Plasminogen: Structure, Activation, and Regulation

Plasminogen: Structure, Activation, and Regulation

Edited by

DAVID M. WAISMAN
Department of Biochemistry & Molecular Biology and Oncology
University of Calgary Health Sciences Centre
Calgary, Alberta, Canada

Springer Science+Business Media, LLC

Library of Congress Cataloging-in-Publication Data

Plasminogen: structure, activation, and regulation/edited by David Morton Waisman.
p. ; cm.
Includes bibliographical references and index.
ISBN 978-0-306-47699-0 ISBN 978-1-4615-0165-7 (eBook)
DOI 10.1007/978-1-4615-0165-7

1. Plasminogen. I. Waisman, David Morton, 1952–

QP93.5 .W356 2003
612′.0151—dc21

2002042762

ISBN 978-0-306-47699-0

Originally published by Kluwer Academic / Plenum Publishers in 2003

To my wife, Mary Sue, and children Meredith and Phillip
for their unfailing support and patience

David M. Waisman

Contributors

Riitta Alitalo
Department of Clinical Chemistry
Hematology and Stem Cell Laboratories and
Helsinki University Central Hospital
Finland

Elizabeth H. Allan
St Vincent's Institute of Medical Research
9 Princes Street
Fitzroy Vic 3065
Australia

Nicholas M. Andronicos
The Scripps Research Institute
Department of Cell Biology
Division of Vascular Biology
La Jolla, California 92037
United States

Toomas Asser
Department of Neurology
University of Tartu
Tartu, Estonia

Felizabel Garcia Bannach
The Scripps Research Institute
Department of Cell Biology
Division of Vascular Biology
La Jolla, California 92037
United States

Francis J. Castellino
W. M. Keck Center for Transgene Research
Department of Chemistry and Biochemistry
University of Notre Dame
Notre Dame, Indiana 46556
United States

Kyu-Sil Choi
Department of Biochemistry and Molecular Biology
University of Calgary
Calgary, Alberta
Canada T2N 4N1

Désiré Collen
Center for Molecular and Vascular Biology
University of Leuven
Campus Gasthuisberg
O & N, Herestraat 49
B-3000 Leuven
Belgium

Evis Daci
Onderwijs en Navorsing
Gasthuisberg
3000 Leuven
Belgium

Paul J. Declerck
Laboratory for Pharmaceutical Biology
Faculty of Pharmaceutical Sciences
Katholieke Universiteit Leuven
Van Evenstraat 4
B-3000 Leuven
Belgium

Yves A. DeClerck
Division of Hematology-Oncology and
Departments of Pediatrics and Biochemistry and
Molecular Biology
Childrens Hospital Los Angeles and USC Keck
School of Medicine
Los Angeles, California
United States

Vincent Ellis
School of Biological Sciences
University of East Anglia
Norwich NR4 7TJ
United Kingdom

Jordi Felez
Institut de Recerca Oncologica
08907, Barcelona
Spain

Darin K. Fogg
Department of Biochemistry and Molecular Biology
University of Calgary
Calgary, Alberta
Canada T2N 4N1

Ann Gils
Laboratory for Pharmaceutical Biology
Faculty of Pharmaceutical Sciences
Katholieke Universiteit Leuven
Van Evenstraat 4
B-3000 Leuven
Belgium

Neill A. Gingles
The Scripps Research Institute
Department of Cell Biology
Division of Vascular Biology
La Jolla, California 92037
United States

Yun Gong
The Scripps Research Institute
Department of Cell Biology
Division of Vascular Biology
La Jolla, California 92037
United States

Davida K. Grella
University of Notre Dame
Notre Dame, Indiana

Ana Gutiérrez-Fernández
The Scripps Research Institute
Department of Cell Biology
Division of Vascular Biology
La Jolla, California 92037
United States

Philip J. Hogg
Centre for Thrombosis and Vascular Research
School of Medical Sciences
University of New South Wales and Department of Haematology
Prince of Wales Hospital
Sydney, NSW 2052
Australia

Jane L. Hoover-Plow
Joseph J. Jacobs Center for Thrombosis and Vascular Biology
Department of Molecular Cardiology
Lerner Research Institute
Cleveland Clinic Foundation
Cleveland, Ohio 44195
United States

G. Ronald Jenkins
The Scripps Research Institute
Department of Cell Biology
Division of Vascular Biology
La Jolla, California 92037
United States

Andres Kulla
Department of Pathology and Neuropathology
Tartu University Clinics
Tartu, Estonia

Mijung Kwon
Cancer Biology Research Group
Department of Biochemistry and Molecular Biology
University of Calgary
Calgary, Alberta
Canada T2N 4N1

Walter E. Laug
Division of Hematology-Oncology and Departments of Pediatrics and Biochemistry and Molecular Biology
Childrens Hospital Los Angeles and USC Keck School of Medicine
Los Angeles, California
United States

H. Roger Lijnen
Center for Molecular and Vascular Biology
University of Leuven
Campus Gasthuisberg
O & N, Herestraat 49
B-3000 Leuven
Belgium

David J. Loskutoff
The Scripps Research Institute
Department of Cell Biology
Division of Vascular Biology
La Jolla, California 92037
United States

T. John Martin
St Vincent's Institute of Medical Research
9 Princes Street
Fitzroy Vic 3065
Australia

Lindsey A. Miles
The Scripps Research Institute
Department of Cell Biology
Division of Vascular Biology
La Jolla, California 92037
United States

Satu Mustjoki
Department of Clinical Chemistry
Hematology and Stem Cell Laboratories and Biomedicum Helsinki
Hematopoietic Stem Cell Research
Helsinki University Central Hospital
Finland

Robert J. Parmer
Department of Medicine
University of California and Veterans Administration Medical Center
San Diego, California 92161
United States

Michael S. Pepper
Department of Morphology
University Medical Center
Geneva, Switzerland

Victoria A. Ploplis
W. M. Keck Center for Transgene Research
Department of Chemistry and Biochemistry
University of Notre Dame
Notre Dame, Indiana 46556
United States

Aadu Simisker
Haartman Institute
University of Helsinki
POB 21, FIN-00014 Helsinki
Finland
Department of Neurology
University of Tartu
Tartu, Estonia

Tambet Teesalu
Haartman Institute,
University of Helsinki
POB 21, FIN-00014 Helsinki
Finland
Department of Pathology and Neuropathology
Tartu University clinics
Tartu, Estonia
Department of Neurology
University of Tartu
Tartu, Estonia

Antti Vaheri
Haartman Institute,
University of Helsinki
POB 21, FIN-00014 Helsinki
Finland

David M. Waisman
Cancer Biology Research Group
Department of Biochemistry and Molecular Biology
University of Calgary
Calgary, Alberta
Canada T2N 4N1

Lu Zhang
The Scripps Research Institute
Department of Cell Biology
Division of Vascular Biology
La Jolla, California 92037
United States

Preface

The observation that a blood clot spontaneously dissolves was first described by Denys in 1889. Subsequently, the blood clotting system was shown to be involved in tumor growth. For example, as early as 1925, Fisher reported that avian tissue explants transformed to malignancy by viruses generated high levels of fibrinolytic activity under conditions in which cultures of normal cells did not. In 1958, the concept that an equilibrium existed between the tendency of blood to clot and to remain fluid was proposed by Astrup. At that time, it was believed that this hemostatic balance was explained by the ability of polymerizing fibrin to orchestrate its own clearance by stimulating fibrinolytic activity. Since these pioneering studies, considerable information has accumulated that has defined the components of the coagulation and fibrinolytic systems and how they are involved in physiological and pathophysiological processes. *Plasminogen: Structure, activation, and regulation* focuses on the basic principles and recent developments in the plasminogen/plasmin research field and how these results provide a conceptual framework for an understanding of the physiological role of plasminogen in health and disease.

The enzymatic cascade triggered by activation of plasminogen has been implicated in a variety of normal and pathological events such as fibrinolysis, wound healing, tissue remodeling, embryogenesis, angiogenesis, and the invasion and metastasis of tumor cells. This impressive list of physiological functions for plasminogen reinforces the wide diversity of roles that plasminogen plays in various physiological processes. Productive plasmin generation requires the assembly of both plasminogen activators and plasminogen on a solid support such as the fibrin polymer or the cell surface. The regulation of plasmin production involves a complex interplay between these plasminogen activators, plasminogen activator inhibitors, and plasmin inhibitors. Clearly, the explosive growth in this research field and the many exciting discoveries suggests that the research efforts in the next decade will reveal the mechanisms by which the components of the plasminogen system interact and regulate both plasmin activation and function at a cellular level.

Plasminogen: Structure, activation, and regulation is divided into two sections. The first section deals with the structure and regulation of plasminogen. The chapters in this section range from discussions of the structure of plasminogen and the regulation of the plasminogen gene to discussions of the structure and regulation of plasminogen activators and plasminogen activator inhibitors. Also examined is the relatively new data concerning the generation of anti-angiogenic molecules from plasminogen. The second section deals with the physiological and pathophysiological roles of plasminogen as well as the consequences of plasminogen gene knockout. Discussions in this section include examination of the role of plasminogen in hematopoietic malignancies, tumor cell progression, angiogenesis, mammary gland involution, wound healing, and bone readsorption.

In closing, I would like to thank my administrative assistant, Ms. Vi Sommerfeld, for her invaluable assistance and timeless efforts with the organization and editing of the book. Lastly, I would like to acknowledge the efforts of the authors of the individual chapters, who are authorities in this field, for agreeing to take time from busy schedules to provide these chapters in a timely fashion.

David Morton Waisman

Contents

Part I. Plasminogen: Structure and Regulation

5. Plasminogen Receptors

Darin K. Fogg, Kyu-Sil Choi, and David M. Wåisman

6. The Role of Lys-Plasminogen in Cell-Mediated Plasmin Production

Lindsey A. Miles, Nicholas M. Andronicos, Jordi Felez, Davida K. Grella, Francis J. Castellino, and Yun Gong

7. Plasmin Reductase

Philip J. Hogg

8. Mechanism of Angiostatin Formation from Plasminogen

Mijung Kwon and David M. Waisman

Part II. Physiological Roles of Plasminogen

9. Lessons Learned from the Plg Deficient Mice (Plg−/−)

Jane L. Hoover-Plow

10. Plasminogen-Directed Phenotypes in Mice

Victoria A. Ploplis and Francis J. Castellino

11. Role of the Plasminogen and MMP Systems in Wound Healing

H. Roger Lijnen and Désiré Collen

12. Matrix Metalloproteinases and the Plasminogen System in Tumor Progression

Yves A. DeClerck and Walter E. Laug

13. Role of Plasminogen Activation in Hematopoietic Malignancies and in Normal Hematopoiesis

Satu Mustjoki, Riitta Alitalo, and Antti Vaheri

14. The Role of Plasminogen in Bone Remodeling

Evis Daci, Elizabeth H. Allan, and T. John Martin

15. Plasminogen Activators in CNS Physiology and Disease

Tambet Teesalu, Andres Kulla, Toomas Asser, Aadu Simisker, and Antti Vaheri

16. Role of the Plasminogen Activator–Plasmin System in Angiogenesis

Michael S. Pepper

Plasminogen: Structure, Activation, and Regulation

Part I

Plasminogen: Structure and Regulation

1

Human Plasminogen: Structure, Activation, and Function

Francis J. Castellino and Victoria A. Ploplis

1. Introduction

The conversion of the plasma zymogen, plasminogen (Pg), to the serine protease plasmin (Pm) is a critical event associated with the activation of the fibrinolytic system (Collen, 1980). The primary vascular function of Pm is to maintain patency by degrading fibrin-rich thrombi. Localization of this proteolytic activity at the site of a developing thrombus, as well as the protection of Pm and its activators from circulating inhibitors are essential in regulating fibrinolysis (Collen, 1980). The identification of cell surface receptors for Pg/Pm and its activators and the demonstration that Pm can either directly or indirectly degrade extracellular matrices have implicated Pm in mediating cell migration, a critical event in a number of physiologies, e.g., wound healing, embryogenesis, angiogenesis, and pathologies, e.g., tumor growth and metastasis processes (Mak *et al.*, 1976; Strickland *et al.*, 1976; Gross *et al.*, 1983; Ossowski and Reich, 1983; Dano *et al.*, 1985; Nielsen *et al.*, 1988; Schafer *et al.*, 1994). Therefore, aside from its vascular function in controlling thrombus formation and dissolution, Pg may play a broad and diverse role in a number of diverse biological events.

2. Structure of Human Plasminogen

2.1. Primary Protein Structure

Human plasminogen (HPg) is synthesized as a 810 amino acid single polypeptide protein but the mature form of this protein is 791 amino acids, the result of cleavage of a 19 amino acid leader peptide during secretion (Forsgren *et al.*, 1987). An illustration of

---

Francis J. Castellino and Victoria A. Ploplis • Department of Chemistry and Biochemistry, The University of Notre Dame, Notre Dame, Indiana.

Plasminogen: Structure, Activation, and Regulation, edited by David M. Waisman. Kluwer Academic/Plenum Publishers, New York, 2003.

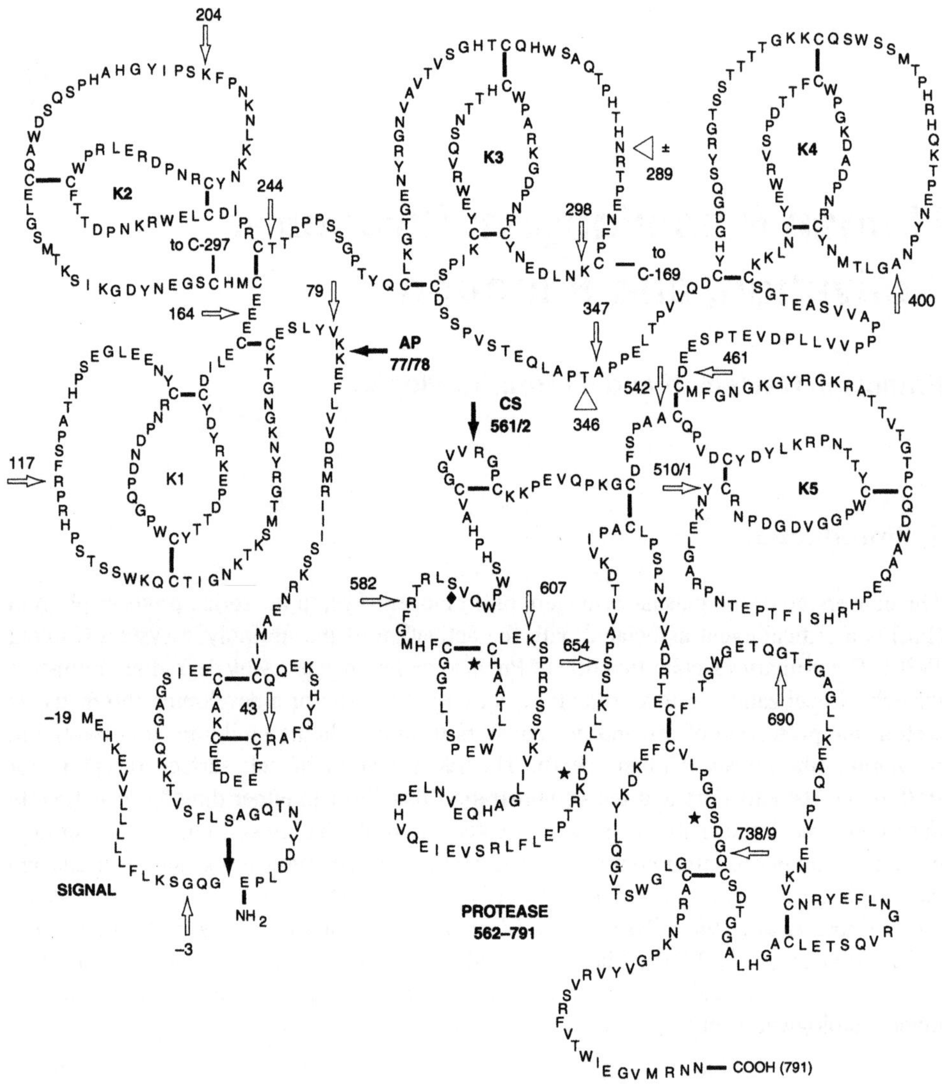

Figure 1.1. The primary structure of HPg. *Filled arrows* identify the cleavage sites for: (a) the release of the signal peptide between residues −1 and 1, which is required for the generation of the mature form of the protein; (b) the release of the activation peptide (Glu^{1}–Lys^{77}) resulting in the conversion of Glu^{1}–Pg to Lys^{78}–Pg or Glu^{1}–Pm to Lys^{78}–Pm; (c) the activation of HPg to HPm (CS) at the Arg^{561}–Val^{562} peptide bond. *Unfilled arrows* identify introns in the gene sequence. *Triangles* locate the N-linked oligosaccharide site at sequence position 289 and the O-linked glycan at position 346. The catalytic triad, His^{603}, Asp^{646}, and Ser^{741}, is also indicated (*). Disulfide bonds are depicted by *heavy bars*. ♦ phosphorylation site

the primary structure of Pg is shown in Figure 1.1. The primary tissue that synthesizes plasminogen is the liver (Bohmfalk and Fuller 1980; Raum *et al.*, 1980; Saito *et al.*, 1980) but other extrahepatic organs have been identified as possible sources of plasminogen, for example, adrenal, kidney, brain, testis, heart, lung, uterus, spleen, thymus, and gut (Zhang *et al.*, 2002). The sequence for HPg was deduced from nucleotide sequences generated from both cDNA (Forsgren *et al.*, 1987) and genomic DNA (Petersen *et al.* 1990) but also confirmed by direct amino acid sequencing (Wiman, 1973, 1977; Wiman and Wallen, 1975; Sottrup-Jensen *et al.*, 1987). The only known posttranslational modifications are N- and O-linked glycosylations (Hayes and Castellino, 1979a,b) and phosphorylation (Wang *et al.*, 1997). The common glycosylation sites within HPg occur at Asn^{289} in approximately one half of the molecules and an O-linked glycan at Thr^{346} in all of the plasminogen molecules (Hayes and Castellino, 1979a,b), and the phosphorylation site exists at Ser^{578}. The N-linked glycoforms can be readily resolved utilizing affinity chromatography by Sepharose–lysine (Brockway and Castellino, 1972). Other minor O-linked sites could potentially exist at Ser^{248} (Pirie-Shepherd *et al.*, 1997) and Ser^{339} (Hortin, 1990). The conversion of the zymogen, Pg, to Pm, a serine protease, is through proteolytic cleavage of Arg^{561}–Val^{562} (Robbins *et al.*, 1967). This results in the generation of Glu^{1}–Pm which now consists of a N-terminus-derived heavy chain (561 amino acids) disulfide linked by two disulfide bonds to a carboxy-terminus-derived light chain (230 amino acids). The typical serine protease catalytic triad is located in the light chain of Pm and consists of His^{603}, Asp^{646}, and Ser^{741}. Another physiologically relevant hydrolytic reaction is cleavage by Glu^{1}–Pm of a 77 amino acid peptide from the N-terminus of Glu^{1}–Pg or the N-terminus of the heavy chain of Glu^{1}–Pm generating, respectively, Lys^{78}–Pg or Lys^{78}–Pm (Violand and Castellino, 1976).

The heavy chain of HPm contains five triple-disulfide-linked peptide regions of approximately 80 amino acids termed kringles (Sottrup-Jensen *et al.*, 1978). These five structures are located within Cys^{84}–Cys^{162}, Cys^{166}–Cys^{243}, Cys^{256}–Cys^{333}, Cys^{358}–Cys^{435}, and Cys^{462}–Cys^{541}. These structures are not unique to HPm but have also been identified in other proteins of hemostasis, for example, prothrombin (Magnusson *et al.*, 1975), factor XII (McMullen and Fujikawa, 1985), tPA (Pennica *et al.*, 1983), urokinase (Steffens *et al.*, 1982) as well as a protein associated with low density lipoprotein, apolipoprotein (a) (McLean *et al.*, 1987).

The kringle domains of Pg serve important functions such as interaction with the substrate of plasminogen, fibrin(ogen) (Suenson and Thorsen, 1981), and binding to bacterial (Berge and Sjorbring, 1993; DiCosta and Boyle, 1998; Wistedt *et al.*, 1998) and mammalian (Miles *et al.*, 1988) cell surfaces. These interactions are inhibited by lysine and its analogues. Additionally, Pg utilizes these structures for interaction with small molecules that affect Pg activation such as Cl^- and ω-amino acids (Urano *et al.*, 1987a,b). Interaction with these small effector molecules occurs with all but kringle 3, with kringle 1 and kringle 4 exhibiting the strongest ω-amino acid binding sites (Menhart *et al.*, 1991, 1993; Sehl *et al.*, 1990), and kringle 2 the weakest (Marti *et al.*, 1999).

2.2. Gene Organization

The gene for HPg is 52.5 kb and has been mapped to chromosome 6q26–6q27 (Murray *et al.*, 1987). Both the cDNA and gDNA have been cloned and sequenced (Forsgren *et al.*, 1987; Petersen *et al.*, 1990). The gene is organized into 19 exons, ranging in size from 75 to 387 base pairs, and 18 introns. The coding sequence consists of a 57 base pair signal sequence and 2,373 nucleotides for the mature protein. Functional domains associated with specific exons are outlined in Table 1.1.

Table 1.1. Exons of Human Plasminogen and Functional Domains

Exon	Amino acids	Function
I	−19 to −3	Signal sequence
II and III	−3 to 78	Preactivation peptide
IV to XIII	79–541	Kringles (K1–5)
XIV	542–582	Activation sequence (Arg^{561}–Val^{562}), Cys^{548}, Cys^{558}, Cys^{566} (involved in disulfide bonds between heavy and light chain)
XV	583–607	Active site residue His^{603}
XVI	608–654	Active site residue Asp^{646}
XVII	655–689	Cys^{666} (involved in disulfide bonds between heavy and light chain)
XVIII	690–738	Cys^{737} (proximal to active center and may be involved in disulfide-mediated stabilization of this site)
XVIX	739–791	Active site residue Ser^{741}

Regulation of expression by sequences within the 5′ and 3′ flanking regions of the coding sequence has been partially characterized (Petersen *et al.*, 1990). Upstream of the signal initiation codon (Met^{-19}), nucleotide sequences (CTGGGA) have been found that are typically identified for acute phase reactant proteins, for example, fibrinogen, human haptoglobin, and transferrin (Fowlkes *et al.*, 1984; Maeda, 1985; Adrian *et al.*, 1986). Additionally, located within nucleotides 2.5 kb upstream of the translational initiation signal, two sequences have been identified as recognition sites for hepatocyte-enriched HNF-1 and AP-3. These two sequences regulate transcript and liver specificity of this gene in a synergistic fashion (Meroni *et al.*, 1996). Within the 5′ flanking region of human plasminogen other potential regulatory sites for transcription have been identified, for example, IL-6, AP-1, DBP, C/EBP, GATA, LF/A-1, and CREB (Kida *et al.*, 1997). Additional studies have confirmed an IL-6 responsive element within a 1,067 base pair region (−914 to +154 relative to transcription initiation site) and a potential glucocorticoid responsive element (GGAACAATGCTTTCT) present between −820 and −806 in the plasminogen promoter 5′ flanking sequence (Jenkins GR *et al.*, 1997). However, regulatory GC boxes have not been identified in the 5′ flanking sequences of this gene.

3. Activation of Human Plasminogen

The conversion of HPg to HPm occurs as a result of the cleavage of the Arg^{561}–Val^{562} peptide bond (Robbins, 1967). Both physiological (urokinase- and tissue-type plasminogen activator) and bacterial-derived (streptokinase and staphylokinase) activators function in this capacity.

In general, the conversion of plasma zymogen Glu^{1}–Pg to ultimately generate the final product, Lys^{78}–Pm, is outlined in Figure 1.2. The first catalytic event is cleavage of the $Arg^{561}Val^{562}$ bond resulting in the plasmin product Glu^{1}–Pm. This initial reaction is slow in the presence of negative effector molecules such as Cl^{-}. In the presence of this

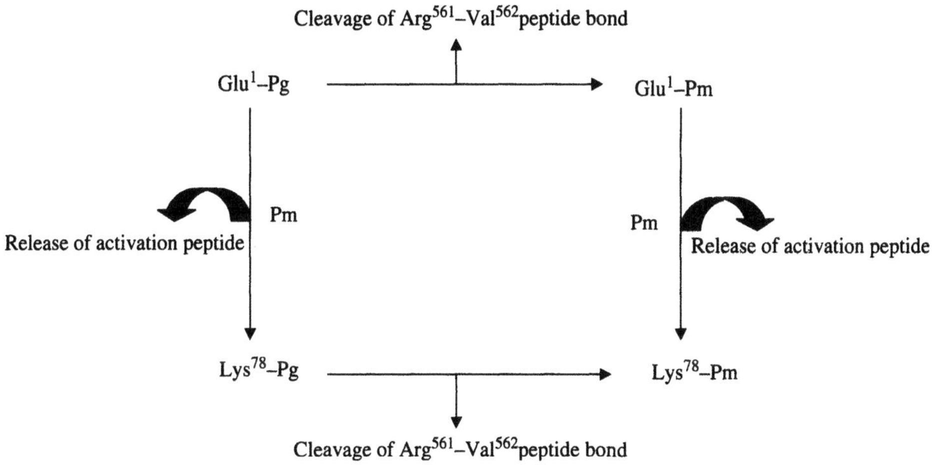

Figure 1.2. Overall scheme for the conversion of the zymogen Glu^{1}–Pg to the serine protease Lys^{78}–Pm.

negative effector anion, Glu^{1}–Pg exists in a compact (T-state), which is poorly activatable (Violand *et al.*, 1975, 1978; Urano *et al.*, 1987a,b, 1988a). The molecular basis of effects of both negative and positive effector molecules on both the structure and activation of Pg has been identified through a number of investigations involving rGlu^{1}–Pg variants (Hoover *et al.*, 1993; McCance *et al.*, 1994). For example, it has been suggested that in the presence of Cl^-, Lys side chains in the 77 amino acid N-terminal activation peptide interact with ω-amino acid binding sites of K1, K4, and to a lesser extent K5. A number of activation peptide residues that interact with these kringle sites and serve to stabilize the T conformation have been identified (Horrevoets *et al.*, 1995). On the other hand, the addition of ω-amino acids, such as *epsilon*-aminocaproic acid (EACA), displaces these interactions and alters the conformation of Glu^{1}–Pg to a more flexible R-state that is readily activatable. Once Glu^{1}–Pm is formed, it acts as a catalysis in releasing the N-terminal activation peptide converting Glu^{1}–Pg to Lys^{78}–Pg or Glu^{1}–Pm to Lys^{78}–Pm (Violand and Castellino, 1976; Gonzalez-Gronow *et al.*, 1977). Release of the activation peptide renders Lys^{78}–Pg a more activatable substrate due to the transformation to the R-state conformation and transition to a protein that is now insensitive to negative and positive effector molecules (Violand *et al.*, 1975).

3.1. Activation by Physiological Activators

3.1.1. Urokinase-type Plasminogen Activator

Urokinase (UK) or urinary plasminogen activator (uPA) is a serine protease whose functional activity was first identified in the 1950s (Williams, 1951; Astrup and Sterndorff, 1952). Major sources of this protease are lung and kidney but a number of tumor cell lines also express uPA (Bernik and Kwaan, 1969; Duffy, 1990). Two molecular weight forms of this protein have been isolated. The full-length form from urine (50 kDa) and a proteolytically processed form from kidney cell cultures (33 kDa), the result of cleavage of Lys^{135}– Lys^{136} (White *et al.*, 1966; Barlow, 1976). Urokinase exists as a zymogen, proUK or single-chain uPA (sc-uPA), and consists of 411 amino acids (DeMunk and Rijken, 1990). The amino-terminal region of sc-uPA consists of 158 amino acids and contains a growth factor-like domain followed by a kringle domain that is homologous to the kringle domain of Pg but does not confer the ability to bind lysine or fibrin (Patthy *et al.*, 1984; Lijnen *et al.*, 1986). The carboxy-terminal region consists of 253 amino acids and the catalytic triad (His^{204}, Arg^{255}, and Ser^{256}).

Debate over whether the zymogen, sc-uPA, has low endogenous activity (<1%) or whether sc-uPA contains low level contamination of the active two chain form, (tc-uPA), the result of Pm-mediated proteolytic processing of sc-uPA, has occurred over the years (Pannell and Gurewich, 1987; Petersen *et al.*, 1988; Urano *et al.*, 1988b). Studies utilizing mutant forms of sc-uPA that are unable to be converted to tc-uPA demonstrated a retention of low activity indicating that sc-uPA has endogenous plasminogen activation function (Fleury *et al.*, 1993). Additional studies utilizing an active site Ser^{740} mutant indicated that sc-uPA can generate the inactive $rSer^{740} \rightarrow$ Ala Pm at a rate consistent with a low level of inherent activity of sc-uPA (Lijnen *et al.*, 1990).

The overall mechanism of sc-uPA-mediated activation of Glu^{1}–Pg initally consists of a low rate of conversion of Glu^{1}–Pg to Glu^{1}–Pm by sc-uPA, or small levels of contaminating tc-uPA, followed by a feedback activation of sc-uPA to tc-uPA by Glu^{1}–Pm. The generation of additional tc-uPA, a better Pg activator relative to sc-uPA, would then accelerate the overall activation of Pg.

3.1.2. Tissue-type Plasminogen Activator

Tissue-type plasminogen activator (tPA) was first identified in melanoma cells in 1980 (Wilson *et al.*, 1980) and has since then been identified as a vascular endothelial cell product from a variety of tissue sources (Levin, 1983). This protein is synthesized as a precursor protein and undergoes post-translation modification (Pennica *et al.*, 1983). Further characterization indicated that it is a serine protease of 68 kDa and is synthesized as a single chain molecule of 527 amino acids. Conversion to the two chain form is the result of Pm-mediated cleavage of Arg^{275}–Ile^{276} bond and results in a conformational change in the protein. The N-terminal derived heavy chain (A chain) is composed of four domains, a finger domain (amino acids 4–50), which is homologous to type I fingers (F) of fibronectin, a domain homologous to epidermal growth factor (EGF) (amino acids 51–87), and two kringle structures, K1 and K2, (amino acids 88–176 and 177–262, respectively). The light chain (B chain) (amino acids 276–527) is homologous to other serine proteases and the catalytic triad consists of His^{322}, Asp^{371}, and Ser^{478}.

Tissue type plasminogen activator has a high affinity for fibrin and Pg activation is enhanced in the presence of this plasmin substrate (Hoylaerts *et al.*, 1982). Single chain tPA and two chain tPA have comparable activities and it is believed that the mechanism involves single chain tPA adopting a two chain tPA conformation, once bound to Fn, thus conferring a similar activation potential (Loscalzo, 1988). Fibrin-mediated enhanced activation is due to a large decrease in the Km of the reaction (Hoylaerts *et al.*, 1982) and fibrin partially digested by Pm is more effective than virgin fibrin (Higgins and Vehar, 1989). The mechanism for this appears to be due to the exposure of COOH-terminal Lys and Arg residues resulting in a tighter association of both tPA and Pg with the clot (de Vries *et al.*, 1989).

3.2. Activation by Bacterial-derived Plasminogen Activators

3.2.1. Streptokinase

Streptokinase (SK), derived from several strains of hemolytic streptococci, can activate Pg from a number of mammalian species (McCoy *et al.*, 1991). The unusual property of this Pg activator is that it functions in a nonhydrolytic manner, the result of SK evolving into two independent domains related to serine proteases and *Streptomyces griseus* protease but containing alterations at critical sites within the catalytic domain (Jackson and Tang, 1982) rendering it enzymatically dormant. Analysis of SK (group C) isolated from infected humans indicated that this protein consists of 415 amino acids in a single polypeptide chain with no disulfide bonds (Jackson and Tang, 1982; Malke *et al.*, 1985).

The activation of HPg by SK initially involves the formation of a stoichiometric complex of SK and HPg (SK–HPg), which induces a conformational alteration in HPg resulting in the formation of an active site within HPg of the complex (SK–HPg*) (Schick and Castellino, 1974). While HPg* in this complex is sensitive to both negative and positive effector molecules the transition of this complex to SK–HPg′ results in a form of Pg that is less sensitive (Chibber *et al.*, 1985, 1986; Chibber and Castellino, 1986). Degradation of both amino-terminal and carboxy-terminal regions of SK results in several forms of SK designated as SK^, which is also generated from SK complexed with HPm (Brockway and Castellino, 1974; Siefring and Castellino, 1976). These three complexes (SK–HPg*, SK–Pg′, and SK^–HPm) are all effective activators of HPg with SK–HPg′ being the most potent (Davidson *et al.*, 1990). At this point, activation of Pg is essentially as described for tPA and uPA. Interestingly, HPm cannot directly activate HPg and none of these complexes can proteolytically release the activation peptide of HPg (Violand *et al.*, 1975; Gonzalez-Gronow *et al.*, 1977).

3.2.2. Staphylokinase

Another bacterial HPg activator, staphylokinase STAR (SakSTAR), has been identified and characterized (Sako *et al.*, 1983; Sako and Tsuchida 1983; Collen *et al.*, 1992). This protein consists of 163 amino acids including a 27 amino acid signal sequence.

The mechanism of Pg activation is similar to SK (Kowalska-Loth and Zakrzewski, 1975) but a major difference is that activation is enhanced in the presence of fibrin (Sakai *et al.*, 1989) or endothelial cells (Ueshima *et al.*, 1996). Binding of SakSTAR to HPg is similar to SKc (Rodriguez *et al.*, 1995) but the initial complex SakSTAR–HPg that is formed does not develop plasminogen activator activity. SakSTAR–HPg needs to be converted to SakSTAR–HPm by other HPg activators, that is, tPA (Collen *et al.*, 1993; Grella and Castellino, 1997), which then converts this complex to an active plasminogen activator (Lijnen *et al.*, 1994).

4. Targets for Plasmin Activity

The primary physiological substrates for Pm are fibrinogen (Fg) and fibrin (Fn). Fg consists of three polypeptide chains identified as Aα (610 amino acids), Bβ (461 amino acids), and γ (410 amino acids) that are intertwined and disulfide bonded together (Shafer *et al.*, 1988). This complex is disulfide bonded at the N-terminal region to another identical complex creating a dimeric trimer. The N-terminal region of the six chain complex is maintained by four disulfide bonds and is the central E domain. Electron microscopic analyses of fibrinogen indicates a trinodular structure consisting of the central E domain and the C-terminal D domains (Hall and Slayter, 1959; Fowler *et al.*, 1980; Telford *et al.*, 1980; Mosesson *et al.*, 1981; Norton and Slayter, 1981; Price *et al.*, 1981). The conversion of Fg to Fn by thrombin is the result of proteolytic cleavage of the Arg^{16}–Gly^{17} in the Aα chain (fibrinopeptide A) and the Arg^{14}–Gly^{15} bond in the Bβ chain (fibrinopeptide B) (Blomback and Vestermark, 1958). This results in the conversion of fibrinogen $(A\alpha B\beta\gamma)_2$ to fibrin $(\alpha\beta\gamma)_2$ which spontaneously polymerizes to form a clot. Stabilization of the clot structure occurs by the formation of covalent crosslinks through the activity of the transglutaminase FXIIIa (Schwartz *et al.*, 1973).

Degradation of Fg by Pm to end products fragment E and fragment D initially involves cleavage of Lys^{583}–Met^{584} in the Aα chain. This is immediately followed by cleavage of Lys^{206}–Met^{207} and Lys^{230}–Ala^{231} in the Aα chain which results in the removal of a 40 kDa fragment from the C terminus of the Aα chain with the generation of the intermediate degradation product fragment X. Plasmin-mediated cleavage of this fragment occurs in all three chains within the coiled-coil region resulting in the generation of fragments E and D (Kwaan, 1980).

Due to the presence of stabilizing crosslinks in Fn, the Pm-mediated degradation products of Fn are somewhat different but it is believed that the same bonds are cleaved. The final end products, in this case, are fragment E and a dimer of fragment D (Aljaersig *et al.*, 1977).

As a result of plasmin's relatively broad substrate specificity, Pm has been shown to both degrade and activate other proteins. Other matrix proteins that can be degraded by Pm include fibronectin, collagen and laminin (Hanbucken *et al.*, 1987; Juranic *et al.*, 1989). Other substrates reported for plasmin include, but not limited to, prekallikrein (Burrowes, 1971), FVa (Zeibdawi and Pryzdial, 2001), FVIII (McKee *et al.*, 1975), tissue factor pathway inhibitor (TFPI) (Li and Wun, 1998), von Willebrand factor (vWF) (Hamilton *et al.*, 1985), thrombospondin (TSP) (Bonnefoy and Legrand, 2000), pro-MMP-9 (Lijnen, 2001), and procarboxypeptidase B (Mao *et al.*, 1999).

5. Dysplasminogenemias and Phenotypic Manifestations

A number of plasminogen variants (dyplasminogenemias) have been identified and most have been characterized. Traditionally, their names have been associated with the city in which the first patient was identified. Pg Chicago I, II, and III were identified by assessing the kinetics of activation of these proteins by SK and uPA (Wohl *et al.*, 1979; Wohl *et al.*, 1982). All three forms demonstrated significantly diminished kcat/Km for uPA. However, the kinetics of activation of Pg Chicago II by SK was similar to the native molecule while it appeared to be altered for Pg Chicago I. Pg Paris I was first identified in a patient who suffered from a thrombotic event. The conversion of this protein to Pm, by either SK or uPA, resulted in a proteolytic inactive protein (Soria *et al.*, 1983). In the case of Pg Frankfurt I, normal kinetics of activation were observed but only a half of the Pm that was formed had enzymatic activity (Scharrer *et al.*, 1986) and the Pm formed was much less stable relative to the native molecule. Pg San Antonio demonstrated reduced activation by SK, uPA, and tPA with differences in isoelectric and immunoelectrophoretic properties (Liu *et al.*, 1988).

The most characterized mutant for Pg was first identified in a patient with recurrent thrombosis (Aoki *et al.*, 1978). The patient presented with a history of thrombotic occurrences with approximately 37% functional plasmin activity. The molecular defect was the result of a replacement of Ala^{601} to Thr near the active His^{603} of the catalytic triad resulting in an inactive Pm molecule (Miyata *et al.*, 1982). This defect was found in a number of other Japanese patients. Genetic analysis indicated that the Ala to Thr mutation was due to a G $\rightarrow$ A transition in exon XV in the DNA sequence (Ichinose *et al.*, 1991). The Ala^{601} to Thr mutation appears at a high frequency in the Japanese population (2%). While a number of these patients have clinical histories of venous thrombosis, many remain asymptomatic. Other Pg mutations identified in the Japanese population involve Ser^{572} to Pro, Asp^{676} to Asn, and Val^{355} to Phe mutations (Ichinose *et al.*, 1991; Azuma *et al.*, 1993; Tsutsumi *et al.*, 1996). The Asp^{676} to Asn mutation leads to a loss of a negative charge at the site and the potential development of a glycosylation site.

A few patients have recently been identified with type I plasminogen deficiency (Mingens *et al.*, 1999; Kraft *et al.*, 2000). These patients present with ligneous conjunctivitus or pseudomembranous disease. Additionally, patients with different homozygote or compound heterozygote mutations have been identified, and also present with ligneous conjunctivitis and, in some cases, occlusive hydrocephalus and gingival overgrowth with ulcerations (Schuster *et al.*, 1997, 1999).

Polymorphisms of HPg have been identified based on isoelectric focusing and zymographic analyses (Hobart 1979; Raum *et al.*, 1980). The two major patterns are PLG1/2 or PLG*A/B. The two most common alleles have been designated PLG*A1 to A3 (acidic forms) and PLGB1 to B3 (basic forms). Less common intermediate forms are PLGM1 to *M5. The pattern is inheritable as an autosomal codominant allele.

6. Conclusions

Through a number of both *in vitro* and *in vivo* studies, plasminogen has been identified as a critical component in hemostasis and a significant player in the biology of other nonhemostasis events. These investigations have facilitated in identifying structure–function relationships within this protein and critical interactions with both effector and substrate molecules. As a result, a better appreciation of its involvement in a number of physiological and pathophysiological processes has developed and has resulted in new targeted approaches in regulating its function toward improved therapies.

References

Adrian, G.S., Korinek, B.W., Bowman, G.H., and Yang, F. (1986). The human transferrin gene: 5′ region contains conserved sequences which match the control elements regulated by heavy metals, glucocorticoids and acute phase reaction. *Gene* **49**, 167–175.

Aljaersig, N., Davies, A., and Fletcher, A. (1977). Fibrin and fibrinogen proteolysis products: Comparison between gel filtration and SDS polyacrylamide electrophoresis analysis. *Thromb. Haemost.* **38**, 524–525.

Aoki, N., Moroi, M., Sakata, Y., Yoshida, N., and Matsuda, M. (1978). Abnormal plasminogen—a hereditary molecular abnormality found in a patient with recurrent thrombosis. *J. Clin. Invest.* **61**, 1186–1195.

Astrup, T. and Sterndorff, I. (1952). An activator of plasminogen in normal urine. *Proc. Soc. Exp. Biol. Med.* **81**, 675–678.

Azuma, H., Uno, Y., Shigekiyo, T., and Saito, S. (1993). Congenital plasminogen deficiency caused by a Ser572 to Pro mutation. *Blood* **82**, 475–480.

Barlow, G.H. (1976). Urinary and kidney cell plasminogen activator (urokinase). In L. Lorand (ed.), *Methods in Enzymology.* Academic Press, San Diego, CA. **45**, 239–244.

Berg, A. and Sjöbring, U. (1993). PAM, a novel plasminogen-binding protein from *Streptococcus pyogenes. J. Biol. Chem.* **268**, 25417–25424.

Bernik, M.B. and Kwaan, H.C. (1969). Plasminogen activator activity in cultures from human tissue. An immunological study. *J. Clin. Invest.* **48**, 1740–1753.

Blomback, B. and Vestermark, A. (1958). Isolation of fibrinopeptides by chromatography. *Arkiv. Kemi.* **12**, 173.

Bohmfalk, J.F. and Fuller, G.M. (1980). Plasminogen is synthesized by primary cultures of rat hepatocytes. *Science* **209**, 408–410.

Bonnefoy, A. and Legrand, C. (2000). Proteolysis of subendothelial adhesive glycoproteins (fibronectin, thrombospondin, and von Willebrand factor) by plasmin, leukocyte cathepsin G, and elastase. *Thromb. Res.* **98**, 323–332.

Brockway, W.J. and Castellino, F.J. (1972). Measurement of the binding of antifibrinolytic amino acids to various plasminogens. *Arch. Biochem. Biophys.* **151**, 194–199.

Brockway, W.J. and Castellino, F.J. (1974). A characterization of native streptokinase and altered streptokinase isolated from a human plasminogen activator complex. *Biochemistry* **13**, 2063–2070.

Chibber, B.A.K. and Castellino, F.J. (1986). Regulation of the streptokinase-mediated activation of human plasminogen by fibrinogen and chloride ions. *J. Biol. Chem.* **261**, 5289–5295.

Chibber, B.A.K., Morris, J.P., and Castellino, F.J. (1985). Effects of human fibrinogen and its cleavage products on activation of human plasminogen by streptokinase. *Biochemistry* **24**, 3429–3434.

Chibber, B.A.K., Radek, J.T., Morris, J.P., and Castellino, F.J. (1986). Rapid formation of an anion sensitive active site in stoichiometric complexes of streptokinase and human [Glu1]plasminogen. *Proc. Natl. Acad. Sci. USA* **83**, 1237–1241.

Collen, D., DeCock, F., Vanlinhout, I., Declerck, P.J., Lijnen, H.R., and Stassen, J.M. (1992). Comparative thrombolytic and immunogenic properties of staphylokinase and streptokinase. *Fibrinolysis* **6**, 232–242.

Collen, D., Schlott, B., Engelborghs, Y., Van Hoef, B., Hartmann, M., Lijnen, H.R., and Behnke, D. (1993). On the mechanism of the activation of human plasminogen by recombinant staphylokinase. *J. Biol. Chem.* **268**, 8284–8289.

Dano, K., Andreasen, P.A., Grondahl-Hansen, J., Kristensen, P., Nielsen, L.S., and Skriver, L. (1985). Plasminogen activators, tissue degradation and cancer. *Adv. Cancer Res.* **44**, 139–266.

Davidson, D.J., Higgins, D.L., and Castellino, F.J. (1990). Plasminogen activator activities of equimolar complexes of streptokinase with variant recombinant plasminogens. *Biochemistry* **29**, 3585–3590.

DeMunk, G.A.W. and Rijken, D.C. (1990). Fibrinolytic properties of single chain urokinase-type plasminogen activator (prourokinase). *Fibrinolysis* **4**, 1–9.

De Vries, C., Vaerman, H., and Pannekoeck, H. (1989). Identification of the domains of tissue-type plasminogen activator involved in augmented binding to fibrin after limited digestion with plasmin. *J. Biol. Chem.* **264**, 12604–12610.

DíCosta, S.S. and Boyle, M.D.P. (1998). Interaction of a group A *Streptococcus* within human plasma results in assembly of a surface plasminogen activator that contributes to occupancy of surface plasmin-binding structures. *Microb. Pathog.* **24**, 341–349.

Duffy, M.J. (1990). Plasminogen activators and cancer. *Blood Coag. Fibrin.* **1**, 681–687.

Fleury, V., Lijnen, H.R., and Angles-Cano, E. (1993). Mechanism of the enhanced intrinsic activity of single-chain urokinase-type plasminogen activator during ongoing fibrinolysis. *J. Biol. Chem.* **268**, 18554–18559.

Forsgren, M., Raden B., Israelsson M., Larsson K., and Hedén, L-O. (1987). Molecular cloning and characterization of a full-length cDNA clone for human plasminogen. *FEBS Lett.* **213**, 254–260.

Fowler, W.E., Fretto, L.J., Erickson, H.P., and McKee, P.A. (1980). Electron microscopy of plasmic fragments human fibrinogen as related to trinodular structure of the intact molecule. *J. Clin. Invest.* **66**, 50–56.

Fowlkes, D.M., Mullis, M.T., Comeau, C.M., and Crabtree, G.R. (1984). Potential basis for regulation of the coordinately expressed fibrinogen genes: Homology in the 5′ flanking regions. *Proc. Natl. Acad. Sci. USA.* **81**, 2313–2316.

Gonzalez-Gronow, M., Violand, B.N., and Castellino, F.J. (1977). Purification and some properties of the glu- and lys- human plasmin heavy chains. *J. Biol. Chem.* **252**, 2175–2177.

Grella, D.K. and Castellino, F.J. (1997). Activation of human plasminogen by staphylokinase. Direct evidence that preformed plasmin is necessary for activation to occur. *Blood* **89**, 1585–1589.

Gross, J.L., Moscatelli, D., and Rifkin, D.B. (1983). Increased capillary endothelial cell protease activity in response to angiogenic stimuli *in vitro*. *Proc. Natl. Acad. Sci. USA.* **80**, 2623–2627.

Hall, C.E. and Slayter, H.S. (1959). The fibrinogen molecule: Its size, shape, and mode of polymerization. *J. Biophys. Biochem. Cytol.* **5**, 11–15.

Hamilton, K.K., Fretto, L.J., Grierson, D.S., and McKee, P.A. (1985). Effects of plasmin on von Willebrand factor multimers. Degradation in vitro and stimulation of release in vivo. *J. Clin. Invest.* **76**, 261–270.

Hanbucken, F.W., Schneider, J., Gunzler, W.A., Friderichs, E., Giertz, H., and Flohe, L. (1987). Selective fibrinolytic activity of recombinant non-glycosylated human pro-urokinase (single chain urokinase-type plasminogen activator) from bacteria. *Arzneimittel Forschung* **37**, 993–997.

Harpel, P.C., Chang, T.S., and Verderber, E. (1985). Tissue plasminogen activator and urokinase mediate the binding of Glu-plasminogen to plasmin fibrin I. Evidence for new binding sites in plasmin-degraded fibrin I. *J. Biol. Chem.* **260**, 4432–4440.

Hayes, M.L. and Castellino, F.J. (1979a). Carbohydrate of human plasminogen variants. II. Structure of the asparagine-linked oligosaccharide unit. *J. Biol. Chem.* **254**, 8772–8776.

Hayes, M.L. and Castellino, F.J. (1979b). Carbohydrate of human plasminogen variants. III. Structure of the O-glycosidically-linked oligosaccharide unit. *J. Biol. Chem.* **254**, 8777–8780.

Higgins, D.L. and Vehar, G.A. (1987). Interaction of one-chain and two-chain tissue plasminogen activator with intact and plasmin-degraded fibrin. *Biochemistry* **26**, 7786–7791.

Hobart, M.J. (1979). Genetic polymorphism of human plasminogen. *Ann. Hum. Genet.* **42**, 419–423.

Hoover, G.J., Menhart, N., Martin, A., Warder, S., and Castellino, F.J. (1993). Amino acids of the recombinant kringle 1 domain of human plasminogen that stabilize its interaction with ω-amino acids. *Biochemistry* **32**, 10936–10943.

Horrevoets, A.J.G., Smilde, A.E., Fredenburgh, J.C., Pannekoek, H., and Nesheim, M.E. (1995). The activation-resistant conformation of recombinant human plasminogen is stabilized by basic residues in the amino-terminal hinge region. *J. Biol. Chem.* **270**, 15770–15776.

Hortin, G.L. (1990). Isolation of glycopeptides containing O-linked oligosaccharides by lectin affinity chromatography on jacalin-agarose. *Anal. Biochem.* **191**, 262–267.

Hoylaerts, M., Rijken, D.C., Lijnen, H.R., and Collen, D. (1982). Kinetics of the activation of plasminogen by human tissue plasminogen activator. Role of fibrin. *J. Biol. Chem.* **257**, 2912–2919.

Ichinose, A., Espling, E.S., Takamatsu, J., Saito, H., Shinmyozu, K., Maruyama, I., Petersen, T.E., and Davie, E.W. (1991). Two types of abnormal genes for plasminogen in families with a predisposition for thrombosis. *Proc. Natl. Acad. Sci. USA.* **88**, 115–119.

Jackson, K.W. and Tang, J. (1982). Complete amino acid sequence of streptokinase and its homology with serine proteases. *Biochemistry* **21**, 6620–6625.

Jenkins, G.R., Seiffart, D., Parmer, R.J., and Miles, L.A. (1997). Regulation of plasminogen gene expression by interleukin-6. *Blood* **89**, 2394–2403.

Juranic, Z., Tomin, R., and Spuzic, I. (1989). Considerations on the role of the plasminogen activators/plasmin system in the generation of conditions for more efficient interaction of malignant cells with large granular lymphocytes. *Res. Immunol.* **140**, 281–283.

Kida, M., Wakabayashi, S., and Ichinose, A. (1997). Expression and induction by IL-6 of the normal and variant genes for human plasminogen. *Biochem. Biophys. Res. Commun.* **230**, 129–132.

Kraft, J., Lieb, W., Zeitler, P., and Schuster, V. (2000). Ligneous conjunctivitis in a girl with severe type I plasminogen deficiency. *Graefe's Arch. Clin. Exp. Ophthalmal.* **238**, 797–800.

Kwaan, H.C. (1980). Fibrinogen–fibrin degradation products. *Ann. Clin. Lab. Sci.* **10**, 234–237.

Levin, E.G. (1983). Latent tissue plasminogen activator produced by human endothelial cells in culture: Evidence for an enzyme-inhibitor complex. *Proc. Natl. Acad. Sci. USA.* **80**, 6804–6808.

Li, A. and Wun, T.C. (1998). Proteolysis of tissue factor pathway inhibitor (TFPI) by plasmin: Effect on TFPI activity. *Thromb. Haemost.* **80**, 423–427.

Lijnen, H.R. (2001). Role of the fibrinolytic and matrix metalloproteinase systems in arterial neointima formation after vascular injury. *Verh. K. Acad. Geneeskd. Belg.* **63**, 605–622.

Lijnen, H.R., De Cook, F., Van Hoef, B., Schlott, B., and Collen, D. (1994). Characterization of the interaction between plasminogen and staphylokinase. *Eur. J. Biochem.* **224**, 143–149.

Lijnen, H.R., Van Hoef, B., Nelles, L., and Collen, D. (1990). Plasminogen activation with single-chain urokinase-type plasminogen activator (scu-PA). Studies with active site mutagenized plasminogen ($Ser^{740} \rightarrow$ Ala) and plasmin-resistant scu-PA ($Lys^{158} \rightarrow$ Glu). *J. Biol. Chem.* **265**, 5232–5236.

Lijnen, H.R., Zamarron, C., Blaber, M., Winkler, M.E., and Collen, D. (1986). Activation of plasminogen by pro-urokinase. I. Mechanism. *J. Biol. Chem.* **261**, 1253–1258.

Liu, Y., Lyons, R.M., and McDonagh, J. (1988). Plasminogen San Antonio: An abnormal plasminogen with a more cathodic migration, decreased activation and associated thrombosis. *Thromb. Haemost.* **59**, 49–53.

Loscalzo, J. (1988). Structural and kinetic comparison of recombinant human single- and two-chain tissue plasminogen activator. *J. Clin. Invest.* **82**, 1391–1397.

Maeda, M. (1985). Nucleotide sequence of the haptoglobin and haptoglobin-related gene pair. *J. Biol. Chem.* **260**, 6698–6709.

Magnusson, S., Petersen, T.E., Sottrup-Jensen, L., and Claeys, H. (1975). Complete primary structure of prothrombin: Isolation and reactivity of ten carboxylated glutamic residues and regulation of prothrombin activation by thrombin. In E. Reich, D.B. Rifkin, and E. Shaw (eds) *Proteases and biological control.* Cold Spring Harbor Laboratories, Cold Spring Harbor, NY. 123–149.

Mak, T.W., Rutledge, G., and Sutherland, D.J. (1976). Androgen-dependent fibrinolytic activity in murine mammary carcinoma (Shionogi SC 115) cells *in vitro*. *Cell* **7**, 223–226.

Malke, H., Roe, B., and Ferretti, J.J. (1985). Nucleotide sequence of the streptokinase gene from *Streptococcus equisimilis* H46A. *Gene* **34**, 357–362.

Mao, S.S., Cooper, C.M., Wood, T., Shafer, J.A., and Gardell, S.J. (1999). Characterization of plasmin-mediated activation of plasma procarboxypeptidase B. Modulation by glycosaminoglycans. *J. Biol. Chem.* **274**, 35046–35052.

Marti, D.N., Schaller, J., and Llinás, M. (1999). Solution structure and dynamics of the plasminogen kringle 2-AMCHA complex: 3_1-helix in homologous domains. *Biochemistry* **38**, 15741–15755.

McCance, S.G., Menhart, N., and Castellino, F.J. (1994). Amino acid residues of the kringle-4 and kringle-5 domains of human plasminogen that stabilize their interactions with omega-amino acid ligands. *J. Biol. Chem.* **269**, 32405–32410.

McCoy, H.F., Broder, C.C., and Lottenberg, R. (1991). Streptokinase produced by pathogenic group C streptococci demonstrate species-specific plasminogen activation. *J. Inf. Dis.* **164**, 515–521.

McKee, P.A., Andersen, J.C., and Switzer, M.E. (1975). Molecular structural studies of human factor VIII. *Ann. N.Y. Acad. Sci.* **240**, 8–33.

McLean, J.W., Tomlinson, J.E., Kuang W-J., Eaton, D.L., Chen, E.Y., Gless, G.M., Scanu, A.M., and Lawn, R.M. (1987). CDNA sequence of human apolipoprotein (a) is homologous to plasminogen. *Nature* **330**, 132–137.

McMullen, B.A. and Fujikawa, K. (1985). Amino acid sequence of the heavy chain of human α-factor XIIa (activated Hageman factor). *J. Biol. Chem.* **260**, 5328–5341.

Menhart, N., McCance, S.G., Sehl, L.C., and Castellino, F.J. (1993). Functional independence of the kringle 4 and kringle 5 regions of human plasminogen. *Biochemistry* **32**, 8799–8806.

Menhart, N., Sehl, L.C., Kelley, R.F., and Castellino, F.J. (1991). Construction, expression and purification of recombinant kringle 1 of human plasminogen and analysis of its interaction with ω-amino acids. *Biochemistry* **30**, 1948–1957.

Meroni, G., Buraggi, G., Mantovani, R., and Taramelli, R. (1996). Motifs resembling hepatocyte nuclear factor 1 and activator protein 3 mediate the tissue specificity of the human plasminogen gene. *Eur. J. Biochem.* **236**, 373–382.

Miles, L.A., Dahlberg, C.M., and Plow, E.F. (1988). The cell binding domains of plasminogen and their function in plasma. *J. Biol. Chem.* **263**, 11928–11934.

Mingers, A.M., Philapitsch, A., Zeitler, P., Schuster, V., Schwarz, H.P., and Kreth, H.W. (1999). Human homozygous type I plasminogen deficiency and ligneous conjunctivitis. *APMIS.* **107**, 62–72.

Miyata, T., Iwanaga, S., Sakata, Y., and Aoki, N. (1982). Plasminogen Tochigi: Inactive plasmin resulting from replacement of alanine 600 by threonine in the active site. *Proc. Natl. Acad. Sci. USA.* **79**, 6132–6136.

Mosesson, M.W., Hainfeld, J., Wall, J., and Haschenmeyer, R. (1981). Identification and mass analysis of human fibrinogen molecules and their domains by scanning transmission electron microscopy. *J. Mol. Biol.* **153**, 695–718.

Murray, J.C., Buetow, K.H., Donovan, M., Hornung, S., Motulsky, A.G., Disteche, C., Dyer, K., Swisshelm, K., Anderson, J., Giblet, E., Sadler, E., Eddy, R., and Shows, T.B. (1987). Linkage disequilibrium of plasminogen polymorphisms and assignment of the gene to human chromosome 6q26–6q27. *Am. J. Hum. Genet.* **40**, 338–350.

Nielsen, L.S., Kallerman, G.M., Behrendt, N., Picone, R., Dano, K., and Blasi, F. (1988). A 55,000–65,000 Mr receptor for urokinase type plasminogen activator. Identification in human tumor cell lines and partial purification. *J. Biol. Chem.* **263**, 2358–2363.

Norton, P.A. and Slayter, H.S. (1981). Immune labeling of the D and E regions of human fibrinogen by electron microscopy. *Proc. Natl. Acad. Sci. USA.* **78**, 1661–1665.

Ossowski, L. and Reich, E. (1983). Antibodies to plasminogen activator inhibit human tumor metastasis. *Cell* **35**, 611–619.

Pannell, R. and Gurewich, V. (1987). The activation of plasminogen by single-chain urokinase or by two-chain urokinase—a demonstration that single-chain urokinase has a low catalytic activity (pro-urokinase). *Blood* **69**, 22–26.

Patthy, L., Trexler, M., Vali Z., Banyai, L., and Varadi, A. (1984). Kringles: Modules specialized for protein binding. Homology of the gelatin-binding region of fibronectin with the kringle structures of proteins. *FEBS Lett.* **171**, 131–136.

Pennica, D., Holmes, W.E., Kohr, W.J., Harkins, R.N., Vehar, G.A., Ward, C.A., Bennett, W.F., Yelverton, E., Seeburg, P.H., Heyneker, H.L., Goeddel, D.V., and Collen, D. (1983). Cloning and expression of human tissue-type plasminogen activator cDNA in *E. coli. Nature* **301**, 214–221.

Petersen, L.C., Lund, L.R., Nielsen, L.S., Dano, K., and Skriver, L. (1988). One-chain urokinase-type plasminogen activator from human sarcoma cells is a proenzyme with little or no intrinsic activity. *J. Biol. Chem.* **263**, 11189–11195.

Petersen, T.E., Martzen, M.R., Ichinose, A., and Davie, E.W. (1990). Characterization of the gene for human plasminogen, a key proenzyme in the fibrinolytic system. *J. Biol. Chem.* **265**, 6104–6111.

Pirie-Sheperd, S.R., Stevens, R.D., Andon, N.L., Enghild, J.J., and Pizzo, S.V. (1997). Evidence for a novel O-linked sialylated trisccharide on Ser-248 of human plasminogen 2. *J. Biol. Chem.* **272**, 7408–7411.

Price, T.M., Strong, D.D., Rudee, M.L., and Doolittle, R.F. (1981). Shadow cast electron microscopy of fibrinogen with antibody fragments bound to specific regions. *Proc. Natl. Acad. Sci. USA.* **78**, 200–204.

Raum, D., Marcus, D., and Alper, C.A. (1980). Genetic polymorphism of human plasminogen. *Am. J. Hum. Genet.* **32**, 681–689.

Raum, D., Marcus, D., Alper, C.A., Levey, R., Taylor, P.D., and Starzl, T.E. (1980). Synthesis of human plasminogen by the liver. *Science* **208**, 1036–1037.

Robbins, K.C., Summaria L., Hsieh, B., and Shah, R.J. (1967). The peptide chains of human plasmin. Mechanism of activation of human plasminogen to plasmin. *J. Biol. Chem.* **242**, 2333–2342.

Rodríguez, P., Collen, D., and Lijnen, H.R. (1995). Binding of streptokinase and staphylokinase to plasminogen. *Fibrinolysis* **9**, 298–303.

Saito, H., Hamilton, S.M., Tavill A.S., Louis L., and Ratnoff, O.D. (1980). Production and release of plasminogen by the liver. *Proc. Natl. Acad. Sci. USA.* **77**, 6837–6840.

Sakai, M., Watanuki, M., and Matsuo, O. (1989). Mechanism of fibrin-specific fibrinolysis by staphylokinase: Participation of α_2-plasmin inhibitor. *Biochem. Biophys. Res. Commun.* **162**, 830–837.

Sako, T., Sawaki, S., Sakurai, T., Ito, S., Yoshizawa, Y., and Kondo, I. (1983). Cloning and expression of the staphylokinase gene of *Staphylococcus aureus* in *E. Coli. Molec. Gen. Genet.* **190**, 271–277.

Sako, T. and Tsuchida, N. (1983). Nucleotide sequence of the staphylokinase gene from *Staphylococcus aureus. Nucleic Acids Res.* **11**, 7679–7693.

Schafer, B.M., Maier, K., Eickhoff, U., Todd, R.F., and Kramer, M.D. (1994). Plasminogen activation in healing human wounds. *Am. J. Pathol.* **144**, 1269–1280.

Scharrer, I.M., Wohl, R.C., Hach, V., Sinio, L., Boreisha, I., and Robbins, K.C. (1986). Investigation of a congenital abnormal plasminogen, Frankfurt I, and its relationship to thrombosis. *Thromb. Haemost.* **55**, 396–401.

Schick, L.A. and Castellino, F.J. (1974). Direct evidence for the generation of an active site in the plasminogen moiety of the streptokinase-human plasminogen activator complex. *Biochem. Biophys. Res. Comm.* **57**, 47–54.

Schuster, V., Mingers, A.M., Seidenspinner, S., Nüssgens, Z., Pukcrop, T., and Kreth, H.W. (1997). Homozygous mutations in the plasminogen gene of two unrelated girls with ligneous conjunctivitis. *Blood* **90**, 958–966.

Schuster, V., Seidenspinner, S., Zeitler, P., Escher, C., Pleyer, U., Bernauer, W., Stiehm, E.R., Isenberg, S., Seregard, S., Olsson, T., Mingers, A.M., Schambeck, C., and Kreth, H.W. (1999). Compound-herozygous mutations in the plasminogen gene predispose to the development of ligneous conjunctivitis. *Blood* **93**, 3457–3466.

Schwartz, M.L., Pizzo, S.V., Hill, R.L., and McKee, P.A. (1973). Human factor XII from plasma and platelets. Molecular weights, subunit structures, proteolytic activation, and cross-linking of fibrinogen and fibrin. *J. Biol. Chem.* **248**, 1395–1407.

Sehl, L.C. and Castellino, F.J. (1990). Thermodynamic properties of the binding of α, ω-amino acids to the isolated kringle 4 region of human plasminogen as determined by high sensitivity titration calorimetry. *J. Biol. Chem.* **265**, 5482–5486.

Shafer, J.A. and Higgins, D.L. (1988). Human fibrinogen. *Crit. Rev. Clin. Lab. Sci.* **26**, 1–41.

Siefring, G.E. and Castellino, F.J. (1976). The interaction of streptokinase and plasminogen. Isolation and characterization of a streptokinase degradation product. *J. Biol. Chem.* **251**, 3913–3921.

Soria, J., Soria, C., Bertrand, O., Dunn, F., Drouet, L., and Caen, J.P. (1983). Plasminogen Paris I: Congenital abnormal plasminogen and its incidence in thrombosis. *Thromb. Res.* **32**, 229–238.

Sottrup-Jensen, L., Claeys, H., Zajdel, M., Petersen, T.E., and Magnusson, S. (1978). The primary structure of human plasminogen: Isolation of two lysine-binding fragments and one "mini" plasminogen (MW, 38000) by elastase-catalyzed-specific limited proteolysis. *Prog. Chem. Fibrinol. Thrombol.* **3**, 191–209.

Steffens, G.J., Günzler, W.A., Ötting, F., Frankus, E., and Flohé, L. (1982). The complete amino acid sequence of low molecular mass urokinase from human urine. *Hoppe-Seyler's Z. Physiol. Chem.* **363**, 1043–1058.

Strickland, S., Reich, E., and Sherman, M.I. (1976). Plasminogen activator in early embryogenesis: Enzyme production by trophoblast and parietal endoderm. *Cell* **9**, 231–240.

Suenson, E. and Thorsen, S. (1981). Secondary-site binding of glu-plasmin, lys-plasmin and miniplasmin to fibrin. *Biochem. J.* **197**, 619–628.

Telford, J.N., Nagy, J.A., Hatcher, P.A., and Scheraga, H.A. (1980). Localization of peptide fragments in the fibrinogen molecule by immunoelectron microscopy. *Proc. Natl. Acad. Sci. USA.* **77**, 2372–2376.

Tsutsumi, S., Saito, T., Sakata, T., Miyata, T., and Ichinose, A. (1996). Genetic diagnosis of dysplasminogenemia: Detection of an Ala601–Thr mutation in 118 out of 125 families and identification of a new Asp676–Asn mutation. *Thromb. Haemost.* **76**, 135–138.

Ueshima, S., Okada, K., Matsumoto, H., Takaishi, T., Fukao, H., and Matsuo, O. (1996). Effects of endothelial cells on activity of staphylokinase. *Blood Coagul. Fibrinol.* **7**, 522–529.

Urano, T., Chibber, B.A.K., and Castellino, F.J. (1987a). The reciprocal effects of ϵ-aminohexanoic acid and chloride ion on the activation of human [Glu^1]plasminogen by human urokinase. *Proc. Natl. Acad. Sci. USA.* **84**, 4031–4034.

Urano, T., de Serrano, V.S., Chibber, B.A.K., and Castellino, F.J. (1987b). The control of the urokinase-catalyzed activation of human glutamic acid 1-plasminogen by positive and negative effectors. *J. Biol. Chem.* **262**, 15959–15964.

Urano, T., de Serrano, V.S., Gaffney, P.J., and Castellino, F.J. (1988a). Effectors of the activation of human [Glu^1]plasminogen by human tissue plasminogen activator. *Biochemistry* **27**, 6522–6528.

Urano, T., de Serrano, V.S., Gaffney, P.J., and Castellino, F.J. (1988b). The activation of human [Glu^1]plasminogen by human single-chain urokinase. *Arch. Biochem. Biophys.* **264**, 222–230.

Violand, B.N., Byrne, R., and Castellino, F.J. (1978). The effect of α-ω-amino acids on human plasminogen structure and activation. *J. Biol. Chem.* **253**, 5395–5401.

Violand, B.N. and Castellino, F.J. (1976). Mechanism of urokinase-catalyzed activation of human plasminogen. *J. Biol. Chem.* **251**, 3906–3912.

Violand, B.N., Sodetz, J.M., and Castellino, F.J. (1975). The effect of ϵ-aminocaproic acid on the gross conformation of plasminogen and plasmin. *Arch. Biochem. Biophys.* **170**, 300–305.

Wang, H., Prorok, M., Bretthauer, R.K., and Castellino, F.J. (1997). Serine-578 is a major phosphorylation locus in human plasma plasminogen. *Biochemistry* **36**, 8100–8106.

White, W.F., Barlow, G.H., and Mozen, M.M. (1966). The isolation and characterization of plasminogen activators (urokinase) from human urine. *Biochemistry* **5**, 2160–2169.

Williams, J.R.B. (1951). The fibrinolytic activity of urine. *Br. J. Exp. Pathol.* **32**, 530–537.

Wilson, E.L., Becker, M.L.B., Hoal, E.G., and Dowdle, E.B. (1980). Molecular species of plasminogen activators secreted by normal and neoplastic cells. *Cancer Res.* **40**, 933–938.

Wiman, B. (1973). Primary structure of peptides released during activation of human plasminogen by urokinase. *Eur. J. Biochem.* **39**, 1–9.

Wiman, B. (1977). The primary structure of the (light) chain of human plasmin. *Eur. J. Biochem.* **76**, 129–137.

Wiman, B. and Wallén, P. (1975). Amino-acid sequence of the cyanogen-bromide fragment from human plasminogen that forms the linkage between the plasmin chains. *Eur. J. Biochem.* **58**, 539–547.

Wistedt, A.C., Kotarsky, H., Marti, D., Ringdahl, U., Castellino, F.J., Schaller, J., and Sjöbring, U. (1998). Kringle 2 mediates high affinity binding of plasminogen to a defined natural sequence in streptococcal surface protein PAM. *J. Biol. Chem.* **273**, 24420–24424.

Wohl, R.C., Summaria, L., Chediak, J., Rosenfeld, S., and Robbins, K.C. (1982). Human plasminogen variant Chicago III. *Thromb. Haemost.* **48**, 146–152.

Wohl, R.C., Summaria, L., and Robbins, K.C. (1979). Physiological activation of the human fibrinolytic system. Isolation and characterization of human plasminogen variants, Chicago I and Chicago II. *J. Biol. Chem.* **254**, 9063–9069.

Zeibdawi, A.R. and Pryzdial, E.L. (2001). Mechanism of factor Va inactivation by plasmin. Loss of A2 and A3 domains from a Ca^{2+}-dependent complex of fragments bound to phospholipid. *J. Biol. Chem.* **276**, 19929–19936.

Zhang, L., Seiffert, D., Fowler, B.J., Jenkins, G.R., Thinnes, T.C., Loskutoff, D.J., Parmer, R.J., and Miles, L.A. (2002). Plasminogen has a broad extrahepatic distribution. *Thromb. Haemost.* **87**, 493–501.

2

Plasminogen Activators: Structure and Function

Vincent Ellis

1. Introduction

Mammalian species have two plasminogen activators, urokinase plasminogen activator (uPA) and tissue-type plasminogen activator (tPA), that are the products of separate genes and members of the trypsin family of serine proteases. Both are mosaic proteins with modular structures similar to the related blood coagulation proteases. The two plasminogen activators provide an excellent example of the regulatory potential made available by such modular construction. Although catalyzing the same reaction, that is, specific hydrolysis of Arg^{560}–Val^{561} of plasminogen, gross differences in the organization of the N-terminal modules of the plasminogen activators, together with subtle differences in the serine protease domain, lead to remarkably different functional properties. These functional properties reflect the differing biological roles of the two plasminogen activators. The activity of tPA is strongly enhanced by fibrin, and tPA is therefore predominantly responsible for plasmin-catalyzed fibrin dissolution in the maintenance of vascular haemostasis. By contrast, the activity of uPA is regulated by interactions at the surface of cells. Therefore, uPA is thought to be primarily responsible for generating pericellular plasmin activity and contributes to the proteolytic activities involved in extracellular matrix degradation and other mechanisms that regulate cellular behavior. Although the two plasminogen activators have well-defined roles in these processes, recent observations have suggested that they also have some overlap of function as well as some previously unsuspected functions.

This chapter will consider how the structure of these two distinct plasminogen activators relates both to their catalytic function in activating plasminogen and to the regulation of this catalytic function by other components of this system, including nonenzymatic cofactors, cellular binding sites, and inhibitors. It will also consider structure–function relationships of these enzymes in the sense of their wider biological roles.

Vincent Ellis • School of Biological Sciences, University of East Anglia, Norwich NR4 7TJ, UK.

Plasminogen: Structure, Activation, and Regulation, edited by David M. Waisman. Kluwer Academic/Plenum Publishers, New York, 2003.

2. Serine Proteases

Serine proteases of the trypsin family (family S1 using the MEROPS classification of Barrett *et al.* (2001), www.merops.co.uk) are the most abundant class of proteases. These proteases are a widely divergent family of enzymes present in animals, bacteria, and viruses, and perform a wide variety of functions, both extracellular and intracellular. They are characterized by a catalytic triad of His^{57}, Asp^{102}, and Ser^{195} (numbered according to the convention of using the sequence of chymotrypsinogen, when comparing sequences the prefix "c" will be used to denote this numbering, e.g., c-His^{57}). All are synthesized as single polypeptide chain precursors (zymogens) with an N-terminal extension of variable length that in virtually all cases acts as a propeptide, requiring proteolytic cleavage to generate the active enzyme. This proteolytic cleavage exposes a new N-terminus which is almost invariably a branched hydrophobic residue, Ile or Val. The free amino group of this Ile^{16} residue forms a salt-bridge with Asp^{194}, the residue preceding the catalytic serine. Conformational changes subsequently occur in four disordered peptide segments (16–19, 142–152, 184–194, and 216–233), collectively called the "activation domain" (Huber and Bode, 1978), resulting in the stabilization and maturation of the oxyanion hole (comprising backbone amides of Gly^{193} and Ser^{195}) and the substrate binding pocket. Only small adjustments in position are made in other regions (including the catalytic residues) during the activation process.

The active site cleft of the trypsin-like serine proteases is located at the interface between the two homologous sub-domains that form the chymotrypsin-fold. Each of these sub-domain consists of an open-ended 6-stranded antiparallel β-barrel, with the two barrels arranged at right-angles to each other. The β-barrel structure shows a crossing pattern of β-strands when viewed from the side and this topology has been termed "Greek key" after a common Greek ornamental motif. Although there is no sequence identity between these sub-domains, it is thought that the two-domain structure is the result of a gene duplication and fusion event, and that the ancestral gene possessed only one β-barrel domain (Lesk and Fordham, 1996). The three catalytic residues are divided between these sub-domains, with His^{57} and Asp^{102} in the first sub-domain and Ser^{195} in the second. The second sub-domain also contributes most of the substrate binding determinants, including the main-chain substrate binding region (Ser^{214}–Gly^{216}), which forms an antiparallel β-sheet with the substrate, and the primary specificity pocket. Asp^{189} is at the bottom of this pocket in trypsin, and interacts with the basic side chains of Lys or Arg P1 residues.* Asp^{189} is conserved in the plasminogen activators, which therefore have a primary specificity for basic residues with a strong preference for Arg.

The major differences between the catalytic domains of the serine proteases are concentrated in the loop regions linking the strands of the various β-sheets. These loops can accommodate quite large insertions and deletions, as well as differences in primary sequence. In many cases these loops are involved in substrate and inhibitor recognition, or provide interaction sites for other proteins. The plasminogen activators contain six significant insertions

*P1 in the nomenclature of Schechter and Berger (1967) refers to the substrate residue on the N-terminal side of the scissile bond and P1′ is the residue on the C-terminal side. Residues distant from the scissile bond in either direction are termed P2, P3, P2′, P3′, etc. The complementary binding sites for these substrate residues on the protease are similarly denoted S1, S1′, etc. Although the P sites constitute a single side-chain, the S sites are each usually composed of several different residues, since they define a binding surface.

(of up to eight residues) in their catalytic domains compared to chymotrypsin. Although both of the plasminogen activators catalyze the same reaction, their catalytic domains have distinct characteristics which play important roles in determining specificity. uPA and tPA have been shown to have distinct profiles for the recognition of small peptide substrates (Ke *et al.*, 1997) and the zymogenic form of tPA has unique properties among the serine proteases.

The catalytic domain of the serine proteases is responsible for the primary recognition of substrate, substrate specificity, and catalysis. However many serine proteases, including the plasminogen activators, have a modular construction with long N-terminal extensions containing multiple additional protein domains or modules (Patthy, 1985). To a large extent these regions dictate biological function, by mediating interactions with other proteins such as regulatory cofactors, cellular binding sites, and biological membranes. In some serine proteases these N-terminal domains are lost after activation, but in the plasminogen activators and plasminogen, as well as many other family members, a disulphide-bridge links the two polypeptide chains and the function of the regulatory domains is maintained. These additional protein modules are thought to have been acquired as a result of exon-shuffling during evolution, a process favored by the presence of introns of the same phase-class flanking the 5′- and 3′-ends of the module (Patthy, 1999).

The additional N-terminal modules present in the plasminogen activators include kringle, EGF-like, and fibronectin type-I modules. Some of these modules appear to have a purely structural role, but more importantly many of them provide interaction sites for other molecules. It is largely these interactions that allow the two plasminogen activators, which of course catalyze the same reaction, to have different mechanisms regulating their activity which is reflected in their differing biological functions.

3. Urokinase Plasminogen Activator, uPA

uPA is the product of the *PLAU* gene, which spans 6.4 kb of human chromosome 10q24. It is composed of 11 exons and 10 introns and the sequence identity between mammalian species is high (Degen *et al.*, 1987). Transcription of the gene produces a single mRNA form of 2.5 kb. uPA gene expression is regulated by a wide variety of cytokines, growth factors, and hormones in many cells in a cell type-specific manner.

uPA was originally known by its alternative name, urokinase, and was first purified from human urine as a 54 kD protein. It has three independent structural domains; an N-terminal EGF-like module (EG), a kringle module (KR) and the serine protease domain. Human uPA is secreted as a single-chain glycoprotein of 411 residues, with a single non-conserved N-linked glycosylation site at Asn^{302} and an unusual O-linked fucosylation at Thr^{18} (Buko *et al.*, 1991). This single-chain pro-enzyme or zymogen form (pro-uPA)* is activated by a single cleavage at Lys^{158}–Ile^{159} giving rise to two disulphide-bridged polypeptides. The N-terminal (A chain) of uPA, containing the EG and KR modules, remains covalently linked to the C-terminal (B chain) serine protease domain by a single disulphide bridge (Cys^{148}–Cys^{279}) (Steffens *et al.*, 1982). The activation cleavage can be catalyzed

*uPA is used here both to describe catalytically active, two-chain uPA and as a generic term also encompassing pro-uPA, the single-chain, zymogen form of uPA.

in vitro by a variety of trypsin-like proteases. Plasmin is probably the most important of these, as is discussed in more detail below.

Activated two-chain uPA can be further cleaved, either by plasmin or uPA itself, at the Lys^{135}–Lys^{136} peptide bond. This is in the extended 16-residue linker region between the KR module and the catalytic domain, and results in the release of the two N-terminal modules. The remaining C-terminal region consists of the catalytic domain (residues 159–411) disulphide bridged to a truncated A-chain fragment (residues 136–158) (Steffens *et al.*, 1982). A similar fragment can also be generated by cleavage at Glu^{143}–Leu^{144} by the matrix metalloproteases matrilysin (MMP-7) (Marcotte *et al.*, 1992) and stromelysin-1 (MMP-3) (Ugwu *et al.*, 1998). This form of the protein has been termed low molecular weight uPA and is the major form found in urine.

3.1. Serine Protease Domain

The structure of the catalytic domain of uPA, in complex with the peptide inhibitor Glu–Gly–Arg–chloromethane, has been determined by X-ray crystallography at a resolution of 2.5 Å (Spraggon *et al.*, 1995). Reengineering of the protease domain to give a more optimal crystal form by removing the largely disordered A-chain fragment and the extended C-terminus has allowed the structure to be refined to 1.5 Å, an ultra-high resolution that may facilitate structure-based therapeutic drug design (Nienaber *et al.*, 2000). The protease domain has an overall structure similar to that of homologous serine proteases, although at six positions insertions of extra residues in loop regions create unique surface areas. The largest difference is a four-residue insertion (Lys^{264}–Glu–Gly–Arg) in the "110-loop" (named from the structure and sequence of chymotrypsin, Figure 2.1). This is termed the "PA loop," as a similar four-residue insertion is also found in tPA. This is highly mobile despite being anchored by the disulphide bridge (Cys^{197}–Cys^{268}) which is characteristic of a small subset of serine proteases including tPA, coagulation Factor XII, and Complement Factor I. This disulphide appears to play a role in stabilizing this region (Davidow *et al.*, 1991). uPA has a disordered 5-residue C-terminal extension compared to other trypsin-like serine proteases after the conserved C-terminal α-helix. This region does not appear to play any functional role and can be modified without affecting the catalytic activity of uPA (Vines *et al.*, 2000).

The crystal structure shows the S1 specificity pocket to be similar to that of trypsin, consistent with the Arg/Lys primary specificity of uPA. The hydrophobic S2 pocket is restricted, and less accessible than that of trypsin, contributing to the narrow substrate specificity of uPA. The S3 pocket is solvent-accessible and appears capable of accommodating a wide range of residues, although substrate residues interacting with this pocket have a key role in determining specificity (Ke *et al.*, 1997). Both of the plasminogen activators have a stringent substrate specificity with plasminogen being the only clearly defined physiological substrate. However, although both activators have the same biological substrate their active sites are sufficiently dissimilar that they recognize small peptide substrates differently. Substrates have been identified by phage-display that are hydrolyzed by the two proteases with up to a 230-fold differential, largely dictated by the residues in the P3 and P4 positions of the substrate (Ke *et al.*, 1997). This differential is maintained when these sequences are introduced into heterologous proteins (Coombs *et al.*, 1998). The stringent substrate specificity of the plasminogen activators is also influenced by specific characteristics of the activation site in plasminogen. The Arg^{561}–Val^{562} scissile bond

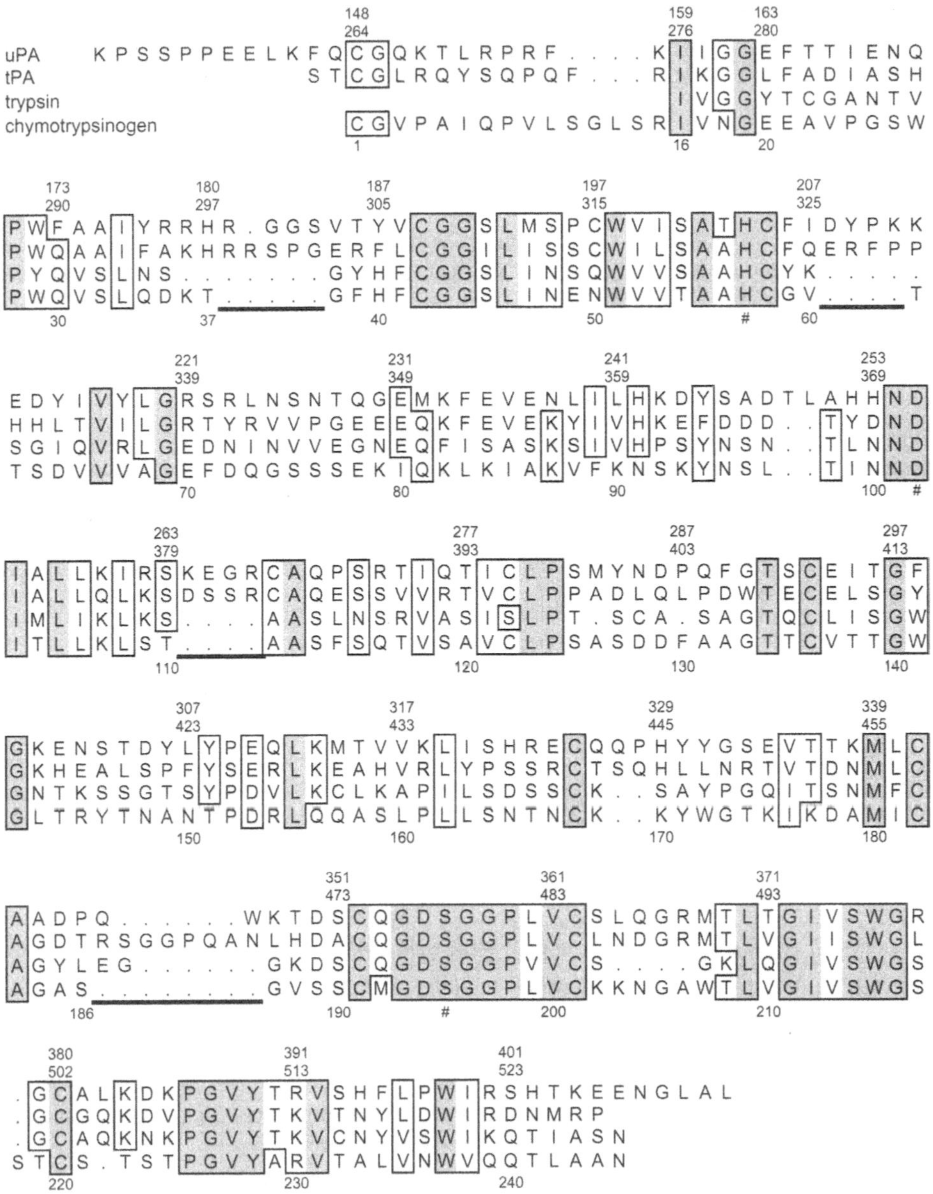

Figure 2.1. Sequence alignment of the catalytic domains of human uPA and tPA with the sequences of bovine trypsin and chymotrypsinogen A. The alignment is based on the topological equivalence of residues (Bode and Renatus, 1997). The sequence numbering of uPA and tPA, respectively, is shown above the alignment and the sequence numbering of chymotrypsinogen below it. Regions of sequence identity and similarity are indicated by gray and white boxes, respectively. Four significant insertion loops in the plasminogen activators, discussed in the text, are underlined in bold. The sequences of uPA and tPA include the linker region preceding c-Cys^{1}, which forms the interchain disulphide bridge with c-Cys^{122} in the activated protease. In the case of uPA the sequence is shown from the proteolytic cleavage site in the linker and therefore lacks four additional N-terminal residues.

is situated in a small disulphide-constrained loop, which is unique among the serine proteases (Wang *et al.*, 2000). Elimination of this Cys^{558}–Cys^{566} disulphide results in reduced activation by both of the plasminogen activators, but increases recognition by nonspecific proteases (Linde *et al.*, 1998).

3.2. N-terminal Domains

The solutions structures of the N-terminal modules of uPA have been determined by NMR spectroscopy. This has been done for the KR module (Li *et al.*, 1994), the EG–KR module pair (Hansen *et al.*, 1994b), and for the entire protein (Nowak *et al.*, 1993). These studies have led to a model of uPA dynamics which involves extensive, but not unrestricted, motion between the different domains. The serine protease domain appears to have extensive motional independence from the adjacent KR module with no intramolecular interactions, consistent with the autonomous unfolding of the protease domain during thermal denaturation of uPA (Novokhatny *et al.*, 1992). The independence of the serine protease domain may be of functional importance as it uncouples catalytic activity from the N-terminal regulatory domains and may be facilitated by the unusually long linker region between the two domains. The KR and EG modules have also been shown to have a high degree of independence (Hansen *et al.*, 1994a).

3.2.1. KR Module

The solution structure of the KR module of uPA determined by NMR spectroscopy (Bokman *et al.*, 1993; Li *et al.*, 1994) resembles the high–resolution crystal structures of KR modules from other proteins, including tPA (de Vos *et al.*, 1992) and plasminogen (Mulichak *et al.*, 1991). The module is named after a type of Danish pastry, having a triple loop structure connected by three disulphide bridges and small sections of anti-parallel β-sheet. The N- and C-termini are linked by one of the disulphide bridges, which forms the top of the laterally flattened, pear-shaped structure for the module. The KR module of uPA is most closely related to tPA KR-2, both structurally and by primary sequence having approximately 46% identity. Despite the close structural similarity, this module does not contain the "lysine-binding site" that characterizes both tPA KR-2 and those of plasminogen. Two of the four aromatic residues that form the hydrophobic cleft of the tPA KR-2 lysine-binding site are substituted by Val^{123} and Glu^{125} in uPA, a lysine residue in the cationic center is replaced by Tyr^{84} and one of the Asp residues in the cationic center is replaced by Arg^{108}. The latter is first of the three consecutive Arg residues that have been identified to be involved in the binding of heparin (Stephens *et al.*, 1992b). This extended heparin, or polyanion, binding site also involves His^{58} and His^{87} at the opposite end of the module (Li *et al.*, 1994). The significance of this site has not been demonstrated, but it may be involved in the binding of uPA to components of the extracellular matrix such as laminin-nidogen (Stephens *et al.*, 1992a).

3.2.2. EG Module

The EG module of uPA is of particular interest as it is responsible for the binding of uPA to its specific cellular receptor uPAR. This module contains all of the structural information required for high-affinity binding to uPAR, and uPA and the isolated module interact with uPAR equally well (Appella *et al.*, 1987; Ploug *et al.*, 1995). This is consistent with the previously discussed structural and dynamic independence of this module from the adjacent KR module of uPA.

This module consists of a major two-stranded antiparallel β-sheet followed by a loop to a minor C-terminal two-stranded β-sheet, a structure typical of many other EG modules (Hansen *et al.*, 1994b). It contains six cysteine residues all of which form disulphide bridges, Cys^{11}–Cys^{19} bridging the N-terminal region to the first strand of the major β-sheet, Cys^{13}–Cys^{31} bridging the N-terminal region to the second strand of the major β-sheet and Cys^{33}–Cys^{42} bridging the C-terminus of the major β-sheet with the connecting loop of the minor β-sheet.

3.2.2a. EG/uPAR Interaction. The two strands of the major β-sheet of the EG module are connected by a seven-residue Ω-loop. This is the most flexible region of the module, and contains the residues Tyr^{24}, Phe^{25}, Ile^{28}, and Trp^{30} that form a continuous exposed hydrophobic patch. Together with Lys^{23}, also in the Ω-loop, these residues have been shown to be important for the binding of uPA to uPAR (Ploug *et al.*, 1995; Magdolen *et al.*, 1996). The binding site on uPAR, although not well-defined at the molecular level, has been shown to be hydrophobic (Ploug *et al.*, 1994). The identification of this binding site has been hampered by the lack of a three-dimensional structure for uPAR. This GPI-anchored protein contains three homologous domains, termed LU-domains (Ly-6/uPAR domain), which are also found in snake venom α-neurotoxins (Ploug and Ellis, 1994). These proteins are structurally well-defined and have a characteristic "three-finger fold," with a three loops projecting from a central β-sheet stabilized by a small highly disulphide-bridged globular core. Mutagenesis of the N-terminal domain of uPAR has shown that Arg^{53}, Leu^{55}, Tyr^{57}, and Leu^{66} in the third loop of this domain are particularly important for uPA binding (Gårdsvoll *et al.*, 1999). However, in contrast to uPA, uPAR needs an intact domain structure for functional activity (Ploug *et al.*, 1994), suggesting the involvement of other domains in binding uPA. Consistent with this, the C-terminal domain of uPAR has been shown to be in close proximity to the N-terminal domain, as a small photo-activatable peptide antagonist of uPA-binding can cross-link these two domains (Ploug, 1998).

The three domains of uPAR are connected by long linker regions, and the linker region between its first two domains is unusually susceptible to proteolysis, containing five cleavage sites for various protease activities. Interestingly, this region contains the sequence Ser^{81}–Gly–Arg–Ala–Val which closely matches the P3–P2′ sequence of the uPA cleavage site of plasminogen (Ser–Gly–Arg–Ile–Val in bovine plasminogen) and is cleaved by uPA (Høyer-Hansen *et al.*, 1992). This makes uPAR the only well-defined substrate for uPA in addition to plasminogen and the plasminogen-related growth factor, HGF/SF. It is possible that this represents a negative regulatory mechanism, and it has been shown that the cleavage of uPAR on the cell surface is enhanced by uPAR-bound uPA (Høyer-Hansen *et al.*, 1997).

4. Mechanisms Regulating uPA Function

4.1. Zymogen Activation

Activation of the uPA zymogen occurs by a single proteolytic cleavage at Lys^{158}–Ile^{159}. The newly exposed N-terminal residue of the B chain, Ile^{159}, is structurally equivalent to the corresponding residue of other trypsin-like serine proteases (c-Ile^{16}), its α-amino group forming a salt bridge with the side-chain carboxylate of Asp^{355}. This cleavage can be catalyzed *in vitro* by a variety of trypsin-like and other proteases. These include trypsin itself (Ichinose *et al.*, 1986), plasmin (Gunzler *et al.*, 1982), plasma kallikrein (Ichinose *et al.*, 1986), mast cell tryptase (Stack and Johnson, 1994) and matriptase/MT-SP1 (Takeuchi *et al.*, 2000), the metalloprotease thermolysin (Marcotte and Henkin, 1993) and the cysteine protease cathepsin B (Kobayashi *et al.*, 1991). Plasmin is presumed to be the primary physiological activator of pro-uPA. However, the only protease that has been formally demonstrated to activate pro-uPA *in vivo* is the glandular kallikrein mGK-6, which was found to be responsible for the presence of active uPA in the urine of plasminogen-deficient mice (List *et al.*, 2000).

The main reason to suppose that plasmin is the primary physiological activator of pro-uPA is because the susceptibility of pro-uPA to activation by plasmin enables a system of "reciprocal zymogen activation," that is, the active protease of each zymogen/protease pair can activate the zymogen of the other protease/zymogen pair. Once proteolytic activity is present in a system of two such zymogens their reciprocal activation will result in the generation of proteolytic activity. Inherent in such a system is an amplification of the generation of proteolytic activity as is observed with pro-uPA (Lijnen *et al.*, 1986; Ellis *et al.*, 1987). The significance of reciprocal zymogen activation will be considered later in the context of the initiation of this proteolytic system.

Other proteases can cleave close to the Lys^{158}–Ile^{159} activation site and give rise to inactive two-chain forms. Thrombin can cleave at Arg^{156}–Phe^{157} leading to an inactive two-chain molecule (Ichinose *et al.*, 1986). This can subsequently be activated only very poorly by plasmin, but can be efficiently activated by removal of the N-terminal dipeptide by dipeptidyl-peptidase I (Nauland and Rijken, 1994). Leukocyte elastase cleaves at Ile^{159}–Ile^{160}, leading to a completely inactive two-chain molecule (Schmitt *et al.*, 1989). Ile^{160} at the exposed N-terminus cannot substitute for Ile^{159} in the activation mechanism, and neither can this molecule subsequently be activated.

4.2. Zymogen Activity

Serine protease zymogens are generally considered to be enzymatically inactive. However, all of these zymogens exist in an equilibrium between a catalytically inactive conformation with a disordered activation domain and an enzymatically active form with a folded activation domain mimicking that of the mature activated enzyme. For most serine protease this equilibrium lies far towards the inactive conformation and in the case of trypsinogen the equilibrium constant has been estimated to be 10^7 (Bode, 1979). This equilibrium can be affected by proteolytic cleavage at the activation site, which forces the zymogen irreversibly into the active state by formation of the Ile^{16} to Asp^{194} salt bridge, or by other mechanisms that affect the stability of these conformations. Some zymogens, most notably that of tPA, can achieve significant catalytic activity by concomitantly destabilizing the inactive conformation and stabilizing the active conformation of the protease. One of the forces responsible for stabilizing inactive zymogen forms is the presence of a "zymogen triad" Ser^{32}–His^{40}–Asp^{194} which antagonizes oxyanion hole formation (Wang *et al.*, 1985; Madison *et al.*, 1993). Apart from the conserved Asp^{194} this triad is absent in both of the plasminogen activators.

Pro-uPA appears to have a low level of intrinsic catalytic activity, although there has been considerable controversy concerning its absolute level. The sensitivity of pro-uPA to plasmin activation means that the determination of this intrinsic proteolytic activity is not trivial. Some investigators have reported an activity similar to that of the two-chain enzyme (Collen *et al.*, 1986), apparently due to a remarkably low K_m for plasminogen, 0.2 μM compared to the 20–50 μM observed in most laboratories with the activated enzyme (Collen *et al.*, 1986; Ellis *et al.*, 1987; Christensen, 1988). These data have not been corroborated by other laboratories, using a variety of strategies to suppress the activation of pro-uPA (Ellis *et al.*, 1987; Pannell and Gurewich, 1987; Petersen *et al.*, 1988b). These studies all show that pro-uPA has an intrinsic activity, but that it is less than 1% that of activated uPA. This conclusion was confirmed in studies using non-activatable pro-uPA with a mutated P1 residue (Lys^{158}→Glu) and plasminogen with a mutated active-site Ser^{740}→Ala, which demonstrated an intrinsic activity of approximately 0.2% (Lijnen *et al.*, 1990).

Despite disagreement concerning the exact magnitude of the intrinsic catalytic activity of pro-uPA, the key question is what the relevance of this intrinsic activity may be and can it represent the initiation mechanism for this proteolytic system? For many serine protease zymogens such a low degree of intrinsic activity may not be of biological relevance, but in this case, where reciprocal zymogen activation can occur, it may be critical for the function of the proteolytic system.

4.3. Reciprocal Zymogen Activation

The degree of amplification achieved in a reciprocal zymogen activation system is dependent on the efficiencies of the proteolytic reactions involved, and in particular the activation of the first zymogen (pro-uPA in this case) which is likely to be present at a much low concentration than the second zymogen (plasminogen). In the absence of contaminating active protease the initiation of the reciprocal activation system will depend on the presence of intrinsic activity in either one of the two zymogens. If pro-uPA and plasminogen are allowed to react in solution reciprocal zymogen activation is apparent, but very slow (Ellis *et al.*, 1989). This is a consequence of both the low intrinsic activity of pro-uPA and, possibly more importantly, the low efficiency of the proteolytic reactions. uPA is a relatively inefficient protease, its K_m for plasminogen is extremely high, in the range 20–50 μM. Even though plasminogen is the most abundant serine protease zymogen in plasma at 2 μM, its concentration is still well below this K_m. Therefore, it must be expected that a mechanism exists either to increase the intrinsic activity of one of the zymogens or to increase the efficiency of the catalytic reactions.

4.4. uPAR Stimulation of Plasminogen Activation

This mechanism is provided by uPAR. If pro-uPA and plasminogen are incubated in the presence of cells expressing uPAR, there is a dramatic increase in plasmin generation (Ellis *et al.*, 1989). The is due to effects on both catalytic reactions, with up to a 100-fold increase in the catalytic efficiency of plasminogen activation by active uPA and a greater than 50-fold increase in the efficiency of pro-uPA activation (Ellis *et al.*, 1989, 1991). The increase in uPA activity is a consequence of a large effect on the K_m for plasminogen which is reduced to 0.1 μM, now approximately 20-fold below its plasma concentration. In addition to requiring

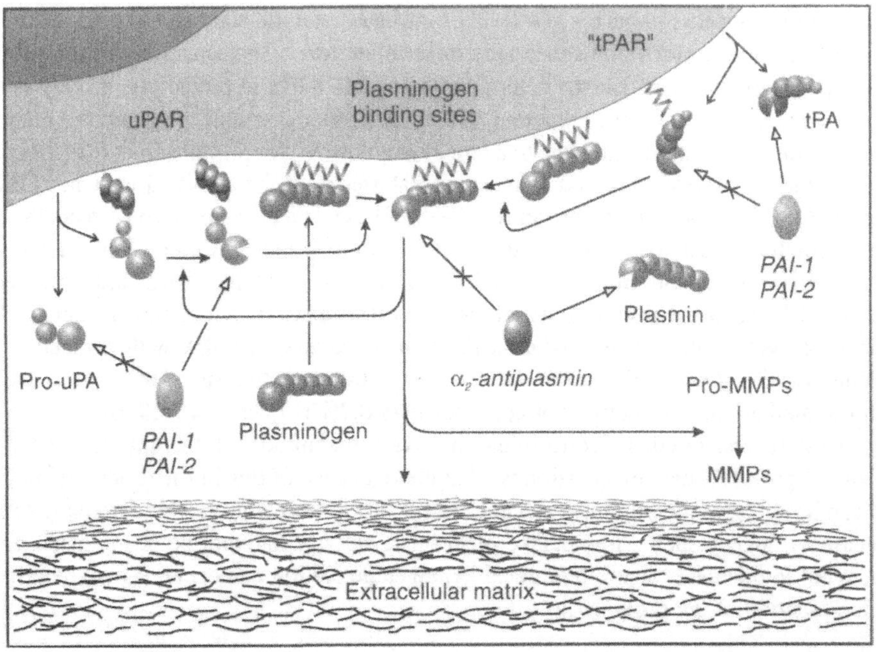

Figure 2.2. Receptor-mediated plasminogen activation. Two independent pathways for pericellular plasmin generation are shown with the principal reactions between membrane-bound components. The binding of secreted pro-uPA to its cellular receptor, uPAR, leads to efficient reciprocal zymogen activation in which uPA preferentially activates cell-bound plasminogen, and cell-bound plasmin can efficiently activate uPAR-bound pro-uPA. Similarly, the binding of secreted tPA to its receptor(s) leads to the efficient activation of cell-associated plasminogen. Generated plasmin can degrade many non-fibrillar proteins of the extracellular matrix, activate certain matrix metalloproteases, and activate or release matrix-bound growth factors (reviewed in (Ellis and Murphy, 2001)). These pathways are modulated by physiological inhibitors. α_2-antiplasmin acts to further focus plasmin activity at the cell surface, as plasmin that is generated is fully protected from inhibition while it remains bound but is rapidly inhibited on dissociation from the cell surface. The plasminogen activator inhibitors, in particular PAI-1, appear to have contrasting functions in the regulation of uPA and tPA activities. Pro-uPA is not inhibited by PAI-1 but once activated to uPA the receptor-bound enzyme is fully available for inhibition (Ellis *et al.*, 1990). In contrast, nascent tPA is susceptible to inhibition by PAI-1, but it can be protected when bound to cellular binding sites (Werner *et al.*, 1999). These and other aspects of the cellular regulation of plasmin generation have recently been reviewed (Bass and Ellis, 2002).

the binding of uPA to uPAR, the overall effect on plasminogen activation also requires the cellular binding of plasminogen, which can be provided by a variety of cell-surface molecules (Plow *et al.*, 1991). The principal interactions of this system are shown in Figure 2.2.

4.4.1. uPA and the Template Mechanism

These increases in catalytic efficiency could be due either to direct molecular effects on the activity of the proteases or their zymogens upon binding, that is, allosteric effects, or to the provision of a catalytic template for the assembly of the reactants. There is compelling experimental evidence to support the latter mechanism. Experiments with a truncated, soluble form of uPAR have shown that this does not potentiate plasminogen activation either in the presence of active uPA or pro-uPA (Ellis *et al.*, 1991, 1996). This is consistent with the observation that binding of uPA or pro-uPA to cellular uPAR has no effect in the absence of plasminogen binding (Ellis *et al.*, 1989; Ellis, 1991), and suggests that it is the concomitant binding of the reactants that is the important event. These observations also suggest that the interaction with uPAR has no effect on the intrinsic activity or zymogenicity of pro-uPA, although it is difficult to demonstrate this definitively in a complex cellular milieu. Further support for the catalytic template mechanism comes from the observation that pro-uPA-initiated plasminogen activation is potentiated by an anti-uPA monoclonal antibody (Ellis and Danø, 1993), and this occurs with kinetics indistinguishable from those determined for cellular uPAR. The mechanism of this effect is the formation of a ternary complex with the antibody recognizing an epitope in the KR module of uPA and binding plasminogen via a C-terminal Lys residue on the IgG heavy chains. In this much less complex system it was clearly shown that no effects other than appropriate juxtaposition of the reactants were necessary to mediate this catalytically efficient system. In addition, this system allowed the construction of a kinetic model which could account for all the experimental observations on uPAR-mediated plasminogen activation. Using this model it can be shown that, as the increased efficiency of the reactions is due to juxtaposition of protease and substrate, increasing the effective concentration of the latter, the zymogen will also experience a similar increase in activity, that is, the activities of the active protease and zymogen will be increased in parallel. Therefore, due to reciprocal zymogen activation and the kinetically favorable conditions provided by the template, whether it be the antibody or cellular uPAR, plasminogen activation will be very efficient and initiated simply by the assembly of this complex, not requiring the presence of activated proteases. This model also predicts that the exact degree of zymogen activity pro-uPA is not critical, and any discrete level of intrinsic activity will lead to efficient initiation of plasminogen activation (Ellis and Danø, 1993).

4.4.2. Plasminogen and the Template Mechanism

The observations with the anti-uPA monoclonal antibody described above further suggest that the catalytic template provided on the cell surface by uPAR and the plasminogen binding sites must in some way be preformed and specific, and not due to random interactions between the cell-bound components. This is highlighted by the observation that plasminogen activation in the presence of soluble uPAR can be stimulated by poly-Lys, which can bind both soluble uPAR and plasminogen and therefore also provide a catalytic template (Petersen, 1997). However, in this case the increase in overall activity was far less than that observed either with cellular uPAR or the monoclonal antibody. What then forms the catalytic template on the cell surface? Unexpectedly, uPAR does not have a direct role in its formation, as uPA directly anchored to the plasma membrane by either a C-terminal glycolipid or transmembrane anchor have catalytic characteristics very similar to those of uPAR-bound uPA (Lee *et al.*, 1994; Vines *et al.*, 2000). Therefore, in terms of the functional activity of this system, uPAR appear to play a rather passive role, localizing uPA at the cell surface, to interact with bound plasminogen. Plasminogen can bind to a variety of molecules on the cell-surface, which could be important for the function of this system (Plow *et al.*, 1991). However, recent observations, which are consistent with the notion of the specificity of the template, suggest that uPA itself may provide the binding site for plasminogen. Plasminogen has been shown to bind to uPA via an "exosite" interaction, that is, independent of the active-site, with a K_d that is close to the K_m for plasminogen activation on the cell surface, 50 nM versus 100 nM, respectively (Ellis *et al.*, 1999). This interaction has not been fully characterized, but may involve the C-terminal region of the A-chain of uPA. In this scenario, uPAR-bound uPA would act as the functional plasminogen binding site with the exosite interaction acting to orientate plasminogen in a catalytically favorable manner. Interestingly, and in support of this model, it has been observed that a significant proportion of plasminogen co-localizes with uPA on the surface of a breast cancer cell line (Andronicos and Ranson, 2001).

4.5. Avian uPA, a Special Case?

The catalytic behavior of uPA in the pericellular environment can be fully explained by the mechanism described above. However, it cannot be excluded that in certain situations other mechanisms could play a role. In this respect there appear to be interesting differences in the functional regulation of uPA in avian species. Chicken uPA has been shown to be capable of plasmin-independent autocatalytic activation (Alexander *et al.*, 1998), a property which mammalian uPA does not have. There is also evidence that this activation is enhanced by interactions with the cell-surface, presumably via uPAR (Berkenpas and Quigley, 1991). The mechanism underlying this autoactivation is not known, but it may involve regions of the protein outside of the catalytic and activation sites as there are significant differences between the avian and mammalian proteins. Chicken uPA has both N- and C-terminal extensions compared to the mammalian proteins, of nine and six residues, respectively, and these regions have no overall similarity to those of the mammalian enzymes. The chicken enzyme additionally lacks the long linker region between the KR module and the catalytic domain, this being only three residues compared to 16–25 residues in mammalian species. Another difference demonstrated by chicken uPA is that the inserted "37-loop" found in the plasminogen activators, and involved in the interaction with PAI-1, contains no basic residues. Consistent with this human PAI-1 inhibits chicken uPA approximately 1,000-fold less efficiently than it does human uPA, but inhibition can be reinstated by substituting the Arg^{178}–Arg–His–Arg motif from the human enzyme (Sipley *et al.*, 1997). Although not well studied, PAI-1 has not been identified in birds. Together these observations suggest that autoactivation may be the initiation mechanism in the avian system, possibly eliminating the need for reciprocal zymogen activation as it occurs in the mammalian system. Such a system would not then need the suppressing effect of the highly efficient inhibitory mechanism provided by PAI-1.

5. Tissue Plasminogen Activator, tPA

tPA is the product of the *PLAT* gene which spans 32.4 kb of human chromosome 8p12–q11.2. This very large gene comprises 14 exons separated by 13 introns which account for 92% of the *PLAT* gene, and is transcribed to a single mRNA form of 2.5 kb. The expression of tPA *in vivo* is markedly different from that of uPA, with a much more restricted distribution. Endothelial cells, vascular smooth muscle cells and keratinocytes all express tPA *in vivo*, and it is also quite widely expressed in the brain in cells of both neuronal and non-neuronal origin.

tPA was first identified as an agent from animal tissues that could activate plasminogen (Astrup and Permin, 1947), although it was not until the 1970s that it was conclusively shown to be identical to various other plasminogen activator activities (e.g., vascular and blood plasminogen activator) and distinct from uPA. This became more significant when tPA was found to have unique properties in relation to plasminogen activation and fibrinolysis, first that tPA had specific affinity for fibrin (Thorsen *et al.*, 1972) and second that fibrin greatly stimulated plasminogen activation (Hoylaerts *et al.*, 1982). Therefore, tPA is a fibrin-specific fibrinolytic protease, in contrast to uPA which will equally lead to the degradation of fibrinogen (Rånby, 1982). It is now apparent that tPA can interact with other molecules that have effects similar to fibrin. These molecules include cell surface proteins that may have a role analogous to that of uPAR.

tPA is a 70 kD mosaic protein composed of five structural domains, an N-terminal fibronectin type-I module (F1), an EGF-like module (EG), two kringle modules (KR), and the C-terminal serine protease domain. The secreted protein contains 530 residues and has three potential N-glycosylation sites (Asn^{117}, Asn^{184}, Asn^{448}). tPA usually comprises a mixture of glycoforms with either all sites glycosylated or with Asn^{184} non-glycosylated. The latter has twice the activity of the fully glycosylated form in fibrinolytic assays (Mori *et al.*, 1995).

The single-chain form of tPA can be proteolytically cleaved by plasmin at a canonical activation site of Arg^{275}–Ile^{276}. As mentioned previously, tPA has an unusually high degree of zymogen activity (10–20% that of the two-chain form) and this cleavage consequently has a much smaller effect on the activity of tPA than it would with uPA or other serine protease zymogens. In addition, binding to fibrin causes single-chain tPA to become equally as active as two-chain tPA. For these reasons the two forms are referred to as single-chain (i.e., rather than pro-tPA) and two-chain tPA.

The interaction of tPA with fibrin is complex and not yet fully understood. The roles of the individual domains of tPA in this interaction have been addressed by site-directed mutagenesis and domain deletion or replacement mutations, but the interpretation of the data from these experiments has proven not to be trivial. The reason for this is that the individual modules of tPA are far less independent than those of uPA. They are connected by short linker peptides, which enforce their close proximity although still allowing some degree of flexibility and independence. For example, the F1 and EG modules are intimately associated (Smith *et al.*, 1995) and the KR-2 module is associated with the protease domain (Hu *et al.*, 1996). Therefore, function may be disrupted when domains are not presented in the correct protein context. More subtle alterations may also indirectly affect function as alanine scanning mutagenesis has shown that mutations in one module of tPA can affect functions assigned to other modules of the protein (Bennett *et al.*, 1991). Nevertheless, the interaction of tPA with fibrin appears to involve at least three discrete sites on the molecule: in the F1-EG modules, in the KR modules, and in the serine protease domain.

5.1. Serine Protease Domain

The structure of the catalytic domain of tPA in complex with the inhibitor benzamidine has been determined by X-ray crystallography at a resolution of 2.3 Å (Lamba *et al.*, 1996). The domain is very similar to that of uPA, but there are notable differences. The largest sequence difference occurs in the "186-loop," which in tPA contains a six-residue insertion (Ser^{463}–Gly–Gly–Pro–Gln–Ala) compared to uPA or trypsin (see Figure 2.1). This loop, adjacent to the proteolytic activation site, is disordered and extremely exposed in the crystal structure. Its importance is not clear, but the residues Asp^{460}–Arg^{462} immediately preceding the insertion, may be involved in the interaction with fibrin (Paoni *et al.*, 1993). The "99-loop," one of the loops surrounding the active-site cleft forming an acidic rim (Asp^{364}–Asp–Asp–Thr–Tyr), is two residues shorter than in uPA. Tyr^{368} in this loop (c-Ile^{99}) restricts the size of the S2 subsite (Gly is the preferred P2 residue for both tPA and uPA) and forms one wall of the hydrophobic S3/S4 groove with the conserved Trp^{498} (c-Trp^{215}) at its base, and bordered on the opposite side by Arg^{449} (a Ser residue in uPA). The comparative shortness of this loop in tPA allows increased access to the S3 and S4 sites, and largely accounts for the previously mentioned ability of the two plasminogen activators to discriminate between peptide substrates. The substrates demonstrating the greatest selectivity for tPA have large P3 and P4 residues, with a preference for Arg and large hydrophobic residues, respectively (Ke *et al.*, 1997). tPA has a four-residue insertion similar to that in uPA, the partially flexible "110-loop" or PA-loop, comprising Asp^{380}–Ser–Ser–Arg. This loop is on the opposite side of the domain to the active-site, and may be involved in interaction with the KR-2 module, as described below.

Another insertion found in the plasminogen activators is the exposed, basic and mobile "37-loop," which in tPA is a five residue insertion compared to chymotrypsin. The first four residues of the Arg^{298}–Arg–Ser–Pro–Gly insertion are disordered in the crystal structure, but modeling suggests that the two Arg residues point away from the molecular surface (Lamba *et al.*, 1996). This loop has been shown to be of importance for the interaction of PAI-1 with tPA (Madison *et al.*, 1989). Deletion of this loop leads to an approximately 500-fold reduction in the rate of inhibition by PAI-1, and with charge-reversal mutations (Arg $>$ Glu) this increased to approximately 3,000-fold (Madison *et al.*, 1990). A similar deletion mutation in uPA has a similar protective effect (Adams *et al.*, 1991). The importance of this region for inhibition by PAI-1 is further highlighted by the observation that introduction of the tPA sequence into the analogous loop of the serine protease thrombin increases its sensitivity to inhibition by PAI-1 by over 2,000-fold (Horrevoets *et al.*, 1993).

The structure of single-chain tPA has also been solved by X-ray crystallography (Renatus *et al.*, 1997a). This structure is largely identical to that of two-chain tPA, with the main differences found in the "activation domain" and surrounding surface loops, and the C-terminal part of the A-chain. The latter is disordered in the two-chain structure, but follows the surface of the molecule in the single-chain structure. This structure has been useful in confirming features underlying the low zymogenicity and high activity of single-chain tPA that had been proposed based on the two-chain structure and mutagenesis experiments, particularly the mechanism of stabilization of Asp^{477} (Asp^{194} in chymotrypsin) in the absence of proteolytic cleavage at Ile^{276} (c-Ile^{16}).

A feature of the tPA serine protease domain is a hydrophobic surface depression, on the opposite side of the domain to the active-site, proposed to accommodate the adjacent KR-2 module. Only two residues separate the protease domain from the adjacent KR module, compared to the long linker in uPA, but, despite its shortness, this linker appears to be non-constrained and therefore flexible. Molecular modeling of interdomain movements around this linker has allowed docking of the KR and serine protease domains (Lamba *et al.*, 1996). The hydrophobic depression identified as the KR-2 interaction site is centered on the strictly conserved Leu^{396} (c-Leu^{123}) and includes Ile^{312} and Val^{394}. This depression is adjacent to Cys^{264} (c-Cys^{1}) in the A-chain which forms the interchain disulphide with Cys^{395} (c-Cys^{122}) and is delimited by the exposed surface loop around Asp^{487}, the segment Pro^{398}–Asp^{400} adjacent to the A-chain attachment site (i.e., Cys^{395}), a segment of the C-terminal α-helix (Leu^{518}–Pro^{527}) and the four residue "PA loop." Biophysical studies of interdomain interactions in tPA are broadly in agreement with such a site on the catalytic domain. Although studies using differential scanning calorimetry failed to identify a stable interaction between KR-2 and catalytic domains of tPA (Novokhatny *et al.*, 1991), ^{1}H NMR studies suggest some mutual stabilization of both domains involving primarily hydrophobic forces (Hu *et al.*, 1994). This interaction appears to be more pronounced in the single-chain form of tPA (Hu *et al.*, 1996), which may reflect changes in the organization of the A-chain subsequent to activation, although the structures of the hydrophobic depression of the two forms of tPA are very similar.

5.2. N-terminal Domains

5.2.1. KR Modules

The two KR modules of tPA are very similar, sharing 54% sequence identity, and conform to the consensus sequence for this module. The crystal structure of KR-2 has been solved and is similar to other solved KR structures (de Vos *et al.*, 1992). Although the structure of KR-1 has not been determined, modeling suggests that it is close to that of KR-2. However, the two modules are functionally dissimilar, as KR-1 lacks the "lysine-binding site" of KR-2. This binding site was first mapped by ^{1}H NMR with 6-aminohexanoic acid as a ligand (Byeon and Llinas, 1991) and subsequently confirmed by the crystal structure in which the site is occupied by a Lys side-chain of another molecule in the crystal packing (de Vos *et al.*, 1992). The binding site in KR-2 is an exposed cleft, at one end of which is a negatively-charged region consisting of the side-chains of Asp^{236} and Asp^{238} that interact with the side-chain amino-group of the ligand. The cleft itself is an aromatic pocket consisting of Trp^{242}, His^{244}, and Trp^{253}, with the indole rings of the two Trp residues forming a V-shaped groove that is in close contact with the methylene backbone of the bound ligand. At the other end of this groove is a site for interaction with the free carboxylate group of C-terminal Lys residues or ω-amino acids such as 6-aminohexanoic acid. This site consists of Lys^{212} with NMR studies suggesting a possible involvement of Arg^{250}, although mutation of this residue appears to have no detectable effect on ligand binding (de Vos *et al.*, 1992). Perhaps surprisingly, all of the charged residues involved in lysine binding are also present in the non-lysine-binding KR-1 of tPA. The difference appears to be due to changes in the hydrophobic interactions, as the residue equivalent to Trp^{253} in the aromatic groove is replaced by a Ser residue. In agreement with this a Trp^{253} to Ser mutation in KR-2 has been shown to lack lysine-binding function (de Vos *et al.*, 1992).

5.2.2. F1-EG Supermodule

The F1 and EG modules of tPA are best considered as a module pair as there are extensive interactions between them. Nevertheless the structure of the isolated F1 module has been determined by ^{1}H NMR (Downing *et al.*, 1992). It shows close similarity to the seventh type 1 repeat of fibronectin with the side-chains of conserved residues lying in similar conformations. This small module (43 residues) consists of a short double-stranded β-sheet stacked on a longer twisted triple-stranded β-sheet with two disulphide bridges, one in the major β-sheet, the other linking this to the minor β-sheet giving an extremely thermodynamically stable structure (Novokhatny *et al.*, 1991). A notable characteristic of the F1 module of tPA is the abundance of hydrophobic residues on the exposed surface of the triple-stranded β-sheet. In the structure of the superpair, which has been solved by heteronuclear NMR (Smith *et al.*, 1995), the residues linking the two modules make up a third strand in the major β-sheet of the EG module, burying the hydrophobic residues of the F1 β-sheet (centered at Leu^{22}/Val^{24}) at an interface with hydrophobic residues in the major β-sheet of the EG module (centered at Ala^{65}/Val^{72}). This stabilizes the module pair, but allows both modules to adopt conformations close to their consensus modular structure.

This supermodule is thought to have the primary role in fibrin binding. The residues involved in this interaction are not well defined, but appear to be situated mainly in the F1 part of the supermodule (Bennett *et al.*, 1991), where strand-connecting loop regions form part of a groove at the interface of the two modules. Thermal unfolding behavior of tPA monitored by differential scanning calorimetry suggests that the F1-EG supermodule interacts with the catalytic domain of tPA (Novokhatny *et al.*, 1991), which would give tPA a compact structure. The reason for this is not known, but it should be noted that plasminogen has a similar behavior (Mangel *et al.*, 1990).

6. Mechanisms Regulating tPA Function

6.1. Zymogenicity

tPA is unusual in that the single-chain zymogen possesses a very high level of intrinsic catalytic activity. This has been measured against plasminogen, synthetic peptide substrates and both natural and synthetic inhibitors. Broadly speaking, the ratio of the activities of two-chain and single-chain tPA (i.e., the "zymogenicity") is similar with all of these, although some differences are apparent. Plasminogen, peptide substrates, and peptide-based inhibitors all show an approximately 10-fold differential between the two forms of tPA (Rånby, 1982; Petersen *et al.*, 1988a; Boose *et al.*, 1989), estimates which have been confirmed using non-cleavable mutants of tPA. Small inhibitors display a smaller differential, 2.4-fold in the case of DFP (Rånby, 1982), whereas some macromolecular inhibitors, notably PAI-2, display a ratio of approximately 100 (Thorsen *et al.*, 1988). By contrast the more specific inhibitor PAI-1 has a differential activity of only 4-fold (Thorsen *et al.*, 1988). Despite these differences the low zymogenicity of tPA is always evident.

The molecular basis of the low zymogenicity of tPA has been addressed both by site-directed mutagenesis and by examination of the structures of the single- and two-chain forms of the enzyme. The central question is what leads to the stabilization of Asp^{477} ($c\text{-}Asp^{194}$) in the absence of the salt-bridge to Ile^{276} ($c\text{-}Ile^{16}$), which is only possible after the activation cleavage and exposure of the new N-terminal α-amino group. An early candidate for this was the side-chain of Lys^{277} ($c\text{-}Lys^{17}$), adjacent to the cleavage site, which in all other serine proteases is an uncharged residue. Lys^{416} ($c\text{-}Lys^{143}$) was also implicated from homology modeling, but site-directed mutagenesis of either of these residues proved to have little or no effect on the zymogenicity of tPA (Petersen *et al.*, 1990).

The determination of the crystal structure of two-chain tPA revealed that another previously suspected candidate, Lys^{429} (c-Lys^{156}) (Nienaber *et al.*, 1992), was potentially responsible for the stabilization of Asp^{477} (Lamba *et al.*, 1996). Mutation of this residue reduced the activity of single-chain tPA by approximately 50-fold, without affecting the activity of two-chain tPA (Tachias and Madison, 1997). The subsequent determination of the crystal structure of single-chain tPA confirmed the presence of a salt-bridge between Lys^{429} and Asp^{477}, definitively demonstrating that the side-chain amino group of Lys^{429} substitutes for the α-amino group of Ile^{276} in the tPA zymogen (Renatus *et al.*, 1997a). c-Lys^{156} is not unique to tPA, but additional structural determinants in tPA shield this residue from solvent and therefore favor salt-bridge formation and all other proteases with Lys at this position have hydrogen bond acceptors in the vicinity which prevent salt-bridge formation and help to stabilize the inactive zymogen (Renatus *et al.*, 1997a).

Stabilization of Asp^{477} (c-Asp^{194}) is a large part of the mechanism underlying the high catalytic activity of single-chain tPA, but the Lys^{429} (c-Lys^{156}) mutation is still significantly more active than most serine protease zymogens. This is probably due to the absence of interactions that stabilize the inactive form in other serine proteases, in particular the previously mentioned "zymogen triad" of c-Ser^{32}, c-His^{40} and c-Asp^{194}. The first two residues of this triad are replaced by Phe^{305} and Ala^{292} in tPA, and mutation of these residues to Ser and His resulted in a 20-fold increase in tPA zymogenicity (Madison *et al.*, 1993). Therefore both the presence of interactions that increase the stability of the "active" tPA zymogen and the absence of interactions that would otherwise stabilize the "inactive" zymogen contribute to the unusual catalytic properties of tPA.

The high intrinsic activity of single-chain tPA appears to be contrary to the fundamental biological principle of zymogens, that is, that proteases are synthesized as inactive precursors to prevent premature proteolysis. The reason that tPA can exist in an active form is that it is a very inefficient enzyme. The catalytic efficiency (defined as the ratio k_{cat}/K_m) of single-chain tPA against plasminogen is in the order of $1 \times 10^3\ M^{-1}s^{-1}$, compared to $1 \times 10^5\ M^{-1}s^{-1}$ for uPA, which itself is a relatively poor enzyme. This means that at physiological concentrations of plasminogen (2 μM) one molecule of tPA will take approximately 10 min to generate one molecule of plasmin. tPA is similarly inefficient against peptide substrates, a peptide corresponding to the natural sequence of plasminogen is cleaved with a catalytic efficiency of only $0.3\ M^{-1}s^{-1}$. For comparison trypsin can cleave this substrate 10^5-fold faster. This is partly due to the unusually poor nature of the plasminogen sequence as a substrate for tPA, as using phage-display techniques much better peptide substrates for tPA have been selected, having catalytic efficiencies up to $1.5 \times 10^3\ M^{-1}s^{-1}$ (Ding *et al.*, 1995). Nevertheless these substrates are still cleaved 1,500-fold more efficiently by trypsin. Inhibitors also have an important role in suppressing tPA activity, as although single-chain tPA is a very poor plasminogen activator, it still reacts efficiently with its inhibitor PAI-1.

6.2. Fibrin Stimulation of Plasminogen Activation

tPA overcomes its inherent inefficiency by using fibrin as a nonenzymatic cofactor. In the presence of fibrin plasminogen activation is increased by up to three orders of magnitude (Hoylaerts *et al.*, 1982; Rånby, 1982). The activity of single-chain tPA is increased to a greater extent than that of two-chain tPA, meaning that both forms of the protease are equally active in the presence of fibrin. Fibrin affects the activity of only single-chain tPA against small peptide substrates, and again the activity is increased to equal that of two-chain tPA (Petersen *et al.*, 1988a). This suggests that binding to fibrin shifts the conformational equilibrium of the protease domain of tPA to the fully active form, although this only accounts for a small part of the fibrin-mediated enhancement.

Plasminogen also binds to fibrin and the major mechanism responsible for the enhancement of plasminogen activation can be attributed to the assembly of tPA and plasminogen on fibrin as a cofactor "template." This acts to increase local concentrations of tPA and plasminogen and to position the active-site of tPA to interact with the plasminogen activation site in a catalytically productive manner (Hoylaerts *et al.*, 1982). More complex models have also been proposed in which the catalytic efficiency of the reaction is determined by the overall stability of this ternary complex, rather than the affinities of the individual binding interactions (Horrevoets *et al.*, 1997). The molecular basis of the interaction between tPA and fibrin underlying these catalytic effects has been studied extensively but is still not fully understood. This is due to various factors including the complexity of the interaction, the complex relationship between tPA binding and effects on catalytic activity, difficulties presented by using fibrin, an insoluble polymer, in biochemical systems and its degradation by plasmin which generates additional binding sites for both tPA and plasminogen.

6.2.1. tPA/fibrin Interaction

The interaction between tPA and fibrin is thought to involve at least three sites on the protease. The most important of these is the site on the F1/EG module pair (Verheijen *et al.*, 1986; van Zonneveld *et al.*, 1986). tPA mutants with these modules deleted have an approximately 10-fold lower affinity for fibrin and a similar reduction in activity in the presence of fibrin (Horrevoets *et al.*, 1997). Much emphasis has been placed on the role of KR-2 and its lysine-binding site, which is thought to have an important role in binding to C-terminal Lys residues exposed by plasmin activity during fibrin dissolution. Consistent with this, mutants of tPA containing only KR-2 and the catalytic domain have significant fibrin-stimulated activity (de Vries *et al.*, 1991). However, others have shown that mutations in KR-2 that abolish lysine binding have no effect on either fibrin affinity or fibrin-stimulated activity (Bennett *et al.*, 1991). Some investigators have suggested that KR-1, despite it lack of a lysine-binding site, can substitute for KR-2 when the latter is deleted (Gething *et al.*, 1988). A site on the protease domain is also implicated, both from the observation that fibrin directly affects the catalytic activity of single-chain tPA against peptide substrates and from the effect of point mutations in the protease domain (Paoni *et al.*, 1993). This fibrin attachment site is thought to involve charged residues from four segments; Glu^{408}, residues His^{432}–Arg^{434}, residues Asp^{460}–Arg^{462} of the inserted "186-loop" of tPA, and possibly Arg^{275} (c-Arg^{15}) in the activation loop of single-chain tPA (Bennett *et al.*, 1991). This attachment site is directly adjacent to the activation domain, which is therefore potentially influenced by fibrin binding. Molecular models have been made in which the modules of tPA can be connected to the protease domain in such a way as to align the various fibrin interaction sites (Bode and Renatus, 1997).

Considering the presence of multiple binding sites, tPA does not bind fibrin with particularly high-affinity, estimates vary widely but a K_d of 0.1–0.5 μM is commonly found (Nesheim *et al.*, 1990; Horrevoets *et al.*, 1994). There are multiple tPA binding sites on fibrin and they are heterogeneous. Conformational changes in the terminal globular domains of fibrin upon polymerization reveal cryptic binding sites for tPA on both the Aα and γ chains (Schielen *et al.*, 1989, 1991), and other sites have also been identified in the same region of the molecule (Tsurupa and Medved, 2001). Plasmin cleavage of fibrin may either generate new sites or destroy existing ones. Therefore, it is possible that the binding sites on tPA are not optimal for any one of the heterogeneous sites on fibrin, but a compromise that allows maximum interaction with fibrin and that under different circumstances the relative importance of the individual sites on tPA may vary.

6.2.2. Vampire Bat Plasminogen Activator

Many of the features of tPA described here are also present, but to a much higher degree, in a plasminogen activator isolated from the vampire bat, *Desmodus rotundus*. This enzyme is closely related to tPA, having a similar modular structure but lacking the lysine-binding KR-2 module. This enzyme is unique amongst the serine proteases in lacking a protease-sensitive activation site, having a His rather than Arg or Lys at P1. This protease is much less active than single-chain tPA and can be considered a true zymogen. In the absence of proteolytic cleavage to act as the "switch" for conversion to the active form, this instead occurs when the protease binds to fibrin (Bringmann *et al.*, 1995). Determination of the crystal structure of this plasminogen activator has again revealed that, like tPA, it uses c-Lys^{156} to stabilize the active conformation (Renatus *et al.*, 1997b). However, in this case interaction of the catalytic domain with fibrin is thought to stabilize the structure of the activation domain, allowing formation of the critical salt-bridge between c-Lys^{156} and c-Asp^{194}. The consequence of these unusual characteristics is that fibrin enhances the activity of vampire bat plasminogen activator to a much greater extent than with tPA, the stimulation factor being in the region of 10^5-fold (Bringmann *et al.*, 1995). Although this is the only known serine protease without an activation cleavage site, there are precedents for the generation of catalytic activity in the absence of cleavage at the activation site. The most well-studied example of this is the generation of catalytic activity in plasminogen when complexed to the nonenzymatic protein streptokinase.

6.3. Cellular Mechanisms Regulating tPA Activity

An intriguing observation regarding the fibrin stimulation of tPA is that mutants with a fibrin specificity much superior to wild-type tPA have proven relatively easy to design, and mutations at multiple sites on the protease domain have this effect (Paoni *et al.*, 1993). As mutations are usually expected to have negative effects, it could be argued that tPA has evolved to have a sub-optimal fibrin specificity. This could suggest that tPA has important catalytic functions in the absence of fibrin, and recent evidence of a role for tPA in pericellular proteolysis appears to substantiate this.

Two independent lines of evidence support the notion that tPA has fibrin-independent roles. First, the identification of cellular mechanisms regulating the functional activity of tPA, similar to those regulating uPA activity, and involving cellular binding sites for tPA. A variety of different sites have been identified, although most do not appear to have the high level of affinity and specificity displayed by uPAR. Second, the observation that tPA is expressed in the brain and, at least in some regions, it is the most abundant serine protease (Davies *et al.*, 1998). The brain is an organ usually devoid of fibrinogen or fibrin, but tPA has nevertheless been shown to activate locally synthesized plasminogen which can then act on non-fibrin substrates (Tsirka *et al.*, 1997).

6.3.1. Endothelial Cells

The first cellular binding site for tPA was identified on endothelial cells, cells that express tPA and secrete it in a regulated manner (Emeis *et al.*, 1997). This was shown to be due to the presence of the Ca^{2+}/phospholipid binding protein annexin II, peripherally associated with the plasma membrane (Hajjar *et al.*, 1994, 1996). This protein stimulated plasminogen activation by up to 20-fold, but only after the generation of an internal C-terminal Lys residue by proteolytic cleavage (Cesarman *et al.*, 1994), which presumably interacts with KR-2 of tPA. This Lys-dependent binding does not discriminate between tPA and plasminogen, and annexin II binds both. Annexin II has also been shown to stimulate plasminogen activation by another mechanism, involving formation of a heterotetramer of two molecules of annexin II and two molecules of a subunit referred to as p11, which has a native C-terminal Lys residue (Kassam *et al.*, 1998a). Plasminogen binding to p11 in the tetramer stimulates tPA activity more efficiently than annexin II alone (Kassam *et al.*, 1998b). Annexin II appears to be quite promiscuous as it also binds a number of other proteins including the cysteine protease cathepsin B (Mai *et al.*, 2000).

6.3.2. Vascular Smooth Muscle Cells

Vascular smooth muscle cells are another cell type that express tPA, particularly in response to vascular injury (Clowes *et al.*, 1990). These cells have been shown to bind tPA in a specific manner leading to a greater than 100-fold increase in plasminogen activation (Ellis and Whawell, 1997). This binding is not mediated by annexin II, any other known tPA binding protein or proteins with C-terminal Lys residues and did not involve the KR-2 module of tPA. Consistent with this is the recent identification of the binding protein as a 63 kD type-II transmembrane protein, which has yet to be fully characterized (unpublished observations). Binding of tPA to this protein appears to have a dual functional role as, in addition to increasing plasminogen activation, it also decreases inhibition of tPA by PAI-1 (see Figure 2.2) and certain low molecular weight inhibitors (Werner *et al.*, 1999). These effects were found to correlate with conformational changes in the protease domain of tPA, indicating that this domain is intimately involved in the binding interaction.

6.3.3. Prion Protein

The brain has been mentioned as an organ with an unexpectedly widespread expression of tPA. Various experiments have shown that it has an important functional role there, both in normal brain function (Seeds *et al.*, 1995, 1999; Huang *et al.*, 1996) and in neurodegenerative events (Tsirka *et al.*, 1995). The mechanisms by which tPA activity is regulated in these situations is not known, but it possibly involves mechanisms similar to those described above, as it has been shown that tPA binds specifically to the surface of neuronal cells (Verrall and Seeds, 1989). One potential candidate for a functional tPA binding site on neuronal cells is the prion protein (PrP). This is a glycolipid-anchored membrane protein, which subsequent to a conformational transition becomes the infectious agent in the group of disease known as the transmissible spongiform encephalopathies or prion diseases (Prusiner, 1998). It has recently been shown that a recombinant soluble form of PrP, having the disease-associated conformation, can specifically bind tPA with very high affinity having a K_d lower than 2.5 nM (Ellis *et al.*, 2002). This leads to a greater than 300-fold increase in plasminogen activation, and requires the binding of plasminogen to an independent site on the same PrP molecule. The latter is consistent with the observation that plasminogen can bind to the disease-associated form of PrP isolated from the brains of infected animals (Fischer *et al.*, 2000). It has yet to be demonstrated whether membrane-bound PrP can also mediate similar effects.

7. Concluding Remarks

Advances over the last decade have increased our understanding of the relationship between the structure and the function of the two plasminogen activators. Structural analysis and mutagenesis of the catalytic domains of these enzymes has revealed much about the interactions with substrates and inhibitors, and what regulates the anomalous catalytic activity of their zymogen forms. Similar analyses have revealed important information about specific characteristics of the non-catalytic N-terminal modules of these proteins that play critical roles in their biological behavior. Enzymological analysis has also played an important role in elucidating the mechanisms regulating the activity of these proteases, both in purified and more complex biological systems, and emphasized the importance of nonenzymatic cofactors in regulating their biological functions. What is currently lacking is structural information on the three-dimensional arrangement of the modules that constitute these relatively complex mosaic proteins. This is unlikely to be achieved for the enzymes in isolation, largely due to their inherent structural mobility, but it may be achieved for the enzymes in complex with their cofactor molecules, which of course would be the ultimate goal of such studies.

Further challenges include acquiring a detailed understanding of how the more recently identified cofactors for the plasminogen activators, such as uPAR, cellular tPA binding sites, and the prion protein, interact with the plasminogen activators to facilitate plasmin generation. The diversity of the mechanisms potentially regulating plasminogen activation strongly suggests that these systems have differing biological functions, despite each generating plasmin activity. The identification of the specific substrates that plasmin is acting on in these different biological situations, amongst the multitude of substrates that have been identified *in vitro*, will help to further define these various functions.

The activity of the plasminogen activators is implicated in a variety of degenerative diseases. Understanding the mechanisms regulating the function of these proteases at the molecular level may help both in defining potential therapeutic targets and in the rational design of therapeutic agents.

References

Adams, D.S., Griffin, L.A., Nachajko, W.R., Reddy, V.B., and Wei, C.M. (1991). A synthetic DNA encoding a modified human urokinase resistant to inhibition by serum plasminogen activator inhibitor. *J. Biol. Chem.* **266**, 8476–8482.

Alexander, D.S., Sipley, J.D., and Quigley, J.P. (1998). Autoactivation of avian urokinase-type plasminogen activator (uPA). A novel mode of initiation of the uPA/plasmin cascade. *J. Biol. Chem.* **273**, 7457–7461.

Andronicos, N.M. and Ranson, M. (2001). The topology of plasminogen binding and activation on the surface of human breast cancer cells. *Br. J. Cancer* **85**, 909–916.

Appella, E., Robinson, E.A., Ullrich, S.J., Stoppelli, M.P., Corti, A., Cassani, G., and Blasi, F. (1987). The receptor-binding sequence of urokinase. A biological function for the growth-factor module of proteases. *J. Biol. Chem.* **262**, 4437–4440.

Astrup, T. and Permin, P.M. (1947). Fibrinolysis in animal organism. *Nature* **159**, 681–682.

Barrett, A.J., Rawlings, N.D., and O'Brien, E.A. (2001). The MEROPS database as a protease information system. *J. Struct. Biol.* **134**, 95–102.

Bass, R. and Ellis, V. (2002). Cellular mechanisms regulating non-haemostatic plasmin generation. *Biochem. Soc. Trans.* **30**, 189–194.

Bennett, W.F., Paoni, N.F., Keyt, B.A., Botstein, D., Jones, A.J., Presta, L., Wurm, F.M., and Zoller, M.J. (1991). High resolution analysis of functional determinants on human tissue-type plasminogen activator. *J. Biol. Chem.* **266**, 5191–5201.

Berkenpas, M.B. and Quigley, J.P. (1991). Transformation-dependent activation of urokinase-type plasminogen activator by a plasmin-independent mechanism: Involvement of cell surface membranes. *Proc. Natl. Acad. Sci. USA* **88**, 7768–7772.

Bode, W. (1979). The transition of bovine trypsinogen to a trypsin-like state upon strong ligand binding. II. The binding of the pancreatic trypsin inhibitor and of isoleucine-valine and of sequentially related peptides to trypsinogen and to p-guanidinobenzoate-trypsinogen. *J. Mol. Biol.* **127**, 357–374.

Bode, W. and Renatus, M. (1997). Tissue-type plasminogen activator: variants and crystal/solution structures demarcate structural determinants of function. *Curr. Opin. Struct. Biol.* **7**, 865–872.

Bokman, A.M., Jimenez-Barbero, J., and Llinas, M. (1993). 1H NMR characterization of the urokinase kringle module. Structural, but not functional, relatedness to homologous domains. *J. Biol. Chem.* **268**, 13858–13868.

Boose, J.A., Kuismanen, E., Gerard, R., Sambrook, J., and Gething, M.J. (1989). The single-chain form of tissue-type plasminogen activator has catalytic activity: Studies with a mutant enzyme that lacks the cleavage site. *Biochemistry* **28**, 635–643.

Bringmann, P., Gruber, D., Liese, A., Toschi, L., Krätzschmar, J., Schleuning, W.D., and Donner, P. (1995). Structural features mediating fibrin selectivity of vampire bat plasminogen activators. *J. Biol. Chem.* **270**, 25596–25603.

Buko, A.M., Kentzer, E.J., Petros, A., Menon, G., Zuiderweg, E.R.P., and Sarin, V.K. (1991). Characterization of a posttranslational fucosylation in the growth factor domain of urinary plasminogen activator. *Proc. Natl. Acad. Sci. USA* **88**, 3992–3996.

Byeon, I.J. and Llinas, M. (1991). Solution structure of the tissue-type plasminogen activator kringle 2 domain complexed to 6-aminohexanoic acid an antifibrinolytic drug. *J. Mol. Biol.* **222**, 1035–1051.

Cesarman, G.M., Guevara, C.A., and Hajjar, K.A. (1994). An endothelial cell receptor for plasminogen/tissue plasminogen activator (t-PA). II. Annexin II-mediated enhancement of t-PA-dependent plasminogen activation. *J. Biol. Chem.* **269**, 21198–21203.

Christensen, U. (1988). Urokinase-catalysed plasminogen activation. Effects of ligands binding to the AH-site of plasminogen. *Biochim. Biophys. Acta* **957**, 258–265.

Clowes, A.W., Clowes, M.M., Au, Y.P.T., Reidy, M.A., and Belin, D. (1990). Smooth muscle cells express urokinase during mitogenesis and tissue-type plasminogen activator during migration in injured rat carotid artery. *Circ. Res.* **67**, 61–67.

Collen, D., Zamarron, C., Lijnen, H.R., and Hoylaerts, M. (1986). Activation of plasminogen by pro-urokinase. II. Kinetics. *J. Biol. Chem.* **261**, 1259–1266.

Coombs, G.S., Bergstrom, R.C., Madison, E.L., and Corey, D.R. (1998). Directing sequence-specific proteolysis to new targets. The influence of loop size and target sequence on selective proteolysis by tissue-type plasminogen activator and urokinase-type plasminogen activator. *J. Biol. Chem.* **273**, 4323–4328.

Davidow, L.S., Dumais, D.R., Smyth, A.P., Greer, J., and Moir, D.T. (1991). Mutations affecting the activity of urokinase-type plasminogen activator. *Protein Eng.* **4**, 923–928.

Davies, B.J., Pickard, B.S., Steel, M., Morris, R.G., and Lathe, R. (1998). Serine proteases in rodent hippocampus. *J. Biol. Chem.* **273**, 23004–23011.

de Vos, A.M., Ultsch, M.H., Kelley, R.F., Padmanabhan, K., Tulinsky, A., Westbrook, M.L., and Kossiakoff, A.A. (1992). Crystal structure of the kringle 2 domain of tissue plasminogen activator at 2.4-A resolution. *Biochemistry* **31**, 270–279.

de Vries, C., Veerman, H., Nesheim, M.E., and Pannekoek, H. (1991). Kinetic characterization of tissue-type plasminogen activator (t-PA) and t-PA deletion mutants. *Thromb. Haemost.* **65**, 280–285.

Degen, S.J., Heckel, J.L., Reich, E., and Degen, J.L. (1987). The murine urokinase-type plasminogen activator gene. *Biochemistry* **26**, 8270–8279.

Ding, L., Coombs, G.S., Strandberg, L., Navre, M., Corey, D.R., and Madison, E.L. (1995). Origins of the specificity of tissue-type plasminogen activator. *Proc. Natl. Acad. Sci. USA* **92**, 7627–7631.

Downing, A.K., Driscoll, P.C., Harvey, T.S., Dudgeon, T.J., Smith, B.O., Baron, M., and Campbell, I.D. (1992). Solution structure of the fibrin binding finger domain of tissue-type plasminogen activator determined by 1H nuclear magnetic resonance. *J. Mol. Biol.* **225**, 821–833.

Ellis, V. (1996). Functional analysis of the cellular receptor for urokinase in plasminogen activation. *J. Biol. Chem.* **271**, 14779–14784.

Ellis, V., Behrendt, N., and Danø, K. (1991). Plasminogen activation by receptor-bound urokinase. A kinetic study with both cell-associated and isolated receptor. *J. Biol. Chem.* **266**, 12752–12758.

Ellis, V., Daniels, M., Misra, R., and Brown, D.R. (2002). Plasminogen activation is stimulated by prion protein and regulated in a copper-dependent manner. *Biochemistry* **41**, 6891–6896.

Ellis, V. and Danø, K. (1993). Potentiation of plasminogen activation by an anti-urokinase monoclonal antibody due to ternary complex formation. A mechanistic model for receptor-mediated plasminogen activation. *J. Biol. Chem.* **268**, 4806–4813.

Ellis, V. and Murphy, G. (2001). Cellular strategies for proteolytic targeting during migration and invasion. *FEBS Lett.* **506**, 1–5.

Ellis, V., Scully, M.F., and Kakkar, V.V. (1987). Plasminogen activation by single-chain urokinase in functional isolation. A kinetic study. *J. Biol. Chem.* **262**, 14998–15003.

Ellis, V., Scully, M.F., and Kakkar, V.V. (1989). Plasminogen activation initiated by single-chain urokinase-type plasminogen activator. Potentiation by U937 monocytes. *J. Biol. Chem.* **264**, 2185–2188.

Ellis, V. and Whawell, S.A. (1997). Vascular smooth muscle cells potentiate plasmin generation by both urokinase and tissue plasminogen activator dependent mechanisms: Evidence for a specific tPA receptor on these cells. *Blood* **90**, 2312–2322.

Ellis, V., Whawell, S.A., Werner, F., and Deadman, J.J. (1999). Assembly of urokinase receptor-mediated plasminogen activation complexes involves direct, non-active-site interactions between urokinase and plasminogen. *Biochemistry* **38**, 651–659.

Ellis, V., Wun, T.-C., Behrendt, N., Rønne, E., and Danø, K. (1990). Inhibition of receptor-bound urokinase by plasminogen-activator inhibitors. *J. Biol. Chem.* **265**, 9904–9908.

Emeis, J.J., Eijnden-Schrauwen, Y., Van den Hoogen, C.M., De Priester, W., Westmuckett, A., and Lupu, F. (1997). An endothelial storage granule for tissue-type plasminogen activator. *J. Cell Biol.* **139**, 245–256.

Fischer, M.B., Roeckl, C., Parizek, P., Schwarz, H.P., and Aguzzi, A. (2000). Binding of disease-associated prion protein to plasminogen. *Nature* **408**: 479–483.

Gårdsvoll, H., Dano, K., and Ploug, M. (1999). Mapping part of the functional epitope for ligand binding on the receptor for urokinase-type plasminogen activator by site-directed mutagenesis. *J. Biol. Chem.* **274**, 37995–38003.

Gething, M.J., Adler, B., Boose, J.A., Gerard, R.D., Madison, E.L., McGookey, D., Meidell, R.S., Roman, L.M., and Sambrook, J. (1988). Variants of human tissue-type plasminogen activator that lack specific structural domains of the heavy chain. *EMBO J.* **7**, 2731–2740.

Gunzler, W.A., Steffens, G.J., Otting, F., Buse, G., and Flohe, L. (1982). Structural relationship between human high and low molecular mass urokinase. *Hoppe Seylers. Z. Physiol. Chem.* **363**, 133–141.

Hajjar, K.A., Guevara, C.A., Lev, E., Dowling, K., and Chacko, J. (1996). Interaction of the fibrinolytic receptor, annexin II, with the endothelial cell surface. Essential role of endonexin repeat 2. *J. Biol. Chem.* **271**, 21652–21659.

Hajjar, K.A., Jacovina, A.T., and Chacko, J. (1994). An endothelial cell receptor for plasminogen/tissue plasminogen activator. I. Identity with annexin II. *J. Biol. Chem.* **269**, 21191–21197.

Hansen, A.P., Petros, A.M., Meadows, R.P., and Fesik, S.W. (1994a). Backbone dynamics of a two-domain protein: ^{15}N relaxation studies of the amino-terminal fragment of urokinase-type plasminogen activator. *Biochemistry* **33**, 15418–15424.

Hansen, A.P., Petros, A.M., Meadows, R.P., Nettesheim, D.G., Mazar, A.P., Olejniczak, E.T., Xu, R.X., Pederson, T.M., Henkin, J., and Fesik, S.W. (1994b). Solution structure of the amino-terminal fragment of urokinase-type plasminogen activator. *Biochemistry* **33**, 4847–4864.

Horrevoets, A.J., Smilde, A., de Vries, C., and Pannekoek, H. (1994). The specific roles of finger and kringle 2 domains of tissue-type plasminogen activator during in vitro fibrinolysis. *J. Biol. Chem.* **269**, 12639–12644.

Horrevoets, A.J., Tans, G., Smilde, A.E., van Zonneveld, A.J., and Pannekoek, H. (1993). Thrombin-variable region 1 (VR1). Evidence for the dominant contribution of VR1 of serine proteases to their interaction with plasminogen activator inhibitor 1. *J. Biol. Chem.* **268**, 779–782.

Horrevoets, A.J.G., Pannekoek, H., and Nesheim, M.E. (1997). A steady-state template model that describes the kinetics of fibrin-stimulated [Glu^1]- and [Lys^{78}]plasminogen activation by native tissue-type plasminogen activator and variants that lack either the finger or kringle-2 domain. *J. Biol. Chem.* **272**, 2183–2191.

Høyer-Hansen, G., Ploug, M., Behrendt, N., Ronne, E., and Danø, K. (1997). Cell-surface acceleration of urokinase-catalyzed receptor cleavage. *Eur. J. Biochem.* **243**, 21–26.

Hoylaerts, M., Rijken, D.C., Lijnen, H.R., and Collen, D. (1982). Kinetics of the activation of plasminogen by human tissue plasminogen activator. Role of fibrin. *J. Biol. Chem.* **257**, 2912–2919.

Høyer-Hansen, G., Rønne, E., Solberg, H., Behrendt, N., Ploug, M., Lund, L.R., Ellis, V., and Danø, K. (1992). Urokinase plasminogen activator cleaves its cell surface receptor releasing the ligand-binding domain. *J. Biol. Chem.* **267**, 18224–18229.

Hu, C.-K., Kohnert, U., Sturzebecher, J., Fischer, S., and Llinas, M. (1996). Complexation of the tissue plasminogen activator protease with benzamidine-type inhibitors: interference by the kringle 2 module. *Biochemistry* **35**, 3270–3276.

Hu, C.-K., Kohnert, U., Wilhelm, O., Fischer, S., and Llinás, M. (1994). Tissue-type plasminogen activator domain-deletion mutant BM 06.022: Modular stability, inhibitor binding, and activation cleavage. *Biochemistry* **33**, 11760–11766.

Huang, Y.Y., Bach, M.E., Lipp, H.P., Zhuo, M., Wolfer, D.P., Hawkins, R.D., Schoonjans, L., Kandel, E.R., Godfraind, J.M., Mulligan, R., Collen, D., and Carmeliet, P. (1996). Mice lacking the gene encoding tissue-type plasminogen activator show a selective interference with late-phase long-term potentiation in both Schaffer collateral and mossy fiber pathways. *Proc. Natl. Acad. Sci. USA* **93**, 8699–8704.

Huber, R. and Bode, W. (1978). Structural basis of the activation and action of trypsin. *Acc. Chem. Res.* **11**, 114–122.

Ichinose, A., Fujikawa, K., and Suyama, T. (1986). The activation of pro-urokinase by plasma kallikrein and its inactivation by thrombin. *J. Biol. Chem.* **261**, 3486–3489.

Kassam, G., Choi, K.S., Ghuman, J., Kang, H.M., Fitzpatrick, S.L., Zackson, T., Zackson, S., Toba, M., Shinomiya, A., and Waisman, D.M. (1998a). The role of annexin II tetramer in the activation of plasminogen. *J. Biol. Chem.* **273**, 4790–4799.

Kassam, G., Le, B.H., Choi, K.S., Kang, H.M., Fitzpatrick, S.L., Louie, P., and Waisman, D.M. (1998b). The p11 subunit of the annexin II tetramer plays a key role in the stimulation of t-PA-dependent plasminogen activation. *Biochemistry* **37**, 16958–16966.

Ke, S.H., Coombs, G.S., Tachias, K., Navre, M., Corey, D.R., and Madison, E.L. (1997). Distinguishing the specificities of closely related proteases. Role of P3 in substrate and inhibitor discrimination between tissue-type plasminogen activator and urokinase. *J. Biol. Chem.* **272**, 16603–16609.

Kobayashi, H., Schmitt, M., Goretzki, L., Chucholowski, N., Calvete, J., Kramer, M., Gunzler, W.A., Janicke, F., and Graeff, H. (1991). Cathepsin B efficiently activates the soluble and the tumor cell receptor-bound form of the proenzyme urokinase-type plasminogen activator (Pro-uPA). *J. Biol. Chem.* **266**, 5147–5152.

Lamba, D., Bauer, M., Huber, R., Fischer, S., Rudolph, R., Kohnert, U., and Bode, W. (1996). The 2.3 Å crystal structure of the catalytic domain of recombinant two-chain human tissue-type plasminogen activator. *J. Mol. Biol.* **258**, 117–135.

Lee, S.W., Ellis, V., and Dichek, D.A. (1994). Characterization of plasminogen activation by glycosylphosphatidylinositol-anchored urokinase. *J. Biol. Chem.* **269**, 2411–2418.

Lesk, A.M. and Fordham, W.D. (1996). Conservation and variability in the structures of serine proteinases of the chymotrypsin family. *J. Mol. Biol.* **258**, 501–537.

Li, X., Bokman, A.M., Llinas, M., Smith, R.A., and Dobson, C.M. (1994). Solution structure of the kringle domain from urokinase-type plasminogen activator. *J. Mol. Biol.* **235**, 1548–1559.

Lijnen, H.R., Van Hoef, B., Nelles, L., and Collen, D. (1990). Plasminogen activation with single-chain urokinase-type plasminogen activator (scuPA). Studies with active site mutagenized plasminogen (Ser740>Ala) and plasmin-resistant scu-PA (Lys158 > Glu). *J. Biol. Chem.* **265**, 5232–5236.

Lijnen, H.R., Zamarron, C., Blaber, M., Winkler, M.E., and Collen, D. (1986). Activation of plasminogen by pro-urokinase. I. Mechanism. *J. Biol. Chem.* **261**, 1253–1258.

Linde, V., Nielsen, L.S., Foster, D.C., and Petersen, L.C. (1998). Elimination of the Cys558-Cys566 bond in Lys78-plasminogen—effect on activation and fibrin interaction. *Eur. J. Biochem.* **251**, 472–479.

List, K., Jensen, O.N., Bugge, T.H., Lund, L.R., Ploug, M., Dano, K., and Behrendt, N. (2000). Plasminogen-independent initiation of the pro-urokinase activation cascade in vivo. Activation of pro-urokinase by glandular kallikrein (mGK-6) in plasminogen-deficient mice. *Biochemistry* **39**, 508–515.

Madison, E.L., Goldsmith, E.J., Gerard, R.D., Gething, M.J., and Sambrook, J.F. (1989). Serpin-resistant mutants of human tissue-type plasminogen activator. *Nature* **339**, 721–724.

Madison, E.L., Goldsmith, E.J., Gerard, R.D., Gething, M.J., Sambrook, J.F., and Bassel-Duby, R.S. (1990). Amino acid residues that affect interaction of tissue-type plasminogen activator with plasminogen activator inhibitor 1. *Proc. Natl. Acad. Sci. USA* **87**, 3530–3533.

Madison, E.L., Kobe, A., Gething, M.J., Sambrook, J.F., and Goldsmith, E.J. (1993). Converting tissue plasminogen activator to a zymogen: A regulatory triad of Asp-His-Ser. *Science* **262**, 419–421.

Magdolen, V., Rettenberger, P., Koppitz, M., Goretzki, L., Kessler, H., Weidle, U.H., König, B., Graeff, H., Schmitt, M., and Wilhelm, O. (1996). Systematic mutational analysis of the receptor-binding region of the human urokinase-type plasminogen activator. *Eur. J. Biochem.* **237**, 743–751.

Mai, J., Finley, R.L., Jr., Waisman, D.M., and Sloane, B.F. (2000). Human procathepsin B interacts with the annexin II tetramer on the surface of tumor cells. *J. Biol. Chem.* **275**, 12806–12812.

Mangel, W.F., Lin, B.H., and Ramakrishnan, V. (1990). Characterization of an extremely large, ligand-induced conformational change in plasminogen. *Science* **248**, 69–73.

Marcotte, P.A. and Henkin, J. (1993). Characterization of the activation of pro-urokinase by thermolysin. *Biochim. Biophys. Acta* **1161**, 105–112.

Marcotte, P.A., Kozan, I.M., Dorwin, S.A., and Ryan, J.M. (1992). The matrix metalloproteinase pump-1 catalyzes formation of low molecular weight (pro)urokinase in cultures of normal human kidney cells. *J. Biol. Chem.* **267**, 13803–13806.

Mori, K., Dwek, R.A., Downing, A.K., Opdenakker, G., and Rudd, P.M. (1995). The activation of type 1 and type 2 plasminogen by type I and type II tissue plasminogen activator. *J. Biol. Chem.* **270**, 3261–3267.

Mulichak, A.M., Tulinsky, A., and Ravichandran, K.G. (1991). Crystal and molecular structure of human plasminogen kringle 4 refined at 1.9-A resolution. *Biochemistry* **30**, 10576–10588.

Nauland, U. and Rijken, D.C. (1994). Activation of thrombin-inactivated single-chain urokinase-type plasminogen activator by dipeptidyl peptidase I (cathepsin C). *Eur. J. Biochem.* **223**, 497–501.

Nesheim, M., Fredenburgh, J.C., and Larsen, G.R. (1990). The dissociation constants and stoichiometries of the interactions of Lys-plasminogen and chloromethyl ketone derivatives of tissue plasminogen activator and the variant delta FEIX with intact fibrin. *J. Biol. Chem.* **265**, 21541–21548.

Nienaber, V., Wang, J., Davidson, D., and Henkin, J. (2000). Re-engineering of human urokinase provides a system for structure-based drug design at high resolution and reveals a novel structural subsite. *J. Biol. Chem.* **275**, 7239–7248.

Nienaber, V.L., Young, S.L., Birktoft, J.J., Higgins, D.L., and Berliner, L.J. (1992). Conformational similarities between one-chain and two-chain tissue plasminogen activator (t-PA): Implications to the activation mechanism on one-chain t-PA. *Biochemistry* **31**, 3852–3861.

Novokhatny, V., Medved, L., Mazar, A., Marcotte, P., Henkin, J., and Ingham, K. (1992). Domain structure and interactions of recombinant urokinase-type plasminogen activator. *J. Biol. Chem.* **267**, 3878–3885.

Novokhatny, V.V., Ingham, K.C., and Medved, L.V. (1991). Domain structure and domain-domain interactions of recombinant tissue plasminogen activator. *J. Biol. Chem.* **266**, 12994–13002.

Nowak, U.K., Li, X., Teuten, A.J., Smith, R.A.G., and Dobson, C.M. (1993). NMR studies of the dynamics of the multidomain protein urokinase-type plasminogen activator. *Biochemistry* **32**, 298–309.

Pannell, R. and Gurewich, V. (1987). Activation of plasminogen by single-chain urokinase or by two-chain urokinase—a demonstration that single-chain urokinase has a low catalytic activity. *Blood* **69**, 22–26.

Paoni, N.F., Chow, A.M., Pena, L.C., Keyt, B.A., Zoller, M.J., and Bennett, W.F. (1993). Making tissue-type plasminogen activator more fibrin specific. *Protein Eng.* **6**, 529–534.

Patthy, L. (1985). Evolution of the proteases of blood coagulation and fibrinolysis by assembly from modules. *Cell* **41**, 657–663.

Patthy, L. (1999). Genome evolution and the evolution of exon-shuffling—a review. *Gene* **238**, 103–114.
Petersen, L.C. (1997). Kinetics of reciprocal pro-urokinase/plasminogen activation—stimulation by a template formed by the urokinase receptor bound to poly(D-lysine). *Eur. J. Biochem.* **245**, 316–323.
Petersen, L.C., Boel, E., Johannessen, M., and Foster, D. (1990). Quenching of the amidolytic activity of one-chain tissue-type plasminogen activator by mutation of lysine-416. *Biochemistry* **29**, 3451–3457.
Petersen, L.C., Johannessen, M., Foster, D., Kumar, A., and Mulvihill, E. (1988a). The effect of polymerised fibrin on the catalytic activities of one-chain tissue-type plasminogen activator as revealed by an analogue resistant to plasmin cleavage. *Biochim. Biophys. Acta* **952**, 245–254.
Petersen, L.C., Lund, L.R., Nielsen, L.S., Danø, K., and Skriver, L. (1988b). One-chain urokinase-type plasminogen activator from human sarcoma cells is a proenzyme with little or no intrinsic activity. *J. Biol. Chem.* **263**, 11189–11195.
Ploug, M. (1998). Identification of specific sites involved in ligand binding by photoaffinity labeling of the receptor for the urokinase-type plasminogen activator. Residues located at equivalent positions in uPAR domains I and III participate in the assembly of a composite ligand-binding site. *Biochemistry* **37**, 16494–16505.
Ploug, M. and Ellis, V. (1994). Structure-function relationships in the receptor for urokinase-type plasminogen activator. Comparison to other members of the Ly-6 family and snake venom a-neurotoxins. [Review]. *FEBS Lett.* **349**, 163–168.
Ploug, M., Ellis, V., and Danø, K. (1994). Ligand interaction between urokinase-type plasminogen activator and its receptor probed with 8-anilino-1-naphthalenesulfonate. Evidence for a hydrophobic binding site exposed only on the intact receptor. *Biochemistry* **33**, 8991–8997.
Ploug, M., Rahbek-Nielsen, H., Ellis, V., Roepstorff, P., and Danø, K. (1995). Chemical modification of the urokinase-type plasminogen activator and its receptor using tetranitromethane. Evidence for the involvement of specific tyrosine residues in both molecules during receptor-ligand interaction. *Biochemistry* **34**, 12524–12534.
Plow, E.F., Felez, J., and Miles, L.A. (1991). Cellular regulation of fibrinolysis. [Review]. *Thromb. Haemost.* **66**, 32–36.
Prusiner, S.B. (1998). Prions. *Proc. Natl. Acad. Sci. USA* **95**, 13363–13383.
Rånby, M. (1982). Studies on the kinetics of plasminogen activation by tissue plasminogen activator. *Biochim. Biophys. Acta* **704**, 461–469.
Renatus, M., Engh, R.A., Stubbs, M.T., Huber, R., Fischer, S., Kohnert, U., and Bode, W. (1997a). Lysine 156 promotes the anomalous proenzyme activity of tPA: X-ray crystal structure of single-chain human tPA. *EMBO J.* **16**, 4797–4805.
Renatus, M., Stubbs, M.T., Huber, R., Bringmann, P., Donner, P., Schleuning, W.D., and Bode, W. (1997b). Catalytic domain structure of vampire bat plasminogen activator: A molecular paradigm for proteolysis without activation cleavage. *Biochemistry* **36**, 13483–13493.
Schechter, I. and Berger, A. (1967). On the size of the active site in proteases. I. Papain. *Biochem. Biophys. Res. Commun.* **27**, 157–162.
Schielen, W.J., Adams, H.P., van Leuven, K., Voskuilen, M., Tesser, G.I., and Nieuwenhuizen, W. (1991). The sequence gamma-(312–324) is a fibrin-specific epitope. *Blood* **77**, 2169–2173.
Schielen, W.J., Voskuilen, M., Tesser, G.I., and Nieuwenhuizen, W. (1989). The sequence A alpha-(148–160) in fibrin, but not in fibrinogen, is accessible to monoclonal antibodies. *Proc. Natl. Acad. Sci. USA* **86**, 8951–8954.
Schmitt, M., Kanayama, N., Henschen, A., Hollrieder, A., Hafter, R., Gulba, D., Janicke, F., and Graeff, H. (1989). Elastase released from human granulocytes stimulated with *N*-formyl-chemotactic peptide prevents activation of tumor cell prourokinase (pro-uPA). *FEBS Lett.* **255**, 83–88.
Seeds, N.W., Basham, M.E., and Haffke, S.P. (1999). Neuronal migration is retarded in mice lacking the tissue plasminogen activator gene. *Proc. Natl. Acad. Sci. USA* **96**, 14118–14123.
Seeds, N.W., Williams, B.L., and Bickford, P.C. (1995). Tissue plasminogen activator induction in Purkinje neurons after cerebellar motor learning. *Science* **270**, 1992–1994.
Sipley, J.D., Alexander, D.S., Testa, J.E., and Quigley, J.P. (1997). Introduction of an RRHR motif into chicken urokinase-type plasminogen activator (ch-uPA) confers sensitivity to plasminogen activator inhibitor (PAI)-1 and PAI-2 and allows ch-uPA-mediated extracellular matrix degradation to be controlled by PAI-1. *Proc. Natl. Acad. Sci. USA* **94**, 2933–2938.
Smith, B.O., Downing, A.K., Driscoll, P.C., Dudgeon, T.J., and Campbell, I.D. (1995). The solution structure and backbone dynamics of the fibronectin type I and epidermal growth factor-like pair of modules of tissue-type plasminogen activator. *Structure* **3**, 823–833.

Spraggon, G., Phillips, C., Nowak, U.K., Ponting, C.P., Saunders, D., Dobson, C.M., Stuart, D.I., and Jones, E.Y. (1995). The crystal structure of the catalytic domain of human urokinase-type plasminogen activator. *Structure* **3**, 681–691.

Stack, M.S. and Johnson, D.A. (1994). Human mast cell tryptase activates single-chain urinary-type plasminogen activator (pro-urokinase). *J. Biol. Chem.* **269**, 9416–9419.

Steffens, G.J., Gunzler, W.A., Otting, F., Frankus, E., and Flohe, L. (1982). The complete amino acid sequence of low molecular mass urokinase from human urine. *Hoppe Seylers. Z. Physiol. Chem.* **363**, 1043–1058.

Stephens, R.W., Aumailley, M., Timpl, R., Reisberg, T., Tapiovaara, H., Myohanen, H., Murphy-Ullrich, J., and Vaheri, A. (1992a). Urokinase binding to laminin-nidogen. Structural requirements and interactions with heparin. *Eur. J. Biochem.* **207**, 937–942.

Stephens, R.W., Bokman, A.M., Myohanen, H.T., Reisberg, T., Tapiovaara, H., Pedersen, N., Grøndahl-Hansen, J., Llinas, M., and Vaheri, A. (1992b). Heparin binding to the urokinase kringle domain. *Biochemistry* **31**, 7572–7579.

Tachias, K. and Madison, E.L. (1997). Converting tissue type plasminogen activator into a zymogen—Important role of Lys^{156}. *J. Biol. Chem.* **272**, 28–31.

Takeuchi, T., Harris, J.L., Huang, W., Yan, K.W., Coughlin, S.R., and Craik, C.S. (2000). Cellular localization of membrane-type serine protease 1 and identification of protease-activated receptor-2 and single-chain urokinase-type plasminogen activator as substrates. *J. Biol. Chem.* **275**, 26333–26342.

Thorsen, S., Glas-Greenwalt, P., and Astrup, T. (1972). Differences in the binding to fibrin of urokinase and tissue plasminogen activator. *Thromb. Haemost.* **28**, 65–74.

Thorsen, S., Philips, M., Selmer, J., Lecander, I., and Astedt, B. (1988). Kinetics of inhibition of tissue-type and urokinase-type plasminogen activator by plasminogen-activator inhibitor type 1 and type 2. *Eur. J. Biochem.* **175**, 33–39.

Tsirka, S.E., Bugge, T.H., Degen, J.L., and Strickland, S. (1997). Neuronal death in the central nervous system demonstrates a non-fibrin substrate for plasmin. *Proc. Natl. Acad. Sci. USA* **94**, 9779–9781.

Tsirka, S.E., Gualandris, A., Amaral, D.G., and Strickland, S. (1995). Excitotoxin-induced neuronal degeneration and seizure are mediated by tissue plasminogen activator. *Nature* **377**, 340–344.

Tsurupa, G. and Medved, L. (2001). Identification and characterization of novel tPA- and plasminogen-binding sites within fibrin(ogen) alpha C-domains. *Biochemistry* **40**, 801–808.

Ugwu, F., Van Hoef, B., Bini, A., Collen, D., and Lijnen, H.R. (1998). Proteolytic cleavage of urokinase-type plasminogen activator by stromelysin-1 (MMP-3). *Biochemistry* **37**, 7231–7236.

van Zonneveld, A.J., Veerman, H., and Pannekoek, H. (1986). On the interaction of the finger and the kringle-2 domain of tissue-type plasminogen activator with fibrin. Inhibition of kringle-2 binding to fibrin by epsilon-amino caproic acid. *J. Biol. Chem.* **261**, 14214–14218.

Verheijen, J.H., Caspers, M.P., Chang, G.T., de Munk, G.A., Pouwels, P.H., and Enger-Valk, B.E. (1986). Involvement of finger domain and kringle 2 domain of tissue-type plasminogen activator in fibrin binding and stimulation of activity by fibrin. *EMBO J.* **5**, 3525–3530.

Verrall, S. and Seeds, N.W. (1989). Characterization of ^{125}I-tissue plasminogen activator binding to cerebellar granule neurons. *J. Cell Biol.* **109**, 265–271.

Vines, D.J., Lee, S.W., Dichek, D.A., and Ellis, V. (2000). Receptor-mediated regulation of plasminogen activator function: plasminogen activation by two directly membrane-anchored forms of urokinase. *J. Pept. Sci.* **6**, 432–439.

Wang, D., Bode, W., and Huber, R. (1985). Bovine chymotrypsinogen A X-ray crystal structure analysis and refinement of a new crystal form at 1.8 A resolution. *J. Mol. Biol.* **185**, 595–624.

Wang, X., Terzyan, S., Tang, J., Loy, J.A., Lin, X., and Zhang, X.C. (2000). Human plasminogen catalytic domain undergoes an unusual conformational change upon activation. *J. Mol. Biol.* **295**, 903–914.

Werner, F., Razzaq, T.M., and Ellis, V. (1999). Tissue plasminogen activator binds to human vascular smooth muscle cells by a novel mechanism. Evidence for a reciprocal linkage between inhibition of catalytic activity and cellular binding. *J. Biol. Chem.* **274**, 21555–21561.

3

Plasminogen Activators Inhibitors

Ann Gils and Paul J. Declerck

The activation of the fibrinolytic system occurs when plasminogen is converted into plasmin by plasminogen activators (PA) such as tissue-type plasminogen activator (t-PA) and urokinase-type plasminogen activator (u-PA). Inhibition of the fibrinolytic system may occur through inhibition of these plasminogen activators by plasminogen activator inhibitors (PAIs) or through inhibition of plasmin by α_2-antiplasmin. In this chapter, the PAIs are discussed. The two most important PAIs are PAI-1 and PAI-2 (Loskutoff *et al.*, 1983; Lecander and Astedt, 1986). In urine, a complex of u-PA and an inhibitor was found and, therefore, this inhibitor was named PAI-3, which was subsequently found to be identical to the protein C inhibitor (PCI) (Stump *et al.*, 1986; Heeb *et al.*, 1987).

All three PAIs belong to the *ser*ine *p*roteinase *in*hibitors (serpin) superfamily (Carrell and Travis, 1985). Serpins represents about 10% of the total protein in plasma. All serpins consist of about 400 residues with molecular masses in the range of 38–70 kDa (dependent on the degree of glycosylation) and an overall amino acid homology of approximately 35% (Carrell and Boswell, 1986; Huber and Carrell, 1989). All serpins have the same highly ordered tertiary structure consisting of 3 β-pleated A, B, and C sheets, α-helices A through I, and a reactive site loop containing residues P16 to P10′ (Figure 3.1) (Huber and Carrell, 1989). The reactive site is located in the reactive site loop situated 30–40 amino acids from the carboxy terminal end and providing a "bait" peptide bond (P1P1′) that mimics the normal substrate of the target proteinase (Laskowski, Jr. and Kato, 1980). In general, serpins occur as a stressed conformation (active) and a relaxed conformation. The transition from the stressed to the relaxed form occurs when the serpin inhibits its target proteinase and is accompanied by insertion of the reactive site loop and conformational shifts involving β-sheet A and C, and helix F (Stein and Carrell, 1995; Huntington *et al.*, 2000).

1. Plasminogen Activator Inhibitor-1

PAI-1 is a single-chain glycoprotein consisting of 379 or 381 amino acids (N-terminal heterogeneity) and a 23 amino acid signal peptide indicating that it is a secreted protein

---

Ann Gils and Paul J. Declerck • Laboratory for Pharmaceutical Biology, Katholieke Universiteit Leuven, Leuven, Belgium.

Plasminogen: Structure, Activation, and Regulation, edited by David M. Waisman. Kluwer Academic/Plenum Publishers, New York, 2003.

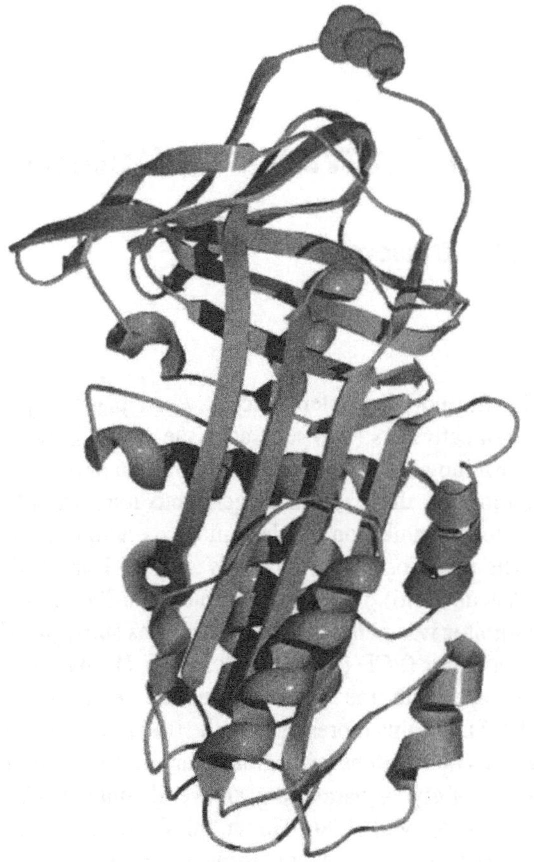

Figure 3.1. Active conformation of PAI-1 (Nar *et al.*, 2000). P1 and P1′ residues, located on the reactive site loop, are indicated as spheres.

(Andreasen *et al.*, 1986; Ginsburg *et al.*, 1986; Ny *et al.*, 1986; Pannekoek *et al.*, 1986). It has an apparent molecular weight of 45 kDa. The reactive site P1P1′ comprises Arg^{345} and Met^{346}. The isoelectric point of PAI-1 is 4.5–5.0 (van Mourik *et al.*, 1984). PAI-1 harbors three potential glycosylation sites, that is, Asn^{209}, Asn^{265}, and Asn^{329}, but no cysteines. The PAI-1 gene is localized on chromosome 7q (Ginsburg *et al.*, 1986).

1.1. Natural and Recombinant PAI-1

Natural PAI-1 has been identified in cultured rabbit endothelial (Loskutoff and Edgington, 1981), bovine endothelial (Loskutoff *et al.*, 1983), and rat hepatoma cells (Coleman *et al.*, 1982). In humans, PAI-1 was identified in conditioned medium of human endothelial cells (Booth *et al.*, 1987; Lambers *et al.*, 1987), of a melanoma cell line (Wagner *et al.*, 1986), of a fibrosarcoma cell line (HT-1080) (Andreasen *et al.*, 1986; Chmielewska *et al.*, 1987), and in human plasma (Juhan Vague *et al.*, 1984; Kruithof *et al.*, 1984; Verheijen *et al.*, 1984) and platelets (Kruithof *et al.*, 1987; Booth *et al.*, 1988;

Declerck *et al.*, 1988). PAI-1 is produced by endothelial cells, smooth muscle cells, adipocytes, spleen cells, and liver cells (Alessi *et al.*, 1997). PAI-1 gene expression is induced by endotoxin, TNF-alpha, TGF-beta, and other growth factors, cytokines, insulin and other hormones, proteinases and hypoxia (Uchiyama *et al.*, 2000).

In 1986, full-length human PAI-1 cDNA was cloned (Andreasen *et al.*, 1986; Ginsburg *et al.*, 1986; Ny *et al.*, 1986; Pannekoek *et al.*, 1986) (EMBL accession number (AC M160006). Between 1988 and 1997, cDNA encoding PAI-1 from different species has been isolated, that is, rat PAI-1 cDNA (AC M24067) from a HTC rat hepatoma cell cDNA library (Zeheb and Gelehrter, 1988), bovine PAI-1 cDNA (AC X16383) from a bovine aortic cell cDNA library (Mimuro *et al.*, 1989), murine PAI-1 cDNA (AC M33960) from a cDNA library from BALB/c 3T3 cells (Prendergast *et al.*, 1990), rabbit PAI-1 cDNA from a rabbit corneal cell cDNA library (Hofmann *et al.*, 1992) and porcine PAI-1 cDNA (AC Y11347) from a porcine aortic cell cDNA library (Bijnens *et al.*, 1997).

Currently, recombinant non-glycosylated human PAI-1 is expressed in *Escherichia coli* (Lawrence *et al.*, 1989; Audenaert *et al.*, 1994; Sancho *et al.*, 1994; Gils *et al.*, 1996) and recombinant glycosylated PAI-1 in Chinese Hamster Ovary cells (Alessi *et al.*, 1988; Lawrence *et al.*, 1989), insect cells (Kjoller *et al.*, 1996), *Pichia pastoris* (Rodenburg *et al.*, 1998), and HEK293 cells (Hansen *et al.*, 2001). The data of Lawrence *et al.* (1989) suggested that the only difference between glycosylated and non-glycosylated PAI-1 is that recombinant prokaryotic PAI-1 forms less stable complexes with plasminogen activators than recombinant eukaryotic PAI-1 or natural PAI-1 (Lawrence *et al.*, 1989). However, other studies suggested that under certain conditions (i.e., presence of the non-ionic detergent Triton X-100) different conformational conversions are observed for non-glycosylated vs glycosylated PAI-1 suggesting that precautions should be taken when evaluating PAI-1 inhibitors (Gils and Declerck, 1998a; Andreasen *et al.*, 1999).

1.2. Distinct Conformations of PAI-1

PAI-1 is unique among the serpins because of its conformational and functional flexibility. PAI-1 can occur in an *active inhibitory* form (Figure 3.1), which can inhibit both t-PA and u-PA by forming a 1:1 stoichiometric complex with it. The active form is not stable and converts spontaneously into the *latent* form (half-life of approximately 2 hr) that can be reactivated *in vitro* (Hekman and Loskutoff, 1985). This *non-reactive latent* form (Figure 3.2) does not react with the target proteinase. Furthermore, a *non-inhibitory substrate* form of PAI-1 has been isolated (Declerck *et al.*, 1992). This *substrate* form is cleaved by t-PA as well as u-PA without the formation of a covalent complex (Declerck *et al.*, 1992; Urano *et al.*, 1992; Munch *et al.*, 1993).

Several crystal structures of PAI-1 forms have been solved. Elucidation of the three-dimensional structure of the active form of a stable PAI-1 mutant revealed that the N-terminal side of the reactive site loop is exposed and accessible for the target proteinase (Sharp *et al.*, 1999; Nar *et al.*, 2000). The C-terminal side of the reactive site loop (P4′–P13′) forms strand s1C in β-sheet C. Conversion to the latent state implies the insertion of the N-terminal side of the reactive site loop into β-sheet A, the loss of strand s1C from β-sheet C, and the formation of an unusual extended loop by the C-terminal side of the reactive site loop, resulting in the distortion of the P1–P1′ peptide bond (Mottonen *et al.*, 1992). The crystal structure of a cleaved non-inhibitory substrate variant of PAI-1

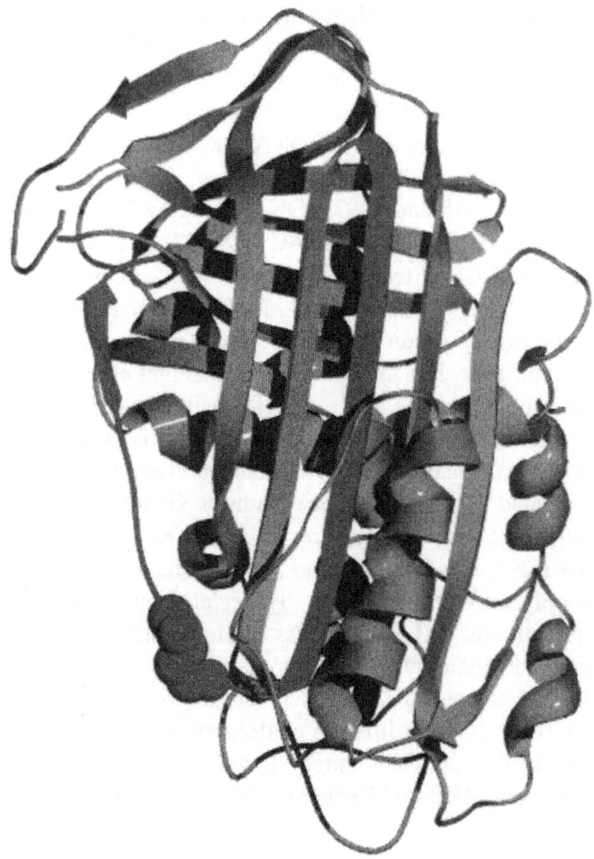

Figure 3.2. Latent conformation of PAI-1 (Mottonen *et al.*, 1992). P1 and P1′ residues, located on the reactive site loop, are indicated as spheres.

revealed the insertion of the P1–P16 portion of the reactive site loop into β-sheet A (Aertgeerts *et al.*, 1995).

1.3. Target Specificity of PAI-1

PAI-1 inhibits t-PA and u-PA very rapidly with second-order rate constants of $0.5–1.5\times10^7\ M^{-1}\ s^{-1}$ (single-chain t-PA), $2.5–4\times10^7\ M^{-1}\ s^{-1}$ (two-chain t-PA), $1\times10^7\ M^{-1}\ s^{-1}$ (u-PA), and $1.1\times10^3\ M^{-1}\ s^{-1}$ (thrombin) (Alessi *et al.*, 1988; Lawrence *et al.*, 1989; Keijer *et al.*, 1991a). Moreover, PAI-1 inhibits plasmin and trypsin. Bovine PAI-1 inhibits plasmin and trypsin with second-order rate constants of $6.6\times10^5\ M^{-1}\ s^{-1}$ and $7.0\times10^6\ M^{-1}\ s^{-1}$, respectively (Hekman and Loskutoff, 1988). It is generally accepted that the P1 residue is the major determinant of the proteinase specificity of serpins (Carrell and Travis, 1985). The active site serine hydroxyl group of the proteinases is critical for generating a high affinity interaction with the serpins (Olson *et al.*, 1995). More than 40–60% of the binding free energy of the serpin–proteinase interaction is lost when the active site serine of the proteinase is altered. Changes at the P1 residue of the serpins through natural mutation, *in vitro* site-directed mutagenesis or species variation, result in an altered target specificity. In PAI-1, a basic residue (Lys or Arg) is required at the P1 position for inhibition of u-PA (Sherman *et al.*, 1992), whereas the presence of neutral or hydrophobic residues at this position does not affect t-PA inhibition properties (Sherman *et al.*, 1995). Inhibition of u-PA still occurs when P1′ is replaced by any amino acid except proline (Sherman *et al.*, 1992).

Substituting the residues at positions P2 and P3 of PAI-1 revealed the possibility for target specific inhibition, with t-PA being more tolerant than u-PA for structural diversity at the P2 and P3 positions (York *et al.*, 1991). Moreover, replacing the residue at position P6 or P10 in PAI-1 by a proline results in a preferential inhibition of t-PA or u-PA, respectively, without affecting the inhibition rate constants, whereas replacing the residue at position P18 in PAI-1 by a proline results in a PAI-1 variant exhibiting exclusively u-PA inhibitory properties (Gils and Declerck, 1997).

It is of interest to note that the inhibition rate constant at which serpins inhibit their target proteinases is also influenced by cofactors. Both heparin and vitronectin enhance the inhibition rate constant of thrombin by PAI-1 up to 200-fold (Ehrlich *et al.*, 1990; Keijer *et al.*, 1991b). The vitronectin-induced acceleration of thrombin inhibition by PAI-1 has been suggested not to be due to a template mechanism (Ehrlich *et al.*, 1990). The latter observation may be related to the conformational change in the reactive site loop of PAI-1, induced upon binding to vitronectin (Fa *et al.*, 1995) and possibly resulting in an increased accessibility toward the active site of thrombin.

Some serpins require secondary interactions for efficient inhibition of the target proteinase. Additional interaction sites between PAI-1 and t-PA that stabilize the reversible complex have been suggested (Chmielewska *et al.*, 1988; Hekman and Loskutoff, 1988; Madison *et al.*, 1989; Lawrence *et al.*, 1990). The residues P4′, P5′, and P9′ of PAI-1 were assumed to be important for this interaction (Madison *et al.*, 1989). However, replacement of residue P9′(Asp) with an Arg resulted in a mutant that was completely inactive toward t-PA and u-PA (Madison *et al.*, 1990). On the other hand, substitution of the P4′(Glu) and P5′(Glu) with an Arg resulted only in a slight decrease of the inhibition rate constants toward t-PA (i.e., $k_2 = 9.2\times10^5\ M^{-1}\ s^{-1}$ and $7.2\times10^5\ M^{-1}\ s^{-1}$, respectively, vs $1.4\times10^6\ M^{-1}\ s^{-1}$ for wtPAI-1) (Madison *et al.*, 1990; Tucker and Gerard, 1996).

1.4. PAI-1 in Patho-Physiological Processes

1.4.1. PAI-1 and Cardiovascular Disease

The plasma concentration of PAI-1 is low (i.e., 20 ng/ml). However, the PAI-1 concentration can increase 10-fold at sites of injury when platelets are activated (Erickson *et al.*, 1984; Juhan Vague *et al.*, 1984; Booth *et al.*, 1988). PAI-1 plays a regulatory role in fibrinolysis by limiting the production of plasmin. In humans, deficiency of PAI-1, either by a functionally abnormal protein (Schleef *et al.*, 1989) or by a lack of plasma PAI-1 antigen (Dieval *et al.*, 1991; Lee *et al.*, 1993), resulted in a hyperfibrinolytic state. Moreover, patients with a complete PAI-1 deficiency caused by a homozygous frameshift mutation manifested an abnormal bleeding, but only after trauma or surgery, without any further abnormalities (Fay *et al.*, 1997). In contrast, high levels of PAI-1 have been suggested to be correlated with myocardial infarction (Hamsten *et al.*, 1985, 1987). Several sequence variations in the promotor region of the PAI-1 gene, especially the 4G/5G polymorphism, have been described. 4G/5G polymorphism refers to a guanosine insertion/deletion polymorphism 675 bp upstream from the start of transcription. *In vitro* studies have suggested that the 4G allele is associated with higher PAI-1 activity compared to the 5G allele due to an additional binding site (onto the 5G allele) for a DNA-binding protein that could be a transcriptional repressor (Eriksson *et al.*, 1995). However, no association between PAI-1 genotype (4G/5G polymorphism at position −675 in the PAI-1 promotor region) and arterial or venous thrombosis was found (Ridker *et al.*, 1997) except in patients with combined protein S deficiency (Zoller *et al.*, 1998). Recently, it was shown that metabolic determinants may be of greater importance than genetic determinants (Henry *et al.*, 1998).

Although the ECAT study demonstrated an association between higher baseline PAI-1 concentrations and the risk for a cardiovascular event, this association disappeared when other parameters reflecting insulin resistance were adjusted (Juhan Vague *et al.*, 1996). The relevance of PAI-1 as an independent risk factor for thrombotic events remains controversial.

In transgenic mouse models, it was found that PAI-1 overexpressing mice developed necrotic tips and swollen hind feet (Erickson *et al.*, 1990), whereas PAI-1 deficiency in mice accelerated the clot-lysis time without induction of moderate bleeding (Carmeliet *et al.*, 1993; Eitzman *et al.*, 1996a; Farrehi *et al.*, 1998; Zhu *et al.*, 1999; Eitzman *et al.*, 2000).

Elevated PAI-1 levels have also been associated with an unfavorable outcome in the cases of sepsis (Kornelisse *et al.*, 1996; Mesters *et al.*, 1996). Severe sepsis can lead to disseminated intravascular coagulation (DIC) (Glauser *et al.*, 1991; Bone, 1992; Levi *et al.*, 1993). In case of sepsis, the 4G/5G polymorphism seems to be of importance since the 4G allele results in a 6-fold higher transcription level than the 5G allele in response to interleukin-1 (Dawson *et al.*, 1993). This higher transcription leads to overproduction of PAI-1 and this may be an important event contributing to DIC in patients suffering from septic shock. Possibly, PAI-1 is the link between the renin–angiotensin–aldosterone system (RAAS) and thrombosis but this is still controversial (Ridker *et al.*, 1993; Vaughan *et al.*, 1995; Margaglione *et al.*, 1997).

Studies have shown that increased PAI-1 levels reduce the efficacy of thrombolytic therapy by preventing or retarding clot dissolution or by supporting early thrombotic reocclusion (Booth *et al.*, 1992). Therefore, several PAI-1 resistant t-PA mutants have been constructed (Madison *et al.*, 1989; Paoni *et al.*, 1993). The most interesting mutant is TNK-t-PA. This t-PA variant is resistant to inactivation by PAI-1 because of a tetra-alanine substitution in the protease domain (K296A/H297A/R298A/R299A, abbreviated K). Moreover, this variant has a longer half-life because the glycosylation site at amino acid 117 (N117Q) has been shifted to amino acid 103 (T103N) (Bennett *et al.*, 1991) and is more fibrin-specific (Stewart *et al.*, 2000). *In vitro* studies have shown that TNK-t-PA retained more than 80% of its lytic activity even at 50% procent molar excess of PAI-1. Animal studies revealed that TNK-t-PA is 13.5 times more potent compared to t-PA. However, studies in high-risk patients with elevated PAI-1 levels still have to confirm the results of the *in vitro* studies (Huber, 2001).

1.4.2. PAI-1 and Cancer

Degradation of extracellular matrix is a prerequisite for invasion and metastasis of cancer cells. Enzymes of the plasmin system (especially u-PA) as well as matrix metalloproteinases (MMPs) are involved in this matrix degradation. u-PA has been shown to be a prognostic marker in many types of cancer (reviewed in Andreasen *et al.* (1997). However, PAI-1 also has been shown to be a prognostic marker in breast cancer (Janicke *et al.*, 1993), gastric cancer (Nekarda *et al.*, 1994), in lung adenocarcinoma (Pedersen *et al.*, 1994), and in ovarian cancer (Kuhn *et al.*, 1994). However, the increased PAI-1 levels may be a protective response to the elevated u-PA. Therefore, it is unclear whether or not high PAI-1 levels contribute directly to poor prognosis. Moreover, PAI-1 overexpressing and PAI-1 deficient mice demonstrated no significant difference in primary tumor size, number of pulmonary metastases, or overall survival after intravenous inoculation of B16 melanoma cells (Eitzman *et al.*, 1996b).

The adhesive glycoprotein vitronectin plays an important role in the attachment of cells to their surrounding matrix and may participate in the regulation of cell differentiation, proliferation, and morphogenesis. The cell-binding properties are mediated through interactions between cellular integrins and arginine–glycine–aspartic acid sequence in the N-terminal region of vitronectin (Hynes, 1992). Active PAI-1 (but not latent PAI-1) has been shown to inhibit cell migration through interference with the binding of integrin $\alpha_v\beta_3$ to vitronectin (Stefansson and Lawrence, 1996). Thus, PAI-1 plays a dual role in extravascular processes, on the one hand via its capacity to inhibit plasminogen activators, and on the other hand via its capacity to prevent cell adhesion. Moreover, a recent study demonstrated that PAI-1 is proangiogenic at physiological concentrations (through its antiproteolytic activity rather than by interacting with vitronectin) and antiangiogenic at higher concentrations (Devy *et al.*, 2002). Interestingly, a monoclonal antibody, that is, MA-124K1 was shown to inhibit rat PAI-1 activity and simultaneously increase the binding of latent PAI-1 to vitronectin subsequently resulting in a decreased cell migration (Ngo *et al.*, 2001).

1.5. PAI-1 Inhibitors and their Binding Sites

As PAI-1 is considered to be a risk factor in cardiovascular diseases, many efforts have been focused at strategies to interfere with its inhibitory activity. In general, the activity of PAI-1 can be inhibited by either conversion to the latent form or to a substrate-behaving form as evidenced by the use of monoclonal antibodies (Declerck *et al.*, 1988; Levi *et al.*, 1992; Perrie *et al.*, 1993; Biemond *et al.*, 1995; Abrahamsson *et al.*, 1996; Bjorquist *et al.*, 1997; Debrock and Declerck, 1997; Berry *et al.*, 1998; Ngo and Declerck, 1998; Muehlenweg *et al.*, 2000; Stoop *et al.*, 2000; Bijnens *et al.*, 2001; Wind *et al.*, 2001), peptides composed of residues corresponding to (or part of) the P14-P1 region of the reactive site loop of PAI-1 (Eitzman *et al.*, 1995; Kvassman *et al.*, 1995; Xue *et al.*, 1998), other peptides (Gardsvoll *et al.*, 1998), low molecular weight compounds (Bryans *et al.*, 1996; Charlton *et al.*, 1996, 1997; Friederich *et al.*, 1997; Bjorquist *et al.*, 1998; Chikanishi *et al.*, 1999; Folkes *et al.*, 2001; Egelund *et al.*, 2001) or detergents (Ehnebom *et al.*, 1997; Gils and Declerck, 1998a; Andreasen *et al.*, 1999). Alternatively, antisense oligonucleotides can inhibit PAI-1 synthesis (Stec *et al.*, 1997; Kobylanska *et al.*, 1999).

Using overlapping peptides, Perrie *et al.* (1993) demonstrated that ESPI-12, recognized an epitope spanning residues 342–349 of PAI-1 (P5–P3′), including the bait peptide bond, and thereby explaining the inhibitory properties of the antibody through a direct interaction with the reactive center residue. The binding region of MAI-12 (=MA-7D4B7; Declerck *et al.*, 1988) has been suggested to be situated between amino acid residues 320 and 379 of PAI-1 (Keijer *et al.*, 1991b), and is overlapping or possibly identical to that of ESPI-12 (Bjorquist *et al.*, 1997). Several other inhibitory monoclonal antibodies, inducing substrate behavior of PAI-1, were mapped to the region of α-helix F (Debrock and Declerck, 1998; Bijnens *et al.*, 2000; Muehlenweg *et al.*, 2000; Stoop *et al.*, 2000; Wind *et al.*, 2001). Epitope mapping of the inhibitory MA-124K1 revealed the major contribution of residues Glu^{212} and Glu^{220} localized on strand s1B and s2B (Ngo *et al.*, 2001). For the monoclonal antibodies I-201 (non-inhibitory) and M-5 (inhibitory), the dominant residues were found to be Q^{56} (located in helix C) and D^{181} (ts3As4C), respectively (Stoop *et al.*, 2000). The epitope of three other inhibitory monoclonal antibodies (i.e., MA-44E4, MA-42A2F6, and MA-56A7C10) were mapped to the distal hinge of the reactive site loop and its proximity (Bijnens *et al.*, 2001). The epitope of the non-inhibiting antibody CB5B10 was localized in α-helix E and the turn connecting helix E and strand s1A (Hoess *et al.*, 1994).

The insertion of two TVASS peptides (P14-P10) into strands 3A and 5A was confirmed by crystal structure determination (Xue *et al.*, 1998). For the low molecular weight compound AR-H029953XX, the residues Arg^{76}, Arg^{115}, and Arg^{188} were suggested to be the binding sites (Bjorquist *et al.*, 1998) whereas binding of XR5118 to PAI-1 was related to the residues 110–145 comprising hE, s1A, and hF (Friederich *et al.*, 1997). Surprisingly, a recent study (Egelund *et al.*, 2001) identified a hydrophobic area in the flexible joint region (regions around hD and hE) of PAI-1 as the binding site of four negatively charged low molecular weight compounds ANS, AR-H029953XX, *bis*-ANS, and 1-dodecyl sulfuric acid as well as one positively charged compound namely XR-5118. The negatively charged compounds were shown to induce a polymerization in PAI-1.

Three different types of polymerization of PAI-1 have been reported: first, polymerization by insertion of the reactive site loop of one molecule as s4A in another molecule (reviewed in Gils and Declerck, 1998b; Carrell and Stein, 1996); second, by insertion of the reactive site loop of one molecule as s1C in another molecule (Carrell *et al.*, 1994; Schreuder *et al.*, 1994; Jin *et al.*, 1997); and third, when the reactive site loop forms hydrogen bonds with s6A resulting in the formation of s7A (Sharp *et al.*, 1999; Nar *et al.*, 2000). A s1C β-linkage as well as a s7A β-linkage was also suggested for native recombinant and latent PAI-1, respectively, when PAI-1 was incubated under certain conditions (Zhou *et al.*, 2001).

2. Plasminogen Activator Inhibitor-2

PAI-2 is a glycoprotein consisting of 415 amino acids that was originally isolated from human placenta. Due to the lack of a propeptide leader sequence, the secretion of PAI-2 (glycosylated form of 60 kDa) is not efficient and, therefore, PAI-2 can also accumulate intracellularly (non-glycosylated form of 47 kDa). The reactive site P1P1′ comprises Arg^{380} and Thr^{381}. The isoelectric point of PAI-2 is 4.5 (glycosylated)–5.0 (non-glycosylated). PAI-2 harbors three potential glycosylation sites, that is, Asn^{75}, Asn^{115}, and Asn^{339} and 6 cysteines at positions 5, 79, 145, 161, 406, 414 (Kruithof *et al.*, 1995). The PAI-2 gene is localized on chromosome 18q21.2–22 (Schneider *et al.*, 1995).

2.1. Function of PAI-2

Since PAI-2 harbors only an internal signal sequence, the secretion of PAI-2 is not efficient, and therefore PAI-2 can accumulate intracellularly. As a consequence, the majority of PAI-2 is in a non-glycosylated form. PAI-2 is produced by epithelial cells, monocytes/macrophages, and keratinocytes (Kruithof *et al.*, 1986; Schleuning *et al.*, 1987; Webb *et al.*, 1987; Wohlwend *et al.*, 1987; Dougherty *et al.*, 1999a). Secretion vs non-secretion of PAI-2 is cell-type specific and culture condition dependent.

PAI-2 has an intracellular function, that is, preventing cell death (Dickinson *et al.*, 1995) as well as an extracellular function, that is, inhibition of t-PA and u-PA. Second-order rate constants are $2.4–2.7 \times 10^6\ M^{-1}\ s^{-1}$ for inhibition of u-PA, $2.5–2.7 \times 10^5\ M^{-1}\ s^{-1}$ for inhibition of single-chain t-PA, and $0.8–1.2 \times 10^4\ M^{-1}\ s^{-1}$ for the inhibition of two-chain t-PA (Mikus *et al.*, 1993). PAI-2 does not bind to vitronectin (Mikus *et al.*, 1993).

PAI-2 levels are normally less than 0.005 μg/ml in plasma but increase during the third trimester of pregnancy (Lecander and Astedt, 1986). Thus, PAI-2 may have a role in maintenance of the placenta or embryonic development (Lecander and Astedt, 1986). However, PAI-2 deficient mice revealed normal development, survival, and fertility (Dougherty *et al.*, 1999b).

A number of transglutaminases (i.e., tissue transglutaminase and FXIIIa) are responsible for the cross-linking of fibrinogen and consequently for the generation of a stable fibrin clot. Some inhibitors of fibrinolysis (i.e., α_2-antiplasmin and PAI-2) can be cross-linked to fibrinogen. This cross-linking occurs via Gln^2 of α_2-antiplasmin and Lys^{303} of the Aα chain of fibrinogen (Kimura and Aoki, 1986) or via Gln^{83} and Gln^{86} (localized on an exposed loop between helices C and D = CD loop) of PAI-2 and Lys^{148}, Lys^{176}, Lys^{183}, Lys^{230}, Lys^{413}, and Lys^{457} of the Aα chain of fibrinogen (Ritchie *et al.*, 2000), respectively. Cross-linked PAI-2 can still inhibit proteinases (Ritchie *et al.*, 1999) and can, therefore, still function as an inhibitor of fibrinolysis. The CD-loop of PAI-2 harbors a potential glycosylation site at position 75 and, therefore, it is hypothesized that glycosylated PAI-2 cannot cross-link with fibrinogen (Kruithof *et al.*, 1995). However, activation of monocytes results in the release of non-glycosylated PAI-2, via an ER-Golgi independent pathway, making it available for cross-linking with fibrinogen in inflammatory lesions (Ritchie and Booth, 1998).

This CD loop is also of importance in the linking of PAI-2 to trophoblast membranes and to extracellular matrix (Jensen *et al.*, 1994). Morever, this loop seems to be of importance in inhibiting programmed cell death in cells (Dickinson *et al.*, 1995).

2.2. PAI-2 Polymerization

PAI-2 is the only serpin that polymerizes spontaneously. Mikus *et al.* demonstrated that glycosylated as well as non-glycosylated PAI-2 can polymerize and that this process can take place in different cellular compartments (Mikus and Ny, 1996). This loop-sheet polymerization, that is, insertion of the reactive site loop of one molecule into β-sheet A of another molecule, is caused by accumulation of PAI-2 and is hindered by a peptide corresponding to residues P2–P14 of the reactive site loop of PAI-2. PAI-2 polymers are found in human placenta cells and in phorbol ester-stimulated U937 cells (Mikus and Ny, 1996).

As a member of the serpin superfamily, PAI-2 occurs in at least two conformations, that is, a stressed and a relaxed conformation (Stein and Carrell, 1995). Moreover, the relaxed form is also induced by insertion of a synthetic reactive site loop peptide resulting in a substrate behavior of PAI-2 for u-PA (Saunders *et al.*, 1998). The same study reported a neoantigenic epitope for MA-2H5 that specifically detects the relaxed conformation of

PAI-2. Comparison of the crystal structure of a stressed PAI-2 form (PAI-2 mutant in which the CD loop is deleted) and a relaxed PAI-2 form (PAI-2 mutant in which the CD loop is deleted in complex with a reactive site loop peptide) suggested that the hinge formed by Gly^{206} (β-strand 3A) plays a crucial role in the stressed to relaxed transition (Harrop *et al.*, 1999; Jankova *et al.*, 2001).

3. Plasminogen Activator Inhibitor-3

PAI-3 also known as PCI is a single-chain glycoprotein consisting of 387 amino acids and a 19-amino-acids signal peptide (Suzuki *et al.*, 1987). The reactive site P1P1′ comprises Arg^{354} and Ser^{355} (Suzuki *et al.*, 1987). The isoelectric point of PAI-3 is 4.5–6.0 (Suzuki, 1993). PAI-3 harbors five potential glycosylation sites, that is, Asn^{230}, Asn^{243},Asn^{319}, Thr^{20}, and Ser^{39} and one cysteine. The PAI-3 gene is localized on chromosome 14q32.1 (Billingsley *et al.*, 1993).

3.1. Target Specificity of PAI-3

PCI (PAI-3) has a broad target specificity and inhibits u-PA (rate constants $8\times10^3\,M^{-1}\,s^{-1}$ (in the absence of heparin (−)) and $9\times10^4\,M^{-1}\,s^{-1}$ (in the presence of heparin (+)) at 37°C (Stump *et al.*, 1986), activated protein C (APC) (rate constants $5.2\times10^3\,M^{-1}\,s^{-1}$ (−) and $2.3\times10^5\,M^{-1}\,s^{-1}$ (+) at 37°C (Hermans and Stone, 1993)), thrombin (rate constants $3.3\times10^4\,M^{-1}\,s^{-1}$ (−) and $2\times10^5\,M^{-1}\,s^{-1}$ (+) at 37°C (Hermans and Stone, 1993)), thrombin–thrombomodulin (rate constant $2.4\times10^6\,M^{-1}\,s^{-1}$ (−) at 23°C (Rezaie *et al.*, 1995)), factor Xa (rate constants $3.2\times10^3\,M^{-1}\,s^{-1}$ (−) and $3.5\times10^3\,M^{-1}\,s^{-1}$ (+) at 23°C (Pratt and Church, 1992)), factor XIa (rate constants $9.4\times10^3\,M^{-1}\,s^{-1}$ (−) and $9.1\times10^4\,M^{-1}\,s^{-1}$ (+) at 37°C (Meijers *et al.*, 1988)), plasma kallikrein (rate constants $1.1\times10^5\,M^{-1}\,s^{-1}$ (−) and $1\times10^5\,M^{-1}\,s^{-1}$ (+) at 37°C (Meijers *et al.*, 1988)), t-PA (rate constant $<10^3\,M^{-1}\,s^{-1}$ (−) at 37°C (Stump *et al.*, 1986)), trypsin (rate constant $1.7\times10^4\,M^{-1}\,s^{-1}$ (+) at 23°C (Phillips *et al.*, 1994)), chymotrypsin (rate constants $1.1\times10^6\,M^{-1}\,s^{-1}$ (−) and $3.3\times10^6\,M^{-1}\,s^{-1}$ (+) at 23°C (Pratt and Church, 1992)), and acrosin (rate constants $2.4\times10^5\,M^{-1}\,s^{-1}$ (−) and $5.6\times10^7\,M^{-1}\,s^{-1}$ (+) at 37°C (Hermans *et al.*, 1994)). It does not inhibit factor XIIa and plasmin. The P3–P3′ residues determine this target specificity of PCI (Phillips *et al.*, 1994; Cooper and Church, 1995; Cooper *et al.*, 1995).

The rate constants for inhibition of serine proteinases by PCI are 30–100-fold increased in the presence of heparin and negatively charged dextran sulfate. Probably heparin induces a conformational change in the structure of PCI by binding to the positively charged H-helix (Neese *et al.*, 1998).

PCI from four other species have been cloned, that is, rhesus monkey PCI (Radtke *et al.*, 1995), bovine PCI (Yuasa *et al.*, 2000), rat PCI (Wakita *et al.*, 1998), and mouse PCI (Zechmeister-Machhart *et al.*, 1997).

3.2. Physiological Role of PAI-3

PCI is found in plasma (2–5 μg/ml), urine (0.2–0.5 μg/ml), and seminal plasma (60–150 μg/ml) (Suzuki *et al.*, 1983; Stump *et al.*, 1986; Espana *et al.*, 1991; Laurell *et al.*, 1992; Espana *et al.*, 1993). In plasma PCI is the primary inhibitor of APC (Espana *et al.*, 1990; Minamikawa *et al.*, 1994) (procoagulant effect) and APC–PCI complexes are considered to be a marker for intravascular coagulation. PCI also inhibits thrombin in plasma (anticoagulant effect) but in the presence of thrombomodulin, PCI inhibits the thrombin–thrombomodulin complex (Rezaie *et al.*, 1995) (procoagulant effect). Thus, in the presence of thrombomodulin, PCI is converted from an anticoagulant into a procoagulant factor. Moreover, urokinase has also been shown to be a target for PCI in plasma (Geiger *et al.*, 1989). Because of its broad target specificity and the contradictory pro- and anticoagulant effects, the physiological role of PCI remains unclear (Mosnier *et al.*, 2001).

In seminal plasma, the primary target for PCI is acrosin (Hermans *et al.*, 1994) and, therefore, a functionally inactive PCI may be associated with infertility. In the kidney tubules and in urine, PCI regulates the urokinase activity (Radtke *et al.*, 1994) and consequently the uPA-mediated invasion and metastasis in case of renal cell carcinoma.

References

Abrahamsson T., Nerme V., Stromqvist M., Akerblom B., Legnehed A., Pettersson K., and Westin Eriksson A. (1996). Anti-thrombotic effect of a PAI-1 inhibitor in rats given endotoxin. *Thromb. Haemost.* **75**, 118–126.

Aertgeerts K., De Bondt H.L., De Ranter C.J., and Declerck P.J. (1995). Mechanisms contributing to the conformational and functional flexibility of plasminogen activator inhibitor-1. *Nat. Struct. Biol.* **2**, 891–897.

Alessi M.C., Declerck P.J., De Mol M., Nelles L., and Collen D. (1988). Purification and characterization of natural and recombinant human plasminogen activator inhibitor-1 (PAI-1). *Eur. J. Biochem.* **175**, 531–540.

Alessi M.C., Peiretti F., Morange P., Henry M., Nalbone G., and Juhan-Vague I. (1997). Production of plasminogen activator inhibitor 1 by human adipose tissue: Possible link between visceral fat accumulation and vascular disease. *Diabetes* **46**, 860–867.

Andreasen P.A., Egelund R., Jensen S., and Rodenburg K.W. (1999). Solvent effects on activity and conformation of plasminogen activator inhibitor-1. *Thromb. Haemost.* **81**, 407–414.

Andreasen P.A., Kjoller L., Christensen L., and Duffy M.J. (1997). The urokinase type plasminogen activator system in cancer metastasis: A review. *Int. J. Of Cancer* **72**, 1–22.

Andreasen P.A., Nielsen L.S., Kristensen P., Grondahl Hansen J., Skriver L., and Dano K. (1986). Plasminogen activator inhibitor from human fibrosarcoma cells binds urokinase-type plasminogen activator, but not its proenzyme. *J. Biol. Chem.* **261**, 7644–7651.

Audenaert A.M., Knockaert I., Collen D., and Declerck P.J. (1994). Conversion of plasminogen activator inhibitor-1 from inhibitor to substrate by point mutations in the reactive-site loop. *J. Biol. Chem.* **269**, 19559–19564.

Bennett W.F., Paoni N.F., Keyt B.A., Botstein D., Jones A.J., Presta L., Wurm F.M., and Zoller M.J. (1991). High resolution analysis of functional determinants on human tissue-type plasminogen activator. *J. Biol. Chem.* **266**, 5191–5201.

Berry C.N., Lunven C., Lechaire I., Girardot C., and OConnor S.E. (1998). Antithrombotic activity of a monoclonal antibody inducing the substrate form of plasminogen activator inhibitor type 1 in rat models of venous and arterial thrombosis. *Br. J. Pharmacol* **125**, 29–34.

Biemond B.J., Levi M., Coronel R., Janse M.J., ten Cate J.W., and Pannekoek H. (1995). Thrombolysis and reocclusion in experimental jugular vein and coronary artery thrombosis. Effects of a plasminogen activator inhibitor type 1-neutralizing monoclonal antibody. *Circulation* **91**, 1175–1181.

Bijnens A.P., Gils A., Knockaert I., Stassen J.M., and Declerck P.J. (2000). Importance of the hinge region between α-helix F and the main part of serpins, based upon identification of the epitope of plasminogen activator inhibitor type 1 neutralizing antibodies. *J. Biol. Chem.* **275**, 6375–6380.

Bijnens A.P., Gils A., Stassen J.M., Komissarov A.A., Knockaert I., Brouwers E., Shore J.D., and Declerck P.J. (2001). The distal hinge of the reactive site loop and its proximity: a target to modulate plasminogen activator inhibitor-1 activity. *J. Biol. Chem.* **276**, 44912–44918.

Bijnens A.P., Knockaert I., Cousin E., Kruithof E.K., and Declerck P.J. (1997). Expression and characterization of recombinant porcine plasminogen activator inhibitor-1. *Thromb. Haemost.* **77**, 350–356.

Billingsley G.D., Walter M.A., Hammond G.L., and Cox D.W. (1993). Physical mapping of four serpin genes: Alpha 1-antitrypsin, alpha 1-antichymotrypsin, corticosteroid-binding globulin, and protein C inhibitor, within a 280-kb region on chromosome I4q32.1. *Am. J. Hum. Genet.* **52**, 343–353.

Bjorquist P., Ehnebom J., Inghardt T., and Deinum J.R. (1997). Epitopes on plasminogen activator inhibitor type 1 important for binding to tissue plasminogen activator. *Biochim. Biophys.* **1341**, 87–98.

Bjorquist P., Ehnebom J., Inghardt T., Hansson L., Lindberg M., Linschoten M., Stromqvist M., and Deinum J. (1998). Identification of the binding site for a low-molecular-weight inhibitor of plasminogen activator inhibitor type 1 by site- directed mutagenesis. *Biochemistry* **37**, 1227–1234.

Bone R.C. (1992). Modulators of coagulation. A critical appraisal of their role in sepsis. *Arch. Intern. Med.* **152**, 1381–1389.

Booth N.A., MacGregor I.R., Hunter N.R., and Bennett B. (1987). Plasminogen activator inhibitor from human endothelial cells. Purification and partial characterization. *Eur. J. Biochem.* **165**, 595–600.

Booth N.A., Robbie L.A., Croll A.M., and Bennett B. (1992). Lysis of platelet-rich thrombi: The role of PAI-1. *Ann. NY Acad. Sci.* **667**, 70–80.

Booth N.A., Simpson A.J., Croll A., Bennett B., and MacGregor I.R. (1988). Plasminogen activator inhibitor (PAI-1) in plasma and platelets. *Br. J. Haematol.* **70**, 327–333.

Bryans J., Charlton P.R., Collins M., Faint R., Latham C., Shaw I., and Trew S. (1996). Inhibition of plasminogen activator inhibitor 1 activity by two diketopiperazines, xr330 and xr334 produced by Streptomyces sp. *J. Antibiot.* **49**, 1014–1021.

Carmeliet P., Kieckens L., Schoonjans L., Ream B., Van Nuffelen A., Prendergast G., Cole M., Bronson R., Collen D., and Mulligan R.C. (1993). Plasminogen activator inhibitor-1 gene-deficient mice. I. Generation by homologous recombination and characterization. *J. Clin. Invest.* **92**, 2746–2755.

Carrell R. and Travis J. (1985). α_1-Antitrypsin and the serpins: Variation and countervariation. *TIBS* 20–24.

Carrell R.W. and Boswell D.R. (1986). Serpins: The superfamily of plasma serine proteinase inhibitors. In Barrett, Salvesen (eds) *Proteinase Inhibitors*. Elsevier Science, pp. 403–420.

Carrell R.W. and Stein P.E. (1996). The biostructural pathology of the serpins: Critical function of sheet opening mechanism. *Biol. Chem. Hoppe Seyler* **377**, 1–17.

Carrell R.W., Stein P.E., Fermi G., and Wardell M.R. (1994). Biological implications of a 3 A structure of dimeric antithrombin [see comments]. *Structure* **2**, 257–270.

Charlton P., Faint R., Barnes C., Bent F., Folkes A., Templeton D., Mackie I., Machin S., and Bevan P. (1997). Xr5118, a novel modulator of plasminogen activator inhibitor 1 (pai 1), increases endogenous tpa activity in the rat. *Fibrinolysis Proteolysis* **11**, 51–56.

Charlton P.A., Faint R.W., Bent F., Bryans J., Chicarelli-Robinson I., Mackie I., Machin S., and Bevan P. (1996). Evaluation of a low molecular weight modulator of human plasminogen activator inhibitor 1 activity. *Thromb. Haemost.* **75**, 808–815.

Chikanishi T., Shinohara C., Kikuchi T., Endo A., and Hasumi K. (1999). Inhibition of plasminogen activator inhibitor-1 by 11-keto-9(E),12(E)-octadecadienoic acid, a novel fatty acid produced by Trichoderma sp. *J. Antibiot.* **52**, 797–802.

Chmielewska J., Carlsson T., Urden G., and Wiman B. (1987). On the relationship between different molecular forms of the fast inhibitor of tissue plasminogen activator. *Fibrinolysis* **1**, 67–73.

Chmielewska J., Ranby M., and Wiman B. (1988). Kinetics of the inhibition of plasminogen activators by the plasminogen-activator inhibitor. Evidence for "second-site" interactions. *Biochem. J.* **251**, 327–332.

Coleman P.L., Barouski P.A., and Gelehrter T.D. (1982). The dexamethasone-induced inhibitor of fibrinolytic activity in hepatoma cells. A cellular product which specifically inhibits plasminogen activation. *J. Biol. Chem.* **257**, 4260–4264.

Cooper S.T. and Church F.C. (1995). Reactive site mutants of recombinant protein C inhibitor. *Biochim. Biophys. Acta* **1246**, 29–33.

Cooper S.T., Whinna H.C., Jackson T.P., Boyd J.M., and Church F.C. (1995). Intermolecular interactions between protein C inhibitor and coagulation proteases. *Biochemistry* **34**, 12991–12997.

Dawson S.J., Wiman B., Hamsten A., Green F., Humphries S., and Henney A.M. (1993). The two allele sequences of a common polymorphism in the promoter of the plasminogen activator inhibitor-1 (PAI-1) gene respond differently to interleukin-1 in HepG2 cells. *J. Biol. Chem.* **268**, 10739–10745.

Debrock S. and Declerck P.J. (1997). Neutralization of plasminogen activator inhibitor-1 inhibitory properties: Identification of two different mechanisms. *Biochim. Biophys. Acta* **1337**, 257–266.

Debrock S. and Declerck P.J. (1998). Identification of a functional epitope in plasminogen activator inhibitor-1, not localized in the reactive center loop. *Thromb. Haemost.* **79**, 597–601.

Declerck P.J., Alessi M.C., Verstreken M., Kruithof E.K., Juhan Vague I., and Collen D. (1988). Measurement of plasminogen activator inhibitor 1 in biologic fluids with a murine monoclonal antibody-based enzyme-linked immunosorbent assay. *Blood* **71**, 220–225.

Declerck P.J., De Mol M., Vaughan D.E., and Collen D. (1992). Identification of a conformationally distinct form of plasminogen activator inhibitor-1, acting as a noninhibitory substrate for tissue-type plasminogen activator. *J. Biol. Chem.* **267**, 11693–11696.

Devy L., Blacher S., Grignet-Debrus C., Bajou K., Masson V., Gerard R.D., Gils A., Carmeliet G., Carmeliet P., Declerck P.J., Noel A., and Foidart J.M. (2002). The pro- or antiangiogenic effect of plasminogen activator inhibitor 1 is dose dependent. *FASEB J.* **16**, 147–154.

Dickinson J.L., Bates E.J., Ferrante A., and Antalis T.M. (1995). Plasminogen activator inhibitor type 2 inhibits tumor necrosis factor alpha-induced apoptosis. Evidence for an alternate biological function. *J. Biol. Chem.* **270**, 27894–27904.

Dieval J., Nguyen G., Gross S., Delobel J., and Kruithof E.K. (1991). A lifelong bleeding disorder associated with a deficiency of plasminogen activator inhibitor type 1. *Blood* **77**, 528–532.

Dougherty K.M., Pearson J.M., Yang A.Y., Westrick R.J., Baker M.S., and Ginsburg D. (1999a) The plasminogen activator inhibitor-2 gene is not required for normal murine development or survival. *Proc. Natl. Acad. Sci. USA* **96**, 686–691.

Dougherty K.M., Pearson J.M., Yang A.Y., Westrick R.J., Baker M.S., and Ginsburg D. (1999b) The plasminogen activator inhibitor-2 gene is not required for normal murine development or survival. *Proc. Natl. Acad. Sci. USA* **96**, 686–691.

Egelund R., Einholm A.P., Pedersen K.E., Nielsen R.W., Christensen A., Deinum J., and Andreasen P.A. (2001). A regulatory hydrophobic area in the flexible joint region of plasminogen activator inhibitor-1, defined with fluorescent activity-neutralizing ligands Ligand induced serpin polymerization. *J. Biol. Chem.* **276**, 13077–13086.

Ehnebom J., Bjorquist P., Andersson J.O., Johansson T., and Deinum J.R. (1997). Detergent tween 80 modifies the specific activity of pai 1. *Fibrinolysis Proteolysis* **11**, 165–170.

Ehrlich A.J., Gebbink R.K., Keijer J., Linders M., Preissner K.T., and Pannekoek H. (1990). Alteration of serpin specificity by a protein cofactor. Vitronectin endows plasminogen activator inhibitor 1 with thrombin inhibitory properties. *J. Biol. Chem.* **265**, 13029–13035.

Eitzman D.T., Fay W.P., Lawrence D.A., Francis Chmura A.M., Shore J.D., Olson S.T., and Ginsburg D. (1995). Peptide-mediated inactivation of recombinant and platelet plasminogen activator inhibitor-1 *in vitro*. *J. Clin. Invest.* **95**, 2416–2420.

Eitzman D.T., Krauss J.C., Shen T., Cui J., and Ginsburg (1996b). Lack of plasminogen activator inhibitor-1 effect in a transgenic mouse model of metastatic melanoma. *Blood* **87**, 4718–4722.

Eitzman D.T., McCoy R.D., Zheng X., Fay W.P., Shen T., Ginsburg D., and Simon R.H. (1996a). Bleomycin-induced pulmonary fibrosis in transgenic mice that either lack or overexpress the murine plasminogen activator inhibitor-1 gene. *J. Clin. Invest.* **97**, 232–237.

Eitzman D.T., Westrick R.J., Nabel E.G., and Ginsburg D. (2000). Plasminogen activator inhibitor-1 and vitronectin promote vascular thrombosis in mice. *Blood* **95**, 577–580.

Erickson L.A., Fici G.J., Lund J.E., Boyle T.P., Polites H.G., and Marotti K.R. (1990). Development of venous occlusions in mice transgenic for the plasminogen activator inhibitor-1 gene. *Nature* **346**, 74–76.

Erickson L.A., Ginsberg M.H., and Loskutoff D.J. (1984). Detection and partial characterization of an inhibitor of plasminogen activator in human platelets. *J. Clin. Invest.* **74**, 1465–1472.

Eriksson P., Kallin B., van 't Hooft F.M., Bavenholm P., and Hamsten A. (1995). Allele-specific increase in basal transcription of the plasminogen-activator inhibitor 1 gene is associated with myocardial infarction. *Proc. Natl. Acad. Sci. USA* **92**, 1851–1855.

Espana F., Estelles A., Fernandez P.J., Gilabert J., Sanchez-Cuenca J., and Griffin J.H. (1993). Evidence for the regulation of urokinase and tissue type plasminogen activators by the serpin, protein C inhibitor, in semen and blood plasma. *Thromb. Haemost.* **70**, 989–994.

Espana F., Gilabert J., Estelles A., Romeu A., Aznar J., and Cabo A. (1991). Functionally active protein C inhibitor/plasminogen activator inhibitor-3 (PCI/PAI-3) is secreted in seminal vesicles., occurs at high concentrations in human seminal plasma and complexes with prostate-specific antigen. *Thromb. Res.* **64**, 309–320.

Espana F., Vicente V., Tabernero D., Scharrer I., and Griffin J.H. (1990). Determination of plasma protein C inhibitor and of two activated protein C-inhibitor complexes in normals and in patients with intravascular coagulation and thrombotic disease. *Thromb. Res.* **59**, 593–608.

Fa M., Karolin J., Aleshkov S., Strandberg L., Johansson L.B., and Ny T. (1995). Time-resolved polarized fluorescence spectroscopy studies of plasminogen activator inhibitor type 1: Conformational changes of the reactive center upon interactions with target proteases, vitronectin and heparin. *Biochemistry* **34**, 13833–13840.

Farrehi P.M., Ozaki C.K., Carmeliet P., and Fay W.P. (1998). Regulation of arterial thrombolysis by plasminogen activator inhibitor-1 in mice. *Circulation* **97**, 1002–1008.

Fay W.P., Parker A.C., Condrey L.R., and Shapiro A.D. (1997). Human plasminogen activator inhibitor 1 (pai 1) deficiency: Characterization of a large kindred with a null mutation in the pai 1 gene. *Blood* **90**, 204–208.

Folkes A., Roe M.B., Sohal S., Golec J., Faint R., Brooks T., and Charlton P. (2001). Synthesis and in vitro evaluation of a series of diketopiperazine inhibitors of plasminogen activator inhibitor-1. *Bioorg. Med. Chem. Lett.* **11**, 2589–2592.

Friederich P.W., Levi M.R., Biemond B.J., Charlton P., Templeton D., Vanzonneveld A.J., Bevan P., Pannekoek H., and Tencate J.W. (1997). Novel low molecular weight inhibitor of pai 1 (xr5118) promotes endogenous fibrinolysis and reduces postthrombolysis thrombus growth in rabbits. *Circulation* **96**, 916–921.

Gardsvoll H., van-Zonneveld A.J., Holm A., Eldering E., van-Meijer M., Dano K., and Pannekoek H. (1998). Selection of peptides that bind to plasminogen activator inhibitor 1 (PAI-1) using random peptide phage-display libraries. *FEBS Lett.* **431**, 170–174.

Geiger M., Huber K., Wojta J., Stingl L., Espana F., Griffin J.H., and Binder B.R. (1989). Complex formation between urokinase and plasma protein C inhibitor in vitro and in vivo. *Blood* **74**, 722–728.

Gils A. and Declerck P.J. (1997). Proteinase specificity and functional diversity in point mutants of plasminogen activator inhibitor 1. *J. Biol. Chem.* **272**, 12662–12666.

Gils A. and Declerck P.J. (1998a). Modulation of plasminogen activator inhibitor 1 by Triton X-100—Identification of two consecutive conformational transitions. *Thromb. Haemost.* **80**, 286–291.

Gils A. and Declerck P.J. (1998b). Structure-function relationship in serpins: Current concepts and controversies. *Thromb. Haemost.* **80**, 531–541.

Gils A., Knockaert I., and Declerck P.J. (1996). Substrate behavior of plasminogen activator inhibitor-1 is not associated with a lack of insertion of the reactive site loop. *Biochemistry* **35**, 7474–7481.

Ginsburg D., Zeheb R., Yang A.Y., Rafferty U.M., Andreasen P.A., Nielsen L., Dano K., Lebo R.V., and Gelehrter T.D. (1986). cDNA cloning of human plasminogen activator-inhibitor from endothelial cells. *J. Clin. Invest.* **78**, 1673–1680.

Glauser M.P., Zanetti G., Baumgartner J.D., and Cohen J. (1991). Septic shock: Pathogenesis. *Lancet* **338**, 732–736.

Hamsten A., de Faire U., Walldius G., Dahlen G., Szamosi A., Landou C., Blomback M., and Wiman B. (1987). Plasminogen activator inhibitor in plasma: Risk factor for recurrent myocardial infarction. *Lancet* **2**, 3–9.

Hamsten A., Wiman B., de Faire U., and Blomback M. (1985). Increased plasma levels of a rapid inhibitor of tissue plasminogen activator in young survivors of myocardial infarction. *N. Engl. J. Med.* **313**, 1557–1563.

Hansen M., Busse M.N., and Andreasen P.A. (2001). Importance of the amino-acid composition of the shutter region of plasminogen activator inhibitor-1 for its transitions to latent and substrate forms. *Eur. J. Biochem.* **268**, 6274–6283.

Harrop S.J., Jankova L., Coles M., Jardine D., Whittaker J.S., Gould A.R., Meister A., King G.C., Mabbutt B.C., and Curmi P.M. (1999). The crystal structure of plasminogen activator inhibitor 2 at 2.0 A resolution: implications for serpin function. *Struct. Fold. Des.* **7**, 43–54.

Heeb M.J., Espana F., Geiger M., Collen D., Stump D.C., and Griffin J.H. (1987). Immunological identity of heparin-dependent plasma and urinary protein C inhibitor and plasminogen activator inhibitor-3. *J. Biol. Chem.* **262**, 15813–15816.

Hekman C.M. and Loskutoff D.J. (1985). Endothelial cells produce a latent inhibitor of plasminogen activators that can be activated by denaturants. *J. Biol. Chem.* **260**, 11581–11587.

Hekman C.M. and Loskutoff D.J. (1988). Bovine plasminogen activator inhibitor 1: Specificity determinations and comparison of the active, latent, and guanidine-activated forms. *Biochemistry* **27**, 2911–2918.

Henry M., Tregouet D.A., Alessi M.C., Aillaud M.F., Visvikis S., Siest G., Tiret L., and Juhan-Vague I. (1998). Metabolic determinants are much more important than genetic polymorphisms in determining the PAI-1 activity and antigen plasma concentrations: A family study with part of the Stanislas Cohort. *Arterioscler. Thromb. Vasc. Biol.* **18**, 84–91.

Hermans J.M., Jones R., and Stone S.R. (1994). Rapid inhibition of the sperm protease acrosin by protein C inhibitor. *Biochemistry* **33**, 5440–5444.

Hermans J.M. and Stone S.R. (1993). Interaction of activated protein C with serpins. *Biochem J.* **295**, 45.

Hoess R.H., Mack A.J., Walton H., and Reilly T.M. (1994). Identification of a structural epitope by using a peptide library displayed on filamentous bacteriophage. *J. Immunol.* **153**, 724–729.

Hofmann K.J., Mayer E.J., Schultz L.D., Socher S.H., and Reilly C.F. (1992). Purification and characterisation of recombinant rabbit plasminogen activator inhibitor-1 expressed in *Saccharomyces cerevisiae. Fibrinolysis* **6**, 263–272.

Huber K. (2001). Plasminogen activator inhibitor type-1 (part two): Role for failure of thrombolytic therapy. PAI-1 resistance as a potential benefit for new fibrinolytic agents. *J. Thromb. Thrombol.* **11**, 195–202.

Huber R. and Carrell R.W. (1989). Implications of the three-dimensional structure of alpha 1-antitrypsin for structure and function of serpins. *Biochemistry* **28**, 8951–8966.

Huntington J.A., Read R.J., and Carrell R.W. (2000). Structure of a serpin-protease complex shows inhibition by deformation. *Nature* **407**, 923–926.

Hynes R.O. (1992). Integrins: Versatility, modulation, and signaling in cell adhesion. *Cell* **69**, 11–25.

Janicke F., Schmitt M., Pache L., Ulm K., Harbeck N., Hofler H., and Graeff H. (1993). Urokinase (uPA) and its inhibitor PAI-1 are strong and independent prognostic factors in node-negative breast cancer. *Breast Cancer Res. Treat.* **24**, 195–208.

Jankova L., Harrop S.J., Saunders D.N., Andrews J.L., Bertram K.C., Gould A.R., Baker M.S., and Curmi P.M. (2001). Crystal structure of the complex of plasminogen activator inhibitor 2 with a peptide mimicking the reactive center loop. *J. Biol. Chem.* **276**, 43374–43382.

Jensen P.H., Schuler E., Woodrow G., Richardson M., Goss N., Hojrup P., Petersen T.E., and Rasmussen L.K. (1994). A unique interhelical insertion in plasminogen activator inhibitor-2 contains three glutamines, Gln83, Gln84, Gln86, essential for transglutaminase-mediated cross-linking. *J. Biol. Chem.* **269**, 15394–15398.

Jin L., Abrahams J.P., Skinner R., Petitou M., Pike R.N., and Carrell R.W. (1997). The anticoagulant activation of antithrombin by heparin. *Proc. Natl. Acad. Sci. USA* **94**, 14683–14688.

Juhan Vague I., Moerman B., De Cock F., Aillaud M.F., and Collen D. (1984). Plasma levels of a specific inhibitor of tissue-type plasminogen activator (and urokinase) in normal and pathological conditions. *Thromb. Res.* **33**, 523–530.

Juhan Vague I., Pyke S.D.M., Alessi M.C., Jespersen J., Haverkate F., and Thompson S.G. (1996). Fibrinolytic factors and the risk of myocardial infarction or sudden death in patients with angina pectoris. *Circulation* **94**, 2057–2063.

Keijer J., Linders M., van Zonneveld A.J., Ehrlich H.J., de Boer J.P., and Pannekoek H. (1991b). The interaction of plasminogen activator inhibitor 1 with plasminogen activators (tissue-type and urokinase-type) and fibrin: Localization of interaction sites and physiologic relevance. *Blood* **78**, 401–409.

Keijer J., Linders M., Wegman J.J., Ehrlich H.J., Mertens K., and Pannekoek H. (1991a). On the target specificity of plasminogen activator inhibitor 1: The role of heparin, vitronectin, and the reactive site. *Blood* **78**, 1254–1261.

Kimura S. and Aoki N. (1986). Cross-linking site in fibrinogen for alpha 2-plasmin inhibitor. *J. Biol. Chem.* **261**, 15591–15595

Kjoller L., Martensen P.M., Sottrupjensen L., Justesen J., Rodenburg K.W., and Andreasen P.A. (1996). Conformational changes of the reactive centre loop and beta strand 5a accompany temperature dependent inhibitor substrate transition of plasminogen activator inhibitor 1. *Eur. J. Biochem.* **241**, 38–46.

Kobylanska A., Pluskota E., Pawlowska Z.., Okruszek A., Cierniewski C.S., and Stec W.J. (1999). The cytotoxicity of anti-PAI-I oligonucleotides and their conjugates. *Nucleosides Nucleotides* **18**, 1709–1710.

Kornelisse R.F., Hazelzet J.A., Savelkoul H.F., Hop W.C., Suur M.H., Borsboom A.N., Risseeuw Appel I.M., van der Voort E., and de Groot R. (1996). The relationship between plasminogen activator inhibitor-1 and proinflammatory and counterinflammatory mediators in children with meningococcal septic shock. *J. Infect. Dis.* **173**, 1148–1156.

Kruithof E.K., Baker M.S., and Bunn C.L. (1995). Biological and clinical aspects of plasminogen activator inhibitor type 2. *Blood* **86**, 4007–4024.

Kruithof E.K., Nicolosa G., and Bachmann F. (1987). Plasminogen activator inhibitor 1: Development of a radioimmunoassay and observations on its plasma concentration during venous occlusion and after platelet aggregation. *Blood* **70**, 1645–1653.

Kruithof E.K., Tran Thang C., Ransijn A., and Bachmann F. (1984). Demonstration of a fast-acting inhibitor of plasminogen activators in human plasma. *Blood* **64**, 907–913.

Kruithof E.K., Vassalli J.D., Schleuning W.D., Mattaliano R.J., and Bachmann F. (1986). Purification and characterization of a plasminogen activator inhibitor from the histiocytic lymphoma cell line U-937. *J. Biol. Chem.* **261**, 11207–11213.

Kuhn W., Pache L., Schmalfeldt B., Dettmar P., Schmitt M., Janicke F., and Graeff H. (1994). Urokinase (uPA) and PAI-1 predict survival in advanced ovarian cancer patients (FIGO III) after radical surgery and platinum-based chemotherapy. *Gynecol. Oncol.* **55**, 401–409.

Kvassman J.O., Lawrence D.A., and Shore J.D. (1995). The acid stabilization of plasminogen activator inhibitor-1 depends on protonation of a single group that affects loop insertion into beta-sheet A. *J. Biol. Chem.* **270**, 27942–27947.

Lambers J.W., Cammenga M., Konig B.W., Mertens K., Pannekoek H., and van Mourik J.A. (1987). Activation of human endothelial cell-type plasminogen activator inhibitor (PAI-1) by negatively charged phospholipids. *J. Biol. Chem.* **262**, 17492–17496.

Laskowski M., Jr., Kato I. (1980). Protein inhibitors of proteinases. *Annu. Rev. Biochem.* **49**, 593–626.

Laurell M., Christensson A., Abrahamsson P.A., Stenflo J., and Lilja H. (1992). Protein C inhibitor in human body fluids. Seminal plasma is rich in inhibitor antigen deriving from cells throughout the male reproductive system. *J. Clin. Invest.* **89**, 1094–1101.

Lawrence D., Strandberg L., Grundstrom T., and Ny T. (1989). Purification of active human plasminogen activator inhibitor 1 from Escherichia coli. Comparison with natural and recombinant forms purified from eucaryotic cells. *Eur. J. Biochem.* **186**, 523–533.

Lawrence D.A., Strandberg L., Ericson J., and Ny T. (1990). Structure-function studies of the SERPIN plasminogen activator inhibitor type 1. Analysis of chimeric strained loop mutants. *J. Biol. Chem.* **265**, 20293–20301.

Lecander I., Astedt B. (1986). Isolation of a new specific plasminogen activator inhibitor from pregnancy plasma. *Br. J. Haematol.* **62**, 221–228.

Lee M.H., Vosburgh E., Anderson K., and McDonagh J. (1993). Deficiency of plasma plasminogen activator inhibitor 1 results in hyperfibrinolytic bleeding. *Blood* **81**, 2357–2362.

Levi M., Biemond B.J., van Zonneveld A.J., ten Cate J.W., and Pannekoek H. (1992). Inhibition of plasminogen activator inhibitor-1 activity results in promotion of endogenous thrombolysis and inhibition of thrombus extension in models of experimental thrombosis [see comments]. *Circulation* **85**, 305–312.

Levi M., ten-Cate H., van-der-Poll T., and van-Deventer S.J. (1993). Pathogenesis of disseminated intravascular coagulation in sepsis. *JAMA* **270**, 975–979.

Loskutoff D.J., Edgington T.S. (1981). An inhibitor of plasminogen activator in rabbit endothelial cells. *J. Biol. Chem.* **256**, 4142–4145.

Loskutoff D.J., van Mourik J.A., Erickson L.A., and Lawrence D. (1983). Detection of an unusually stable fibrinolytic inhibitor produced by bovine endothelial cells. *Proc. Natl. Acad. Sci. USA* **80**, 2956–2960.

Madison E.L., Goldsmith E.J., Gerard R.D., Gething M.J., and Sambrook J.F. (1989). Serpin-resistant mutants of human tissue-type plasminogen activator. *Nature* **339**, 721–724.

Madison E.L., Goldsmith E.J., Gething M.J., Sambrook J.F., and Gerard R.D. (1990). Restoration of serine protease-inhibitor interaction by protein engineering. *J. Biol. Chem.* **265**, 21423–21426.

Margaglione M., Grandone E., Vecchione G., Cappucci G., Giuliani N., Colaizzo D., Celentano E., Panico S., and Diminno G.R. (1997). Plasminogen activator inhibitor 1 (pai 1) antigen plasma levels in subjects attending a metabolic ward: Relation to polymorphisms of pai 1 and angiotensin converting enzyme (ace) genes. *Arterioscler. Thromb. Vasc. Biol.* **17**, 2082–2087.

Meijers J.C., Kanters D.H., Vlooswijk R.A., van-Erp H.E., Hessing M., and Bouma B.N. (1988). Inactivation of human plasma kallikrein and factor XIa by protein C inhibitor. *Biochemistry* **27**, 4231–4237.

Mesters R.M., Florke N., Ostermann H., and Kienast J. (1996). Increase of plasminogen activator inhibitor levels predicts outcome of leukocytopenic patients with sepsis. *Thromb. Haemost.* **75**, 902–907.

Mikus P., Ny T. (1996). Intracellular polymerization of the serpin plasminogen activator inhibitor type 2. *J. Biol. Chem.* **271**, 10048–10053.

Mikus P., Urano T., Liljestrom P., and Ny T. (1993). Plasminogen-activator inhibitor type 2 (PAI-2) is a spontaneously polymerising SERPIN. Biochemical characterisation of the recombinant intracellular and extracellular forms. *Eur. J. Biochem.* **218**, 1071–1082.

Mimuro J., Sawdey M., Hattori M., and Luskutoff D.J. (1989). cDNA for bovine type 1 plasminogen activator inhibitor (PAI-1). *Nucleic Acids Res.* **17**, 8872–8872.

Minamikawa K., Wada H., Wakita Y., Ohiwa M., Tanigawa M., Deguchi K., Hiraoka N., Huzioka H., Nishioka J., and Hayashi T., *et al.* (1994). Increased activated protein C-protein C inhibitor complex levels in patients with pulmonary embolism. *Thromb. Haemost.* **71**, 192–194.

Mosnier L.O., Elisen M.G., Bouma B.N., and Meijers J.C. (2001). Protein C inhibitor regulates the thrombin–thrombomodulin complex in the up- and down regulation of TAFI activation. *Thromb. Haemost.* **86**, 1057–1064.

Mottonen J., Strand A., Symersky J., Sweet R.M., Danley D.E., Geoghegan K.F., Gerard R.D., and Goldsmith E.J. (1992). Structural basis of latency in plasminogen activator inhibitor-1. *Nature* **355**, 270–273.

Muehlenweg B., Guthaus E., de P.N., Schmitt M., Schmiedeberg N., Kotzsch M., Creutzburg S., Kramer M.D., Kessler H., Wilhelm O.G., and Magdolen V. (2000). Epitope mapping of monoclonal antibodies directed to PAI-1 using PAI-1/PAI-2 chimera and PAI-1-derived synthetic peptides. *Thromb. Res.* **98**, 73–81.

Munch M., Heegaard C.W., and Andreasen P.A. (1993). Interconversions between active, inert and substrate forms of denatured/refolded type-1 plasminogen activator inhibitor. *Biochim. Biophys. Acta* **1202**, 29–37.

Nar H., Bauer M., Stassen J.M., Lang D., Gils A., and Declerck P.J. (2000). Plasminogen activator inhibitor 1. Structure of the native serpin, comparison to its other conformers and implications for serpin inactivation. *J. Mol. Biol.* **297**, 683–695

Neese L.L., Wolfe C.A., and Church F.C. (1998). Contribution of basic residues of the d and h helices in heparin binding to protein c inhibitor. *Arch. Biochem. Biophy.* **355**, 101–108.

Nekarda H., Siewert J.R., Schmitt M., and Ulm K. (1994). Tumour-associated proteolytic factors uPA and PAI-1 and survival in totally resected gastric cancer. *Lancet* **343**, 117.

Ngo T.H., Declerck P.J. (1998). Suppression of plasminogen activator inhibitor 1 (PAI-1) activity levels in rats by monoclonal antibodies. *Fibrinolysis Proteolysis* **12**, 335–339.

Ngo T.H., Hoylaerts M.F., Knockaert I., Brouwers E., and Declerck P.J. (2001). Identification of a target site in plasminogen activator inhibitor-1 that allows neutralization of its inhibitory properties concomitant with an allosteric up-regulation of its antiadhesive properties. *J. Biol. Chem.* **276**, 26243–26248.

Ny T., Sawdey M., Lawrence D., Millan J.L., and Loskutoff D.J. (1986). Cloning and sequence of a cDNA coding for the human beta- migrating endothelial-cell-type plasminogen activator inhibitor. *Proc. Natl. Acad. Sci. USA* **83**, 6776–6780.

Olson S.T., Bock P.E., Kvassman J., Shore J.D., Lawrence D.A., Ginsburg D., and Bjork I. (1995). Role of the catalytic serine in the interactions of serine proteinases with protein inhibitors of the serpin family. Contribution of a covalent interaction to the binding energy of serpin–proteinase complexes. *J. Biol. Chem.* **270**, 30007–30017.

Pannekoek H., Veerman H., Lambers H., Diergaarde P., Verweij C.L., van Zonneveld A.J., and van Mourik J.A. (1986). Endothelial plasminogen activator inhibitor (PAI): A new member of the Serpin gene family. *EMBO J.* **5**, 2539–2544.

Paoni N.F., Keyt B.A., Refino C.J., Chow A.M., Nguyen H.V., Berleau L.T., Badillo J., Pena L.C., Brady K., and Wurm F.M., *et al.* (1993). A slow clearing, fibrin-specific, PAI-1 resistant variant of t-PA (T103N, KHRR 296-299 AAAA). *Thromb. Haemost.* **70**, 307–312.

Pedersen H., Grondahl Hansen J., Francis D., Osterlind K., Hansen H.H., Dano K., and Brunner N. (1994). Urokinase and plasminogen activator inhibitor type 1 in pulmonary adenocarcinoma. *Cancer Res.* **54**, 120–123.

Perrie A.M., MacGregor I.R., and Booth N.A. (1993). Definition of epitopes within plasminogen activator inhibitor type-1 (PAI-1) using multiple peptide synthesis. *Fibrinolysis* **7**, 257–263.

Phillips J.E., Cooper S.T., Potter E.E., and Church F.C. (1994). Mutagenesis of recombinant protein C inhibitor reactive site residues alters target proteinase specificity. *J. Biol. Chem.* **269**, 16696–16700.

Pratt C.W., Church F.C. (1992). Heparin binding to protein C inhibitor. *J. Biol. Chem.* **267**, 8789–8794.

Prendergast G.C., Diamond L.E., Dahl D., and Cole M.D. (1990). The c-myc-regulated gene mrl encodes plasminogen activator inhibitor 1. *Mol. Cell. Biol.* **10**, 1265–1269.

Radtke K.P., Fernandez J.A., Greengard J.S., Tang W.W., Wilson C.B., Loskutoff D.J., Scharrer I., and Griffin J.H. (1994). Protein C inhibitor is expressed in tubular cells of human kidney. *J. Clin. Invest.* **94**, 2117–2124.

Radtke K.P., Fernandez J.A., Villoutreix B.O., Greengard J.S., and Griffin J.H. (1995). Characterization of a cDNA for rhesus monkey protein C inhibitor—evidence for N-terminal involvement in heparin stimulation. *Thromb. Haemost.* **74**, 1079–1087.

Rezaie A.R., Cooper S.T., Church F.C., and Esmon C.T. (1995). Protein C inhibitor is a potent inhibitor of the thrombin–thrombomodulin complex. *J. Biol. Chem.* **270**, 25336–25339.

Ridker P.M., Gaboury C.L., Conlin P.R., Seely E.W., Williams G.H., and Vaughan D.E. (1993). Stimulation of plasminogen activator inhibitor in vivo by infusion of angiotensin II. Evidence of a potential interaction between the renin–angiotensin system and fibrinolytic function. *Circulation* **87**, 1969–1973.

Ridker P.M., Hennekens C.H., Lindpaintner K., Stampfer M.J., and Miletich J.P. (1997). Arterial and venous thrombosis is not associated with the 4g/5g polymorphism in the promoter of the plasminogen activator inhibitor gene in a large cohort of us men. *Circulation* **95**, 59–62.

Ritchie H., Booth N.A. (1998). Secretion of plasminogen activator inhibitor 2 by human peripheral blood monocytes occurs via an endoplasmic reticulum-golgi-independent pathway. *Exp. Cell Res.* **242**, 439–450.

Ritchie H., Lawrie L.C., Crombie P.W., Mosesson M.W., and Booth N.A. (2000). Cross-linking of plasminogen activator inhibitor 2 and alpha(2)-antiplasmin to fibrin(ogen). *J. Biol. Chem.* **275**, 24915–24920.

Ritchie H., Robbie L.A., Kinghorn S., Exley R., and Booth N.A. (1999). Monocyte plasminogen activator inhibitor 2 (PAI-2) inhibits u-PA-mediated fibrin clot lysis and is cross-linked to fibrin. *Thromb. Haemost.* **81**, 96–103.

Rodenburg K.W., Kjoller L., Petersen H.H., and Andreasen P.A. (1998). Binding of urokinase-type plasminogen activator plasminogen activator inhibitor-1 complex to the endocytosis receptors alpha(2)-macroglobulin receptor low-density lipoprotein receptor- related protein and very-low-density lipoprotein receptor involves basic residues in the inhibitor. *Biochem. J.* **329** (Part 1); 55–63.

Sancho E., Tonge D.W., Hockney R.C., and Booth N.A. (1994). Purification and characterization of active and stable recombinant plasminogen-activator inhibitor accumulated at high levels in Escherichia coli. *Eur. J. Biochem.* **224**, 125–134.

Saunders D.N., Buttigieg K.M., Gould A., McPhun V., and Baker M.S. (1998). Immunological detection of conformational neoepitopes associated with the serpin activity of plasminogen activator inhibitor type-2. *J. Biol. Chem.* **273**, 10965–10971.

Schleef R.R., Higgins D.L., Pillemer E., and Levitt L.J. (1989). Bleeding diathesis due to decreased functional activity of type 1 plasminogen activator inhibitor. *J. Clin. Invest.* **83**, 1747–1752.

Schleuning W.D., Medcalf R.L., Hession C., Rothenbuhler R., Shaw A., and Kruithof E.K. (1987). Plasminogen activator inhibitor 2: Regulation of gene transcription during phorbol ester-mediated differentiation of U-937 human histiocytic lymphoma cells. *Mol. Cell. Biol.* **7**, 4564–4567.

Schneider S.S., Schick C., Fish K.E., Miller E., Pena J.C., Treter S.D., Hui S.M., and Silverman G.A. (1995). A serine proteinase inhibitor locus at 18q21.3 contains a tandem duplication of the human squamous cell carcinoma antigen gene. *Proc. Natl. Acad. Sci. USA* **92**, 3147–3151.

Schreuder H.A., de Boer B., Dijkema R., Mulders J., Theunissen H.J., Grootenhuis P.D., and Hol W.G. (1994). The intact and cleaved human antithrombin III complex as a model for serpin-proteinase interactions. *Nat. Struct. Biol.* **1**, 48–54.

Sharp A.M., Stein P.E., Pannu N.S., Carrell R.W., Berkenpas M.B., Ginsburg D., Lawrence D.A., and Read R.J. (1999). The active conformation of plasminogen activator inhibitor 1, a target for drugs to control fibrinolysis and cell adhesion. *Struc. Fold. Des.* **7**, 111–118.

Sherman P.M., Lawrence D.A., Verhamme I.M., Paielli D., Shore J.D., Ginsburg D. (1995). Identification of tissue-type plasminogen activator-specific plasminogen activator inhibitor-1 mutants. Evidence that second sites of interaction contribute to target specificity. *J. Biol. Chem.* **270**, 9301–9306.

Sherman P.M., Lawrence D.A., Yang A.Y., Vandenberg E.T., Paielli D., Olson S.T., Shore J.D., and Ginsburg D. (1992). Saturation mutagenesis of the plasminogen activator inhibitor-1 reactive center. *J. Biol. Chem.* **267**, 7588–7595.

Stec W.J., Cierniewski C.S., Okruszek A., Kobylanska A., Pawlowska Z., Koziolkiewicz M., Pluskota E., Maciaszek A., Rebowska B., and Stasiak M. (1997). Stereodependent inhibition of plasminogen activator inhibitor type 1 by phosphorothioate oligonucleotides: Proof of sequence specificity in cell culture and *in vivo* rat experiments. *Antisense Nucleic Acid Drug Dev.* **7**, 567–573.

Stefansson S., Lawrence D.A. (1996). The serpin PAI-1 inhibits cell migration by blocking integrin alpha V beta3 binding to vitronectin [see comments]. *Nature* **383**, 441–443.

Stein P.E., Carrell R.W. (1995). What do dysfunctional serpins tell us about molecular mobility and disease? *Nat. Struct. Biol.* **2**, 96–113.

Stewart R.J., Fredenburgh J.C., Leslie B.A., Keyt B.A., Rischke J.A., and Weitz J.I. (2000). Identification of the mechanism responsible for the increased fibrin specificity of TNK-tissue plasminogen activator relative to tissue plasminogen activator. *J. Biol. Chem.* **275**, 10112–10120.

Stoop A.A., Jespers L., Lasters I., Eldering E., and Pannekoek H. (2000). High-density mutagenesis by combined DMA shuffling and phage display to assign essential amino acid residues in protein–protein interactions: Application to study structure-function of plasminogen activation inhibitor 1 (PAI-I). *J. Mol. Biol.* **301**, 1135–1147.

Stump D.C., Thienpont M., and Collen D. (1986). Purification and characterization of a novel inhibitor of urokinase from human urine. Quantitation and preliminary characterization in plasma. *J. Biol. Chem.* **261**, 12759–12766.

Suzuki K. (1993). Protein C inhibitor. *Methods Enzymol.* **222**, 385–399.

Suzuki K., Deyashiki Y., Nishioka J., Kurachi K., Akira M., Yamamoto S., and Hashimoto S. (1987). Characterization of a cDNA for human protein C inhibitor. A new member of the plasma serine protease inhibitor superfamily. *J. Biol. Chem.* **262**, 611–616.

Suzuki K., Nishioka J., Hashimoto S., Kamiya T., and Saito H. (1983). Normal titer of functional and immunoreactive protein-C inhibitor in plasma of patients with congenital combined deficiency of factor V and factor VIII. *Blood* **62**, 1266–1270.

Tucker H.M., Gerard R.D. (1996). Sequence requirements in the reactive-center loop of plasminogen activator inhibitor-1 for recognition of plasminogen activators. *Eur. J. Biochem.* **237**, 180–187.

Uchiyama T., Kurabayashi M., Ohyama Y., Utsugi T., Akuzawa N., Sato M., Tomono S., Kawazu S., and Nagai R. (2000). Hypoxia induces transcription of the plasminogen activator inhibitor-1 gene through genistein-sensitive tyrosine kinase pathways in vascular endothelial cells. *Arterioscler. Thromb. Vasc. Biol.* **20**, 1155–1161.

Urano T., Strandberg L., Johansson L.B., and Ny T. (1992). A substrate-like form of plasminogen-activator-inhibitor type 1. Conversions between different forms by sodium dodecyl sulphate. *Eur. J. Biochem.* **209**, 985–992.

van Mourik J.A., Lawrence D.A., and Loskutoff D.J. (1984). Purification of an inhibitor of plasminogen activator (antiactivator) synthesized by endothelial cells. *J. Biol. Chem.* **259**, 14914–14921.

Vaughan D.E., Lazos S.A., and Tong K. (1995). Angiotensin II regulates the expression of plasminogen activator inhibitor-1 in cultured endothelial cells. A potential link between the renin–angiotensin system and thrombosis. *J. Clin. Invest.* **95**, 995–1001.

Verheijen J.H., Chang G.T., and Kluft C. (1984). Evidence for the occurrence of a fast-acting inhibitor for tissue-type plasminogen activator in human plasma. *Thromb. Haemost.* **51**, 392–395.

Wagner O.F., Vetterlein M., and Binder B.R. (1986). Purification of an active plasminogen activator inhibitor immunologically related to the endothelial type plasminogen activator inhibitor from the conditioned media of a human melanoma cell line [published erratum appears in *J. Biol. Chem.* 1988 Jan 25; **263**(3), 1593]. *J. Biol. Chem.* **261**, 14474–14481.

Wakita T., Hayashi T., Yuasa H., Nishioka J., Kawamura J., and Suzuki K. (1998). Molecular cloning, tissue distribution and androgen regulation of rat protein C inhibitor. *FEBS Lett.* **429**, 263–268.

Webb A.C., Collins K.L., Snyder S.E., Alexander S.J., Rosenwasser L.J., Eddy R.L., Shows T.B., and Auron P.E. (1987). Human monocyte Arg-Serpin cDNA. Sequence, chromosomal assignment, and homology to plasminogen activator-inhibitor. *J. Exp. Med.* **166**, 77–94.

Wind T., Jensen M.A., and Andreasen P.A. (2001). Epitope mapping for four monoclonal antibodies against human plasminogen activator inhibitor type-1—Implications for antibody-mediated PAI-1-neutralization and vitronectin-binding. *Eur. J. Biochem.* **268**, 1095–1106.

Wohlwend A., Belin D., and Vassalli J.D. (1987). Plasminogen activator-specific inhibitors produced by human monocytes/macrophages. *J. Exp. Med.* **165**, 320–339.

Xue Y., Björquist P., Inghardt T., Linschoten M., Musil D., Sjölin L., and Deinum J. (1998). Interfering with the inhibitory mechanism of serpins: Crystal structure of a complex formed between cleaved plasminogen activator inhibitor type 1 and a reactive-centre loop peptide. *Structure* **6**, 627–636.

York J.D., Li P., and Gardell S.J. (1991). Combinatorial mutagenesis of the reactive site region in plasminogen activator inhibitor I. *J. Biol. Chem.* **266**, 8495–8500.

Yuasa H., Tanaka H., Hayashi T., Wakita T., Nakamura H., Nishioka J., Kawarada Y., and Suzuki K. (2000). Bovine protein C inhibitor has a unique reactive site and can transiently inhibit plasmin. *Thromb. Haemost.* **83**, 262–267.

Zechmeister-Machhart M., Hufnagl P., Uhrin P., Korschineck I., Binder B.R., and Geiger M. (1997). Molecular cloning and sequence analysis of the mouse protein C inhibitor gene. *Gene* **186**, 61–66.

Zeheb R., Gelehrter T.D. (1988). Cloning and sequencing of cDNA for the rat plasminogen activator inhibitor-1. *Gene* **73**, 459–468.

Zhou A., Faint R., Charlton P., Dafforn T.R., Carrell R.W., and Lomas D.A. (2001). Polymerization of plasminogen activator inhibitor-1. *J. Biol. Chem.* **276**, 9115–9122.

Zhu Y.H., Carmeliet P., and Fay W.P. (1999). Plasminogen activator inhibitor-1 is a major determinant of arterial thrombolysis resistance. *Circulation* **99**, 3050–3055.

Zoller B., Garcia-de-Frutos P., and Dahlback B. (1998). A common 4G allele in the promoter of the plasminogen activator inhibitor-1 (PAI-1) gene as a risk factor for pulmonary embolism and arterial thrombosis in hereditary protein S deficiency. *Thromb. Haemost.* **79**, 802–807.

4

Regulation of Plasminogen Gene Expression

Ana Gutiérrez-Fernández, Neill A. Gingles, Lu Zhang, Felizabel Garcia Bannach, G. Ronald Jenkins, David J. Loskutoff, Robert J. Parmer, and Lindsey A. Miles

1. Introduction

Plasminogen is the zymogen of the active serine protease plasmin. Plasminogen circulates in plasma at a concentration of 2 μM (Collen and Verstraete, 1975) and is also present in the extravascular fluid. Plasmin is the major enzyme responsible for the degradation of fibrin in blood clots within the vascular tree (Carmeliet and Collen, 1996; Petersen *et al.*, 1990; Ploplis *et al.*, 1995). In addition to its established role in intravascular thrombolysis, plasmin has been implicated in such diverse processes as angiogenesis (Pepper *et al.*, 1987; Baker *et al.*, 2000; van Hinsbergh *et al.*, 2001; Pepper, 2001;), tumor progression (Chappuis *et al.*, 2001; van der Pluijm *et al.*, 2001), degradation of extracellular matrices, neurite outgrowth (Tsirka *et al.*, 1995; Wu *et al.*, 2000; Jacovina *et al.*, 2001), and inflammatory reactions important in host defense and wound repair.

Because plasmin is a broad spectrum serine protease, its activity is tightly regulated by the presence of inhibitors of plasmin and of plasminogen activators and by the synthesis and secretion of plasminogen activators in response to specific stimulation by growth factors and cytokines. Recent data from our laboratory and others suggest that the expression of the plasminogen gene is regulated, also, in response to cytokines and glucocorticoids at the protein, mRNA, and transcriptional levels. Furthermore, although liver is the major site of plasminogen synthesis (Bohmfalk and Fuller, 1980; Saito *et al.*, 1980; Zhang *et al.*, 2002.), we have shown recently that plasminogen synthesis is exhibited at low levels in many extrahepatic tissues, separated from the circulation by anatomical barriers. Thus, regulation of plasminogen gene expression should contribute to the proteolytic profile within these tissues.

Ana Gutiérrez-Fernández, Neill A. Gingles, Lu Zhang, Felizabel Garcia Bannach, G. Ronald Jenkins, David J. Loskutoff, and Lindsey A. Miles • The Scripps Research Institute, Department of Cell Biology, Division of Vascular Biology, La Jolla, California 92037, USA **Robert J. Parmer** • Department of Medicine, University of California and Veterans Administration Medical Center, San Diego, CA 92161, USA.

Plasminogen: Structure, Activation, and Regulation, edited by David M. Waisman. Kluwer Academic/Plenum Publishers, New York, 2003.

This chapter will review the extrahepatic distribution of the plasminogen gene, mechanisms by which the plasminogen gene is constitutively regulated, and recent studies demonstrating regulation of the plasminogen gene in response to specific cytokines and glucocorticoids. The regulation of other fibrinolytic proteins by mediators that regulate plasminogen gene expression will be compared also.

2. The Plasminogen Gene

The human plasminogen gene spans 52.5 kb of DNA and consists of 19 exons separated by 18 introns on chromosome 6q26. The promoter contains two clusters of regulatory elements for transcription, including forward and reverse CCAAT boxes and TATAA sequences (Petersen *et al.*, 1990). Each of the five kringle structures of plasminogen is encoded by two separate exons with a single intron inserted in the middle of the DNA sequence encoding each kringle.

The human genome contains another plasminogen related gene, apolipoprotein(a), that has a high degree of similarity to plasminogen and contains up to 37 tandem repeats of plasminogen kringle 4, a kringle 5 homologue and a protease domain (McLean *et al.*, 1987; Tomlinson *et al.*, 1989). However, although the apo(a) cDNA encodes the serine protease catalytic triad, apo(a) cannot be cleaved by plasminogen activators to generate protease activity. It has been speculated that apolipoprotein(a) may have evolved via exon shuffling (Petersen *et al.*, 1990).

In addition, three genes belonging to the plasminogen-apolipoprotein(a) gene family are present in the human genome and have been designated plasminogen related genes A (PRGA), B (PRGB), and C (PRGC) (Ichinose, 1992). PRGA and PRGB correspond to exons I–V of the plasminogen gene. Their nucleotide sequences are ~95% identical to the plasminogen gene and differ from each other by 1–2%. The expected gene products of PRGA and PRGB are a peptide of 8.8 kDA, corresponding to the preactivation peptide of plasminogen. PRGC corresponds to exons VII–XVII of plasminogen. To date, only mRNA corresponding to PRGA (Lewis *et al.*, 1999) and PRGB (Tateno and Ichinose, 1999) have been detected in tissues.

2.1. Mechanisms of Constitutive Regulation of Plasminogen Gene Expression

Cis-acting elements responsible for constitutive expression of the plasminogen gene in hepatocytic cell lines have been identified. Deletional analysis has implicated an HNF-1 recognition site and an AP-3 recognition site that participate in constitutive plasminogen gene expression (Meroni *et al.*, 1996). The HNF-1 site is in an unusual position, within the 5′ untranslated region. The AP-3 recognition site is present in a region between −2200 and −2100 from the plasminogen mRNA cap site (Meroni *et al.*, 1996). In addition, inhibitory sequences are present between −700 and −300 bp (Meroni *et al.*, 1996; Garcia Bannach *et al.*, 2002).

2.2. Plasminogen Distribution in Tissues

Plasminogen circulates in plasma at a high concentration, 2 μM (Collen and Verstraete, 1975) and is produced predominately in liver (Bohmfalk and Fuller, 1980;

Saito *et al.*, 1980). Recent studies carried out in our laboratory and in others, have demonstrated that plasminogen is expressed at low levels extrahepatically in a number of tissues, including tissues separated from the circulation by anatomic barriers (Sappino *et al.*, 1993; Tateno and Ichinose, 1999; Pepper, 2001; Zhang *et al.*, 2002). Using a sensitive and specific RT-PCR assay, we found that plasminogen mRNA was present in murine adrenal, kidney, brain, testis, heart, lung, uterus, spleen, thymus, and gut (Figure 4.1) (Zhang *et al.*, 2002). Plasminogen mRNA was not detected in skeletal muscle or in adipose tissue. Using an isotopic quantitative RT-PCR assay, we quantified the plasminogen mRNA contents of the extrahepatic tissues providing the strongest signals in the RT-PCR and compared the values with liver. As anticipated, liver contained the highest plasminogen mRNA level. There were wide variations in plasminogen mRNA expression in the extrahepatic tissues (Table 4.1). The highest plasminogen mRNA level in the extrahepatic tissues was found in adrenal (1.3% of hepatic mRNA content) followed by kidney (0.3% of liver plasminogen

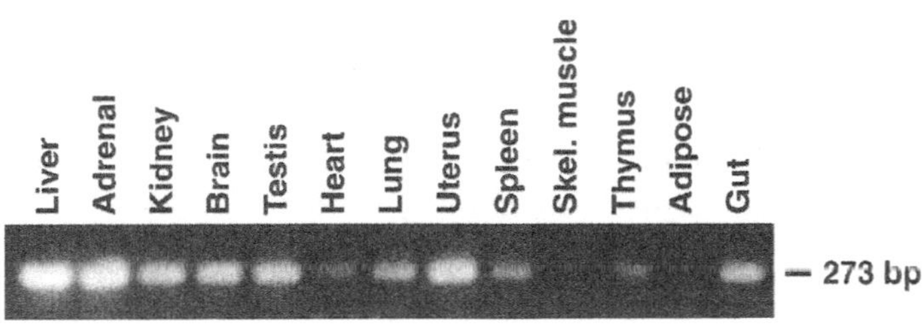

Figure 4.1. Distribution of Pg mRNA in murine organs. RT-PCR was performed using murine total RNA from liver (2 ng), adrenal and kidney (200 ng each), brain and testis (400 ng each), heart, lung, uterus, spleen, skeletal muscle, thymus, adipose, and gut (1 μg each) as templates. Plasminogen specific oligonucleotide primers were used. The RT-PCR products were fractioned on 2.3% NuSieve 3:1 of the aragose gel and stained with ethidium bromide. The size of the product (273 bp) is indicated to the right (from Zhang, L., Seiffert, D., Fowler, B.J., Jenkins, G.R., Thinnes T.C.D., Loskutoff, D.J., Parmer, R.J., Miles, L.A. Plasminogen has a broad extrahepatic distribution, *Thrombosis Haemostasis*, 2002, 87: 493–501, reprinted with permission).

Table 4.1. Quantitative Analysis of Plasminogen Gene Expression in Murine Tissues

	Pg mRNA fg/μg Total RNA	Percentage of liver Pg mRNA
Liver	2240 ± 157.48	100
Adrenal	28.45 ± 5.05	1.3
Kidney	6.98 ± 1.84	0.3
Testis	1.74 ± 0.28	0.08
Brain	0.59 ± 0.29	0.03
Uterus	0.19	0.01

The concentration of plasminogen competitor required for equivalent incorporation of ^{32}P dCTP along with plasmingen cDNA was determined using serial dilutions of the standard. Samples are from 2 to 3 different animals (from Zhang, L., Seiffert, D., Fowler B.J., Jenkins, G.R., Thinnes, T.C.D., Loskutoff, D.J., Parmer, R.J., Miles, L.A. Plasminogen has a broad extrahepatic distribution, *Thrombosis Haemostasis*, 2002, 87: 493–501, reprinted with permission).

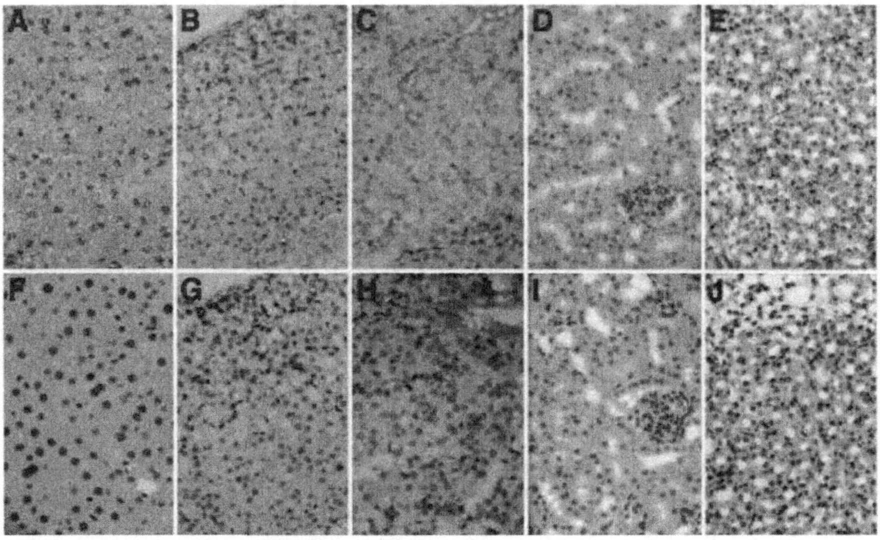

Figure 4.2. Localization of plasminogen mRNA in murine liver, adrenal, and kidney. Representative sections from liver (Panels A and F), adrenal cortex (Panels B and G), adrenal medulla (Panels C and H), kidney cortex (Panels D and I), and kidney medulla (Panels E and J) were subjected to *in situ* hybridization using either a murine plasminogen antisense riboprobe (Panels A–E) or Pg sense riboprobe (Panels F–J). The slides were exposed at 4°C for either 4 weeks (Panels A and F), 20 weeks (Panels B, C, G, and H) or 16 weeks (Panels D, E, I, and J) and stained with hematoxylin and eosin. Magnification is 400× (from Zhang, L., Seiffert, D., Fowler, B.J., Jenkins, G.R., Thinnes, T.C.D., Loskutoff, D.J., Parmer, R.J., Miles, L.A. Plasminogen has a broad extrahepatic distribution, *Thrombosis Haemostasis*, 2002, 87: 493–501, reprinted with permission).

mRNA content). Brain and testis expressed 0.03–0.08% respectively of the liver plasminogen mRNA content, and the plasminogen mRNA content of uterus was 0.01% of that of liver. Cell types expressing plasminogen mRNA were analyzed by *in situ* hybridization using an antisense riboprobe specific for murine plasminogen (Zhang *et al.*, 2002). In the adrenal, plasminogen mRNA was expressed in cells throughout the cortex (Figure 4.2, Panel B) and medulla (Figure 4.2, Panel C). In the kidney, plasminogen mRNA was distributed throughout the cortex (Figure 4.2, Panel D) and medulla with the most intense signal in the nephron (Figure 4.2, Panel E). In spleen, plasminogen mRNA was expressed in the white and red pulps. Of note, plasminogen was expressed also in tissues separated from the circulation by anatomic barriers. In testis, plasminogen mRNA was detected in the spermatogonia and in primary and secondary spermatozones in the semineferous tubules. In thymus, the major plasminogen mRNA signal was detected in the cortex with weak signals in the medulla. Plasminogen mRNA was present in several distinct areas within the central nervous system (CNS) (Figure 4.3): neurons in the cortex (Figure 4.3 Panel A), pyramidal cells in the hippocampus (Figure 4.3, Panels B and C) and in granular cells within the dentate gyrus. In the cerebellum, plasminogen was expressed in Purkinje cells (Figure 4.3, Panel D, light arrow) and in the granular cells (Figure 4.3, Panel D heavy arrow and Panel E dark field). The olfactory bulb is also rich in granular cells and we detected plasminogen mRNA in this region as well (Zhang *et al.*, 2002).

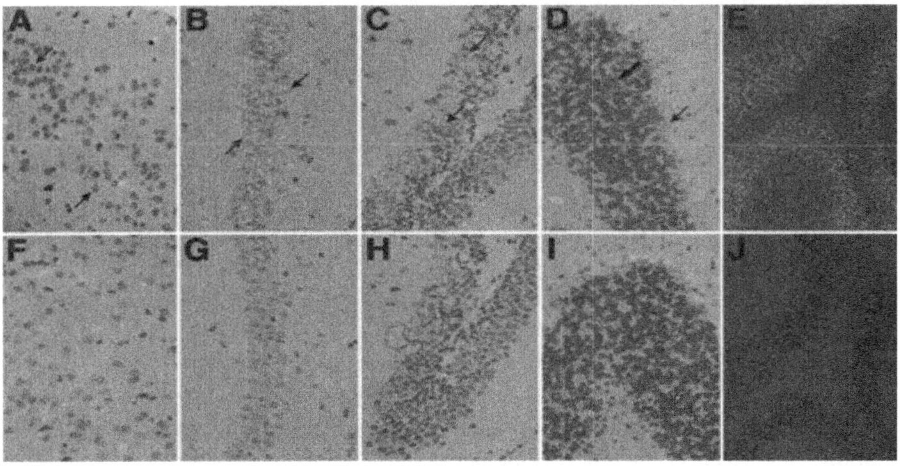

Figure 4.3. Localization of plasminogen mRNA in murine brain. Representative sections from brain cortex (Panels A and F), hippocampus (Panels B, C, G, H) or cerebellum (D, E, I, J) were subjected to *in situ* hybridization using a murine plasminogen antisense riboprobe (Panels A–E) or a plasminogen sense riboprobe (F–J). Arrows indicate representative neurons (panel A), pyramidal cells (Panel B), granular cells (Panel C), or Purkinje cells (Panel D, light arrow) or granular cells (Panel D, heavy arrow). The slides were exposed at 4°C for 16 weeks and then strained with hematoxylin and eosin. Panels E and J are dark field exposures. Magnification is 400× for Panels A–D, F–I, and 200× for Panels E and J (from Zhang, L., Seiffert, D., Fowler, B.J., Jenkins, G.R., Thinnes, T.C.D., Loskutoff, D.J., Parmer, R.J., Miles, L.A. Plasminogen has a broad extrahepatic distribution. *Thrombosis Haemostasis*, 2002, 87: 493–501, reprinted with permission).

The expression of plasminogen and plasminogen mRNA have been demonstrated, also, in hippocampus as well as in the external granule cell, molecular and internal layers of the cerebellum and in the cortex (Tsirka *et al.*, 1997; Basham and Seeds, 2001). Plasminogen has been detected also in the avascular cornea (Twining *et al.*, 1999). Thus, at these sites separated from the circulation, tissue synthesis should provide the only source of plasmin for performance of proteolytic functions in these tissues.

Plasminogen function, specifically in the hippocampus, has been demonstrated using plasminogen deficient mice generated by targeted gene inactivation. Plasminogen-/-mice are resistant to excitotoxin-induced neurodegeneration that is observed in normal mice (Chen *et al.*, 1999). The basis for the protection from cell death in the plasminogen-/- mice is that, in wild-type mice, plasmin degrades the neuronal extracellular matrix, laminin, thus depriving the cells of a necessary survival signal (Chen and Strickland, 1997).

It is important to note that plasminogen activators are expressed in many of the tissues in which plasminogen mRNA is present, especially within the brain tissues. t-PA has been detected by RT-PCR in adrenal, kidney, heart, lung, brain, spleen, thymus, gut, testis, and liver (Sappino *et al.*, 1993). u-PA is localized in the nephron and collecting tubules of the kidney, the gut, ductus deferens, and epididymis (Yamamoto and Loskutoff, 1996). Thus, at many tissue sites, local sources of plasminogen and its activators are present. The colocalization of these fibrinolytic proteases at tissues sites, is likely to result in generation of the relatively broad spectrum proteolytic activity of plasmin independently of circulating plasminogen.

3. Regulation of Plasminogen Gene Expression *in vitro* and *in vivo*

A number of inflammatory mediators, growth factors, and other agonists have been shown recently to regulate plasminogen expression levels in distinct cell types *in vitro* and *in vivo*. Herein, we will review the effect of the molecules shown to regulate plasminogen expression, interleukin-6, glucocorticoids, LPS, tissue necrosis factor-α, transforming growth factor-β, interleukin-1, and kainic acid. The effects of these effector molecules on plasminogen mRNA expression in different tissues are summarized in Table 4.2. The regulation of expression of other fibrinolytic proteins by these agents will be compared.

3.1. Interleukin-6

Interleukin-6 (IL-6) is a cytokine that plays a role in inflammation, bone metabolism, immunity, reproduction, neural development, and haematopoiesis. IL-6 is produced by macrophages, fibroblasts, endothelial cells, and activated T-helper cells and acts synergistically with IL-1 and TNF-α. IL-6 is the primary inducer of the acute-phase response in liver. IL-6 also enhances the differentiation of B-cells and their consequent production of immunoglobulin and affects platelet production and reactivity as well as endothelial function (Heinrich *et al.*, 1990).

Several reports have suggested that plasminogen behaves as an acute phase reactant and therefore may respond to cytokines such as IL-6 (Lackner and Javid, 1973). Moreover, plasminogen levels are positively correlated with C-reactive protein and consequently with an acute phase response in patients with angina pectoris (Juhan-Vague *et al.*, 1989).

Based on these earlier reports, we examined whether IL-6 could modulate plasminogen expression. We found that IL-6 increased plasminogen mRNA expression in primary murine hepatocytes in a time-dependent fashion, reaching an ~8-fold level of stimulation at 24 hr (Figure 4.4) (Jenkins *et al.*, 1997). IL-6 also increased plasminogen mRNA expression in human Hep3B cells. In *in vivo* studies, injection of mice with IL-6 increased steady state levels of hepatic plasminogen mRNA ~4-fold, compared to saline-injected mice (Figure 4.5). Consequently, circulating plasminogen concentrations were increased, also.

Table 4.2. Regulation of Plasminogen by Various Agonists

Agonist	Plasminogen	References
IL-1 α	↑ Human cornea	Twining (1999)
IL-1 β	↑ Human cornea	
TGF-β1	↓ Monkey hepatocytes	Ramharack (1998)
TNF- α	↓ Monkey hepatocytes	
IL-6	↑ Human hepatocytes	Jenkins (1997)
	↓ Human cornea	Twining (1999)
LPS	↑ Human hepatocytes	
Kainic acid	↑ Rat piramidal neurons	Matsuoka (1998)

Relative increases or decreases in plasminogen mRNA expression are depicted by arrows.

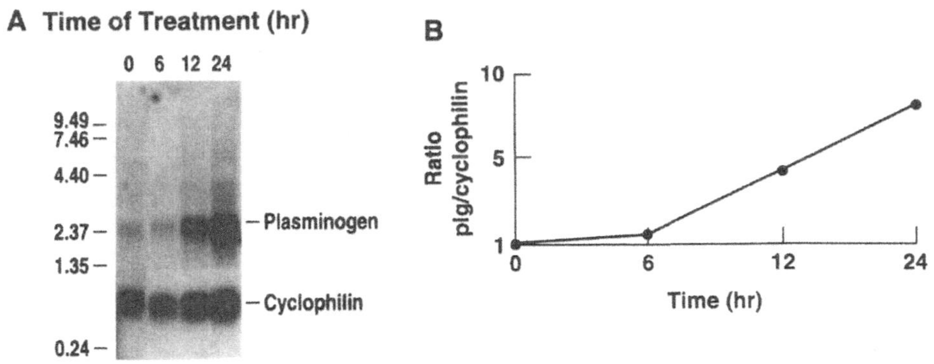

Figure 4.4. Time dependence of the effect of rhIL-6 on plasminogen mRNA expression in primary murine hepatocytes. (A) Northern blotting for plasminogen using total RNA from primary murine hepatocytes treated with 500 U/ml IL-6. (B) Fold stimulation of plasminogen mRNA levels were determined as the ratio of plasminogen/cyclophilin hybridization band intensity determined by laser densiometric scanning of the autoradiogram (from Jenkins, G.R.E., Seiffert, D., Parmer, R.J., Miles, L.A. Regulation of plasminogen gene expression by interleukin-6 in hepatoma and murine liver cells, *Blood*, 1997, 89: 2394, copyright American Society of Hematology, used by permission).

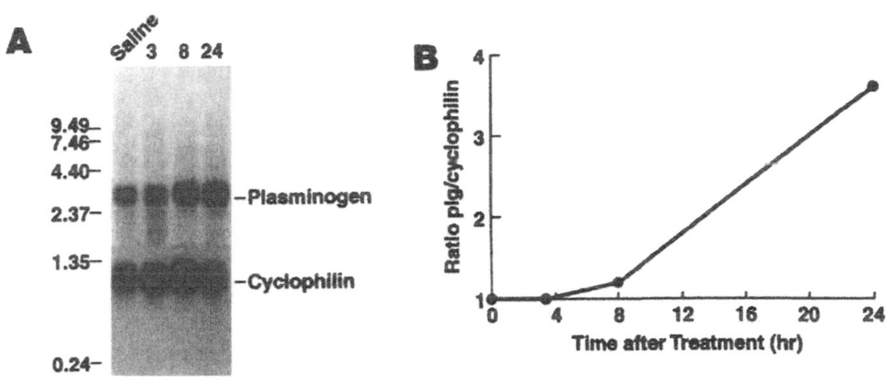

Figure 4.5. Time dependence of the effect of rhIL-6 on hepatic plasminogen mRNA expression in mice. (A) C57BL/6J male mice were injected intraperitoneally with either saline or rhIL-6 and sacrificed at the indicated times. Total RNA was extracted and subjected to Northern blot analysis for murine plasminogen. (B) Fold stimulation was determined as the ratio of plasminogen/cyclophilin hybridization band intensity determined by laser densitometric scanning of the autoradiogram (from Jenkins, G.R.E., Seiffert, D., Parmer, R.J., Miles, L.A. Regulation of plasminogen gene expression by interleukin-6 in hepatoma and murine liver cells, *Blood*, 1997, 89: 2394, copyright American Society of Hematology, used by permission).

We also found that the 5′ flanking region of the plasminogen gene could respond to IL-6 (Figure 4.6). A 1,067-bp fragment (nucleotides −914 to +154) cloned upstream of a luciferase reporter gene provided ~100-fold greater luciferase activity in Hep3B cells, compared to cells transfected with plasmid alone. In addition, a 1067-bp fragment of the 5′ flanking region of the human plasminogen gene drove luciferase expression in HEP3B cells that was increased ~5-fold in response to IL-6 (Jenkins *et al.*, 1997). The positive

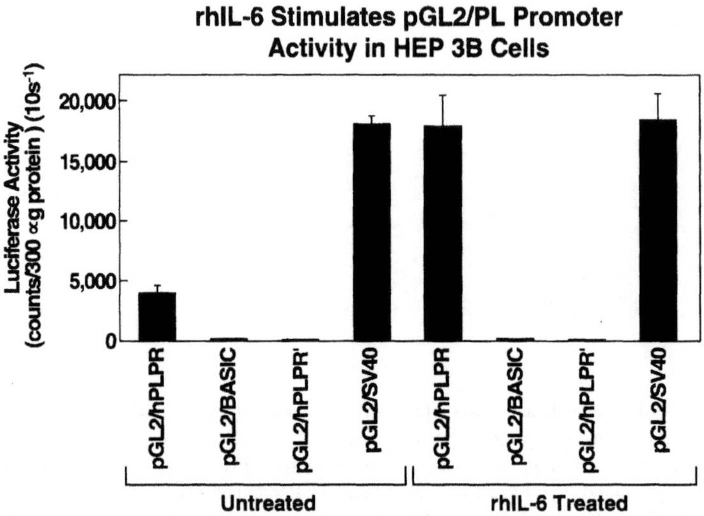

Figure 4.6. rhIL-6 stimulates plasminogen promoter activity in Hep3B cells. Hep3B cells where transiently transfected with the indicated plasmid constructs. (pGL2/hPLPR = a fragment of human plasminogen DNA −914 to +154 cloned in the correct orientation upstream of the promoterless pGL2 basic plasmid; pGL2/hPLPR = the same plasminogen fragment cloned into the pGL2 basic plasmid in the reverse orientation; pGL2/Basic = the empty vector; pGL2/SV40 = an SV40 promoter/enhancer cloned into the pGL2 basic plasmid.) Following transfection, cells were cultured either in the presence or absence of rhIL-6 for 24 hr. Error bars represent standard deviation (from Jenkins, G.R.E., Seiffert, D., Parmer, R.J., Miles, L.A. Regulation of plasminogen gene expression by interleukin-6 in hepatoma and murine liver cells, *Blood*, 1997, 89: 2394, copyright American Society of Hematology, used by permission).

response of the plasminogen promoter to IL-6 is consistent with the presence of functional IL-6 responsive elements in the plasminogen promoter. The 5′ flanking sequences of human plasminogen contains six consensus sequences for IL-6 (Malgaretti *et al.*, 1990; Wade *et al.*, 1993). In recent studies, we have determined the sequence of 2600 bp upstream of murine plasminogen exon I (Garcia Bannach *et al.*, 2002). Thirteen consensus IL-6 responsive elements are present, also, within the 5′ flanking sequence of the murine plasminogen gene. When murine Hepa 1-6 cells were transfected with a 1064-bp 5′ flanking fragment of the murine plasminogen gene, cloned upstream of a luciferase reporter, luciferase activity was stimulated ~8-fold when the cells were treated with IL-6 (Garcia Bannach *et al.*, 2002).

IL-6 also increases plasminogen mRNA expression in primary cultures of monkey hepatocytes (Ramharack *et al.*, 1998). In contrast, in the human cornea, IL-6 inhibits plasminogen synthesis by ~40% (Twining *et al.*, 1999). Thus, the differential expression of specific transcription factors in different tissues must account for the tissue specific responses to IL-6.

3.2. Glucocorticoids

Other mediators of the acute phase response, such as glucocorticoids, are also able to enhance the IL-6–dependent responses of specific genes including α_2-antiplasmin, fibrinogen, thiostatin, and hemopexin (Marinkovic *et al.*, 1989). With plasminogen, the

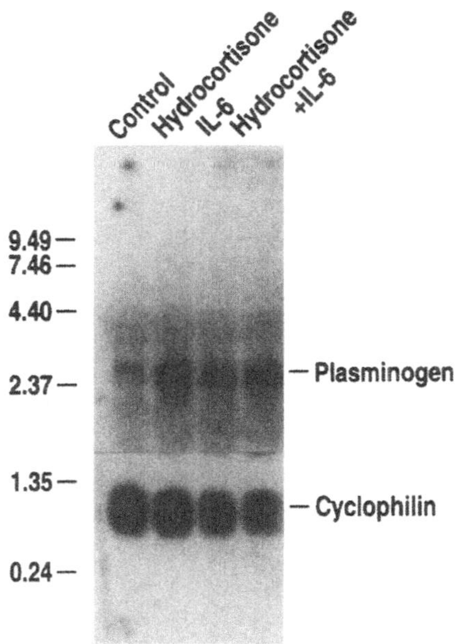

Figure 4.7. Effect of hydrocortisone and IL-6 on plasminogen mRNA expression in primary murine hepatocytes. Primary murine hepatocytes were incubated with either 10 μM hydrocortisone, 500 U/ml rhIL-6 or hydrocortisone + rhIL-6 at the foregoing concentrations. Total RNA was isolated and Northern blot analysis for plasminogen was performed (from Jenkins, G.R.E., Seiffert, D., Parmer, R.J., Miles, L.A. Regulation of plasminogen gene expression by interleukin-6 in hepatoma and murine liver cells, *Blood*, 1997, 89: 2394, copyright American Society of Hematology, used by permission).

addition of hydrocortisone did not markedly enhance plasminogen gene expression in response to IL-6. However, hydrocortisone, alone increased plasminogen mRNA expression in primary hepatocytes ~4–fold (Figure 4.7) (Jenkins *et al.*, 1997). A putative glucocorticoid responsive elements is present also in the human plasminogen 5′ flanking region.

3.3. LPS

Lipopolysaccharide (LPS) induces the biosynthesis of several cytokines including IL-6 (Andus *et al.*, 1988; Haddad and Fahlman, 2002). Therefore, under conditions where LPS is present, circulating plasminogen levels would be expected to increase. Consequently, we found that mice injected with LPS showed increases in hepatic levels of plasminogen mRNA. The increase in plasminogen mRNA occurred in a time dependent manner, achieving a maximum ~3-fold increase at 24 hr and returning to basal levels by 72 hr (Figure 4.8).

In contrast, in PC3 human prostatic cancer cell, LPS induces u-PA gene expression but IL-1, IL-6, and TNF do not affect u-PA expression (Ohta *et al.*, 2000). Thus, the increase in u-PA expression most likely follows a pathway distinct from that of plasminogen gene stimulation. The induction of the u-PA occurs in the presence of cycloheximide and thus, is not dependent upon *de novo* protein synthesis.

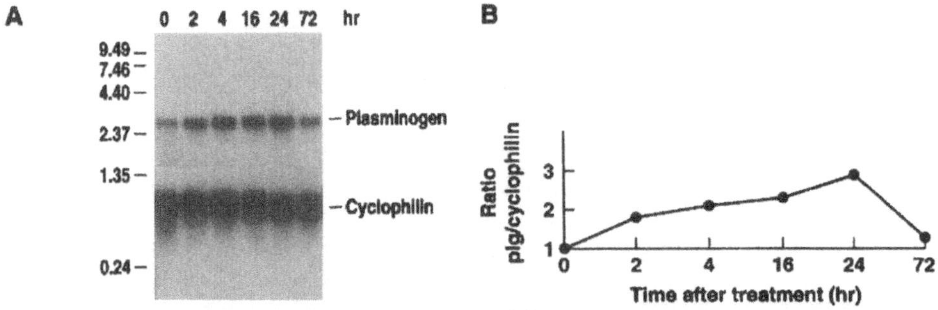

Figure 4.8. Time dependence of the effect of the LPS injection on hepatic plasminogen mRNA expression in mice. (A) Mice were injected intraperitoneally with either saline or 50 μg LPS and sacrificed at the indicated times. Livers were harvested and total RNA was extracted and subjected to Northern blot analysis for plasminogen. (B) Fold-simulation was determined as the ratio of plasminogen/cyclophilin hybridization band intensity determined by laser densiometric scanning of the autoradiogram (from Jenkins, G.R.E., Seiffert, D., Parmer, R.J., Miles, L.A. Regulation of plasminogen gene expression by interleukin-6 in hepatoma and murine liver cells, *Blood*, 1997, 89: 2394, copyright American Society of Hematology, used by permission).

3.4. Tumour Necrosis Factor-α (TNF-α) and Transforming Growth Factor-β (TGF-β)

Other cytokines that participate in the inflammatory response include TNF-α and TGF-β. TNF-α (cachectin) is a major immune response-modifying cytokine produced primarily by activated macrophages. TNF-α induces the expression of additional autocrine growth factors, increases cellular responsiveness to growth factors and induces signaling pathways that lead to proliferation. TGF-β has proliferative effects on many mesenchymal and epithelial cell types. Several members of the TGF-β family are potent inducers of mesodermal differentiation in early embryos.

TNF-α and TGF-β participate in the regulation of several components of the fibrinolytic system. For example, TNF-α and TGF-β decrease plasminogen mRNA expression in primary monkey hepatocytes (Ramharack *et al.*, 1998). In human peritoneal mesangial cells, TGF-β also decreases t-PA gene expression, thus affecting expression of the fibrinolytic system in parallel with plasminogen expression (Falk *et al.*, 2000). In contrast, in these same cells, TGF-β stimulates u-PA expression (Falk *et al.*, 2000).

3.5. Interleukin-1

Interleukin-1 (IL-1) is the prototypic pro-inflammatory cytokine (Dinarello, 1996, 1997a,b) that exists in two forms IL-1α and IL-1β, and in most studies their biological activities are indistinguishable. The IL-1s are secreted primarily by macrophages but also by neutrophils, endothelial cells, smooth muscle cells, glial cells, astrocytes, B- and T-cells, fibroblasts and keratinocytes. Production of IL-1 by these different cell types occurs only in response to cellular stimulation. IL-1 affects nearly every cell type, often in concert with other pro-inflammatory cytokines, such as TNF-α. IL-1 has been reported to participate in human malignancy (Jin *et al.*, 1997). IL-β is expressed by most breast carcinomas and a high IL-1β content is often associated with tumour invasiveness.

IL-1 has diverse actions in the brain and is a mediator of several forms of neurodegeneration (Rothwell *et al.*, 1997; Touzani *et al.*, 1999). IL-1 mediates ischaemic, excitotoxic, and traumatic brain injury, most likely via multiple actions on glia, neurons, and the vasculature. The expression of IL-1α and IL-1β is increased in response to clinical insults such as cerebral ischemia, infusion of excitotoxins, seizures or CNS infections and in Parkinson's disease, multiple sclerosis and Alzheimer's disease.

IL-1α and IL-1β treatment increases plasminogen expression 2–3-fold in human corneal cells (Twining *et al.*, 1999). Thus far, this is the only tissue in which IL-1 has been demonstrated to modulate plasminogen expression. Nonetheless, in view of the recent studies (reviewed above) demonstrating plasminogen expression in the CNS, the possibility that IL-1α and IL-1β and other inflammatory mediators modulate plasminogen expression in the brain is a promising area of investigation.

3.6. Kainic Acid

Kainic acid (KA) has potent excytotoxic effects in the CNS. Injection of kainic acid induces neuronal loss and activation of glial cells in the rat hippocampus (Matsuoka *et al.*, 1998).

KA promotes morphological changes in hippocampal pyramidal neurons, 6 hr after injection. Almost all neurons die 7 days following KA injection. Interestingly, plasminogen mRNA and protein expression in the hippocampus is increased markedly 6 hr after KA injection and subsequently decreases gradually. The changes in plasminogen mRNA levels are most marked in the CA3 pyramidal neurons (Matsuoka *et al.*, 1998). Therefore, the expression of plasminogen appears to be synchronous with the morphological changes in pyramidal neurons preceding cell death.

Plasminogen deficient mice and t-PA deficient mice are resistant to kainate-induced neurodegeneration. Laminin is expressed in the hippocampal pyramidal neuronal layers and disappears after KA injection (Endo *et al.*, 1999). However, laminin degradation is not observed in plasminogen-/-mice, suggesting that neuronal cell death observed in the wild type mice is due to a loss of a necessary survival signal dependent upon laminin adherence (Nagai *et al.*, 1999). The increased plasminogen expression induced after KA injection may contribute also to laminin degradation in the vicinity of the hippocampal neurons.

4. Conclusions

Previously, many studies have addressed the regulation of expression of the plasminogen activators and PAI-1. However, studies of plasminogen gene regulation are an emerging area. Although plasminogen circulates at high concentration, changes in the levels of circulating plasminogen have important pathophysiological consequences. Firstly, congenital deficiencies in plasminogen (both hypoplasminogenmia, dysplasminogenemia, and aplasminogenemia) are associated with thrombosis (Aoki *et al.*, 1978; Robbins, 1992; Schuster *et al.*, 1997). Secondly, mice homozygous for plasminogen deficiency are predisposed to severe thrombosis (Bugge *et al.*, 1995). Heterozygous mice, expressing 50% of the normal circulating level of plasminogen, exhibit decreased spontaneous lysis of fibrin clots compared to control mice (Ploplis *et al.*, 1995). Thirdly, epidemiological studies of racial groups, in the same geographic area with different degrees of expression of coronary heart disease have shown statistically significantly higher plasminogen levels in the populations exhibiting lower incidences of coronary heart disease (Lackner and Johnson, 1967; Dischinger *et al.*, 1980; Szczeklik *et al.*, 1980). These studies suggest that even 2-fold changes in plasminogen levels have profound consequences with regard to changes in fibrinolytic activity and consequent thrombosis.

In tissues separated from the circulation by anatomic barriers (brain, testis, and thymus), the synthesis of plasminogen by cells within these tissues should be the only source of plasminogen. Thus, regulation of expression of the plasminogen gene within these tissues should have profound effects on local proteolysis and function within these tissues.

Further studies are necessary to identify the *cis*- and *trans*-acting elements responsible for regulation of plasminogen gene expression in distinct tissues and cell types and also to elucidate whether the regulation of the plasminogen gene is regulated concomitantly with t-PA and u-PA in specific tissues.

Acknowledgments

Work in the authors' laboratories was supported by grants from the NIH (HL45394 and HL38272 to L. A. M.), (HL50398 to R. J. P.) (HL47819 to D. J. L.) and from the Department of Veterans Affairs (R. J. P.). We thank Kathy Kling for expert editorial assistance.

References

Andus, T., Geiger, T., Hirano, T., Kishimoto, T., and Heinrich, P.C. (1988). Action of recombinant human interleukin 6, interleukin 1 beta and tumor necrosis factor alpha on the mRNA induction of acute-phase proteins. *Eur. J. Immunol.* **18**, 739–746.

Aoki, N., Moroi, M., Sakata, Y., Yoshida, N., and Matsuda, M. (1978). Abnormal plasminogen. A hereditary molecular abnormality found in a patient with recurrent thrombosis. *J. Clin. Invest.* **61**, 1186–1195.

Baker, E.A., Bergin, F.G., and Leaper, D.J. (2000). Plasminogen activator system, vascular endothelial growth factor, and colorectal cancer progression. *Mol. Pathol.* **53**, 307–312.

Basham, M.E. and Seeds, N.W. (2001). Plasminogen expression in the neonatal and adult mouse brain. *J. Neurochem.* **77**, 318–325.

Bohmfalk, J.F. and Fuller, G.M. (1980). Plaminogen is synthesized by primary cultures of rat hepatocytes. *Science* **209**, 408–410.

Bugge, T.H., Flick, M.J., Daugherty, C.C., and Degen, J.L. (1995). Plasminogen deficiency causes severe thrombosis but is compatible with development and reproduction. *Genes Dev.* **9**, 794–807.

Carmeliet, P. and Collen, D. (1996). Genetic analysis of the plasminogen and coagulation system in mice. *Haemostasis* **26**(Suppl. 4), 132–153.

Chappuis, P.O., Dieterich, B., Sciretta, V., Lohse, C., Bonnefoi, H., Remadi, S., and Sappino, A.P. (2001). Functional evaluation of plasmin formation in primary breast cancer. *J. Clin. Oncol.* **19**, 2731–2738.

Chen, Z.L., Indyk, J.A., Bugge, T.H., Kombrinck, K.W., Degen, J.L., and Strickland, S. (1999). Neuronal death and blood-brain barrier breakdown after excitotoxic injury are independent processes. *J. Neurosci.* **19**, 9813–9820.

Chen, Z.L. and Strickland, S. (1997). Neuronal death in the hippocampus is promoted by plasmin-catalyzed degradation of laminin. *Cell* **91**, 917–925.

Collen, D. and Verstraete, M. (1975). Molecular biology of human plasminogen. II. Metabolism in physiological and some pathological conditions in man. *Thromb. Diath. Haemorrh.* **34**, 403–408.

Dinarello, C.A. (1996). Biologic basis for interleukin-1 in disease. *Blood* **87**, 2095–2147.

Dinarello, C.A. (1997a). Blocking interleukin-1 and tumor necrosis factor in disease. *Eur. Cytokine Netw.* **8**, 294–296.

Dinarello, C.A. (1997b). Interleukin-1. *Cytokine Growth Factor Rev.* **8**, 253–265.

Dischinger, P., Tyroler, H.A., McDonagh, R., Jr, and Hames, C.G. (1980). Blood fibrinolytic activity, social class and habitual physical activity—I. A study of black and white men in Evans County, Georgia. *J. Chronic Dis.* **33**, 283–290.

Endo, A., Nagai, N., Urano, T., Takada, Y., Hashimoto, K., and Takada, A. (1999). Proteolysis of neuronal cell adhesion molecule by the tissue plasminogen activator-plasmin system after kainate injection in the mouse hippocampus. *Neurosci. Res.* **33**, 1–8.

Falk, P., Ma, C., Chegini, N., and Holmdahl, L. (2000). Differential regulation of mesothelial cell fibrinolysis by transforming growth factor beta 1. *Scand. J. Clin. Lab. Invest.* **60**, 439–447.

Garcia Bannach, F., Gutierrez, A., Fower, B.J., Bugge, T.H., Degen, J.L., Parmer, R.L., and Miles, L.A. (2002). Localization of regulatory elements mediating constitutive and cytokine-stimulated plasminogen gene expression. *J. Biol. Chem.* **277**: 38579–38588.

Haddad, J.J. and Fahlman, C.S. (2002). Nuclear factor-kappaB-independent regulation of lipopolysaccharide-mediated interleukin-6 biosynthesis. *Biochem. Biophys. Res. Commun.* **291**, 1045–1051.

Heinrich, P.C., Castell, J.V., and Andus, T. (1990). Interleukin-6 and the acute phase response. *Biochem. J.* **265**, 621–636.

Ichinose, A. (1992). Multiple members of the plasminogen-apolipoprotein(a) gene family associated with thrombosis. *Biochemistry* **31**, 3113–3118.

Jacovina, A.T., Zhong, F., Khazanova, E., Lev, E., Deora, A.B., and Hajjar, K.A. (2001). Neuritogenesis and the nerve growth factor-induced differentiation of PC-12 cells requires annexin II-mediated plasmin generation. *J. Biol. Chem.* **276**, 49350–49358.

Jenkins, G.R., Seiffert, D., Parmer, R.J., and Miles, L.A. (1997). Regulation of plasminogen gene expression by interleukin-6. *Blood* **89**, 2394–2403.

Jin, L., Yuan, R.Q., Fuchs, A., Yao, Y., Joseph, A., Schwall, R., Schnitt, S.J., Guida, A., Hastings, H.M., Andres, J., Turkel, G., Polverini, P.J., Goldberg, I.D., and Rosen, E.M. (1997). Expression of interleukin-1beta in human breast carcinoma. *Cancer* **80**, 421–434.

Juhan-Vague, I., Alessi, M.C., Joly, P., Thirion, X., Vague, P., Declerck, P.J., Serradimigni, A., and Collen, D. (1989). Plasma plasminogen activator inhibitor-1 in angina pectoris. Influence of plasma insulin and acute-phase response. *Arteriosclerosis* **9**, 362–367.

Lackner, H. and Javid, J.P. (1973). The clinical significance of the plasminogen level. *Am. J. Clin. Pathol.* **60**, 175–181.

Lackner, H. and Johnson, A.J. (1967). The fibrinolytic system in South African white and Bantu men. *Thromb. Diath. Haemorrh.* **18**, 456–461.

Lewis et al. (1999). Homologous plasminogen N-terminal and plasminogen-related gene A and B peptides. *Eur. J. Biochem.* **259**, 618–625.

Malgaretti, N., Bruno, L., Pontoglio, M., Candiani, G., Meroni, G., Ottolenghi, S., and Taramelli, R. (1990). Definition of the transcription initiation site of human plasminogen gene in liver and non hepatic cell lines. *Biochem. Biophys. Res. Commun.* **173**, 1013–1018.

Marinkovic, S., Jahreis, G.P., Wong, G.G., and Baumann, H. (1989). IL-6 modulates the synthesis of a specific set of acute phase plasma proteins *in vivo*. *J. Immunol.* **142**, 808–812.

Matsuoka, Y., Kitamura, Y., and Taniguchi, T. (1998). Induction of plasminogen in rat hippocampal pyramidal neurons by kainic acid. *Neurosci. Lett.* **252**, 119–122.

McLean, J.W., Tomlinson, J.E., Kuang, W.J., Eaton, D.L., Chen, E.Y., Fless, G.M., Scanu, A.M., and Lawn, R.M. (1987). cDNA sequence of human apolipoprotein(a) is homologous to plasminogen. *Nature* **330**, 132–137.

Meroni, G., Buraggi, G., Mantovani, R., and Taramelli, R. (1996). Motifs resembling hepatocyte nuclear factor 1 and activator protein 3 mediate the tissue specificity of the human plasminogen gene. *Eur. J. Biochem.* **236**, 373–382.

Nagai, N., Urano, T., Endo, A., Takahashi, H., Takada, Y., and Takada, A. (1999). Neuronal degeneration and a decrease in laminin-like immunoreactivity is associated with elevated tissue-type plasminogen activator in the rat hippocampus after kainic acid injection. *Neurosci. Res.* **33**, 147–154.

Nakajima, K., Tsuzaki, N., Nagata, K., Takemoto, N., and Kohsaka, S. (1992). Production and secretion of plasminogen in cultured rat brain microglia. *FEBS Lett.* **308**, 179–182.

Ohta, S., Niiya, K., Sakuragawa, N., and Fuse, H. (2000). Induction of urokinase-type plasminogen activator by lipopolysaccharide in PC-3 human prostatic cancer cells. *Thromb. Res.* **97**, 343–347.

Pepper, M.S. (2001). Extracellular proteolysis and angiogenesis. *Thromb. Haemost.* **86**, 346–355.

Pepper, M.S., Vassalli, J.D., Montesano, R., and Orci, L. (1987). Urokinase-type plasminogen activator is induced in migrating capillary endothelial cells. *J. Cell Biol.* **105**, 2535–2541.

Petersen, T.E., Martzen, M.R., Ichinose, A., and Davie, E.W. (1990). Characterization of the gene for human plasminogen, a key proenzyme in the fibrinolytic system. *J. Biol. Chem.* **265**, 6104–6111.

Ploplis, V.A., Carmeliet, P., Vazirzadeh, S., Van Vlaenderen, I., Moons, L., Plow, E.F., and Collen, D. (1995). Effects of disruption of the plasminogen gene on thrombosis, growth, and health in mice. *Circulation* **92**, 2585–2593.

Ramharack, R., Barkalow, D., and Spahr, M.A. (1998). Dominant negative effect of TGF-beta1 and TNF-alpha on basal and IL-6-induced lipoprotein(a) and apolipoprotein(a) mRNA expression in primary monkey hepatocyte cultures. *Arterioscler. Thromb. Vasc. Biol.* **18**, 984–990.

Robbins, K.C. (1992). Dysplasminogenemias. *Prog. Cardiovasc. Dis.* **34**, 295–308.

Rothwell, N., Allan, S., and Toulmond, S. (1997). The role of interleukin 1 in acute neurodegeneration and stroke: Pathophysiological and therapeutic implications. *J. Clin. Invest.* **100**, 2648–2652.

Saito, H., Hamilton, S.M., Tavill, A.S., Louis, L., and Ratnoff, O.D. (1980). Production and release of plasminogen by isolated perfused rat liver. *Proc. Natl. Acad. Sci. USA* **77**, 6837–6840.

Sappino, A.P., Madani, R., Huarte, J., Belin, D., Kiss, J.Z., Wohlwend, A., and Vassalli, J.D. (1993). Extracellular proteolysis in the adult murine brain. *J. Clin. Invest.* **92**, 679–685.

Schuster, V., Mingers, A.M., Seidenspinner, S., Nüssgens, Z., Pukrop, T., and Kreth, H.W.(1997). Homozygous mutations is the plasminogen gene of two unrelated girls with ligneous conjunctivitis. *Blood* **90**, 958–966.

Szczeklik, A., Dischinger, P., Kueppers, F., Tyroler, H.A., Hames, C.G., Cassel, J.C., and Creagan, S. (1980). Blood fibrinolytic activity, social class and habitual physical activity—II. A study of black and white men in Southern Georgia. *J. Chronic Dis.* **33**, 291–299.

Tateno, T. and Ichinose, A. (1999). Expression of plasminogen-related gene B varies among normal tissues and increases in cancer tissues. *FEBS Lett.* **445**, 31–35.

Tomlinson, J.E., McLean, J.W., and Lawn, R.M. (1989). Rhesus monkey apolipoprotein(a). Sequence, evolution, and sites of synthesis. *J. Biol. Chem.* **264**, 5957–5965.

Touzani, O., Boutin, H., Chuquet, J., and Rothwell, N. (1999). Potential mechanisms of interleukin-1 involvement in cerebral ischaemia. *J. Neuroimmunol.* **100**, 203–215.

Tsirka, S.E., Gualandris, A., Amaral, D.G., and Strickland, S. (1995). Excitotoxin-induced neuronal degeneration and seizure are mediated by tissue plasminogen activator. *Nature* **377**, 340–344.

Tsirka, S.E., Rogove, A.D., Bugge, T.H., Degen, J.L., and Strickland, S. (1997). An extracellular proteolytic cascade promotes neuronal degeneration in the mouse hippocampus. *J. Neurosci.* **17**, 543–552.

Twining, S.S., Wilson, P.M., and Ngamkitidechakul, C. (1999). Extrahepatic synthesis of plasminogen in the human cornea is up-regulated by interleukins-1alpha and -1beta. *Biochem. J.* **339** (Pt 3), 705–712.

van der Pluijm, G., Sijmons, B., Vloedgraven, H., van der Bent, C., Drijfhout, J.W., Verheijen, J., Quax, P., Karperien, M., Papapoulos, S., and Lowik, C. (2001). Urokinase-receptor/integrin complexes are functionally involved in adhesion and progression of human breast cancer *in vivo*. *Am. J. Pathol.* **159**, 971–982.

van Hinsbergh, V.W., Collen, A., and Koolwijk, P. (2001). Role of fibrin matrix in angiogenesis. *Ann. N Y Acad. Sci.* **936**, 426–437.

Wade, D.P., Clarke, J.G., Lindahl, G.E., Liu, A.C., Zysow, B.R., Meer, K., Schwartz, K., and Lawn, R.M. (1993). 5′ control regions of the apolipoprotein(a) gene and members of the related plasminogen gene family. *Proc. Natl. Acad. Sci. USA* **90**, 1369–1373.

Wu, Y.P., Siao, C.J., Lu, W., Sung, T.C., Frohman, M.A., Milev, P., Bugge, T.H., Degen, J.L., Levine, J.M., Margolis, R.U., and Tsirka, S.E. (2000). The tissue plasminogen activator (tPA)/plasmin extracellular proteolytic system regulates seizure-induced hippocampal mossy fiber outgrowth through a proteoglycan substrate. *J. Cell. Biol.* **148**, 1295–1304.

Yamamoto, K. and Loskutoff, D.J. (1996). Fibrin deposition in tissues from endotoxin-treated mice correlates with decreases in the expression of urokinase-type but not tissue-type plasminogen activator. *J. Clin. Invest.* **97**, 2440–2451.

Zhang, L., Seiffert, D., Fowler, B.J., Jenkins, G.R., Thinnes, T.C., Loskutoff, D.J., Parmer, R.J., and Miles, L.A. (2002). Plasminogen has a broad extrahepatic distribution. *Thromb. Haemost.* **87**, 493–501.

5

Plasminogen Receptors

Darin K. Fogg, Kyu-Sil Choi, and David M. Waisman

1. Introduction to Cellular Plasminogen Activation

The broad substrate protease plasmin is a key enzyme in a number of physiological and pathological processes, owing to its ability to degrade fibrin and extracellular matrix (ECM) proteins, and to activate latent matrix metalloproteases and growth factors. Plasmin is formed when the Arg^{561}–Val^{562} bond of plasminogen is hydrolyzed by one of two plasminogen activators, tissue plasminogen activator (tPA) or urokinase plasminogen activator (uPA). This results in the formation of the two-chain, disulfide-linked active serine protease. In general, tPA is considered responsible for activating plasminogen during fibrinolysis, and uPA for activating cell-associated plasminogen during cell migration in such processes as embryo implantation, wound healing, inflammation, angiogenesis, and tumor metastasis. Plasminogen binds to specific cellular receptors with low affinity and high capacity. This cell binding has a number of important consequences. First, the rate of plasminogen activation, that is, formation of active plasmin, is enhanced on the cell-surface compared to the soluble phase. This results in efficient plasmin generation and, because active plasmin remains bound to the cell, leads to a pericellular proteolytic cascade and subsequent remodeling of the ECM or fibrin matrix surrounding the cell. Second, when bound to the cell, plasmin is inaccessible to physiological inhibitors such as its major inhibitor α2-antiplasmin as well as α2-macroglobulin, which rapidly inactivate soluble plasmin. Third, plasminogen binding to specific surface receptors allows the cell to regulate pericellular proteolysis. By altering the expression and/or distribution of specific receptors for plasminogen, the cell can spatially and temporally regulate the potentiation of plasmin activity on its surface.

While plasmin activity is required for cell migration, tight regulation of pericellular proteolysis is absolutely required for proper control of this process. Because of the potency of plasmin and the breadth of its substrates, excess amounts of plasmin activity would lead to loss of cell adhesion to the substratum, thus preventing migration. Regulation of

Darin K. Fogg, Kyu-Sil Choi, and David M. Waisman • Department of Biochemistry and Molecular Biology, University of Calgary, Calgary, Alberta, Canada T2N 4N1

Plasminogen: Structure, Activation, and Regulation, edited by David M. Waisman. Kluwer Academic/Plenum Publishers, New York, 2003.

plasmin generation can occur at a number of levels including (a) expression of receptors for plasminogen on the cell-surface, (b) expression and/or activity of plasminogen activators, (c) expression and/or distribution of plasminogen activator receptors on the cell-surface, and (d) expression of plasminogen activator inhibitors (PAIs). In this chapter, we will review the current state of knowledge of the regulation of plasmin production and activity by cellular plasminogen receptors, with attention to different cell types and model systems. We will also summarize data on the candidate plasminogen receptors that are most likely involved in regulating plasmin production. Other chapters in this volume address regulation at the level of plasminogen activators and their inhibitors.

2. Mechanism of Cellular Plasminogen Binding

2.1. History

It has long been held that proteases can be intimately associated with the surface of cells, and that this feature confers on them the ability to dissolve fibrin clots or to degrade ECM protein during cell migration. When cell culture was in its infancy, fibrin clots were used to culture cells *ex vivo*. It was observed that metastatic cancer cells dissolved the fibrin substratum, while normal cells did not. Therefore, it was proposed that the ability to dissolve the clot was related to the ability of tumor cells to migrate through tissues (Fischer, 1925). The observed fibrinolytic activity of cancer cells is now known to be largely plasmin-mediated, but many different families of proteases are now known to contribute to cell-associated proteolysis during cell migration, including cathepsins and matrix metalloproteases.

The first direct evidence for cell-associated plasmin activity came from studies on platelets, which were shown to possess fibrinolytic activity that was resistant to cell washing (Morse *et al.*, 1965; Thorson *et al.*, 1972) and that increased with increasing platelet concentration (Taylor, Jr. *et al.*, 1981; Carroll *et al.*, 1982; Miles and Plow, 1988). Direct binding of plasminogen to the cell-surface, as well as cell-associated plasminogen activation was also initially demonstrated on platelets (Miles and Plow, 1985). Here, perhaps the most important feature of cellular plasminogen binding was introduced. That is, that binding of plasminogen to specific cellular receptors on the platelet surface leads to an enhanced rate of plasminogen activation compared to that in the soluble phase. This fundamental property is now an essential criterion in the definition of plasminogen-binding proteins as candidate plasminogen receptors. Plasminogen binding has been observed in many other cell types, including B cells (Miles and Plow, 1987), smooth muscle cells (Plow *et al.*, 1986), fibroblasts (Hajjar *et al.*, 1986; Plow *et al.*, 1986), keratinocytes (Reinartz *et al.*, 1993), and metastatic colorectal or mammary carcinoma cells (Burtin *et al.*, 1985, 1987). Today, the cells most extensively used for characterization of plasminogen binding are endothelial cells (Hajjar *et al.*, 1986) and monocytes or monocytoid cell lines (Ellis *et al.*, 1987).

2.2. Kinetics of Plasminogen Binding

The increased rate of plasminogen activation upon ligation to specific cellular receptors is due to a large reduction in the Km for activation by tPA (Felez *et al.*, 1996) or receptor-bound uPA (Ellis *et al.*, 1991), accompanied by a small reduction in the k_{cat}, leading to an overall increase in catalytic efficiency, k_{cat}/Km, of about 5- to 13-fold.

Binding of plasminogen to cells appears to be mediated via a single class of low-affinity receptors. The capacity for plasminogen binding to cells varies from around 10^4 to 10^7 binding sites per cell, and while the affinity of cells for plasminogen is consistent, different binding kinetics are observed for different cell types, implying that plasminogen receptors, or the mechanisms for regulating plasminogen binding, are cell-type specific. For example, Plow *et al.* showed that 6.84×10^4 plasminogen molecules bound to U937 cells at 37°C and this was reduced to one fourth that at 4°C. This was due to a decrease in number of plasminogen receptors, as demonstrated by Scatchard analysis. In contrast, binding of plasminogen to GM1380 fibroblasts at 4°C was only marginally reduced compared to that at 37°C (Plow *et al.*, 1986). Cell specific differences in temperature dependence have also been demonstrated among peripheral blood cells (Miles and Plow, 1987). For example, monocytes and granulocytes bound 33% and 49% less plasminogen, respectively, at 4°C than at 37°C, while plasminogen binding to lymphocytes dropped only 8% from 37°C to 4°C. As revealed by Scatchard analysis, the observed differences were not due to changes in receptor affinity, but the cellular capacity for plasminogen binding, therefore differences exist in the regulation of receptor number among different blood cells.

2.3. Binding of Plasminogen Isoforms

At least two differentially glycosylated forms of Glu-plasminogen are known to occur in blood (Brockway and Castellino, 1972; Hayes and Castellino, 1979a, 1979b, 1979c). Plasminogen type 1 is O-glycosylated at Thr^{346} and N-glycosylated at Asn^{289}, while plasminogen type 2 has only the threonine modification. Although these isozymes behave similarly under most conditions, some differences have been observed, such as binding to α2-antiplasmin (Lijnen *et al.*, 1981)and enhancement of plasminogen activation on fibrin (Takada *et al.*, 1985). Gonzalez-Gronow *et al.* observed a 10-fold greater binding for Glu-plasminogen type 2 compared to type 1 on U937 cells (Gonzalez-Gronow *et al.*, 1989). Moreover, type 1 Glu-plasminogen was unable to displace type 2 from the cell-surface, implying a different subset of receptors for each glycoform. In contrast, Hall *et al.* showed that type 2 Glu-plasminogen bound to rat hepatocytes and C6 glioma cells with slightly higher affinity than type 1, but the capacity was similar among isozymes (Hall *et al.*, 1990). Additionally, the latter study showed that radiolabeled Glu-plasminogen type 1 was displaced by cold Glu-plasminogen type 2 and vice-versa, suggesting that they were competing for the same receptor on those cells. These data suggest that differential binding of the two glycoforms of plasminogen may be cell-type dependent, and thus may depend on the profile of specific plasminogen receptors expressed on each particular cell type.

Glu-plasminogen can be cleaved by plasmin after Lys^{77} or Lys^{78} to yield Lys-plasminogen (Henkin *et al.*, 1991). In some cases differences have been noted in terms of cellular binding and/or activation of the native and proteolyzed forms. For example, binding of Glu-plasminogen, but not Lys-plasminogen, to fibrin is enhanced on the platelet cell-surface (Miles *et al.*, 1988a). On the other hand, U937 cells bound twice as much Lys- as Glu-plasminogen. In addition, Lys-plasminogen was activated on the surface of

U937 cells with a 3-fold greater catalytic efficiency (Felez *et al.*, 1996). This indicates that different receptors may be responsible for binding native versus proteolyzed forms of plasminogen, and this may result in important differences with respect to the rate of activation of plasminogen on the cell-surface.

Recently it has been shown that in order for endothelial cells to potentiate plasminogen activation by tPA or uPA, Glu-plasminogen must first be converted to Lys-plasminogen (Gong *et al.*, 2001). This was shown by using a mutant form of plasminogen (Asp^{646}–Glu) that can be converted to plasmin (can be cleaved at the Arg^{561}–Val^{562} bond) but the resulting two-chain plasmin is catalytically inactive. Compared with a 13-fold stimulation of wild-type Glu-plasminogen, endothelial cells could stimulate plasmin production only 2-fold using the mutant plasminogen protein. This indicated that plasmin-dependent cleavage of Glu-plasminogen, presumably to Lys-plasminogen, was necessary for optimal enhancement of plasmin production (Gong *et al.*, 2001). While conversion to Lys-plasminogen may be necessary for optimal stimulation of plasminogen activation, it is not sufficient. Namiranian *et al.* (1995) showed that upon addition to U937 cells, Glu-plasminogen was completely converted to Lys-plasminogen after 20 min and in another experiment they showed that Lys-plasminogen could be completely converted to plasmin on the cell-surface within 5 min. However, conversion of Glu-plasminogen to plasmin on U937 cells was less than 25% complete by 25 min, indicating that activation of Glu-plasminogen does not occur according to a simple Glu-plasminogen to Lys-plasminogen to plasmin mechanism.

In general, binding of active, two-chain plasmin to cell-surfaces is qualitatively and quantitatively similar to that for plasminogen, suggesting that the mature enzyme can bind to the same receptor(s) as the zymogen. However, some differences have been noted. For example, at low concentrations (10–30 nM) plasmin and plasminogen bound to human umbilical vein endothelial cells (HUVEC) with similar capacity. At 100 nM, however, nearly twice as much plasmin bound to these cells as plasminogen (Miles *et al.*, 1988b). This may be a result of a distinct plasmin receptor with similar affinity, but greater capacity than that for plasminogen. Alternatively, it could be argued that limited cell-surface proteolysis may occur as a result of using plasmin as a ligand, which would expose more carboxy-terminal lysine residues for increased binding. However, if this were the case then increased plasmin binding would be expected even at the lower plasmin concentrations of 10 and 30 nM, based on previous reports (Camacho *et al.*, 1989). The heterogeneity of plasmin(ogen) receptors on different cell types suggests that there may be promiscuous receptors capable of binding plasmin and plasminogen with only slightly different affinities.

Association of plasminogen with cells is mediated through the lysine-binding "kringle" domains of plasminogen, which interact with exposed lysine residues of protein receptors (Miles and Plow, 1985; Hajjar *et al.*, 1986; Plow *et al.*, 1986; Miles and Plow, 1988; Miles *et al.*, 1991; Plow *et al.*, 1986; Ellis *et al.*, 1999). These kringle-dependent interactions result in plasminogen binding to cells with an affinity in the low micromolar range ($\sim$1–2 μM). In all cases described to date, specific plasminogen binding is inhibited by lysine or by lysine analogs such as epsilon aminocaproic acid (ε-ACA). The inhibitory concentration required to disrupt cellular plasminogen binding by 50% (IC_{50}) for ε-ACA is between 2.5 and 65 μM (Miles and Plow, 1985, 1986a; Plow *et al.*, 1986; Miles and Plow, 1987; Miles *et al.*, 1988a). This value closely corresponds with the Kd of ε-ACA binding to the high affinity lysine-binding site of Glu-plasminogen kringle 1 (between

9 and 44 μM) (Markus *et al.*, 1978; Miles and Plow, 1986b, 1988). Indeed, it has been shown that binding of plasminogen to platelets and U937 cells is mediated primarily through interactions involving kringles 1–3, including the high affinity lysine-binding region of kringle 1, with some contribution from kringle 5 (Miles and Plow, 1988). The high-affinity sites are not solely responsible for plasminogen binding, however (Miles and Plow, 1988). Miniplasminogen, composed of part of kringle 5 plus the plasmin serine protease domain (Sottrup-Jensen *et al.*, 1978; Moroz, 1981; Henkin *et al.*, 1991) can bind to cells, albeit with a 10-fold lower affinity than native Glu-plasminogen (Felez *et al.*, 1996; Miles and Plow, 1986b), and therefore may interact with some cellular plasminogen receptors. Because different kringle domains have different affinities for lysine-containing proteins, it is possible that plasminogen binding and orientation on the cell-surface can be heterogeneous (Plow and Miles, 1990). This may translate into more or less favorable interactions with specific receptors and thus plasminogen could display differential binding and activation kinetics unique to certain receptors on the cell-surface.

3. Plasminogen Receptors as Regulators of Plasmin Activity

Plasminogen is produced in the liver and secreted to the blood stream where it circulates at a concentration of around 2 μM in humans (Plow and Miles, 1990). Given their affinity for plasminogen, around half of the receptors on a given cell will have bound plasminogen (Adelman *et al.*, 1988; Miles *et al.*, 1988a; Plow and Miles, 1990). As mentioned above, the capacity for plasminogen binding is remarkably high, between 1×10^4 and 1×10^7 sites per cell, depending on the cell type and condition. This suggests that cell-bound plasminogen should be plentiful and the rate of plasmin activation would depend on the availability of active plasminogen activator, which is in turn dependent on levels of the uPA receptor (uPAR). Surprisingly, it has been demonstrated that it is plasminogen binding, and not uPA binding to cells that is rate limiting for plasminogen activation. This was illustrated with U937 cells, which express around 5000 uPA receptors per cell and the gastric cancer cell line KATO III, which expresses only 400 uPA receptors per cell. Plasminogen activation on the two cell types occurs with similar kinetics (Namiranian *et al.*, 1995) suggesting that the difference in uPA binding alone could not account for cell-type specific differences in plasminogen activation. It was concluded that the rate of plasmin formation on the cell-surface depends on the ability of certain plasminogen receptors to present Glu- or Lys-plasminogen to uPA in a conformation that can be efficiently activated (Namiranian *et al.*, 1995). This implies heterogeneity of plasminogen receptors with respect to binding of plasminogen and subsequent presentation to uPA. The authors speculated that, while the bulk of the cellular plasminogen receptors are not involved in potentiation of plasminogen activation, a small subset of plasminogen receptors have carboxyl-terminal residues that are capable of inducing an activator-susceptible plasminogen conformation and have access to plasminogen activators. These are the receptors that are responsible for the enhancement of plasminogen activation on the cell surface.

This notion is supported by studies that demonstrate heterogeneity among plasminogen receptors with respect to enhancement of plasminogen activation. In particular, receptors with carboxy-terminal lysine residues that are exposed and susceptible to cleavage by basic carboxypeptidases have been implicated in the observed stimulation of plasminogen activation on cells although they account for only a fraction of the cell's total plasminogen-binding capacity. For example, treatment of U937 monocytoid cells with carboxypeptidase B reduced plasminogen binding by 60%, but reduced the enhancement of plasminogen activation by 95%, suggesting that the subset of receptors with carboxy-terminal lysine residues are responsible for the stimulation of cell-associated plasminogen activation (Felez *et al.*, 1996; Hawley *et al.*, 2000). Binding of plasminogen to these types of receptor may alter its conformation, allowing more efficient cleavage by plasminogen activators (Markus, 1996). This idea is supported by the description of a conformational change in plasminogen induced by ε-ACA that leads to a more open form that is perhaps more accessible to cleavage by the plasminogen activators (Mangel *et al.*, 1990). This ligand-induced conformational shift in plasminogen has also been observed with the candidate cellular plasminogen receptor annexin II heterotetramer (see Section 6). Alternatively or in addition, receptors with a carboxy-terminal lysine residue may function as a template for bringing plasminogen and its activator within close proximity, leading to a greater rate of activation (Felez *et al.*, 1996). The template model is supported by the demonstration that on U937 monocytoid cells, tPA and plasminogen compete for the same binding sites (Felez *et al.*, 1993) and that uPA-dependent plasminogen activation is dependent on formation of a uPA–plasminogen complex (Ellis and Dano, 1993; Ellis *et al.*, 1999). A uPA–plasminogen complex formed by an anti-uPA monoclonal antibody with a carboxy-terminal lysine stimulated plasminogen activation 50-fold (Ellis and Dano, 1993). This suggests that ternary complex formation could explain the enhancement of plasminogen activation observed in the presence of cells. However, the effect on the conformation of plasminogen following interaction with a protein containing a carboxy-terminal lysine residue should be considered as well. The effect of the antibody on the conversion of pro-uPA to mature, two-chain uPA by plasmin, which is demonstrative of the advantage of ternary complex formation was relatively small, so it is possible that both complex formation and plasminogen conformation contribute partially to the enhanced plasminogen activation. In addition, plasminogen and tPA compete for binding to gangliosides, suggesting these are capable of ternary complex formation (see Section 6). However, based on carboxypeptidase data, it appears that gangliosides do not contribute significantly to stimulation of plasminogen activation. Therefore, it could be concluded that ternary complex formation cannot be entirely responsible for the increase in plasminogen activation rates observed in the presence of cells.

4. Modulation of Receptor Expression

If plasmin activity can be regulated at the level of plasminogen receptor, it would be expected that the expression of receptors that serve to enhance plasminogen activation on the cell-surface might be altered during cell activation or differentiation. In some cases, the capacity for plasminogen binding to one cell type under different conditions has been observed.

4.1. Plasminogen Receptor Expression on Platelets

Miles *et al.* first showed plasminogen receptor upregulation on platelets in response to stimulation with thrombin (Miles and Plow, 1985). A 3-fold increase in plasminogen binding was induced by thrombin, which was correlated with an increase in the rate of tPA, uPA, and streptokinase-dependent plasminogen activation. It was later determined that there existed at least two mechanisms for plasminogen binding to platelets; one was insensitive, and one sensitive to induced upregulation by thrombin (Miles *et al.*, 1986). Plasminogen binding to thrombin-stimulated platelets was later demonstrated in plasma, indicating that the initial observations made on plasminogen binding to platelets were physiologically relevant (Adelman *et al.*, 1988).

4.2. Plasminogen Receptor Expression on Endothelial Cells

Reports of differing binding kinetics for plasminogen on endothelial cells (Hajjar *et al.*, 1985, 1986) prompted Miles *et al.* to investigate binding of plasminogen and uPA to endothelial cells before and after stimulation with thrombin (Miles *et al.*, 1988b). The average dissociation constant (Kd) for plasminogen binding to HUVEC was found to be in the order of 2 μM with plasminogen binding to around 1–2 $\times 10^7$ molecules per cell. The binding capacity dropped approximately 19% upon stimulation of HUVEC with thrombin. This finding was supported by the more recent observation that a reduction in generation of active plasmin on the surface of porcine aortic endothelial cells (PAEC) occurred following thrombin treatment (Li *et al.*, 1995).

4.3. Plasminogen Receptor Expression on Peripheral Blood Cells

Felez *et al.* (1990) have shown that plasminogen receptor expression can be regulated on the surface of monocytoid cells and peripheral mononuclear cells. For example, it was demonstrated that treatment of THP-1 monocytoid cells with phorbol myristate acetate (PMA) led to a 17-fold increase in plasminogen binding in a specific subpopulation of cells (those that were nonadherent), accompanied by a measured 8-fold increase in rate of cell-dependent plasmin generation. Because uPA binding was only modestly altered under these conditions, this result emphasizes the notion that plasminogen binding, and not uPAR expression or uPA binding, is rate limiting in plasminogen activation. An increase in plasminogen binding was also observed when freshly isolated human peripheral blood monocytes were cultured for 18 hr, demonstrating modulation of plasminogen receptor expression on primary cells. For each cell type, only the capacity and not the affinity for plasminogen binding was altered, suggesting increased expression of existing plasminogen receptor(s), not expression of a distinct, high affinity receptor. Similar experiments by the same group have since shown an increase in expression or exposure of carboxy-terminal lysine-containing plasminogen receptors on U937 and THP-1 cells, following adhesion to the β1 integrin ligands fibronectin, laminin, and vitronectin. These findings suggest that when circulating monocytes come into contact with ECM substrates, their plasminogen receptor expression levels and thus cell-associated plasmin activity may be altered in order to allow modification of adhesive and proteolytic states (Kim *et al.*, 1996). Cycling between these two states is a complex process, which is not yet fully understood, but is important for cell migration during inflammation, wound healing, angiogenesis, and tumor metastasis.

Recently, similar observations to those mentioned above for monocytes have been made with peripheral blood neutrophils. Herren *et al.* (2001) reported that, compared with freshly isolated neutrophils, those that have been cultured up to 20 hr increase their plasminogen binding by an average of 31-fold, again as a result of increased capacity and not affinity. Concomitant with the increase in plasminogen-binding capacity was an ~7.7-fold increase in rate of plasminogen activation on these cells. Consistent with reports on other cell types, the increase in plasminogen binding was sensitive to carboxypeptidase B, again implicating proteins containing carboxy-terminal lysine residues.

4.4. Plasminogen Receptor Expression on Tumor Cells

Certain tumor cell types have also been shown to modulate their plasminogen-binding capacity. The highly metastatic breast cancer cell line MDA-MB-231 was shown to bind and activate significantly more plasminogen than the non-metastatic cell lines MCF-7 and T-47D (Ranson *et al.*, 1998). In addition, it was noted that nonviable cancer cells bound more plasminogen than viable ones. This finding was corroborated with subsequent studies showing that plasminogen binding was increased up to 100-fold on the surface of apoptotic and dead U937 cells, and that apoptotic cells had significantly higher plasminogen activation rates compared to viable or dead cells, owing to a concomitant increase in surface uPA expression of the apoptotic cells (O'Mullane and Baker, 1998). The increase in plasminogen binding occurred as a late phase event in apoptosis, suggesting that plasmin activity may play some role in the degradative phase of apoptosis (O'Mullane and Baker, 1999), however the significance of plasmin activity during apoptosis remains to be elucidated. It may be worth noting that the increase in plasminogen binding during apoptosis was shown to follow phosphatidylserine exposure. Because the phosphatidylserine-binding proteins annexin II and annexin II heterotetramer have been implicated in plasminogen binding and activation (see Section 6), the increase in binding could be as a result of exposure of these proteins from the inner leaflet of the plasma membrane. However, the increase in plasminogen binding was delayed compared to phosphatidylserine exposure, implying that if annexin(s) were involved, some modification step may be necessary before plasminogen binding could occur. Also, other studies have implicated cytokeratin 8 and cell-surface actin as the major plasminogen receptor in regards to enhancement of plasminogen activation on MCF-7, BT20, and MDA-MB-157 breast cancer cell lines (Hembrough *et al.*, 1996). It is especially evident among different breast cancer cell lines that those proteins that seem to be the most important plasminogen receptors are different among different cell types.

5. Candidate Plasminogen Receptors

Because the binding of plasminogen to specific cellular receptors is the key event in the regulation of plasmin generation and activity and also because this receptor-associated plasmin activity is central to cell migration in development, immunology, and tumor metastasis, there has been much interest in identifying the receptor(s) that have the greatest influence on plasminogen regulation. A number of candidate plasminogen receptors are presented in Table 5.1. Although there appears to be a single class of binding sites, there may be many distinct receptors on the same cell or on different cell types that bind with this affinity (Miles and Plow, 1988). Much evidence has been gathered in support of individual receptors, however it remains unclear which of these are most important and if the plasminogen receptor profile is different with different physiological processes.

The first receptor identified as a plasminogen receptor that contains a carboxy-terminal lysine and is capable of stimulating plasminogen activation was α-enolase. It was isolated by affinity chromatography as a protein from U937 cells that interacted with plasminogen and was expressed on the surface of these cells. α-enolase was shown to be capable of binding iodinated plasminogen in ligand blotting experiments, and carboxypeptidase B abolished this binding, while reducing U937 cell-surface plasminogen binding up to 40% (Miles *et al.*, 1991). It was later demonstrated that α-enolase could stimulate plasminogen activation *in vitro*, that it was present on peripheral blood monocytes and neutrophils as well as cell lines, and that it accounted for about 1.8×10^6 binding sites on the surface of U937 cells. This is approximately 10% of the total plasminogen-binding capacity of these cells (Redlitz *et al.*, 1995). These results were corroborated by Andronicos *et al.*, who used recombinant human protein to verify the plasminogen-binding kinetics and carboxy-terminal lysine-dependence of α-enolase (Andronicos *et al.*, 1997). Because α-enolase does not contain a signal sequence, there is some debate whether the protein found on the surface of cells is identical or related to the ubiquitous glycolytic enzyme found in the cytoplasm. To our knowledge, there has been no evidence to suggest that surface α-enolase differs from cytoplasmic α-enolase (Redlitz *et al.*, 1995; Andronicos *et al.*, 1997; Arza *et al.*, 1997). It remains to be determined by which leaderless secretion mechanism α enolase may become externalized.

Using two-dimensional electrophoresis and ligand blotting of U937 and peripheral blood monocyte membranes, Hawley *et al.* demonstrated differences between cell types in regard to susceptibility of surface proteins to carboxypeptidase activity (Hawley *et al.*, 2000). Binding of plasminogen to the candidate receptors α-enolase, annexin II, and cell-surface actin from U937 membrane extracts was unaffected by carboxypeptidase B treatment, while treatment of monocyte membranes with carboxypeptidase led to reductions in ligand binding of 71% and 14% for α-enolase and annexin II, respectively. This suggested

Table 5.1. Candidate Cellular Plasminogen Receptors. (Those receptors that have been demonstrated to bind tPA in addition to plasminogen are indicated, and where applicable, the reported dissociation constants (Kd) are given)

Candidate receptor	MW (kDa)	Cell type	Binds tPA (Kd)	Reference
Glycoprotein IIb/IIIa	~80–130	Platelets	ND*	Miles *et al.* (1986)
Gangliosides	~1–3	Various cells	Yes (ND)	Miles *et al.* (1989)
Actin	45	Endothelial cells	Yes (65 nM)	Dudani and Ganz (1996)
gp330	76	Kidney epithelium	Yes (ND)	Kanalas and Makker (1991)
α-enolase	54	Monocytoid cells	Yes (ND)	Miles *et al.* (1991)
Amphoterin	30	Neuronal cells	Yes (ND)	Parkkinen and Rauvala (1991)
45 kDa protein	45	Endothelial cells	Yes (ND)	Dudani *et al.* (1993)
Annexin II	36	Endothelial cells	Yes (25 nM)	Hajjar *et al.* (1994)
Cytokeratin 8	52	Epithelial cells	Yes (250 nM)	Hembrough *et al.* (1996)
Annexin II heterotetramer	94	Endothelial cells	Yes (ND)	Kassam *et al.* (1998a)
Tip 49a	54	Monocytoid cells	ND	Hawley *et al.* (2001)

*ND: Not determined.

that the accessibility of the carboxy-terminal lysine residues on the same plasminogen receptors might differ even among related cell types, adding yet another layer of complexity to the plasminogen receptor story. It remains unclear what physiological role each putative plasminogen receptor may play, given that few experiments have been performed that examine what effect varying expression levels of particular receptors has on cell-surface plasminogen binding and activation. In one study, expression of α-enolase was increased on c-*jun* transformed rat fibroblasts compared to normal (FR3T3) fibroblasts. There was no difference found between these two cell lines in terms of plasminogen binding or activation, suggesting that α-enolase does not serve as an important plasminogen receptor on these cells (Bergman *et al.*, 1997). However, in this report there was no indication that expression of surface-localized α-enolase was increased on the transformed fibroblasts compared to normal, and thus the conclusion that α-enolase does not play a role in plasminogen regulation on these cells is premature. It remains to be seen what role α-enolase plays in plasminogen regulation *in vivo*.

Using ligand blotting of endothelial cell extracts separated by one-dimensional SDS-PAGE and radiolabeled plasminogen, a 45 kDa protein was found that bound a major fraction of the ligand (Dudani *et al.*, 1991, 1993). This was later shown to interact with tPA (Dudani *et al.*, 1994). Using two-dimensional electrophoresis, the 45 kDa band was separated, based on isoelectric point, into two bands, each able to bind plasminogen. One component was identified as actin, and it was shown that actin could bind to plasminogen, tPA, and lipoprotein a (Lp(a)) (Dudani and Ganz, 1996). The other 45 kDa protein has not been identified to date and its role in plasminogen activation awaits clarification. Actin has been observed on the surface of a variety of types of endothelial cells (Shapiro *et al.*, 1987; Pardridge *et al.*, 1989; Hu *et al.*, 1991; Hu and Riordan, 1993; Moroianu *et al.*, 1993), and other cell types (Chen *et al.*, 1978; Owen *et al.*, 1978; Jones *et al.*, 1979; Sanders and Craig, 1983), and therefore could potentially play a role in plasminogen binding. In keeping with this, it has been reported that plasminogen partially colocalizes with actin on the surface of the breast cancer cell line MDA-MB-231, while other candidate plasminogen receptors were either not expressed (α-enolase, cytokeratin 8) or did not significantly colocalize with plasminogen (annexin II) (Andronicos and Ranson, 2001). Actin lacks carboxy-terminal lysine, and actin-dependent enhancement of plasminogen activation has not been demonstrated. The studies performed by Miles *et al.* would argue against a role for actin in enhancement of plasminogen activation on the surface of, at least, monocytes and monocytoid cell lines, because treatment of membrane preparations of these cells with carboxypeptidase did not significantly affect plasminogen binding to actin (Hawley *et al.*, 2000). However, the results by Andronicos *et al.* suggest that on at least one breast cancer cell line, cell-surface actin may play a role in plasminogen binding or localization.

Several other proteins have been identified that are expressed on the surface of various types of cells that bind plasminogen and that stimulate plasminogen activation *in vitro*, including amphoterin (neuronal cells) (Parkkinen and Rauvala, 1991), cytokeratin 8 (breast cancer cells) (Hembrough *et al.*, 1996), and Tip49a (monocytes and monocytoid cell lines) (Hawley *et al.*, 2001). Further investigation of these putative receptors will determine whether they play a role in plasminogen binding or activation *in vivo*. In order to unequivocally demonstrate the importance of each of the receptors mentioned above, the kinetics of plasminogen activation on cells with targeted downregulation (or overexpression) of these proteins must be examined.

In addition to cell-surface proteins, plasminogen can also bind to cell-surface gangliosides; glycosphingolipids that contain variable amounts of sialic acid. Lysine analogues also prevent this plasminogen binding to gangliosides, suggesting that the kringle domains of plasminogen also mediate this interaction. Binding of plasminogen to gangliosides is inhibited by tPA, and the reverse is also true. This suggests that gangliosides can bind both tPA and plasminogen (Felez *et al.*, 1993). Although gangliosides and protein receptors can bind plasminogen with similar affinity and capacity, enhanced plasminogen activation upon binding to gangliosides has not been demonstrated.

6. Annexin II, P11, and Annexin II Heterotetramer as Candidate Plasminogen Receptors

As mentioned above, the Ca^{2+} binding protein, annexin II, has been suggested to be a receptor for plasminogen and tPA on the surface of endothelial cells (Cesarman *et al.*, 1994; Hajjar *et al.*, 1994). In most cells, the majority of annexin II exists in a heterotetrameric form referred to as AIIt, with two copies of annexin II and two copies of a member of the S100 family of proteins, S100A10, also known as p11. AIIt is formed when a p11 homodimer binds to two molecules of annexin II, via the carboxy-terminus of p11 and the amino-terminal region of annexin II. In most cell types so far examined, at least 90% of the total cellular annexin II is in the tetrameric form (Gerke and Weber, 1984; Nilius *et al.*, 1996). Although originally described as cytoplasmic proteins with many proposed intracellular functions, annexin II and AIIt are now also known to exist on the outer leaflet of the plasma membrane of many cell lines (Yeatman *et al.*, 1993; Cesarman *et al.*, 1994; Chung and Erickson, 1994; Wright *et al.*, 1994; Waisman, 1995; Siever and Erickson, 1997; Mai *et al.*, 2000). In many cases, biological functions such as plasminogen binding are ascribed to annexin II but since co-detection of p11 is seldom attempted, this conclusion is premature. Many plasminogen-binding experiments utilize SDS-PAGE techniques and proteins of molecular weight less than 20 kDa are seldom examined. As a result, p11 is often run out of the gel and therefore is overlooked. Our laboratory has extensively examined the effect on plasminogen activation of annexin II compared with AIIt and have reported many lines of evidence to suggest that it is AIIt, and not annexin II, that is an important cellular receptor for plasminogen (Figure 5.1A).

First, while annexin II enhances plasminogen activation by tPA around 6-fold, AIIt does so up to 341-fold (Kassam *et al.*, 1998a).

Second, the binding of AIIt, and not annexin II or other annexins, to a mutant form of plasminogen labeled fluorescently at its active site (Ser^{741}–Cys) caused the fluorescence to be quenched, indicating a large conformation shift upon binding to AIIt (Kassam *et al.*, 1998a). Using this mutant, a Kd for plasminogen binding to AIIt of 1.26 ± 0.04 μM was determined. As mentioned above, this conformational shift in plasminogen has been observed using lysine analogs capable of interacting with plasminogen kringle domains, and results in an enhanced rate of plasminogen activation.

Third, AIIt possesses the prerequisite carboxyl-terminal residues indicative of a plasminogen receptor. The importance of carboxy-terminal lysine residues of plasminogen receptors is well established and it is thought that the enhancement of plasminogen activation by cellular receptors is a result of the conformational shift induced by these lysine

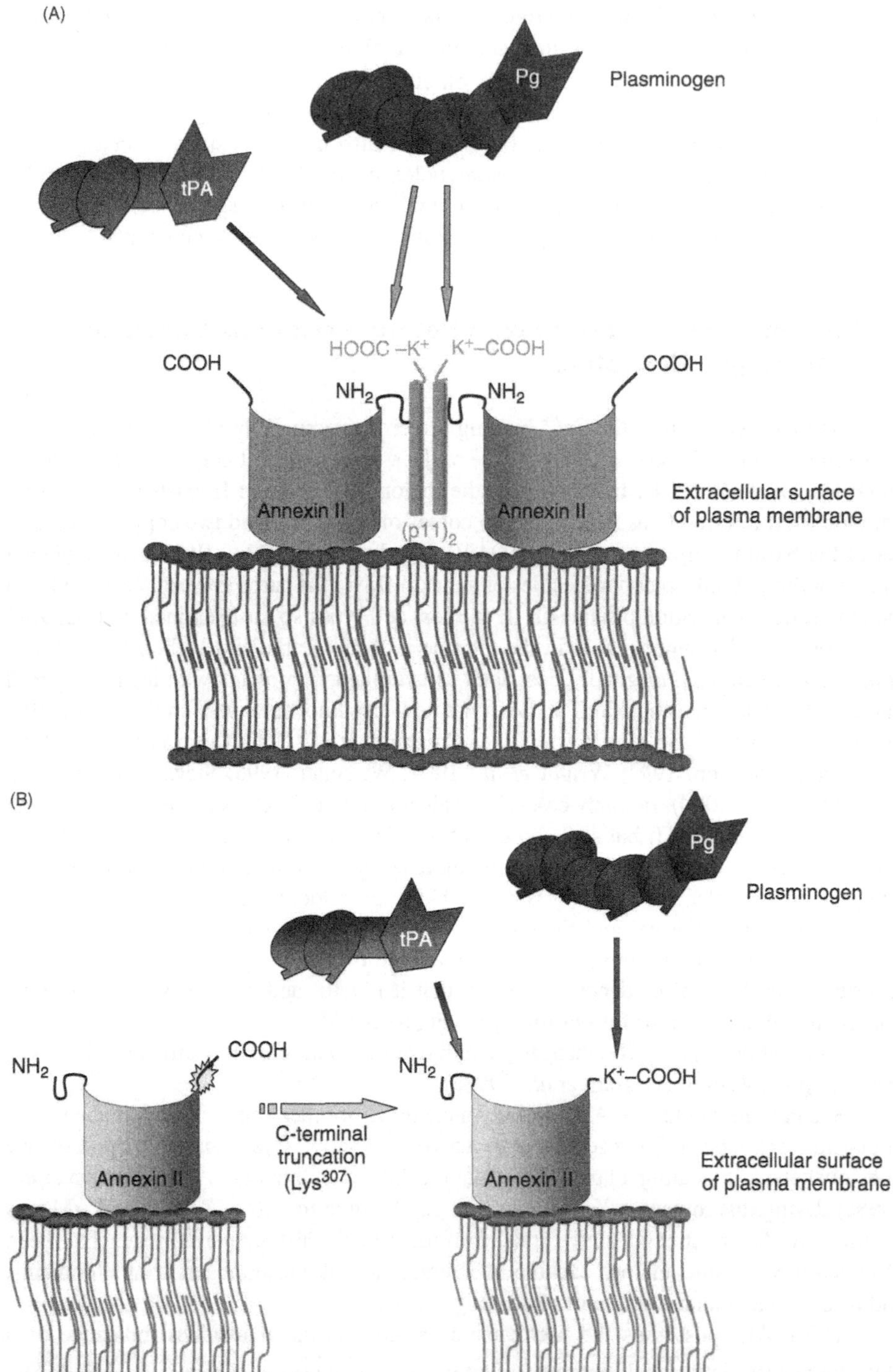
(A)
Pg
Plasminogen
tPA
HOOC –K+
K+–COOH
COOH
NH2
NH2
COOH
Annexin II
(p11)2
Annexin II
Extracellular surface
of plasma membrane
(B)
Pg
Plasminogen
tPA
NH2
COOH
NH2
K+–COOH
C-terminal
truncation
(Lys307)
Annexin II
Annexin II
Extracellular surface
of plasma membrane

(C)

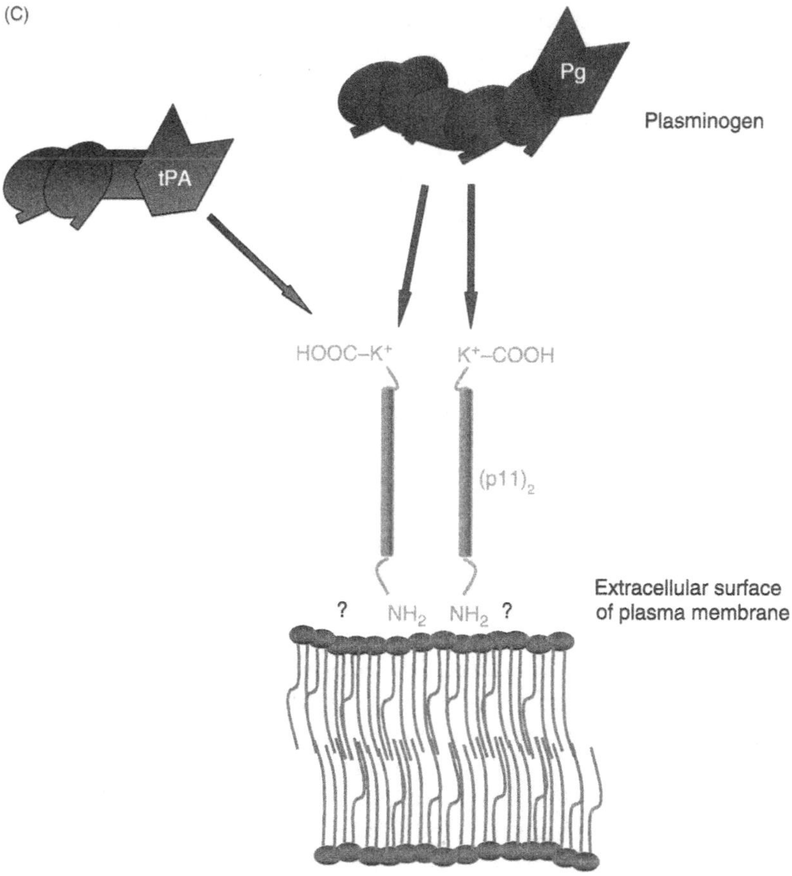

Figure 5.1. Model of plasminogen binding to candidate cell-surface receptors. Binding of plasminogen or plasmin and tPA to annexin II heterotetramer $[(AII)_2\,(p11)_2]$, (A), or its component subunits annexin II monomer (B) and $(p11)_2$ homodimer (referred to simply as p11) (C). While intact p11 contains carboxy-terminal lysine residues, annexin II does not, and therefore must undergo limited proteolysis to expose plasminogen-binding carboxy-terminal lysine ((B), see text). Although it is apparent that p11 alone can interact directly with plasminogen and tPA *in vitro*, it is unclear whether p11 exists at the cell-surface independent of annexin II ((C), question marks).

residues. While annexin II lacks carboxy-terminal lysine, the carboxy terminus of p11 is F^{86}-V-V-H-M-K-Q-K-G-K-K^{96}. Thus plasminogen activation in the presence of purified recombinant p11 is 46-fold above the rate with tPA alone, compared with 77-fold in the presence of recombinant AIIt (Kassam *et al.*, 1998b).

Fourth, the full activity of AIIt can be reconstituted by binding the amino-terminus of annexin II to p11. That is, the carboxyl-terminal plasminogen-binding domain of annexin II does not contribute to the activity of AIIt. The addition of a peptide consisting of the amino-terminal 15 amino acids, representing the p11-binding portion of annexin II (Johnsson *et al.*, 1988) to p11 in the plasminogen activation assay brought the rate of plasmin generation up to the level observed with AIIt. This suggests that only the amino-terminal region of annexin II plays a role in plasminogen binding and that binding to annexin II may influence the conformation of p11 to allow proper interaction with plasminogen (Kassam *et al.*, 1998b). In contrast with p11 and AIIt, the monomeric form of annexin II lacks carboxy-terminal lysine. It has been suggested that in order for plasminogen to bind to monomeric annexin II, the latter must first be proteolyzed to expose Lys^{307} (Figure 5.1B). However, we found that the substitution of this amino acid with threonine does not have a dramatic effect on plasminogen activation rate (Kassam *et al.*, 1998b). Hajjar has reported that the tPA binding site on annexin II resides in its amino terminus, from Leu^{8} to Leu^{13} (Hajjar *et al.*, 1998), and that annexin II enhances plasminogen activation by forming a ternary complex with tPA and plasminogen, thus bringing the enzyme and zymogen in close proximity and allowing efficient activation. The requirement for limited proteolysis of annexin II to expose Lys^{307} at the carboxy-terminus would mean that the protease-sensitive amino-terminal domain of annexin II (Drust and Creutz, 1988) would most likely also be lost, and therefore the proteolyzed protein would be unable to form the proposed ternary complex. It is also interesting that the Hajjar model postulates that annexin II contains distinct binding sites for tPA (amino-terminal region) and plasminogen (carboxyl-terminal region). However, it has been demonstrated that about 90% of the plasminogen bound to cells is blocked by the binding of tPA. That is, plasminogen and tPA share a common binding site (Felez *et al.*, 1993). This observation is incompatible with the Hajjar model.

A fifth important distinction between annexin II and AIIt with regard to candidacy as a plasminogen receptor is the prerequisite ability of plasminogen receptors to protect plasmin from inactivation by α2-antiplasmin. While annexin II is unable to protect plasmin from inactivation (Cesarman *et al.*, 1994), AIIt is capable of protecting plasmin from α2-antiplasmin, as well as tPA from its inhibitor, PAI-1.

Sixth, AIIt but not annexin II stimulates plasmin autoproteolysis. In addition to hydrolyzing several ECM proteins, plasmin is also capable of autoproteolysis (Jespersen *et al.*, 1986). This self-destruct mechanism is thought to be important to prevent collateral tissue damage by the accumulation of plasmin in the tissues and has been observed on the surface of THP-1 monocytoid cells (Falcone *et al.*, 1994). Our *in vitro* studies suggested that in addition to stimulating plasmin production, AIIt and the p11 subunit of AIIt, could also stimulate plasmin autoproteolysis (Fitzpatrick *et al.*, 2000).

Seventh, we have identified a potential mechanism to terminate AIIt-stimulated cellular plasmin production. Since AIIt has such a profound effect on plasminogen activation, we speculated that there may exist a mechanism of downregulating plasmin production and that removal of cellular binding sites by physiologic carboxypeptidases was a reasonable possibility. We have demonstrated that in addition to the pancreatic carboxypeptidase B (CpB), physiologically relevant concentrations of plasma carboxypeptidase N (CpN) and thrombin-activated fibrinolysis inhibitor (TAFI) are capable of completely ablating the enhancement of plasminogen activation by AIIt and p11. We confirmed that the mechanism by which the carboxypeptidases abrogated stimulation of plasminogen activation was by removal of carboxy-terminal lysine residues (Fogg *et al.*, 2002). These results suggest that a physiological mechanism, such as plasma carboxypeptidase-mediated loss of cellular plasminogen binding, may exist to downregulate plasmin production and thereby protect cells from the deleterious effect of plasmin overproduction. It could also be proposed that membrane-bound carboxypeptidase (CpM), which is associated with neutrophils (Herren *et al.*, 2001) and monocytoid cells (Rehli *et al.*, 1995) could play a role in terminating the enhancement of plasminogen activation on the surface of these cell types.

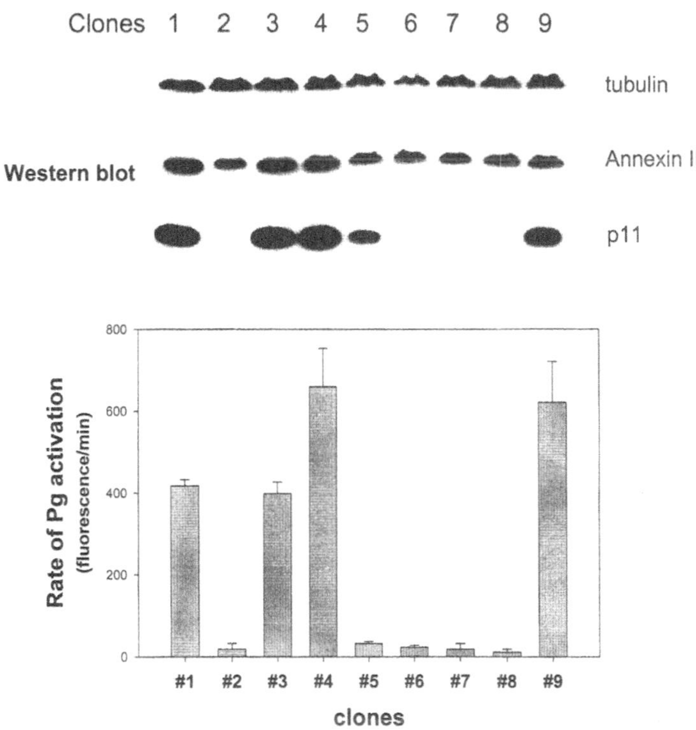

Figure 5.2. Effect of p11 expression on cell surface plasminogen activation. HT 1080 fibrosarcoma cells were transfected with antisense to p11, and clones (1–9) were isolated. Expression of p11 and annexin II was determined by immunoblot analysis. Expression of α-tubulin was also measured as a protein loading control. Plasminogen activation was examined on the surface of fibrosarcoma clones showing different expression levels of p11. Cell-surface plasmin generation for each of clones 1–9 was determined by measuring amidolytic activity of the plasmin generated from plasminogen. The reaction was initiated by the addition of 1 μM [Glu]-plasminogen and a fluorogenic substrate directly to the cells and was monitored with a fluorescence plate reader.

In order to compare the possible roles of AIIt and annexin II as regulators of cellular plasmin production, we established HT1080 cell lines with altered levels of extracellular p11. It was reasoned that the loss of p11 from the cell-surface would have substantial effects on plasmin production if AIIt were an important plasminogen receptor. In contrast, since the loss of p11 from the cell surface would be expected not to affect free annexin II and only to moderately affect annexin II that was normally complexed with p11, it was expected that the loss of p11 would cause little if any effect on cellular plasmin production if annexin II was the important plasminogen regulator. The importance of p11 in plasminogen activation and plasmin-dependent cell migration was dramatically illustrated in these cells. Expression levels of p11 were reduced by transfection of cells with antisense-oriented p11, and increased by transfection of sense-oriented p11. The transfected cells were cloned and cell lines established. In parallel with p11 expression levels, the rates of plasminogen activation rose and fell (Figure 5.2). In fact, in some cell lines, the loss of cell-surface p11 resulted in an astounding 90% loss in cellular plasmin production. Moreover, in individual clones, we observed that cellular plasminogen binding (Figure 5.3A), rate of plasminogen activation (Figure 5.3B), and extent of ECM degradation (Figure 5.3C) all

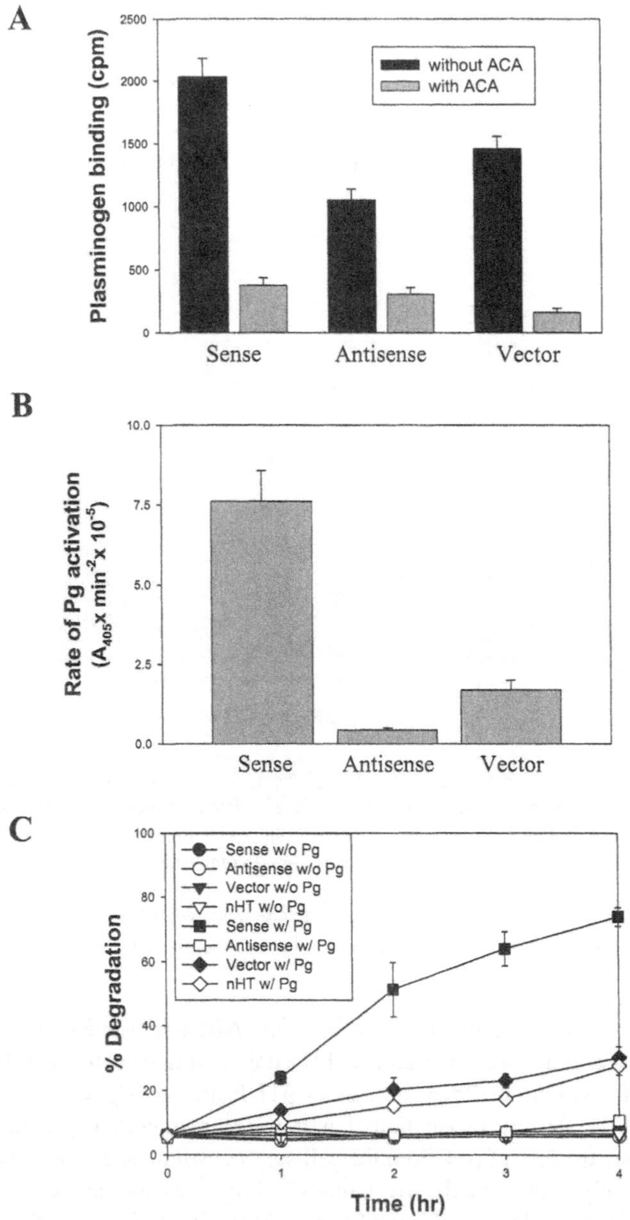

Figure 5.3. p11 regulates plasminogen binding and activation on the cell-surface. HT 1080 cells were transfected with full-length p11 cDNA, either in the sense or antisense orientation, or with empty vector, and stable clones were established. (A) Plasminogen binding to sense-, antisense-, and vector-transfected cells in the absence or presence of ε-ACA was examined with radioiodinated plasminogen. (B) Plasminogen activation on the surface of these cells was measured by using plasmin amidolytic substrate with 1 μM plasminogen. (C) The proteolytic activity of the clones toward an extracellular matrix generated by human aortic smooth muscle cells was also examined without (w/o) or with (w/) 1 μM exogenous plasminogen. nHT means normal, untransfected HT 1080 fibrosarcoma cells.

correlated with the expression of p11 protein on the extracellular surface of the cell (Choi *et al.*, 2003). Interestingly, the extracellular levels of annexin II were identical in all of the transfected HT1080 fibrosarcoma cell lines. The levels of uPA and the PAIs were unaffected; therefore the differences in plasminogen activation rates were due to specific reduction or increase in the level of p11. These results underscore the importance of the p11 subunit of AIIt in plasminogen activation, at least in this tumor cell model system.

Recently, we have used surface plasmon resonance to investigate the binding interactions between plasminogen and tPA and AIIt. These experiments utilized a novel approach in which a phospholipid bilayer was immobilized on the BIAcore biosensor chip. Since AIIt or annexin II bound essentially irreversibly to the bilayer, we were able to measure, in real time, the interactions between plasminogen and tPA and the phospholipid-bound AIIt or annexin II. For these experiments we also amine-couple the p11 protein to the biosensor chip. We found that, while neither plasminogen nor tPA bound to annexin II, plasminogen bound to AIIt with a Kd of 0.11 μM and to p11 with a Kd of 1.81 μM. tPA bound to AIIt with a Kd of 0.68 μM and to p11 with a Kd of 0.45 μM (Macleod *et al.*, *J. Biol. Chem.* in press). Removal of the carboxyl-terminal lysines from the p11 subunit attenuated t-PA and plasminogen binding to AIIt. These results show that the carboxyl-terminal lysines of p11 form t-PA and plasminogen binding sites. Based on these findings, we now have substantial and definitive evidence to propose that AIIt, and not annexin II, is the major plasminogen receptor on the surface of many cells. Given that p11 is itself endowed with many of the qualities ascribed to plasminogen receptors, including specific binding via carboxyl-terminal lysines, enhancement of plasminogen activation (Kassam *et al.*, 1998b), and cell-surface expression (Kassam *et al.*, 1998a), it is clear that within the AIIt complex, p11 is the regulator of plasminogen activation and activity. It is also important to point out that although annexin II and p11 colocalize on the surface of several cell lines, it has not been unequivocally established if p11 can exist in a free, uncomplexed form on the cell-surface. The possibility exists that p11 acts as a plasminogen receptor in the absence of annexin II (Figure 5.1C). To date, this possibility remains purely speculative.

7. Concluding Remarks

uPA has long been considered an important prognostic indicator in many human cancers, in particular, breast cancer (Duffy, 2001). In addition, uPA is thought to play a key role in the immune system, allowing leukocytes to migrate throughout the body and home to sites of inflammation (Gyetko *et al.*, 1996). It is also well established that regulated pericellular proteolysis by normal and tumor cells is fundamental to their ability to migrate through tissues and, in the case of tumor cells, metastasize to distant body sites. Treatments based on blocking uPA binding to tumor cells are currently being designed and tested in clinical trials. At present, clinical studies are being performed in which uPA binding is being targeted as a potential therapy for solid tumors (Muehlenweg *et al.*, 2001). Given the demonstration that plasminogen binding, and not uPA binding, appears to be rate limiting for plasminogen activation associated with a number of different cells, studies directed toward blocking the critical plasminogen receptors should also be seriously considered.

It is obvious that the importance of each individual plasminogen receptor under varying cellular contexts needs to be clarified. While no single receptor can be responsible for the productive binding and activation of plasminogen on all cell surfaces, new evidence is illustrating very dramatic effects on plasminogen activation by manipulation of the expression levels of single receptors. This indicates that under certain conditions, one particular receptor may have a dominant influence on the overall regulation of plasmin production. Once clearly defined, these most important mechanisms of plasmin regulation will serve as novel therapeutic targets for many disorders including inflammation and tumor metastasis.

References

Adelman, B., Rizk, A., and Hanners, E. (1988). Plasminogen interactions with platelets in plasma. *Blood* **72**, 1530–1535.

Andronicos, N.M. and Ranson, M. (2001). The topology of plasminogen binding and activation on the surface of human breast cancer cells. *Br. J. Cancer* **85**, 909–916.

Andronicos, N.M., Ranson, M., Bognacki, J., and Baker, M.S. (1997). The human ENO1 gene product (recombinant human α-enolase) displays characteristics required for a plasminogen binding protein. *Biochim. Biophys. Acta Protein Struct. Mol. Enzymol.* **1337**, 27–39.

Arza, B., Félez, J., Lopez-Alemany, R., Miles, L.A., and Muñoz-Cánoves, P. (1997). Identification of an epitope of α-enolase (a candidate plasminogen receptor) by phage display. *Thromb. Haemost.* **78**, 1097–1103.

Bergman, A. C., Linder, C., Sakaguchi, K., Sten-Linder, M., Alaiya, A.A., Franzen, B., Shoshan, M.C., Bergman, T., Wiman, B., Auer, G., Appella, E., Jornvall, H., and Linder, S. (November 3, 1997). Increased expression of alpha-enolase in c-jun transformed rat fibroblasts without increased activation of plasminogen. *FEBS Lett.* **417**, 17–20.

Brockway, W.J. and Castellino, F.J. (1972). Measurement of the binding of antifibrinolytic amino acids to various plasminogens. *Arch. Biochem. Biophys.* **151**, 194–199.

Burtin, P., Chavanel, G., Andre-Bougaran, J., and Gentile, A. (1987). The plasmin system in human adenocarcinomas and their metastases. A comparative immunofluorescence study. *Int. J. Cancer* **39**, 170–178.

Burtin, P., Chavanel, G., and Andre, J. (1985). The plasmin system in human colonic tumors: An immunofluorescence study. *Int. J. Cancer* **35**, 307–314.

Camacho, M., Fondaneche, M.C., and Burtin, P. (March 13, 1989). Limited proteolysis of tumor cells increases their plasmin-binding ability *FEBS Lett.* **245**, 21–24.

Carroll, R.C., Radcliffe, R.D., Taylor, F.B., Jr., and Gerrard, J.M. (1982). Plasminogen, plasminogen activator, and platelets in the regulation of clot lysis. Reconstitution and patient studies. *J. Lab. Clin. Med.* **100**, 986–996.

Cesarman, G.M., Guevara, C.A., and Hajjar, K.A. (1994). An endothelial cell receptor for plasminogen/tissue plasminogen activator (t-PA). II. Annexin II-mediated enhancement of t-PA-dependent plasminogen activation. *J. Biol. Chem.* **269**, 21198–21203.

Chen, L.B., Murray, A., Segal, R.A., Bushnell, A., and Walsh, M.L. (1978). Studies on intercellular LETS glycoprotein matrices. *Cell* **14**, 377–391.

Choi, K.S., Fogg, D.K., Yoon, C.S., and Waisman, D.M. (2003). P11 regulates extracellular plasmin production and invasiveness of HT1080 fibrosarcoma cells. *FASEB J* 17, 235–246.

Chung, C.Y. and Erickson, H.P. (1994). Cell-surface annexin II is a high affinity receptor for the alternatively spliced segment of tenascin-C. *J. Cell Biol.* **126**, 539–548.

Drust, D.S. and Creutz, C.E. (1988). Aggregation of chromaffin granules by calpactin at micromolar levels of calcium. *Nature* **331**, 88–91.

Dudani, A.K., Cummings, C., Hashemi, S., and Ganz, P.R. (1993). Isolation of a novel 45 kDa plasminogen receptor from human endothelial cells. *Thromb. Res.* **69**, 185–196.

Dudani, A.K. and Ganz, P.R. (1996). Endothelial cell-surface actin serves as a binding site for plasminogen, tissue plasminogen activator and lipoprotein(a). *Br. J. Haematol.* **95**, 168–178.

Dudani, A.K., Hashemi, S., Aye, M.T., and Ganz, P.R. (1991). Identification of an endothelial cell-surface protein that binds plasminogen. *Mol. Cell Biochem.* **108**, 133–139.

Dudani, A.K., Pluskota, A., and Ganz, P.R. (1994). Interaction of tissue plasminogen activator with a human endothelial cell 45-kilodalton plasminogen receptor *Biochem. Cell Biol.* **72**, 126–131.
Duffy, M.J. (2001). Biochemical markers in breast cancer: Which ones are clinically useful? *Clin. Biochem.* **34**, 347–352.
Ellis, V., Behrendt, N., and Dano, K. (1991). Plasminogen activation by receptor-bound urokinase. A kinetic study with both cell-associated and isolated receptor. *J. Biol. Chem.* **266**, 12752–12758.
Ellis, V. and Dano, K. (1993). Potentiation of plasminogen activation by an anti-urokinase monoclonal antibody due to ternary complex formation. A mechanistic model for receptor-mediated plasminogen activation. *J. Biol. Chem.* **268**, 4806–4813.
Ellis, V., Scully, M.F., and Kakkar, V.V. (1987). Plasminogen activation by single-chain urokinase in functional isolation. A kinetic study. *J. Biol. Chem.* **262**, 14998–15003.
Ellis, V., Whawell, S.A., Werner, F., and Deadman, J.J. (January 12, 1999). Assembly of urokinase receptor-mediated plasminogen activation complexes involves direct, non-active-site interactions between urokinase and plasminogen [in process citation]. *Biochemistry* **38**, 651–659.
Falcone, D.J., Borth, W., McCaffrey, T.A., Mathew, J., and McAdam, K. (1994). Regulation of macrophage receptor-bound plasmin by autoproteolysis. *J. Biol. Chem.* **269**, 32660–32666.
Felez, J., Chanquia, C.J., Fabregas, P., Plow, E.F., and Miles, L.A. (1993). Competition between plasminogen and tissue plasminogen activator for cellular binding sites. *Blood* **82**, 2433–2441.
Felez, J., Miles, L.A., Fabregas, P., Jardi, M., Plow, E.F., and Lijnen, R.H. (1996). Characterization of cellular binding sites and interactive regions within reactants required for enhancement of plasminogen activation by tPA on the surface of leukocytic cells. *Thromb. Haemost.* **76**, 577–584.
Felez, J., Miles, L.A., Plescia, J., and Plow, E.F. (1990). Regulation of plasminogen receptor expression on human monocytes and monocytoid cell lines. *J. Cell Biol.* **111**, 1673–1683.
Fischer, A. (1925). Tissue culture. Studies in experimental morphology and general physiology of tissue cells *in vitro*. London. William Heinemann Ltd.
Fitzpatrick, S.L., Kassam, G., Choi, K.S., Kang, H.-M., Fogg, D.K., and Waisman, D.M. (2000). Regulation of plasmin activity by annexin II tetramer. *Biochemistry* **39**, 1021–1028.
Fogg, D.K., Bridges, D.E., Cheung, K.K., Kassam, G., Filipenko, N.R., Choi, K.S., Fitzpatrick, S.L., Nesheim, M., and Waisman, D.M. (April 16, 2002). The p11 subunit of annexin II heterotetramer is regulated by basic carboxypeptidase. *Biochemistry* **41**, 4953–4961.
Gerke, V. and Weber, K. (1984). Identity of p36K phosphorylated upon Rous sarcoma virus transformation with a protein purified from brush borders; calcium-dependent binding to non-erythroid spectrin and F-actin. *EMBO J.* **3**, 227–233.
Gong, Y., Kim, S.O., Felez, J., Grella, D.K., Castellino, F.J., and Miles, L.A. (2001). Conversion of Glu-plasminogen to Lys-plasminogen is necessary for optimal stimulation of plasminogen activation on the endothelial cell-surface. *J. Biol. Chem.* **276**, 19078–19083.
Gonzalez-Gronow, M., Edelberg, J.M., and Pizzo, S.V. (1989). Further characterization of the cellular plasminogen binding site: evidence that plasminogen 2 and lipoprotein a compete for the same site. *Biochemistry* **28**, 2374–2377.
Gyetko, M.R., Chen, G.H., McDonald, R.A., Goodman, R., Huffnagle, G.B., Wilkinson, C.C., Fuller, J.A., and Toews, G.B. (1996). Urokinase is required for the pulmonary inflammatory response to *Cryptococcus neoformans*—a murine transgenic model. *J. Clin. Invest.* **97**, 1818–1826.
Hajjar, K.A., Harpel, P.C., Jaffe, E.A., and Nachman, R.L. (1986). Binding of plasminogen to cultured human endothelial cells. *J. Biol. Chem.* **261**, 11656–11662.
Hajjar, K.A., Harpel, P.C., and Nachman, R.L. (1985). Binding of plasminogen to endothelial cell monolayers. *Blood* **66**, 356A.
Hajjar, K.A., Jacovina, A.T., and Chacko, J. (1994). An endothelial cell receptor for plasminogen/tissue plasminogen activator. I. Identity with annexin II. *J. Biol. Chem.* **269**, 21191–21197.
Hajjar, K.A., Mauri, L., Jacovina, A.T., Zhong, F., Mirza, U.A., Padovan, J.C., and Chait, B.T. (April 17, 1998). Tissue plasminogen activator binding to the annexin II tail domain. Direct modulation by homocysteine. *J. Biol. Chem.* **273**, 9987–9993.
Hall, S.W., Vandenberg, S.R., and Gonias, S.L. (1990). Plasminogen carbohydrate side chains in receptor binding and enzyme activation: A study of C6 glioma cells and primary cultures of rat hepatocytes *J. Cell Biochem.* **43**, 213–227.

Hawley, S.B., Green, M.A., and Miles, L.A. (2000). Discriminating between cell-surface and intracellular plasminogen-binding proteins: Heterogeneity in profibrinolytic plasminogen-binding proteins on monocytoid cells.[in process citation]. *Thromb. Haemost.* **84**, 882–890.

Hawley, S.B., Tamura, T., and Miles, L.A. (January 5, 2001). Purification, cloning, and characterization of a profibrinolytic plasminogen-binding protein, TIP49a. *J. Biol. Chem.* **276**, 179–186.

Hayes, M.L. and Castellino, F.J. (1979a). Carbohydrate of the human plasminogen variants. I. Carbohydrate composition, glycopeptide isolation, and characterization. *J. Biol. Chem.* **254**, 8768–8771.

Hayes, M.L. and Castellino, F.J. (1979b). Carbohydrate of the human plasminogen variants. II. Structure of the asparagine-linked oligosaccharide unit. *J. Biol. Chem.* **254**, 8772–8776.

Hayes, M.L. and Castellino, F.J. (1979c). Carbohydrate of the human plasminogen variants. III. Structure of the *O*-glycosidically linked oligosaccharide unit. *J. Biol. Chem.* **254**, 8777–8780.

Hembrough, T.A., Li, L., and Gonias, S.L. (1996). Cell-surface cytokeratin 8 is the major plasminogen receptor on breast cancer cells and is required for the accelerated activation of cell-associated plasminogen by tissue-type plasminogen activator. *J. Biol. Chem.* **271**, 25684–25691.

Henkin, J., Marcotte, P., and Yang, H.C. (1991). The plasminogen–plasmin system [Review]. *Prog. Cardiovasc. Dis.* **34**, 135–164.

Herren, T., Burke, T.A., Jardi, M., Felez, J., and Plow, E.F. (February 15, 2001). Regulation of plasminogen binding to neutrophils. *Blood* **97**, 1070–1078.

Hu, G.F., Chang, S.I., Riordan, J.F., and Vallee, B.L. (March 15, 1991). An angiogenin-binding protein from endothelial cells. *Proc. Natl. Acad. Sci. USA* **88**, 2227–2231.

Hu, G.F. and Riordan, J.F. (1993). Angiogenin enhances actin acceleration of plasminogen activation. *Biochem. Biophys. Res. Commun.* **197**, 682–687.

Jespersen, J., Gram, J., and Astrup, T. (1986). The autodigestion of human plasmin follows a bimolecular mode of reaction subject to product inhibition. *Thromb. Res.* **41**, 395–404.

Johnsson, N., Marriott, G., and Weber, K. (1988). p36, the major cytoplasmic substrate of src tyrosine protein kinase, binds to its p11 regulatory subunit via a short amino-terminal amphiphatic helix. *EMBO J.* **7**, 2435–2442.

Jones, P.A., Scott-Burden, T., and Gevers, W. (1979). Glycoprotein, elastin, and collagen secretion by rat smooth muscle cells. *Proc. Natl. Acad. Sci. USA* **76**, 353–357.

Kanalas, J.J. and Makker, S.P. (1991). Identification of rat heymann nephritis autoantigen as a receptor site for plasminogen. *J. Biol. Chem.* **266**, 10825–10829.

Kassam, G., Choi, K.S., Ghuman, J., Kang, H.M., Fitzpatrick, S.L., Zackson, T., Zackson, S., Toba, M., Shinomiya, A., and Waisman, D.M. (1998a). The role of annexin II tetramer in the activation of plasminogen. *J. Biol. Chem.* **273**, 4790–4799.

Kassam, G., Le, B.H., Choi, K.S., Kang, H.M., Fitzpatrick, S., Louie, P., and Waisman, D.M. (1998b). The p11 subunit of annexin II tetramer plays a key role in the stimulation of t-PA-dependent plasminogen activation. *Biochemistry* **37**, 16958–16966.

Kim, S.O., Plow, E.F., and Miles, L.A. (1996). Regulation of plasminogen receptor expression on monocytoid cells by beta1-integrin-dependent cellular adherence to extracellular matrix proteins. *J. Biol. Chem.* **271**, 23761–23767.

Li, X.N., Varma, V.K., Parks, J.M., Benza, R.L., Koons, J.C., Grammer, J.R., Grenett, H., Tabengwa, E.M., and Booyse, F.M. (1995). Thrombin decreases the urokinase receptor and surface-localized fibrinolysis in cultured endothelial cells. *Arterioscler. Thromb. Vasc. Biol.* **15**, 410–419.

Lijnen, H.R., Van Hoef, B., and Collen, D. (1981). On the role of the carbohydrate side chains of human plasminogen in its interaction with alpha 2-antiplasmin and fibrin. *Eur. J. Biochem.* **120**, 149–154.

Mai, J., Finley, R.L. Jr., Waisman, D.M., and Sloane, B.F. (2000). Human procathepsin B interacts with the annexin II tetramer on the surface of tumor cells. *J. Biol. Chem.* **275**, 12806–12812.

Mangel, W.F., Lin, B.H., and Ramakrishnan, V. (1990). Characterization of an extremely large, ligand-induced conformational change in plasminogen. *Science* **248**, 69–73.

Markus, G. (1996). Conformational changes in plasminogen, their effect on activation, and the agents that modulate activation rates—a review. *Fibrinolysis* **10**, 75–85.

Markus, G., DePasquale, J.L., and Wissler, F.C. (1978). Quantitative determination of the binding of epsilon-aminocaproic acid to native plasminogen. *J. Biol. Chem.* **253**, 727–732.

Miles, L.A., Dahlberg, C.M., Levin, E.G., and Plow, E.F. (1989). Gangliosides interact directly with plasminogen and urokinase and may mediate binding of these fibrinotylic components to cells. *Biochemistry* **28**, 9337–9343.

Miles, L.A., Dahlberg, C.M., Plescia, J., Felez, J., Kato, K., and Plow, E.F. (1991). Role of cell-surface lysines in plasminogen binding to cells: Identification of alpha-enolase as a candidate plasminogen receptor. *Biochemistry* **30**, 1682–1691.

Miles, L.A., Dahlberg, C.M., and Plow, E.F. (1988a). The cell-binding domains of plasminogen and their function in plasma. *J. Biol. Chem.* **263**, 11928–11934.

Miles, L.A., Ginsberg, M.H., White, J.G., and Plow, E.F. (1986). Plasminogen interacts with human platelets through two distinct mechanisms. *J. Clin. Invest.* **77**, 2001–2009.

Miles, L.A., Levin, E.G., Plescia, J., Collen, D., and Plow, E.F. (1988b). Plasminogen receptors, urokinase receptors, and their modulation on human endothelial cells. *Blood* **72**, 628–635.

Miles, L.A. and Plow, E.F. (1985). Binding and activation of plasminogen on the platelet surface. *J. Biol. Chem.* **260**, 4303–4311.

Miles, L.A. and Plow, E.F. (1986a). Localization of fibrinolytic proteins on peripheral blood cells and endothelial cells. *Circulation* Suppl. **74**, 246 (Abstract).

Miles, L.A. and Plow, E.F. (1986b). Topography of the high-affinity lysine binding site of plasminogen as defined with a specific antibody probe. *Biochemistry* **25**, 6926–6933.

Miles, L.A. and Plow, E.F. (1987). Receptor mediated binding of the fibrinolytic components, plasminogen and urokinase, to peripheral blood cells. *Thromb. Haemost.* **58**, 936–942.

Miles, L.A. and Plow, E.F. (1988). Plasminogen receptors: Ubiquitous sites for cellular regulation of fibrinolysis. *Fibrinolysis* **2**, 61–71.

Moroianu, J., Fett, J.W., Riordan, J.F., and Vallee, B.L. (May 1, 1993). Actin is a surface component of calf pulmonary artery endothelial cells in culture. *Proc. Natl. Acad. Sci. USA* **90**, 3815–3819.

Moroz, L.A. (1981). Mini-plasminogen: A mechanism for leukocyte modulation of plasminogen activation by urokinase. *Blood* **58**, 97–104.

Morse, E.E., Jackson, D.P., and Conley, C.L. (1965). Role of platelet fibrinogen in the reactions of platelets to thrombin. *J. Clin. Invest.* **44**, 809–816.

Muehlenweg, B., Sperl, S., Magdolen, V., Schmitt, M., and Harbeck, N. (2001). Interference with the urokinase plasminogen activator system: A promising therapy concept for solid tumours. *Expert. Opin. Biol. Ther.* **1**, 683–691.

Namiranian, S., Naito, Y., Kakkar, V.V., and Scully, M.F. (1995). Bound plasminogen is rate-limiting for cell-surface-mediated activation of plasminogen by urokinase. *Biochem. J.* **309**, 977–982.

Nilius, B., Gerke, V., Prenen, J., Szücs, G., Heinke, S., Weber, K., and Droogmans, G. (1996). Annexin II modulates volume-activated chloride currents in vascular endothelial cells. *J. Biol. Chem.* **271**, 30631 30636.

O'Mullane, M.J. and Baker, M.S. (July 10, 1998). Loss of cell viability dramatically elevates cell-surface plasminogen binding and activation. *Exp. Cell Res.* **242**, 153–164.

O'Mullane, M.J. and Baker, M.S. (1999). Elevated plasminogen receptor expression occurs as a degradative phase event in cellular apoptosis. *Immunol. Cell Biol.* **77**, 249–255.

Owen, M.J., Auger, J., Barber, B.H., Edwards, A.J., Walsh, F.S., and Crumpton, M.J. (1978). Actin may be present on the lymphocyte surface. *Proc. Natl. Acad. Sci. USA* **75**, 4484–4488.

Pardridge, W.M., Nowlin, D.M., Choi, T.B., Yang, J., Calaycay, J., and Shively, J.E. (1989). Brain capillary 46,000 dalton protein is cytoplasmic actin and is localized to endothelial plasma membrane *J. Cereb. Blood Flow Metab.* **9**, 675–680.

Parkkinen, J. and Rauvala, H. (1991). Interactions of plasminogen and tissue plasminogen activator (t-PA) with amphoterin. Enhancement of t-PA-catalyzed plasminogen activation by amphoterin. *J. Biol. Chem.* **266**, 16730–16735.

Plow, E.F., Freaney, D.E., Plescia, J., and Miles, L.A. (1986). The plasminogen system and cell-surfaces: Evidence for plasminogen and urokinase receptors on the same cell type. *J. Cell Biol.* **103**, 2411–2420.

Plow, E.F. and Miles, L.A. (1990). Plasminogen receptors in the mediation of pericellular proteolysis [Review]. *Cell Differ. Dev.* **32**, 293–298.

Ranson, M., Andronicos, N.M., O'Mullane, M.J., and Baker, M.S. (1998). Increased plasminogen binding is associated with metastatic breast cancer cells: Differential expression of plasminogen binding proteins. *Br. J. Cancer* **77**, 1586–1597.

Redlitz, A., Fowler, B.J., Plow, E.F., and Miles, L.A. (1995). The role of an enolase-related molecule in plasminogen binding to cells. *Eur. J. Biochem.* **227**, 407–415.

Rehli, M., Krause, S.W., Kreutz, M., and Andreesen, R. (June 30, 1995). Carboxypeptidase M is identical to the MAX.1 antigen and its expression is associated with monocyte to macrophage differentiation. *J. Biol. Chem.* **270**, 15644–15649.

Reinartz, J., Batrla, R., Boukamp, P., Fusenig, N., and Kramer, M.D. (1993). Binding and activation of plasminogen at the surface of human keratinocytes. *Exp. Cell Res.* **208**, 197–208.

Sanders, S.K. and Craig, S.W. (1983). A lymphocyte cell-surface molecule that is antigenically related to actin. *J. Immunol.* **131**, 370–377.

Shapiro, R., Strydom, D.J., Olson, K.A., and Vallee, B.L. (August 11, 1987). Isolation of angiogenin from normal human plasma. *Biochemistry* **26**, 5141–5146.

Siever, D.A. and Erickson, H.P. (1997). Extracellular annexin II. *Int. J. Biochem. Cell Biol.* **29**, 1219–1223.

Sottrup-Jensen, L., Claeys, H., Zajdel, M., Petersen, T.E., and Magnusson, S. (1978). Progress in chemical fibrinolysis and thrombosis. Raven Press, New York (in) Davidson J.F., Rowan R.M., Samama M.M., Desnoyers P.C. (eds) **3**, 191–209.

Takada, Y., Makino, Y., and Takada, A. (1985). Glu-plasminogen I and II: Their activation by urokinase and streptokinase in the presence of fibrin and fibrinogen. *Thromb. Res.* **39**, 289–296.

Taylor, F.B., Jr., Carroll, R.C., Gerrard, J., Esmon, C.T., and Radcliffe, R.D. (May 15, 1981). Lysis of clots prepared from whole blood and plasma. *Fed. Proc.* **40**, 2092–2098.

Thorson, S., Brakman, F., and Astrup, T. (1972). Influence of platelets on fibrinolysis: A critical review. *Hematol. Rev.* **3**, 123–179.

Waisman, D.M. (1995). Annexin II tetramer: Structure and function. *Mol. Cell Biochem.* **149/150**, 301–322.

Wright, J.F., Kurosky, A., and Wasi, S. (1994). An endothelial cell-surface form of annexin II binds human cytomegalovirus. *Biochem. Biophys. Res. Commun.* **198**, 983–989.

Yeatman, T.J., Updyke, T.V., Kaetzel, M.A., Dedman, J.R., and Nicolson, G.L. (1993). Expression of annexins on the surfaces of non-metastatic and metastatic human and rodent tumor cells. *Clin. Exp. Metastasis* **11**, 37–44.

6

The Role of Lys-Plasminogen in Cell-Mediated Plasmin Production

Lindsey A. Miles, Nicholas M. Andronicos, Jordi Felez, Davida K. Grella, Francis J. Castellino, and Yun Gong

1. Introduction

A key control point by which the activity of plasmin is positively regulated is by localizing Glu-plasminogen (the native circulating form of the plasminogen molecule) to specific binding sites on cell surfaces. Glu-plasminogen is activated much more efficiently when bound to cells than when in solution. Upon activation, plasmin remains associated with the cell surface where plasmin activity is relatively protected from inhibitors. Furthermore, the enzymatic activity of plasmin is enhanced in the milieu of the cell surface.

Localization of plasmin activity on cells generates a local nidus of protected proteolytic activity, which has distinct functions on different cell types. On endothelial cells, plasmin becomes localized to sites of thrombus formation where plasmin proteolytic activity can contribute to fibrin clot lysis. On cancer cells and leukocytes, the association of plasmin with these cell surfaces arms the cells with the proteolytic activity of plasmin required in processes in which cells must degrade extracellular matrices in order to migrate. Association of plasmin with neuroendocrine cells enhances processing of prohormones (Parmer *et al.*, 2000). Plasmin bound to bacterial plasmin(ogen) receptors serve to enhance the invasive potential of bacteria by increasing penetration of extracellular matrices, basement membranes, and tissues (Boyle and Lottenberg, 2000).

The molecular identity of plasminogen binding sites is also cell-type dependent. Although plasminogen can interact with both protein (Miles *et al.*, 1991) and non-protein (Miles *et al.*, 1989) cell-surface components, proteins exposing carboxyl-terminal lysines on cell surfaces are responsible predominantly for the ability of cells to enhance plasminogen activation (Felez *et al.*, 1996). Cell-surface plasminogen binding proteins synthesized with carboxyl-terminal lysines include TIP49a on monocytoid cells

---

Lindsey A. Miles, Nicholas M. Andronicos, and Yun Gong • The Scripps Research Institute, Department of Cell Biology, Division of Vascular Biology, La Jolla, California 92037. **Jordi Felez** • Institut de Recerca Oncologica, 08907, Barcelona, Spain. **Francis J. Castellino and Davida K. Grella** • The University of Notre Dame, Notre Dame, Indiana.

Plasminogen: Structure, Activation, and Regulation, edited by David M. Waisman. Kluwer Academic/Plenum Publishers, New York, 2003.

(Hawley *et al.*, 2001) and α-enolase on monocytes, (Miles *et al.*, 1991; Redlitz *et al.*, 1995), neurons (Nakajima *et al.*, 1994), carcinoma cells (Lopez-Alemany *et al.*, 1994), and pathogenic streptococci (Pancholi and Fischetti, 1998). Glyceraldehyde-3-phosphate dehydrogenase binds plasminogen on the surface of bacteria (Winram and Lottenberg, 1996) and cytokeratin 8 is a plasminogen binding protein present on the surfaces of hepatocellular and breast carcinoma cells (Hembrough *et al.*, 1995). Other plasminogen binding proteins including annexin II on endothelial cells (Hajjar *et al.*, 1994) and actin on endothelial cells (Dudani and Ganz, 1996), carcinoma cells (Andronicos and Ranson, 2001) and catecholaminergic cells (Miles and Parmer, 2001), amphoterin on various transformed cell lines (Parkkinen *et al.*, 1993), and αIIbβ3 on platelets (Miles *et al.*, 1986) appear to require post-translational processing to generate carboxyl-terminal lysines in order to interact with plasminogen. Most of these cell-surface plasminogen binding proteins are "moonlighting" proteins, members of a growing list of proteins that are recognized as identical gene products exhibiting multiple functions at distinct cellular and extracellular sites through "gene sharing" (Wistow and Piatigorsky, 1987; Jeffery, 1999). With the exception of αIIbβ3, the DNA sequences encoding the previously identified cell-surface plasminogen binding proteins do not predict classical signal sequences. Thus, other mechanisms for release and subsequent membrane localization of these plasminogen binding proteins or direct translocation of these proteins to the cell membrane must exist (Muesch *et al.*, 1990).

In this review, we focus on mechanisms by which cell-surface Glu-plasminogen activation is enhanced with emphasis on our recent studies (Gong *et al.*, 2001) that distinguished between two potential mechanisms as major pathways for stimulation of plasminogen activation on the cell surface: (a) promotion of the direct activation of cell-associated Glu-plasminogen by plasminogen activators; or (b) conversion of Glu-plasminogen to Lys-plasminogen and subsequent activation of Lys-plasminogen by plasminogen activators. [Plasmin catalyzes the cleavage of Glu-plasminogen (Wiman, 1973; Wiman and Wallen, 1975; Violand and Castellino, 1976; Horrevoets *et al.*, 1995) to generate new amino-termini of plasminogen, resulting in plasminogen molecular forms designated as "Lys-plasminogen." *Lys-plasminogen is more readily activated by plasminogen activators* (Markus *et al.*, 1978a, 1979; Hoylaerts *et al.*, 1982).] In the first proposed mechanism, Glu-plasminogen remains uncleaved, yet its direct activation is promoted on the cell surface relative to the solution phase, perhaps through conformational changes induced in the molecule upon its interaction with the cell surface. In the alternative mechanism, conversion of Glu-plasminogen by plasmin to yield the more readily activated Lys-plasminogen is necessary so that Lys-plasminogen becomes the major substrate for plasminogen activators on the cell surface. In addition, we summarize recent data (directly related to the mechanisms above) demonstrating a previously unrecognized pathway by which cells modulate plasminogen activation: promotion of cleavage of cell-associated Glu-plasminogen to Lys-plasminogen by plasmin (Gong *et al.*, 2001).

2. Key Differences between Glu-Plasminogen and Lys-Plasminogen

Plasminogen is secreted constitutively [primarily by hepatocytes (Bohmfalk and Fuller, 1980; Saito *et al.*, 1980; Zhang *et al.*, 2002) with amino-terminal glutamic acid

(Glu-plasminogen)]. Lys-plasminogen is formed when plasmin cleaves Glu-plasminogen at the carboxyl sides of K^{62}, R^{68}, K^{77} (Wiman, 1973; Wiman and Wallen, 1975; Violand and Castellino, 1976), and at additional minor sites (Horrevoets *et al.*, 1995) to remove the amino-terminal peptide (NTP) and to generate new amino-termini resulting in plasminogen molecular forms that are collectively termed "Lys-plasminogen."

There are clear differences in the conformations of Glu-plasminogen and Lys-plasminogen. The crystal structure of Glu-plasminogen has not been solved. However, a combination of physical techniques as well as biochemical studies have advanced the hypothesis that Glu-plasminogen exists in a closed right-handed spiral conformation. Electron microscopy of Glu-plasminogen demonstrates that it has a right-handed spiral conformation with a diameter of 9–11 nm (Weisel *et al.*, 1994). Small angle, electron scattering experiments also suggest that the structure of Glu-plasminogen in solution is compact. Glu-plasminogen has a radius of gyration of 3.05 nm and a maximum intramolecular distance of 9.1 nm (Ponting *et al.*, 1992), which is comparable to electron microscopy data (Weisel *et al.*, 1994). However, the precise dimensions of the zymogen remain controversial since Mangel and coworkers (Mangel *et al.*, 1990) using neutron scattering techniques, described Glu-plasminogen as a prolate ellipsoid with radius of gyration and the maximum intramolecular distance of 3.39 and 15 nm, respectively. Chemical cross-linking analysis of Glu-plasminogen demonstrates that Lys_{204} of K2 and Tyr_{672} of the C-terminal protease domain are approximately 3 Å from each other in the protein (Banyai and Patty, 1984). Taken together these studies suggest that Glu-plasminogen exists in a tight right-hand spiral, known as the closed conformation.

In contrast, Lys-plasminogen, exhibits a more open conformation than Glu-plasminogen. Electron microscopy of Lys-plasminogen demonstrates that it exists in a U-shaped conformation with approximate dimensions of 14 nm × 7 nm (Weisel *et al.*, 1994). This open conformation has also been described as a Debye random coil with a radius of gyration of 56 Å (Ramakrishnan *et al.*, 1991). The N-terminal fragment of Lys-plasminogen (K1–3) is described equally well by either a Debye random coil or an elongated, prolate ellipsoid, shapes with a radius of gyration of 29 Å (Ramakrishnan *et al.*, 1990). This more open conformation of Lys-plasminogen is also associated with an increase in the intrinsic fluorescence of the protein (Markus *et al.*, 1979) and by differences in sedimentation equilibria (Violand *et al.*, 1975, Markus *et al.*, 1978b, 1979). The secondary structures of both Glu- and Lys-plasminogen are very similar (Sjoholm, 1973; Misselwitz *et al.*, 1994), suggesting that a negligible change in secondary structure is associated with a large tertiary conformational change. Thus, the spatial relationship of the individual domains with each other is different between the closed and the open conformations of plasminogen, while the structures of the domains themselves remain constant. Taken together, these studies suggest that the NTP domain may act as a conformational determinant of Glu-plasminogen.

The ability to detect Lys-plasmin(ogen) *in vivo* is dependent upon the local site being evaluated. Using an anti-Lys-plasminogen antibody to examine plasma of patients undergoing thrombolytic therapy, only 3 of 7 patients exhibited evidence of formation of complexes of Lys-plasmin with α_2-antiplasmin (Holvoet *et al.*, 1985). In contrast, Fredenburgh and Nesheim demonstrated that Lys-plasminogen was formed and activated in plasma clots formed *in vitro*, raising the possibility that Lys-plasminogen may be the preferred substrate for tissue plasminogen activator (t-PA) in the environment of the clot.

Lys-plasminogen is detected when activation of Glu-plasminogen is analyzed on U937 cells (Ellis *et al.*, 1991). Furthermore, eluates of umbilical cord vessels contain Lys-plasminogen (Hajjar and Nachman, 1988), further suggesting that Lys-plasminogen is produced and activated at local sites *in vivo*.

The key difference between Glu-plasminogen and Lys-plasminogen, from the standpoint of understanding plasminogen activation on the cell surface, is that Lys-plasminogen is more readily activated by plasminogen activators than is Glu-plasminogen (Claeys and Vermylen, 1974; Thorsen and Astrup, 1974; Thorsen *et al.*, 1974; Thorsen and Mullertz, 1974; Hoylaerts *et al.*, 1982; Fredenburgh and Nesheim, 1992). When t-PA is used as activator in solution, the catalytic efficiency of Lys-plasminogen activation is increased 11-fold over that of Glu-plasminogen (Hoylaerts *et al.*, 1982; Fredenburgh and Nesheim, 1992). Similarly, with urokinase-type plasminogen activator (u-PA) the catalytic efficiency of Lys-plasminogen activation is increased ~9-fold (Ellis *et al.*, 1991).

Of particular relevance to Glu-plasminogen activation on cell surfaces, Glu-plasminogen acquires both the conformation and increased activatability of Lys-plasminogen in the presence of lysine analogs (Markus *et al.*, 1979). The weak lysine binding sites in plasminogen kringle 4 and kringle 5 modulate the conformational change of Glu-plasminogen to a Lys-plasminogen-like conformation, without removal of the NTP (Christensen and Molgaard, 1992). These kringle domains (as well as the kringle 1–3 domain) also appear to modulate the interactions of plasminogen with cells (Miles *et al.*, 1988). Thus, one possibility is that lysine binding sites within the kringle domains of Glu-plasminogen may interact with carboxyl-terminal lysines on cell surfaces to induce conformational changes in Glu-plasminogen that mimic the conformation of Lys-plasminogen without requiring cleavage and removal of the NTP. These considerations were the bases for postulating mechanism 1 (promotion of the direct activation of cell-associated Glu-plasminogen by plasminogen activators) as a key pathway by which activation of cell-associated plasminogen is stimulated.

3. Mechanisms by which Plasminogen Activation is Enhanced on the Cell Surface

When Glu-plasminogen is bound to cells, its activation is markedly enhanced, compared to the reaction in solution, predominantly due to a marked reduction in the Km for Glu-plasminogen activation (Miles and Plow, 1985; Hajjar *et al.*, 1986; Stricker *et al.*, 1986; Loscalzo and Vaughan, 1987; Ellis *et al.*, 1991; Duval-Jobe and Parmely, 1994; Félez *et al.*, 1996; Longstaff *et al.*, 1999; Sinniger *et al.*, 1999). The majority of cells that bind plasminogen also express cell-surface receptors for plasminogen activators, for u-PA (u-PAR) and for t-PA. Colocalization of activator and plasminogen serves to increase the local concentration of reactants, consistent with the decrease in Km for Glu-plasminogen activation on cell surfaces. A clear requirement for the interaction of plasminogen with the cell surface in the enhancement of plasminogen activation has been demonstrated in studies with lysine analogs, which interact with the plasminogen kringles to block the binding of plasminogen to the cells. In reciprocal experiments, inactivation of cellular binding sites by treatment of cell surfaces with carboxypeptidase B, abrogates the enhancement in plasminogen activation on cell surfaces (Félez *et al.*, 1996).

Other mechanisms by which plasminogen activation is promoted on cell surfaces have been delineated in previous studies. The interactions of both plasminogen and the zymogen, single chain u-PA (scu-PA), with cells enhances cleavage of scu-PA to its active form to promote further plasminogen activation (Ellis *et al.*, 1989). Interaction of scu-PA with the u-PAR also results in increased activity of the zymogen itself, without a requirement for cleavage (Manchanda and Schwartz, 1991). Treatment of some cell surfaces with plasmin also serves to generate new plasminogen binding sites for stimulation of plasminogen activation (Gonzalez-Gronow *et al.*, 1991).

3.1. Role of the Glu-Plasminogen to Lys-Plasminogen Conversion in Plasminogen Activation by t-PA on the Cell Surface

Previously, using kinetic assays, it was not possible to distinguish between two potential mechanisms by which plasmin formation is enhanced on the cell surface compared to the soluble phase: either (a) direct activation of Glu-plasminogen forms by plasminogen activators; or (b) initial cleavage of Glu-plasminogen to Lys-plasminogen by plasmin, resulting in Lys-plasminogen (the more readily activated form) becoming the predominant substrate for plasminogen activators on the cell surface. In these earlier studies, cell-associated Glu-plasminogen was converted to Lys-plasminogen (~50–60% conversion) on the surfaces of both human umbilical vein endothelial cells (HUVEC) (Hajjar and Nachman, 1988) and U937 monocytoid cells (Ellis *et al.*, 1991) without the addition of either plasminogen activators or plasmin. Thus, it was not possible to determine whether the *direct* activation of Glu-plasminogen was enhanced on the cell surface. In contrast, in our recent studies, Glu-plasminogen remained in its native form, without proteolytic cleavage to Lys-plasminogen when bound to HUVEC (Gong *et al.*, 2001). This enabled us to assess the role of conversion of Glu-plasminogen to Lys-plasminogen in enhancement of Glu-plasminogen activation on the cell surface. We used a plasminogen recombinant variant, [D(646)E]Glu-Pg, for these studies. This recombinant plasminogen/plasmin variant can be converted to the molecular form of plasmin, but does not exhibit significant plasmin proteolytic activity due to the absence of the necessary D residue in the serine proteolytic catalytic triad. Hence, when [D(646)E]Glu-Pg is incubated with plasminogen activators, neither [D(646)E]Lys-Pg nor [D(646)E]Lys-Pm is formed (Grella and Castellino, 1997). We compared the rates of activation of Glu-plasminogen, Lys-plasminogen, and [D(646)E]Glu-Pg by either t-PA or u-PA when these plasminogen forms were either bound to HUVEC, or in solution (Gong *et al.*, 2001). To further distinguish between potential mechanisms for enhancement of Glu-plasminogen activation, we compared the results obtained with HUVEC with results using U937 cells. These latter experiments provided an opportunity to compare results under conditions where cell-associated Glu-plasminogen and [D(646)E]Glu-plasminogen did exhibit conversion to their "Lys"-forms upon binding to the U937 cells.

To monitor the activation of ^{125}I-Glu-plasminogen (^{125}I-Glu-Pg), ^{125}I-[D(646)E] Glu-Pg, and ^{125}I-Lys-plasminogen (^{125}I-Lys-Pg) by t-PA, sodium dodecyl sulfate polyacrylamide gel electrophoresis (SDS-PAGE) was employed because kinetic assays could not be performed on [D(646)E]Glu-Pg since its plasmin form is proteolytically inactive. The cell-bound ligand was recovered and subjected to SDS-PAGE under reducing conditions, which distinguish native Glu-plasminogen from Lys-plasminogen and distinguish the heavy chains of Glu- and Lys-plasmin (Figure 6.1). We found that in the absence of

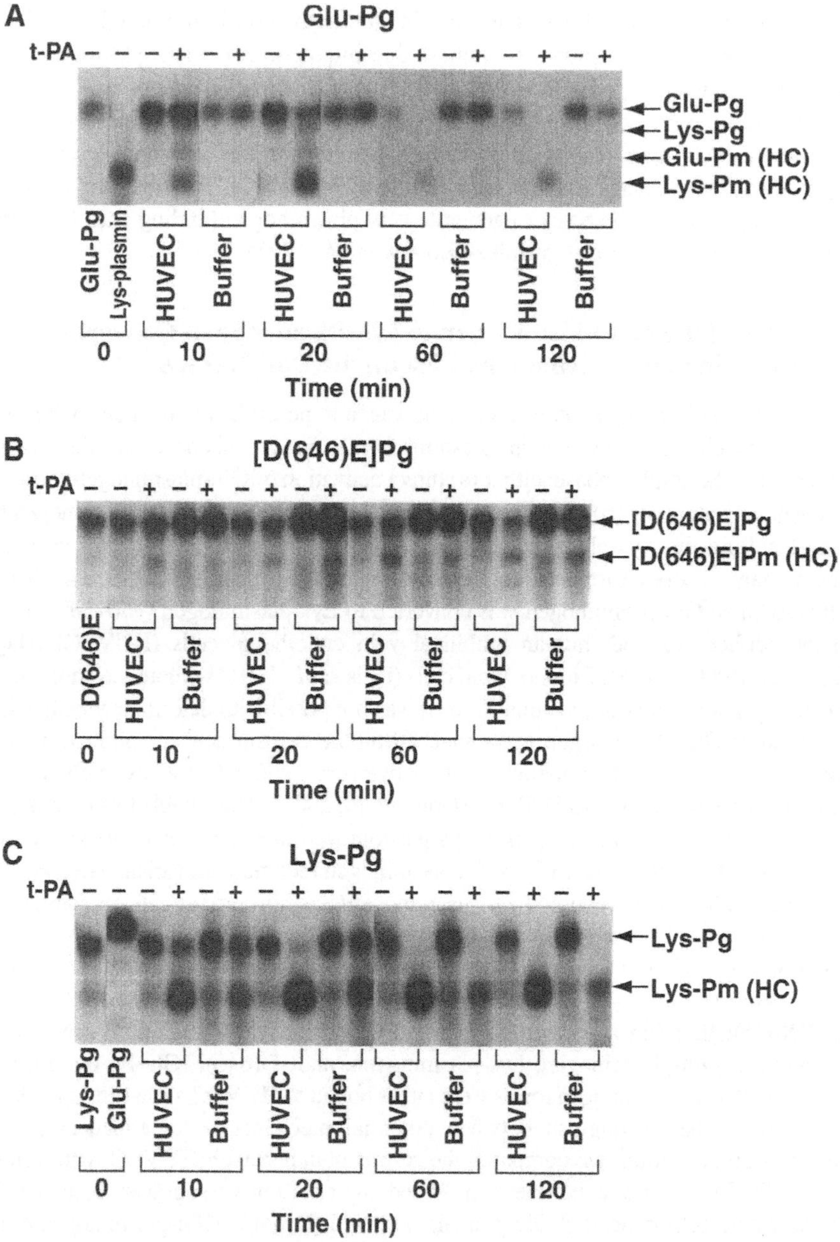

Figure 6.1. Role of the Glu-plasminogen to Lys-plasminogen conversion in plasminogen activation on the cell surface. A concentration of 25 nM of either ^{125}I-Glu-Pg (Panel A), ^{125}I-[D(646)E]Glu-Pg (Panel B), or ^{125}I-Lys-Pg (Panel C) was incubated with buffer (HBSS-0.4% BSA) or HUVEC (5×10^4) in the presence or absence of 20 nM t-PA for the indicated times at 37°C. Cell-bound ligand was eluted and greater than 90% of the cell-bound ligand was recovered by the elution procedure. The cell-bound ligand was subjected to electrophoresis on 7% SDS-PAGE under reducing conditions (from Yun Gong, Sun-OK Kim, Jordi Felez, Davida K. Grella, Francis J. Castellino, and Lindsey A. Miles. Conversion of Glu-plasminogen to Lys-plasminogen is necessary for optimal stimulation of plasminogen activation on the endothelial cell surface, *Journal of Biological Chemistry*, May 2001, 276: 19078–19083, reprinted with permission).

added t-PA, >95% of the added ^{125}I-Glu-Pg and ^{125}I-[D(646)E]Glu-Pg was present as Glu-plasminogen on the cell surface and plasminogen activation was not detected on these cells (Figure 6.1). Activation of the three plasminogen forms by the addition of 20 nM t-PA was compared when the ligands were bound to HUVEC with the activation reaction in the solution phase, in the absence of cells. The activation of ^{125}I-Glu-Pg was enhanced ~13-fold at 10 min compared to the reaction in solution (Figure 6.1, panel A and Figure 6.2) (Gong *et al.*, 2001). The predominant form of the plasmin heavy chain corresponded to the Lys-plasmin heavy chain consistent with the presence of plasmin activity (Figure 6.1, Panel A). Following activation with t-PA, both Glu-plasminogen and Lys-plasminogen were present on the cell surface. *In contrast, Lys-plasminogen formation was not detected when the reaction was carried out in solution.*

From the foregoing analyses, as in previous studies with plasminogen activation by u-PA on U937 cell (Ellis *et al.*, 1991), it was not possible to determine whether Glu-plasminogen or Lys-plasminogen was the predominant t-PA substrate. To address this issue, we analyzed cleavage of ^{125}I-[D(646)E]Glu-Pg by t-PA. Cleavage of cell-associated ^{125}I-[D(646)E]Glu-Pg by t-PA was only 2-fold greater than in the absence of cells at 10 min (Figure 6.1, panel B and Figure 6.2) (Gong *et al.*, 2001). The predominant form of the plasmin heavy chain was the ^{125}I-[D(646)E]Glu-plasmin form and no conversion of ^{125}I-[D(646)E]Glu-Pg to ^{125}I-[D(646)E]Lys-Pg was observed on the cell surface, consistent with the absence of plasmin. Notably, the extent of cleavage of cell-bound ^{125}I-[D(646)E]Glu-Pg was markedly less than when ^{125}I Glu Pg was bound to the cells (Figures 6.1 and 6.2). In contrast, activation of cell-associated ^{125}I-Lys-Pg by t-PA was

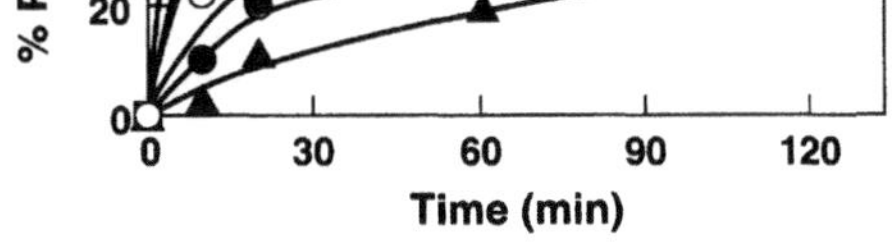

Figure 6.2. Comparison of cleavage of plasminogen forms by t-PA on HUVEC and in solution. The autoradiograms in Figure 6.1 were scanned. The % plasmin formation was calculated by dividing the sum of the densities of the Glu-plasmin heavy chain and Lys-plasmin heavy chain bands by the sum of the densities of the Glu-Pg and Lys-Pg bands and the Glu-plasmin and Lys-plasmin heavy chain bands. (The light chain of plasmin does not incorporate ^{125}I in proportion to the heavy chain and was not used in the calculation of 100% cell-associated ligand). The % plasmin in the starting material was subtracted. Symbols are as follows: ▲ ^{125}I-Glu-Pg + t-PA; △ ^{125}I-Glu-Pg + t-PA + HUVEC; ● ^{125}I-[D(646)E)Glu-Pg + t-PA; ○ ^{125}I-[D(646)E)Glu-Pg + t-PA HUVEC; ■ ^{125}I-Lys-Pg + t-PA; □ ^{125}I-Lys-Pg + t-PA + HUVEC (from Yun Gong, Sun-OK Kim, Jordi Felez, Davida K. Grella, Francis J. Castellino, and Lindsey A. Miles. Conversion of Glu-plasminogen to Lys-plasminogen is necessary for optimal stimulation of plasminogen activation on the endothelial cell surface, *Journal of Biological Chemistry*, May 2001, 276: 19078–19083, reprinted with permission).

similar to that of ^{125}I-Glu-Pg (Figure 6.1, Panels A and C and Figure 6.2). However, at 10 min, activation of Lys-plasminogen bound to HUVEC was only 1.3-fold greater than the reaction in solution. The cells provided only this small enhancement of Lys-plasminogen activation by t-PA, compared to the solution phase, because Lys-plasminogen was activated more rapidly in solution than was Glu-plasminogen. (At 10 min, 51% of the Lys-plasminogen in solution was Lys-plasmin while only 3% of the Glu-plasminogen in solution was activated to plasmin).

We ruled out the possibility that the increase in plasmin associated with the cells could be caused by either a higher affinity or capacity of the cells for plasmin versus plasminogen. The numbers of molecules of cell-bound ^{125}I-Glu-plasmin(ogen) did not increase in the presence, compared to the absence of t-PA (Table 6.1). In addition, in the presence of t-PA, the number of molecules of ligand bound to the cell was not increased when cell-bound ^{125}I-Glu-Pg was activated by t-PA compared to treatment of cell-bound ^{125}I-[D(646)E]Glu-Pg with t-PA. These data suggest that the increase in plasmin on the cell surface could not be accounted for by enhanced binding of plasmin to the cells.

To further explore the role of the Glu-plasminogen to Lys-plasminogen conversion in plasminogen activation, we examined plasminogen activation on U937 cells, under conditions where localization of Glu-plasminogen on the cell surface does result in conversion to Lys-plasminogen. In an earlier study by Ellis and coworkers, addition of either Glu-plasminogen or an active site mutant plasminogen ([S(740)A]Pg) to U937 cells resulted in conversion of these ligands to their "Lys-plasminogen" forms (Ellis *et al.*, 1991). Consistent with the results of Ellis and coworkers, and in contrast to our data with endothelial cells, 67% conversion of Glu-plasminogen to either Lys-plasminogen or to Lys-plasmin and 64% conversion of [D(646)E]Pg to its Lys-plasminogen and Lys-plasmin forms occurred upon binding to the U937 cells (data not shown). Under these conditions, the extent of activation of [D(646)E]Pg to its two chain (inactive plasmin) form was similar to that of Glu-plasminogen (Figure 6.3). *Thus, under conditions where cleavage of*

Table 6.1. Effect of t-PA on Plasminogen Binding to HUVEC

	Plasmin bound (Molecules/cell $\times 10^5$)	
Added radiolabeled ligand	Buffer	PA
Glu-plasminogen	1.01 ± 0.03	1.04 ± 0.16
[D(646)E]Glu-Pg	1.02 ± 0.10	1.20 ± 0.06
Lys-plasminogen	1.93 ± 0.70	2.09 ± 0.60

Confluent HUVEC in wells of 24-well culture dishes were incubated with 25 nM each of either ^{125}I-Glu-plasminogen, ^{125}I-[D(646)E]Glu-Pg, or ^{125}I-Lys-plasminogen in either the presence or absence of 20 nM t-PA for 120 min at 37°C. Reactions were terminated by aspirating the unbound radiolabeled ligands, rapidly washing the culture twice with HBSS-BSA, and extracting the cell-bound ligands. Values are mean ± S.E.M. of two experiments (from Yun Gong, Sun-OK Kim, Jordi Felez, Davida K. Grella, Francis J. Castellino, and Lindsey A. Miles. Conversion of Glu-plasminogen to Lys-plasminogen is necessary for optimal stimulation of plasminogen activation on the endothelial cell surface, *Journal of Biological Chemistry*, May 2001, 276: 19078–19083, reprinted with permission).

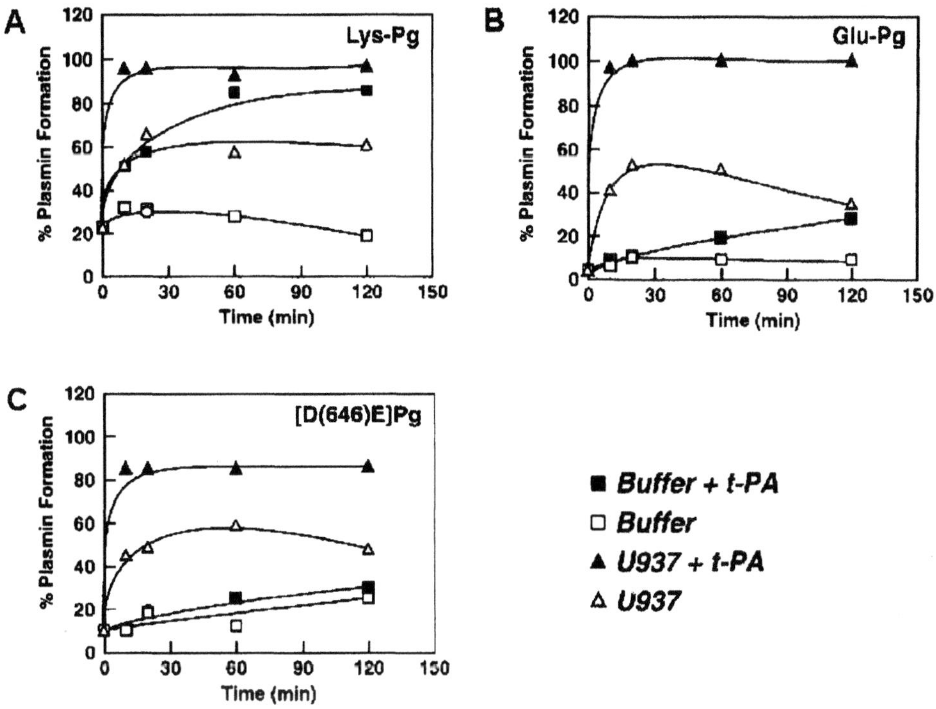

Figure 6.3. Comparison of cleavage of plasminogen forms by t-PA on U937 cells and in solution. A concentration of 25 nM of either ^{125}I-Lys-Pg (Panel A), ^{125}I-Glu-Pg (Panel B), ^{125}I-[D(646)E]Glu-Pg (Panel C), was incubated with buffer (HBSS–0.4% BSA) or HUVEC (5×10^4) in the presence or absence of 20 nM t-PA for the indicated times at 37°C. Cell-bound ligand was subjected to electrophoresis on 7% SDS PAGE under reducing conditions.

[D(646)E]Pg to its Lys-form did occur, its activation by t-PA was markedly promoted on the cell surface. With all three forms of plasminogen, some activation was observed in the absence of added t-PA, due to the secretion of u-PA by these cells. In the case of Lys-plasminogen, the extent of cleavage in the presence of cells alone added to the extent of activation by t-PA in the absence of cells, was equivalent to the total cleavage observed in the presence of cells plus t-PA, suggesting, that, as with the endothelial cells, the interaction of Lys-plasminogen with the cells did not markedly enhance Lys-Pg activation, compared to the reaction in the soluble phase. However, cleavage of both Glu-Pg and [D(646)E]Pg by t-PA was markedly enhanced in the presence of cells compared to the reaction in solution. These results suggest that under conditions where the transition to the Lys-plasminogen form is permitted, the ability of Glu-plasminogen and [D(646)E]Pg to be activated by plasminogen activators is markedly enhanced. *Thus, analysis of the cleavage of [D(646)E]Pg on both HUVEC and U937 cells, showed that no stimulation of activation was observed under conditions where the mutant could not be cleaved to its Lys-plasminogen form while significant stimulation of activation was observed when the mutant was cleaved to its Lys-plasminogen form.*

3.2. Role of the Glu-Plasminogen to Lys-Plasminogen Conversion in Plasminogen Activation by u-PA on the Cell Surface

The requirement for conversion of Glu-plasminogen to Lys-plasminogen for enhanced activation on the cell surface was also observed, using both high molecular weight u-PA (hmw u-PA) and low molecular weight u-PA (lmw u-PA) as the plasminogen activators (Gong *et al.*, 2001). On the HUVEC surface, cleavage of Glu-plasminogen to plasmin by hmw u-PA (10 nM) was stimulated 4-fold at 10 min, compared to the reaction in the solution phase (Figure 6.4, Panel A). The predominant form of the plasmin heavy chain was the Lys-Pm heavy chain (Figure 6.4, Panel A). Furthermore, Lys-plasminogen

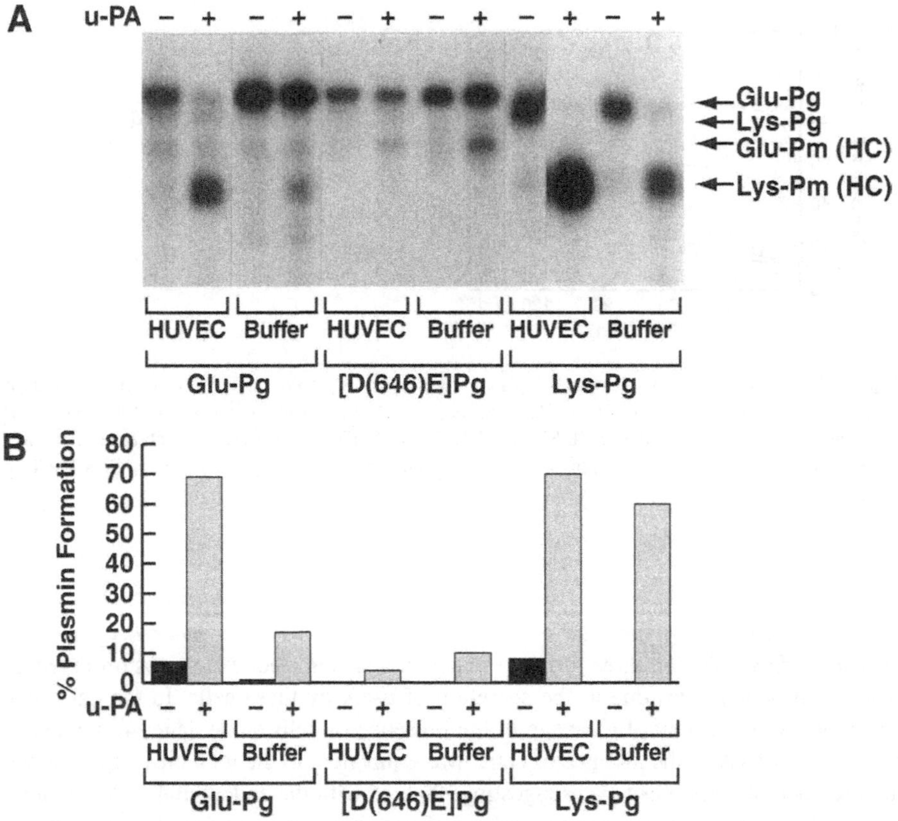

Figure 6.4. Role of the Glu-Pg to Lys-Pg conversion in plasminogen activation by hmw u-PA on the cell surface. (Panel A) A concentration of 25 nM of either ^{125}I-Glu-Pg, ^{125}I-[D(646)E]Glu-Pg, or ^{125}I-Lys-Pg was incubated with buffer (HBSS–0.4% BSA) or HUVEC (5×10^4) for 10 min at 37°C in the presence or absence of 10 nM hmw u-PA. Cell-bound ligand was electrophoresed on 7% SDS-PAGE under reducing conditions. (Panel B) The autoradiograms in Panel A were scanned. The percent plasmin formation was calculated by dividing the sum of the densities of the Glu-plasmin heavy chain and Lys-plasmin heavy chain bands by the sum of the densities of the Glu-Pg and Lys-Pg bands and the Glu-plasmin and Lys-plasmin heavy chain bands. The % plasmin in the starting material was subtracted (from Yun Gong, Sun-OK Kim, Jordi Felez, Davida K. Grella, Francis J. Castellino, and Lindsey A. Miles. Conversion of Glu-plasminogen to Lys-plasminogen is necessary for optimal stimulation of plasminogen activation on the endothelial cell surface, *Journal of Biological Chemistry*, May 2001, 276: 19078–19083, reprinted with permission).

accounted for 43% of the uncleaved plasminogen on the cell surface. In contrast, cleavage of cell-associated [D(646)E]Pg by u-PA was still markedly less than cleavage of cell-associated ^{125}I-Glu-Pg and cleavage of cell-associated [D(646)E]Pg was not enhanced compared to its cleavage in solution (Figure 6.4, Panel B). Formation of [D(646)E]Lys-Pg was not detected on the cell surface. The percentage activation of Lys-plasminogen on the cell surface was similar to that of Glu-plasminogen, but, again, activation of Lys-plasminogen in solution was markedly greater than that of Glu-plasminogen. (At 10 min 71% of the Lys-plasminogen in solution was activated to plasmin while only 17% of the Glu-plasminogen in solution was cleaved to plasmin.) Hence, activation of Lys-plasminogen on the HUVEC surface, compared to the solution phase, was enhanced only 1.1-fold.

Results using lmw u-PA as activator were similar to those obtained when hmw u-PA was used [(compare Figures 6.4 and 6.5) (Gong *et al.*, 2001)]. When cell-associated Glu-plasminogen was activated with lmw u-PA, the predominant form of the plasmin heavy chain was Lys-plasmin and Lys-plasminogen accounted for 41% of the unactivated plasminogen

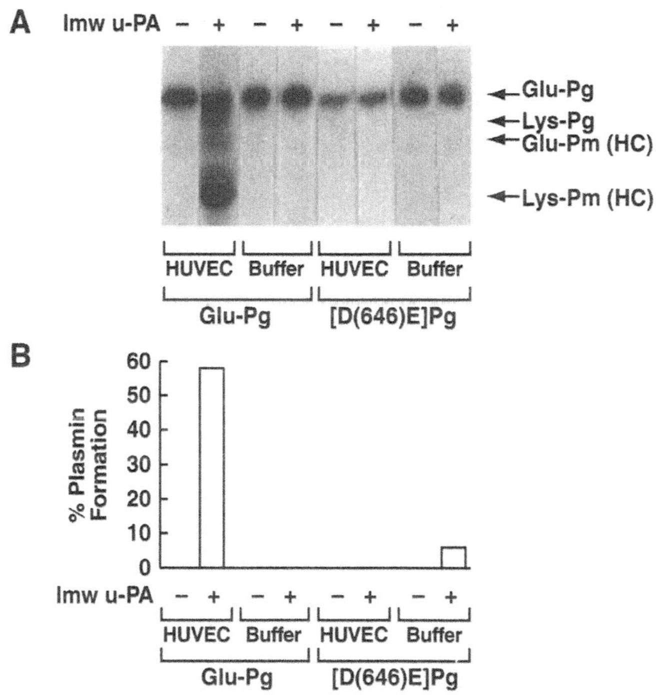

Figure 6.5. Role of the Glu-Pg to Lys-Pg conversion in plasminogen activation by lmw u-PA on the cell surface. (Panel A) A concentration of 25 nM of either ^{125}I-Glu-Pg or ^{125}I-[D(646)E]Glu-Pg was incubated with buffer (HBSS–0.4% BSA) or HUVEC (5×10^4) for 10 min at 37°C in the presence or absence of 15 nM lmw u-PA. Cell-bound ligand was subjected to electrophoresis on 7% SDS-PAGE under reducing conditions. (Panel B) The autoradiograms in Panel A were scanned. The % plasmin formation was calculated by dividing the sum of the densities of the Glu-plasmin heavy band and Lys-plasmin heavy chain band by the sum of the densities of the Glu-Pg and Lys-Pg bands and the Glu-plasmin and Lys-plasmin heavy chain bands. The % plasmin in the starting material was subtracted (from Yun Gong, Sun-OK Kim, Jordi Felez, Davida K. Grella, Francis J. Castellino, and Lindsey A. Miles. Conversion of Glu-plasminogen to Lys-plasminogen is necessary for optimal stimulation of plasminogen activation on the endothelial cell surface, *Journal of Biological Chemistry*, May 2001, 276: 19078–19083, reprinted with permission).

on the cell surface. Under these conditions, activation of Glu-plasminogen in solution was not detected and cleavage of ^{125}I-[D(646)E]Pg bound to HUVEC was not detected.

3.3. Effect of Cells on Conversion of [D(646)E]Glu-Pg to [D(646)E]Lys-Pg by Exogenous Plasmin

The foregoing results suggested that conversion of Glu-plasminogen to Lys-plasminogen was necessary for the optimal stimulation of plasminogen activation on the cell surfaces. Consequently, we examined whether localization of Glu-plasminogen on the cell surface could enhance its conversion to Lys-plasminogen by exogenous plasmin,

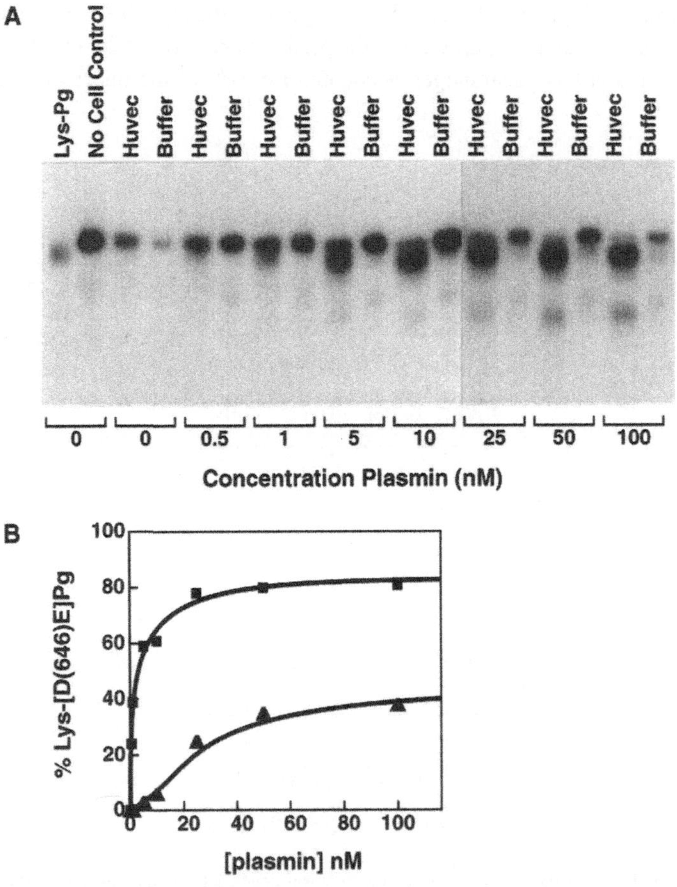

Figure 6.6. Effect of HUVEC on conversion of [D(646)E]Glu-Pg to [D(646)E]Lys-Pg by exogenous plasmin. ^{125}I-[D(646)E]Glu-Pg (25 nM) was incubated with either HUVEC (5×10^4) (■) or buffer (▲) in the presence of the indicated concentration of plasmin for 20 min at 37°C. The cell-bound ligand was electrophoresed in parallel with the solution phase samples on 7% SDS-PAGE under reducing conditions (Panel A). The autoradiograms were scanned (Panel B) (from Yun Gong, Sun-OK Kim, Jordi Felez, Davida K. Grella, Francis J. Castellino, and Lindsey A. Miles. Conversion of Glu-plasminogen to Lys-plasminogen is necessary for optimal stimulation of plasminogen activation on the endothelial cell surface, *Journal of Biological Chemistry*, May 2001, 276: 19078–19083, reprinted with permission).

compared to the solution phase. ^{125}I-[D(646)E]Glu-Pg was used in this analysis to avoid a possible contribution of the added ligand. ^{125}I-[D(646)E]Glu-Pg was incubated with either buffer or adherent HUVEC and increasing concentrations of plasmin were added and incubated for 20 min at 37°C. Conversion of the cell bound and solution phase ^{125}I-[D(646)E]Glu-Pg to [D(646)E]Lys-Pg was monitored by SDS-PAGE. Marked enhancement in conversion of [D(646)E]Glu-Pg to its Lys-form was observed when the ligand was bound to the HUVEC, compared to the solution phase (Figure 6.6) (Gong *et al.*, 2001). Under conditions where 59% of the cell-bound ligand was in the Lys-form, only 3% of the ligand in the solution phase was in the Lys-form. Therefore, the interaction of Glu-plasminogen with the cell surface promoted cleavage to Lys-plasminogen by plasmin.

4. Conclusions

In this review, we have summarized our recent data demonstrating that activation of cell-associated Glu-plasminogen is markedly enhanced only when its conversion to Lys-plasminogen on the cell surface is permitted. The predominant mechanism by which Glu-plasminogen activation is enhanced on the cell surface thus appears to be an increase in cleavage of Glu-plasminogen to Lys-plasminogen by plasmin. In this respect, the interaction of Glu-plasminogen with cells mimics the interaction with partly digested fibrin on which plasmin converts Glu-plasminogen to Lys-plasminogen 200-fold more efficiently than the reaction solution (Suenson and Thorsen, 1988; Fredenburgh and Nesheim, 1992).

Differences in activation mechanisms when Glu-plasminogen is bound to cells, compared with the reaction in solution are illustrated in Figure 6.7. On the cell surface

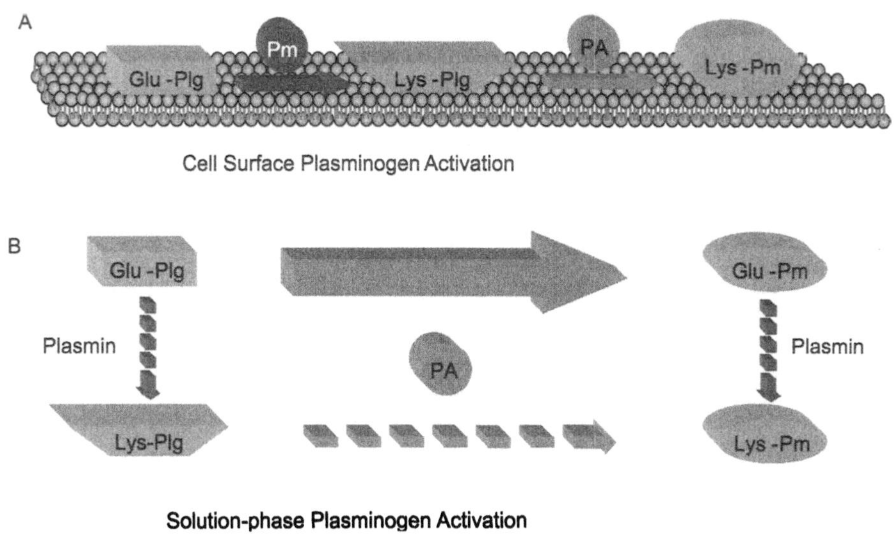

Figure 6.7. Comparison of Glu-plasminogen activation on the cell surface with the reaction in solution. On the cell surface (Panel A) the conversion of Glu-plasminogen to Lys-plasminogen is enhanced so that Lys-plasminogen becomes the major substrate for plasminogen activators. In contrast, in the solution phase (Panel B), the major substrate for plasminogen activators is Glu-plasminogen. Thus, on cells, the conversion to the more readily activated Lys-plasminogen is necessary for optimal stimulation of plasminogen activation.

(Panel A) conversion of Glu-plasminogen to Lys-plasminogen is enhanced so that Lys-plasminogen is the major substrate for plasminogen activators. In contrast, in solution (Panel B), the major substrate for plasminogen activators is Glu-plasminogen, consistent with earlier studies (Holvoet *et al.*, 1985). Thus on the cell surface, the more readily activated Lys-plasminogen is the substrate for the plasminogen activators, while in solution, the more poorly activated Glu-plasminogen is the substrate, accounting for the difference in kinetics of activation on the cell surface compared to the reaction in solution.

Small amounts of stimulation of activation of Lys-plasminogen or of [D646E]Glu-plasminogen were observed when bound to the HUVEC surface. Other mechanisms could account for this small amount of stimulation, including concentration of reactants, as these cells express binding sites for both t-PA and u-PA. Alternatively, or in addition, conformational changes induced in both Glu-plasminogen and Lys-plasminogen upon binding to cells, may contribute to the enhancement in activation. Nonetheless, the major effect of the interaction of Glu-plasminogen with the cell surface appears to be an increase in cleavage to the Lys-plasminogen form so that the more readily activated Lys-plasminogen becomes the major substrate for plasminogen activators on the cell surface. Further studies are necessary to address whether this mechanism is due to conformational changes in the cell-associated Glu-plasminogen to enhance its cleavage by plasmin and/or concentration of plasmin on the cell surface.

Acknowledgments

Research in the authors' laboratories was supported by National Institutes of Health Grants HL38272, HL45934 (to L.A.M.), HL13423 (to F.J.C.), and American Heart Grant-in-aid 9650636N (to L.A.M.). We thank Kathy Kling for excellent editorial assistance.

References

Andronicos, N.M. and Ranson, M. (2001). The topology of plasminogen binding and activation on the surface of human breast cancer cells. *Br. J. Cancer* **85**, 909–916.

Banyai, L. and Patty, L. (1984). Importance of intramolecular interactions in the control of the fibrin affinity and activation of human plasminogen. *J. Biol. Chem.* **259**, 6466–6471.

Bohmfalk, J.F. and Fuller, G.M. (1980). Plasminogen is synthesized by primary cultures of rat hepatocytes. *Science* **209**, 408–410.

Boyle, M.D. and Lottenberg, R. (2000). The interaction of pathogens with humans. *Methods* **21**, 99–102.

Christensen, U. and Molgaard, L. (1992). Positive co-operative binding at two weak lysine-binding sites governs the Glu-plasminogen conformational change. *Biochem. J.* **285**, 419–425.

Claeys, H. and Vermylen, J. (1974). Physico-chemical and proenzyme properties of NH2-terminal glutamic acid and NH2-terminal lysine human plasminogen. Influence of 6-aminohexanoic acid. *Biochim. Biophys. Acta* **342**, 351–359.

Dudani, A.K. and Ganz, P.R. (1996). Endothelial cell surface actin serves as a binding site for plasminogen, tissue plasminogen activator and lipoprotein(a). *Br. J. Haematol.* **95**, 168–178.

Duval-Jobe, C. and Parmely, M.J. (1994). Regulation of plasminogen activation by human U937 promonocytic cells. *J. Biol. Chem.* **269**, 21353–21357.

Ellis, V., Behrendt, N., and Dano, K. (1991). Plasminogen activation by receptor-bound urokinase. A kinetic study with both cell-associated and isolated receptor. *J. Biol. Chem.* **266**, 12752–12758.

Ellis, V., Scully, M.F., and Kakkar, V.V. (1989). Plasminogen activation initiated by single-chain urokinase-type plasminogen activator. *J. Biol. Chem.* **264**, 2185–2188.

Félez, J., Miles, L.A., Fábregas, P., Jardi, M., Plow, E.F., and Lijnen, R.H. (1996). Characterization of cellular binding sites and interactive regions within reactants required for enhancement of plasminogen activation by tPA on the surface of leukocytic cells. *Thromb. Haemost.* **76**, 577–584.

Fredenburgh, J.C. and Nesheim, M.E. (1992). Lys-plasminogen is a significant intermediate in the activation of Glu-plasminogen during fibrinolysis *in vitro*. *J. Biol. Chem.* **267**, 26150–26156.

Gong, Y., Kim, S.-O., Felez, J., Grella, D.K., Castellino, F.J., and Miles, L.A. (2001). Conversion of glu-plasminogen to lys-plasminogen is necessary for optimal stimulation of plasminogen activation on the endothelial cell surface. *J. Biol. Chem.* **276**, 19078–19083.

Gonzalez-Gronow, M., Stack, S., and Pizzo, S.V. (1991). Plasmin binding to the plasminogen receptor enhances catalytic efficiency and activates the receptor for subsequent ligand binding. *Arch. Biochem. Biophys.* **286**, 625–628.

Grella, D.K. and Castellino, F.J. (1997). Activation of human plasminogen by staphylokinase. Direct evidence that preformed plasmin is necessary for activation to occur. *Blood* **89**, 1585–1589.

Hajjar, K.A., Harpel, P.C., Jaffe, E.A., and Nachman, R.L. (1986). Binding of plasminogen to cultured human endothelial cells. *J. Biol. Chem.* **261**, 11656–11662.

Hajjar, K.A., Jacovina, A.T., and Chacko, J. (1994). An endothelial cell receptor for plasminogen/tissue plasminogen activator. I. Identity with annexin II. *J. Biol. Chem.* **269**, 21191–21197.

Hajjar, K.A. and Nachman, R.L. (1988). Endothelial cell-mediated conversion of glu-plasminogen to lys-plasminogen. *J. Clin. Invest.* **82**, 1769–1778.

Hawley, S.B., Tamura, T., and Miles, L.A. (2001). Purification, cloning, and characterization of a profibrinolytic plasminogen-binding protein, TIP49a. *J. Biol. Chem.* **276**, 179–186.

Hembrough, T.A., Vasudevan, J., Allietta, M.M., Glass, W.F., and Gonias, S.L. (1995). A cytokeratin 8-like protein with plasminogen-binding activity is present on the external surfaces of hepatocytes, HepG2 cells and breast carcinoma cell lines. *J. Cell Sci.* **108** (Pt. 3), 1071–1082.

Holvoet, P., Lijnen, H.R., and Collen, D. (1985). A monoclonal antibody specific for Lys-plasminogen. Application to the study of the activation pathways of plasminogen *in vivo*. *J. Biol. Chem.* **260**, 12106–12111.

Horrevoets, A.J.G., Smilde, A.E., Fredenburgh, J.C., Pannekoek, H., and Nesheim, M.E. (1995). The activation-resistant conformation of recombinant human plasminogen is stabilized by basic residues in the amino-terminal hinge region. *J. Biol. Chem.* **270**, 15770–15776.

Hoylaerts, M., Rijken, D.C., Lijnen, H.R., and Collen, D. (1982). Kinetics of the activation of plasminogen by human tissue plasminogen activator. *J. Biol. Chem.* **257**, 2912–2919.

Jeffery, C.J. (1999). Moonlighting proteins. *Trends Biochem. Sci.* **24**, 8–11.

Longstaff, C., Merton, R.E., Fabregas, P., and Felez, J. (1999). Characterization of cell-associated plasminogen activation catalyzed by urokinase-type plasminogen activator, but independent of urokinase receptor (uPAR, CD87). *Blood* **93**, 3839–3846.

Lopez-Alemany, R., Correc, P., Camoin, L., and Burtin, P. (1994). Purification of the plasmin receptor from human carcinoma cells and comparison to α-enolase. *Thromb. Res.* **75**, 371–381.

Loscalzo, J. and Vaughan, D.E. (1987). Tissue plasminogen activator promotes platelet disaggregation in plasma. *J. Clin. Invest.* **79**, 1749–1755.

Manchanda, N. and Schwartz, B.S. (1991). Single chain urokinase: Augmentation of enzymatic activity upon binding to monocytes. *J. Biol. Chem.* **266**, 14580–14584.

Mangel, W.F., Lin, B., and Ramakrishnan, V. (1990). Characterization of an extremely large, ligand-induced conformational change in plasminogen. *Science* **248**, 69–73.

Markus, G., De Pasquale, J.L., and Wissler, F.C. (1978a). Quantitative determination of the binding of epsilon-aminocaproic acid to native plasminogen. *J. Biol. Chem.* **253**, 727–732.

Markus, G., Evers, J.L., and Hobika, G.H. (1978b). Comparison of some properties of natiave (glu) and modified (lys) human plasminogen. *J. Biol. Chem.* **253**, 733–739.

Markus, G., Priore, R.L., and Wissler, F.C. (1979). The binding of tranexamic acid to native (Glu) and modified (Lys) human plasminogen and its effect on conformation. *J. Biol. Chem.* **254**, 1211–1216.

Miles, L.A., Dahlberg, C.M., Levin, E.G., and Plow, E.F. (1989). Gangliosides interact directly with plasminogen and urokinase and may mediate binding of these components to cells. *Biochemistry* **280**, 9337–9343.

Miles, L.A., Dahlberg, C.M., Plescia, J., Felez, J., Kato, K., and Plow, E.F. (1991). Role of cell-surface lysines in plasminogen binding to cells: Identification of alpha-Enolase as a candidate plasminogen receptor. *Biochemistry* **30**, 1682–1691.

Miles, L.A., Dahlberg, C.M., and Plow, E.F., (1988). The cell-binding domains of plasminogen and their function in plasma. *J. Biol. Chem.* **263**, 11928–11934.

Miles, L.A., Ginsberg, M.H., White, J.G., and Plow, E.F. (1986). Plasminogen interacts with human platelets through two distinct mechanisms. *J. Clin. Invest.* **77**, 2001–2009.

Miles, L.A. and Parmer, R.J. (2001). Neuroendocrine cell plasminogen receptors promote prohormone processing. *Thromb. Haemost.* **86**(Suppl. I).

Miles, L.A. and Plow, E.F. (1985). Binding and activation of plasminogen on the platelet surface. *J. Biol. Chem.* **260**, 4303–4311.

Misselwitz, R., Welfle, K., and Welfle, H. (1994). Conformations and stabilities of human Glu1- and Lys78-plasminogen and of the fragments mini- and microplasminogen, analysed by circular dichroism and differential scanning calorimetry. *Int. J. Biol. Macromol.* **16**, 187–194.

Muesch, A., Hartmann, E., and Rohde, K. (1990). A novel pathway for secretory proteins? No journal or journal lost.

Nakajima, K., Hamanoue, M., Takemoto, N., Hattori, T., Kato, K., and Kohsaka, S. (1994). Plasminogen binds specifically to alpha-Enolase on rat neuronal plasma membrane. *J. Neurochem.* **63**, 2048–2057.

Pancholi, V. and Fischetti, V.A. (1998). α-enolase, a novel strong plasmin(ogen) binding protein on the surface of pathogenic streptococci. *J. Biol. Chem.* **273**, 14503–14515.

Parkkinen, J., Raulo, E., Merenmies, J., Nolo, R., Kajander, E.O., Baumann, M., and Rauvala, H. (1993). Amphoterin, the 30-kDa protein in a family of HMG1-type polypeptides. *J. Biol. Chem.* **268**, 19726–19738.

Parmer, R.J., Mahata, M., Gong, Y., Mahata, S.K., Jiang, Q., O'Connor, D.T., Xi, X.-P., and Miles, L.A. (2000). Processing of chromogranin A by plasmin provides a novel mechanism for regulating catecholamine secretion. *J. Clin. Invest.* **106**, 907–915.

Ponting, C.P., Holland, S.K., Cederholm-Williams, S.A., Marshall, J.M., Brown, A.J., Spraggon, G., and Blake, C.C. (1992). The compact domain conformation of human Glu-plasminogen in solution. *Biochim. Biophys. Acta* **1159**, 155–161.

Ramakrishnan, V., Patthy, L., and Mangel, W.F. (1991). Conformation of Lys-plasminogen and the kringle 1–3 fragment of plasminogen analyzed by small-angle neutron scattering. *Biochemistry* **30**, 3963–3969.

Ramakrishnan, V., Sinicropi, D.V., Dere, R., Darbonne, W.C., Bechtol, K.B., and Baker, J.B. (1990). Interaction of wild-type and catalytically inactive mutant forms of tissue-type plasminogen activator with human umbilical vein endothelial cell monolayers. *J. Biol. Chem.* **265**, 2755–2761.

Redlitz, A., Fowler, B.J., Plow, E.F., and Miles, L.A. (1995). The role of an enolase-related molecule in plasminogen binding to cells. *Eur. J. Biochem.* **227**, 407–415.

Saito, H., Hamilton, S.M., Tavill, A.S., Louis, L., and Ratnoff, O.D. (1980). Production and release of plasminogen by isolated perfused rat liver. *Proc. Natl. Acad. Sci. USA* **77**, 6837–6840.

Sinniger, V., Merton, R.E., Fabregas, P., Felez, J., and Longstaff, C. (1999). Regulation of tissue plasminogen activator activity by cells. *J. Biol. Chem.* **274**, 12414–12422.

Sjoholm, I. (1973). Studies on the conformational changes of plasminogen induced during activation to plasmin and by 6-aminohexanoic acid. *Eur. J. Biochem.* **39**, 471–479.

Suenson, E. and Thorsen, S. (1998). The course and prerequisites of Lys-plasminogen formation during fibrinolysis, *Biochemistry* **27**, 2435–2443.

Stricker, R.B., Wong, D., Tak Shiu, D., Reyes, P.T., and Shuman, M.A. (1986). Activation of plasminogen by tissue plasminoagen activator on normal and thrombasthenic platelets: Effects on surface proteins and platelet aggregation. *Blood* **68**, 275–280.

Thorsen, S. and Astrup, T. (1974). Substrate composition and the effect of epsilon-aminocaproic acid on tissue plasminogen activator and urokinase-induced fibrinolysis. *Thromb. Diath. Haemorrh.* **32**, 306–324.

Thorsen, S., Kok, P., and Astrup, T. (1974). Reversible and irreversible alterations of human plasminogen indicated by changes in susceptibility to plasminogen activators and in response to epsilon-aminocaproic acid. *Thromb. Diath. Haemorrh.* **32**, 325–340.

Thorsen, S. and Mullertz, S. (1974). Rate of activation and electrophoretic mobility of unmodified and partially degraded plasminogen. Effects of 6-aminohexanoic acid and related compounds. *Scand. J. Clin. Lab. Invest.* **34**, 167–176.

Violand, B.N. and Castellino, F.J. (1976). Mechanism of the urokinase-catalyzed activation of human plasminogen. *J. Biol. Chem.* **251**, 3906–3912.

Violand, B.N., Sodetz, J.M., and Castellino, F.J. (1975). The effect of epsilon-amino caproic acid on the gross conformation of plasminogen and plasmin. *Arch. Biochom. Biophys.* **170**, 300–305.

Weisel, J.W., Nagaswami, C., Korsholm, B., Petersen, L.C., and Suenson, E. (1994). Interactions of plasminogen with polymerizing fibrin and its derivatives, monitored with a photoaffinity cross-linker and electron microscopy. *J. Mol. Biol.* **235**, 1117–1135.

Wiman, B. (1973). Primary structure of peptides released during activation of human plasminogen by urokinase. *Eur. J. Biochem.* **39**, 1–9.

Wiman, B. and Wallen, P. (1975). Structural relationship between "glutamic acid" and "lysine" forms of human plasminogen and their interaction with the NH_2-terminal activation peptide as studied by affinity chromatography. *Eur. J. Biochem.* **50**, 489–494.

Winram, S.B. and Lottenberg, R. (1996). The plasmin-binding protein Plr of group A streptococci is identified as glyceraldehyde-3-phosphate dehydrogenase. *Microbiology* **142**, 2311–2320.

Wistow, G. and Piatigorsky, J. (1987). Recruitment of enzymes as lens structural proteins. *Science* **236**, 1554–1556.

Zhang, L., Seiffert, D., Fowler, B., Jenkins, G.R., Thinnes, T., Loskutoff, D.J., Parmer, R.J., and Miles, L.A. (2002). Plasminogen has a broad extrahepatic distribution. *Thromb. Haemost.* **87**, 493–501.

7

Plasmin Reductase

Philip J. Hogg

1. Introduction

The formation of new blood vessels from pre-existing vessels is an important factor in a broad spectrum of diseases, including tumor growth and spread (Folkman, 1995). New blood vessel growth by the process of angiogenesis is balanced by several protein activators and inhibitors. One such inhibitor is angiostatin, which accumulated in the murine circulation in the presence of a growing Lewis lung tumor and disappeared when the tumor was removed (O'Reilly *et al.*, 1994). The angiostatin produced by the primary tumor was found to inhibit the neovascularization and growth of its remote metastases. Angiostatin has been shown to inhibit the growth of a number of murine and human primary carcinomas in mice (e.g., see O'Reilly *et al.*, 1996; Lannutti *et al.*, 1997; Sim *et al.*, 1997). The mechanism of action of angiostatin is not known but may relate to induction of endothelial cell apoptosis (Claesson-Welsh *et al.*, 1998), perhaps through binding to cell–surface ATP synthase (Moser *et al.*, 2001).

Angiostatin is an internal fragment of plasmin(ogen), and has been variously defined in the literature as consisting of approximately 1–4 kringles or smaller fragments thereof. We define angiostatin as a protein consisting of kringle 1–4 and part of kringle 5 of plasmin (Stathakis *et al.*, 1999), and suggest that the smaller fragments derive from this precursor. Both serine- and metallo-proteinase activity has been implicated in the cleavage of plasmin(ogen) to form angiostatin.

Angiostatin formation from plasmin involves not only peptide-bond cleavage but cleavage of disulfide-bonds. We have suggested that disulfide-bond cleavage precedes and makes possible the peptide-bond cleavage that results in the angiostatin fragments. This chapter discusses the nature and mechanism of the disulfide-bond cleavage in plasmin. In particular, the chapter focuses on the identification and characterization of a plasmin reductase that facilitates cleavage of the plasmin disulfide-bonds. The role of plasmin reductase in tumor angiogenesis is also discussed.

Philip J. Hogg • Centre for Vascular Research, University of New South Wales and Department of Haematology, Prince of Wales Hospital, Sydney, NSW 2052, Australia.

Plasminogen: Structure, Activation, and Regulation, edited by David M. Waisman. Kluwer Academic/Plenum Publishers, New York, 2003.

2. Identification of Plasmin Reductase Activity

Stathakis *et al.* (1997) observed that incubation of plasminogen or plasmin with adherent Chinese hamster ovary (CHO) cells or CHO conditioned medium resulted in the formation of ~45, ~41 and ~38 kDa plasmin fragments with Lys78 N-termini. The fragments encompassed approximately the first four kringle domains. The angiostatin fragments derived from plasmin, not plasminogen, as plasminogen activator inhibitor 1 (PAI-1) prevented the formation of both plasmin and the angiostatin fragments by CHO cells, whereas PAI-1 did not influence production of angiostatin from exogenous plasmin.

Stathakis *et al.* (1997) also showed that the tight-binding plasmin inhibitor aprotinin, but not the metalloproteinase inhibitor EDTA, inhibited the formation of angiostatin from both plasmin derived from plasminogen and from plasmin added exogenously, implying that proteolysis (perhaps auto-proteolysis) of plasmin is required for the production of angiostatin by CHO cells. These findings are in agreement with those of Gately *et al.* (1996), who observed that serine proteinase activity was necessary for angiostatin production by human prostate cancer cells. In addition, Stathakis *et al.* (1997) showed that the lysine analogue, ϵ-amino caproic acid, did not perturb angiostatin formation by CHO cells, implying that the lysine-binding activity of the plasmin(ogen) kringles is not required for angiostatin formation.

A key feature of the mechanism of formation of the angiostatin fragments was provided by the observation that they contained free sulfhydryl group(s) (Stathakis *et al.*, 1997). This was demonstrated by the incorporation of the biotin-linked maleimide, *N*-(3-(*N*-maleimidyl)propionyl)biocytin (MPB) (Figure 7.1). Maleimides react specifically and rapidly (rate constant of ~10^4 M^{-1} s^{-1}) with free thiols at neutral pH. Plasminogen and plasmin did not incorporate MPB, which is consistent with the absence of free

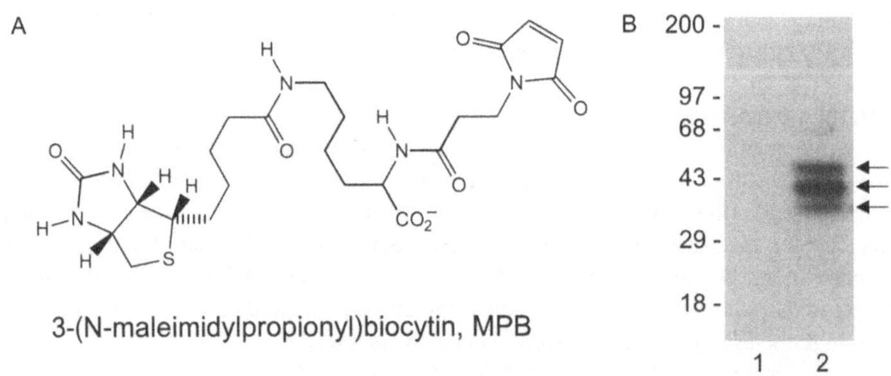

Figure 7.1. Measurement of cleavage of plasmin disulfide-bond(s). **A** Structure of MPB. **B** Plasmin was incubated without (lane 1) or with (lane 2) HT1080 conditioned medium and the angiostatin fragments labeled with MPB as described by Stathakis *et al.* (1997, 1999). The plasmin products were collected on lysine-Sepharose beads and eluted with ϵ-amino-caproic acid. The MPB-labeled fragments were separated on SDS–PAGE under non-reducing conditions, blotted with streptavidin-peroxidase and visualized using chemiluminescence. The positions of molecular weight markers in kDa are shown at left. The arrows indicate the three angiostatin fragments.

sulfhydryl groups in these proteins (Stathakis *et al.*, 1997). The demonstration of free sulfhydryl group(s) implied that formation of the angiostatin fragments involves disulfide-bond cleavage in plasmin.

Cleavage of protein disulfide-bonds usually requires a catalyst. Two proteins that are known to catalyze reduction of protein disulfide-bonds are protein disulfide isomerase (PDI) (Novia, 1999) and thioredoxin (Holmgren, 1989). Incubation of plasmin with PDI or thioredoxin resulted in the formation of three angiostatin fragments that were identical to those made by CHO cells (Stathakis *et al.*, 1997). In contrast, incubation of plasminogen with either PDI or thioredoxin did not perturb plasminogen structure by sodium dodecyl sulfate–polyacrylamide gel electrophoresis (SDS–PAGE), or generate free sulfhydryl groups in plasminogen that could be alkylated by MPB (Stathakis *et al.*, 1997). Therefore, the fragments are derived from plasmin by reduction and auto-proteolysis of the enzyme. These findings were consistent with the results using CHO cells or conditioned medium and implied that auto-proteolysis of reduced plasmin is sufficient for angiostatin formation. The ability of PDI or thioredoxin to generate angiostatin from plasmin is not based on thermodynamic considerations alone, as dithiothreitol, which is a better reductant than either PDI or thioredoxin, did not generate angiostatin (Stathakis *et al.*, 1997). These results supported the contention that the mechanism of formation of angiostatin by CHO cells involves cleavage of one or more plasmin disulfide-bonds, which triggers proteolysis of the enzyme releasing fragment(s) containing approximately the first four kringle domains.

Although PDI and thioredoxin could produce angiostatin from plasmin, these enzymes were relatively inefficient. Immunodepletion of CHO conditioned medium of PDI or thioredoxin did not effect angiostatin generating activity, confirming that neither of these enzymes are used by CHO cells (Stathakis *et al.*, 1997). These observations implied that CHO cells secrete a "plasmin reductase" which facilitates the cleavage of plasmin disulfide-bond(s).

Plasmin reductase activity was also secreted by the human fibrosarcoma cell line, HT1080, and the human mammary carcinoma cell lines, MCF-7, MDA231, and BT20 (Stathakis *et al.*, 1999). In contrast, the monocyte/macrophage cell line, THP-1, did not secrete significant levels of plasmin reductase before or after stimulation with phorbol ester. Neither primary foreskin fibroblasts, blood monocyte/macrophages, nor macrovascular or microvascular endothelial cells secreted detectable plasmin reductase (Stathakis *et al.*, 1999). In contrast, cultured bovine and rat vascular smooth muscle cells secrete small but reproducible levels of plasmin reductase (Stathakis *et al.*, 1999). In general terms, cellular transformation appears to be associated with the secretion of plasmin reductase and angiostatin formation. This finding suggests that angiostatin formation is driven by tumor cells *in vivo*. Production of angiostatin by vascular smooth muscle cells is an interesting observation and suggests that angiostatin may function in the blood vessel wall.

Gateley *et al.* (1997) showed that sufficient concentration of small thiols alone can generate angiostatin from plasmin. The concentration of small thiols in HT1080 conditioned medium was determined to be to be ~1 μM (Stathakis *et al.*, 1999). The major small thiol was GSH. The level of secretion of GSH by HT1080 cells is similar to other cultured cells (Bannai and Tsukeda, 1979; Dethmers and Meister, 1981). Incubation of plasmin with 1 μM GSH, the concentration in HT1080 conditioned medium, had no angiostatin-generating activity, while 10 μM GSH had ~6% of the angiostatin generating activity of

HT1080 conditioned medium (Stathakis *et al.*, 1999). The authors concluded that high concentrations (~100 μM) of small thiols have enough reducing power to reduce plasmin, but that these concentrations are not achievable in cell culture.

3. Disulfide-Bonds Cleaved by Plasmin Reductase

Stathakis *et al.* (1999) made fragments of plasminogen and tested them as substrates for plasmin reductase in HT1080 conditioned medium. One or more disulfide-bonds in the kringle 5-serine proteinase fragment of plasminogen was reduced by plasmin reductase. In contrast, neither kringle 4, kringle 1–3, nor kringle 1–4 were substrates for plasmin reductase. This finding implied that the target disulfide-bond(s) for plasmin reductase are in kringle 5.

Plasmin undergoes auto-proteolysis in alkaline pH producing a catalytically active microplasmin fragment with a Lys531 N-terminus (Wu *et al.*, 1987a,b). Microplasmin consists of the last 10 amino acids of kringle 5, the remaining 21 amino acids of the A chain, and the serine proteinase domain. Wu *et al.* (1987a,b) noticed that both the Cys462–Cys541 and Cys512–Cys536 disulfide-bonds in kringle 5 must have been cleaved to release microplasmin from kringle 1–4. They proposed that the increased hydroxide ion concentration at pH 11 was responsible for cleaving the two disulfide-bonds. Stathakis *et al.* (1999) observed that the proteinase fragments produced from plasmin in pH 11 buffer, apparent molecular weights of ~40, ~30, and ~29, were of identical molecular weight to the proteinase fragments generated from plasmin in HT1080 conditioned medium. The 29 molecular weight fragment is the same size as microplasmin. They suggested that the mechanism of plasmin proteolysis at pH 11 is similar to, if not the same, as the mechanism of proteolysis in HT1080 conditioned medium. More recently, Kassam *et al.* (2001) described the formation of a Lys78–Lys468 plasmin fragment that resulted from auto-digestion of plasmin in pH 9 buffer. It is likely that the plasmin auto-proteolytic events described by Kassam *et al.* (2001) in pH 9 buffer were similar, if not the same, as the events described by Wu *et al.* (1987a,b) in pH 11 buffer.

Three major angiostatin fragments and three major serine proteinase fragments were produced in HT1080 conditioned medium. Based on the close similarity in plasmin fragments produced at alkaline pH and in HT1080 conditioned medium, Stathakis *et al.* (1999) concluded that at least the Cys462–Cys541 and Cys512–Cys536 disulfide-bonds in kringle 5 were cleaved by plasmin reductase (see Figure 7.2). For instance, cleavage of only the kringle 5 Cys483–Cys524 and Cys512–Cys536 disulfide-bonds would not have resulted in the release of angiostatin from plasmin. Similarly, angiostatin would have remained covalently linked to the remaining kringles if the Cys462–Cys541 and Cys483–Cys524 disulfide-bonds were cleaved, but not the disulfide-bond at Cys512–Cys536. Cleavage of the Cys483–Cys524 disulfide-bond in kringle 5 or other disulfide-bonds in kringle 1–4, however, could not be excluded.

It has been proposed that the reduction of plasmin precedes proteolysis (Stathakis *et al.*, 1999). This was supported by an experiment that examined generation of angiostatin from plasmin that had been inactivated by Val-Phe-Lys-chloromethyl ketone. MPB labeled both intact inactivated plasmin and angiostatin fragments derived from the inactivated plasmin. This finding demonstrated the cleavage of disulfide-bond(s) in intact plasmin, which supports the contention that plasmin reduction precedes proteolysis.

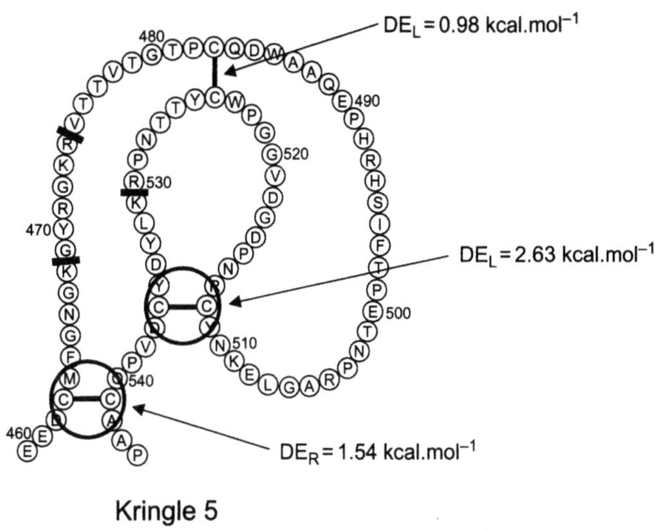

Figure 7.2. Cartoon of kringle 5 of plasmin(ogen) showing the dihedral strain energies of the disulfide-bonds and sites of disulfide- and peptide-bond cleavage. The handedness (left or right) of the disulfide-bonds are indicated with the dihedral strain energies (DE). We suggest that both the Cys462–Cys541 and Cys512–Cys536 disulfide-bonds in plasmin kringle 5 are cleaved (large circles). The reduced kringle 5 is then subject to proteolysis at the Arg530–Lys531 peptide bond or two other unidentified peptide bonds (solid black lines). The other peptide bonds are likely to be Arg474–Val475 and Lys468–Gly469. Numbering is based on the sequence of human plasminogen (791 residues) beginning at Glu1, excluding the 19 amino acid signal peptide that ends at Met.

4. Proteolysis of Reduced Plasmin

Cleavage of both the kringle 5 Cys462–Cys541 and Cys512–Cys536 disulfide-bonds renders kringle 5 susceptible to proteolysis at either the Arg530–Lys531 peptide bond or two other unidentified peptide bonds (Stathakis *et al.*, 1999). The other peptide bonds are likely to be Arg474–Val475 (Stathakis *et al.*, 1999) and Lys468–Gly469 (Kassam *et al.*, 2001).

Gately *et al.* (1996) observed that serine proteinase activity was required for the generation of angiostatin from plasminogen or plasmin by cultured human prostate carcinoma cells and Stathakis *et al.* (1999) reported that serine proteinase activity was necessary for angiostatin generation by cultured CHO and HT1080 cells. Incubation of purified plasmin with thioredoxin or protein disulfide isomerase (Stathakis *et al.*, 1997) or high concentrations of small thiols (Gately *et al.*, 1997) resulted in the formation of angiostatin. This indicated that plasmin auto-proteolysis could account for angiostatin formation. Angiostatin was generated almost equally well, however, from either active or inactive plasmin in HT1080 conditioned medium (Stathakis *et al.*, 1999), which implied that a proteinase(s) secreted by HT1080 cells was mostly responsible for plasmin proteolysis. The enzyme(s) was a serine proteinase as it was inactivated by soybean trypsin inhibitor. Depletion of serine proteinases in HT1080 conditioned medium by soybean trypsin inhibitor-agarose reduced the angiostatin generating activity to 34% of control (Stathakis *et al.*, 1999). This result implied that the rate of angiostatin formation is slowed when auto-proteolysis of reduced plasmin is the operative mechanism. There were three major serine proteinases in HT1080 conditioned medium with apparent molecular weights of ~70, ~50, and ~39 that cleaved gelatin. One or more of these enzymes may contribute to the generation of angiostatin from reduced plasmin. Therefore, although plasmin autoproteolysis can account for angiostatin formation, HT1080 cells do not rely entirely on this mechanism and secreted other serine proteinase(s) that cleave reduced plasmin.

5. Plasmin Reductase is Phosphoglycerate Kinase

Lay *et al.* (2000) observed that plasmin reductase activity in HT1080 conditioned medium was inhibited by the adenine nucleotides, NAD(H) or ATP, and inactivated by the thiol alkylating reagents, iodoacetamide, or *N*-ethylmaleimide. These results implied that plasmin reductase would have affinity for the chomatography matrices, Cibachron Blue-Sepharose and activated thiopropyl-Sepharose. Lay *et al.* (2000) collected 20 L of conditioned medium from HT1080 cells, which was resolved on Cibachron Blue-Sepharose and activated thiopropyl-Sepharose and then gel filtered. The plasmin reductase activity was associated with a single polypeptide chain with a molecular mass of ~42 kDa by SDS–PAGE. The protein was found to have 100% sequence identity to phosphoglycerate kinase (PGK; ATP:3-phospho-D-glycerate 1-phosphotransferase, EC 2.7.2.3).

Recombinant human PGK resolved at the same size as the HT1080 protein on SDS-PAGE and had the same plasmin reductase activity as the HT1080 protein (Lay *et al.*, 2000). These findings demonstrated that plasmin reductase is PGK. Rabbit muscle PGK had comparable specific plasmin reductase activity to recombinant human PGK or HT1080 PGK while yeast PGK had negligible activity (Lay *et al.*, 2000). Human and rat PGK share 97% sequence identity while the human and yeast enzyme share only 65% sequence identity.

Rate of secretion of PGK by carcinoma cells varies considerably. Cultured HT1080 cells secrete ~0.4 μg of PGK per 10^6 cells in 24 hr (Lay *et al.*, 2000). This rate of secretion was compared with rates of secretion of PGK from four human pancreatic, three human breast, and three human colon carcinoma cell lines. The secretion rates varied 26-fold among the cell lines. Importantly, the rate of secretion of PGK by the HT1080, breast and colon carcinoma cells correlated with the plasmin reductase activity secreted by these cells (Lay *et al.*, 2000).

6. Structure/Function Aspects of Plasmin Reduction by Phosphoglycerate Kinase

PGK is the sixth enzyme of the glycolytic pathway where it catalyzes the high-energy phosphoryl transfer reaction from the acid anhydride bond of 1,3-bisphosphoglycerate to the β-phosphate of MgADP. PGK also influences DNA replication and repair in mammalian cell nuclei (Vishwanatha *et al.*, 1992; Popanda *et al.*, 1998), stimulates viral mRNA synthesis in the cytosol (Ogino *et al.*, 1999), and extends through the cell wall of *C. albicans* (Alloush *et al.*, 1997). The human enzyme has a molecular mass of ~45 kDa and consists of a single polypeptide chain of 418 residues. Crystallographic studies of the yeast (Banks *et al.*, 1979; Watson *et al.*, 1982), horse (Blake and Evans, 1974), and pig (Harlos *et al.*, 1992; May *et al.*, 1996; Szilágyi *et al.*, 2001) enzyme revealed that the molecule is composed of two domains of similar size, which corresponds to the N- and C-terminal halves of the chain, separated by a hinge region.

Protein reductant active sites typically contain a redox active dithiol/disulfide with the sequence, CysGlyXCys (Holmgren, 1989; Novia, 1999). The Cys thiols cycle between the reduced dithiol and oxidized disulfide-bond in coordination with a dithiol or disulfide of a protein substrate. This can lead to the net formation, the net rearrangement, or the net reduction of protein disulfide-bonds depending on the nature of the protein substrate, the redox conditions, and the presence of other thiols and disulfides. Mammalian PGK's contain 7 Cys and only 2 of the 7 are nearby in the primary or tertiary structure (Szilágyi *et al.*, 2001). Cys379, 380 are close to the hinge region and have been referred to as "fast-reacting" as they are amenable to alkylation by several thiol-reactive compounds (Cserpán and Vas, 1983; Minard *et al.*, 1989). This arrangement of Cys indicates that the mechanism by which PGK reduces disulfide-bonds in plasmin is unconventional.

Lay *et al.* (2002) mutated the seven Cys in PGK to Ala in a cumulative fashion in the order, Cys379, 380, 367, 316, 108, 99, and 50. The two fast-reacting Cys, Cys379, 380, were mutated first and the resulting cDNA was then used to mutate Cys367, and so forth. The specific plasmin reductase activity of the mutant PGK's, with the exception of the Cys-less PGK, was similar to that of the wild-type protein. Some mutations were shown to change the tertiary structure of PGK as measured by susceptibility to proteolysis by plasmin (Lay *et al.*, 2002). For instance, the Cys-less PGK was rapidly degraded by plasmin which probably accounted for the loss of plasmin reductase activity. These results implied that PGK Cys were not directly involved in plasmin reduction. The question remained, therefore, why alkylation of Cys379, 380 inhibited plasmin reduction (Stathakis *et al.*, 1999). Lay *et al.* explored this question by reacting the Cys residues with different alkylating/oxidizing agents and examining the consequences for plasmin reductase activity.

Carboxymethylation, but not methylation, of the fast-reacting Cys of pig PGK inactivates the kinase activity (Dékány and Vas, 1984). Reaction of the pig enzyme with 5,5′-dithiobis(2-nitrobenzoic acid) also inactivates the kinase activity (Cserpán and Vas, 1983), while reaction with $HgCl_2$ or dibromobimane reduces kinase activity by up to 80% (Vas and Csanády, 1987). Reaction of PGK with four different alkylating/oxidizing agents, including $HgCl_2$ and dibromobimane, reduced the number of reactive thiols per mol of PGK to ~0.2–0.5 (Lay *et al.*, 2002). The plasmin reductase activity of the modified proteins was reduced to 7–35% of control. These results indicate that alkylation of the fast-reacting thiols perturb plasmin reductase activity. Neither changes in pH, ionic strength, nor temperature markedly affected plasmin reductase activity, which implies that the reaction of PGK and plasmin involves predominantly hydrophobic interactions.

Kinase substrate binding studies have shown that MgATP and MgADP bind to the inner surface of the C-domain (Banks *et al.*, 1979; Watson *et al.*, 1982; Minard *et al.*, 1989; Szilágyi *et al.*, 2001) while 3-phosphoglycerate (3-PG) binds to the inner surface of the N-domain (Harlos *et al.*, 1992; May *et al.*, 1996; Szilágyi *et al.*, 2001). The bound substrates are ≥ 10 Å from each other, too large a distance for direct in-line phosphoryl transfer. The enzyme overcomes this distance by undergoing a "hinge-bending" conformational change that brings the two substrates closer together.

The effect of 3-PG/MgADP-induced conformational changes in PGK for plasmin reductase activity was evaluated by Lay *et al.* (2002). Incubation of PGK with saturating concentrations of 3-PG or MgATP completely inhibited angiostatin-generating activity. The half-maximal inhibitory effects of 3-PG and MgATP on angiostatin-generating activity were ~1 mM, which is in the range of the Michaelis constants for these substrates in the kinase reaction (McHarg *et al.*, 1999). These findings indicate that hinge-bending conformational change in PGK negates the plasmin reductase activity of the protein.

7. Putative Mechanism of Action of Phosphoglycerate Kinase

A disulfide-bond in a protein usually results in an increase in stability through entropic effects, and decrease in stability due to bond enthalpy effects and strain. A method for calculating the strain of a disulfide-bond uses dihedral angles. A dihedral angle is the angle of rotation about a certain bond. The dihedral strain energy of a disulfide-bond can be calculated from the five dihedral angles of a disulfide-bond as described by Weiner *et al.* (1984) and Katz and Kossiakoff (1986). The calculated strain energies only consider the dihedral angles of the disulfide-bond and do not include other factors such as bond lengths, bond angles, and van der Waals contacts in calculating energy. The findings of Pjura *et al.* (1990), however, indicate that such calculations can give useful semi-quantitative insights into the amount of strain in a disulfide-bond.

There is a correlation between the redox potentials of disulfide-bonds and their calculated dihedral energies. Those bonds with higher dihedral energies are more easily cleaved (Katz and Kossiakoff, 1986; Wells and Powers, 1986; Wetzel, 1987). Calculation of the dihedral strain energies of the kringle 5 disulfide-bonds from the crystal structure described by Chang *et al.* (1998) revealed that the left-handed Cys512–Cys536 bond has a high strain energy (2.63 kcal.mol^{-1}) compared to the right-handed Cys462–Cys541 (1.54 kcal.mol^{-1}) and left-handed Cys483–Cys524 (0.98 kcal.mol^{-1}) disulfide-bonds (Figure 7.2). The average dihedral strain energies for left- and right-handed disulfide-bonds are 1.68 and 3.19 kcal.mol^{-1}, respectively (Weiner *et al.*, 1984; Katz and Kossiakoff, 1986).

These calculations led Lay *et al.* (2002) to propose that PGK facilitates cleavage of the Cys512–Cys536 disulfide-bond by hydroxide ion, which results in the formation of a sulfenic acid at position 512 and a free thiol at Cys536. The Cys536 thiol is then available to exchange with the Cys462–Cys541 disulfide-bond resulting in the formation of a new disulfide at Cys536–Cys541 and a free thiol at Cys462. Kringle 5 is then susceptible to proteolysis at Arg530–Lys531, Arg474–Val475, and/or Lys468–Gly469. These events are summarized in Figure 7.3 and are the simplest sequence of events that can explain all the available data (Stathakis *et al.*, 1997, 1999; Lay *et al.*, 2000, 2002). The free thiol that is labeled by MPB in the three angiostatin fragments would be Cys462.

Lay *et al.* (2002) concluded that PGK presumably binds to plasmin and induces a conformational change in kringle 5 that facilitates hydroxide ion attack on the Cys512–Cys536 disulfide-bond. This suggests that other molecules that interact with plasmin might also facilitate cleavage of the kringle 5 disulfides. There is recent evidence to support this hypothesis. Interaction of a truncated porcine PAI-1 (residues 80–265), but not full length protein, with plasmin has been shown to result in the generation of kringle-containing angiostatin fragments (Mulligan-Kehoe *et al.*, 2001). It is possible that

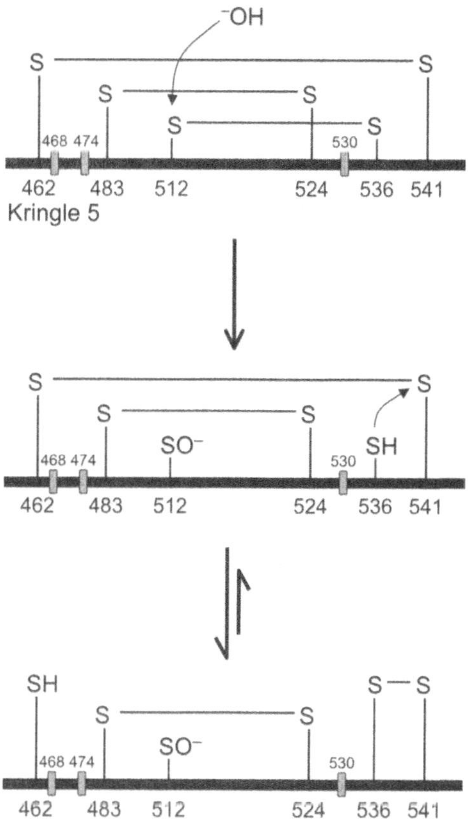

Figure 7.3. Model of the molecular events facilitated by PGK in plasmin kringle 5.

the truncated protein is facilitating the same sequence of events in plasmin that are achieved by PGK.

8. Phosphoglycerate Kinase in Tumor Angiogenesis

The plasma of mice bearing HT1080 tumors contained several-fold more PGK than mice without tumors (Lay *et al.*, 2000). This finding implied that relatively small amounts of PGK are secreted from normal tissues, which is in accordance with the observation that transformed cells secrete much more PGK than primary cells (Stathakis *et al.*, 1999). Moreover, administration of PGK to mice with tumors resulted in a significant increase in the plasma levels of MPB-labeled angiostatin (Lay *et al.*, 2000). PGK administration to mice without tumors did not change plasma levels of MPB-angiostatin. This result implied that PGK facilitated reduction of plasmin produced by the HT1080 tumors, which supports the proposed role for PGK in angiostatin formation.

Lay *et al.* (2000) tested whether treatment of tumor-bearing mice with PGK might inhibit tumor angiogenesis and growth via production of angiostatin. Indeed, the growth of human fibrosarcoma (HT1080) and pancreatic carcinoma (AsPC-1) tumors in immunocompromised mice was suppressed by systemic administration of rhPGK. Intraperitoneal administration of 5 mg rhPGK per kg per day caused a ~50% inhibition of the rate of HT1080 tumor growth and ~70% inhibition of the rate of AsPC-1 tumor growth (Lay *et al.*, 2000). Immunohistochemical staining of the AsPC-1 tumors for vascularity indicated a significant reduction in angiogenesis in the rhPGK-treated tumors.

Lay *et al.* (2000) noted that AsPC-1 tumorigenesis was more susceptible to suppression by systemic administration of PGK than HT1080 tumorigenesis. They suggested that this may reflect the 5-fold reduced secretion of PGK by AsPC-1 cells compared to HT1080 cells (Lay *et al.*, 2000). They also noted that the rate of secretion of PGK by three different human pancreatic carcinoma cells was inversely correlated with the tumorigenicity of the cells grown subcutaneously in the proximal midline dorsum of SCID mice.

The hypoxic nature of solid tumors triggers expression of vascular endothelial growth factor, which stimulates angiogenesis, and glycolytic enzymes including PGK, which facilitate anaerobic production of ATP. We have measured the effect of hypoxia on production and secretion of PGK and angiostatin formation by human tumor cells. As anticipated, hypoxia resulted in an increase in the total PGK protein in the tumor cells. In contrast, hypoxia markedly inhibits secretion of PGK and angiostatin production by the tumor cells (unpublished observations). These results indicate that secretion of PGK is regulated independently and inversely of its production. Our findings are consistent with the correlation between hypoxia and angiogenesis. That is, hypoxia triggers production of vascular endothelial growth factor which stimulates angiogenesis and inhibits secretion of PGK which reduces angiostatin formation. This is the first example of the regulation of an angiogenesis inhibitor by hypoxia.

The findings discussed in this chapter support the model for angiostatin formation by tumors shown in Figure 7.4. This model is premised on the observations that plasmin is relatively resistant to proteolysis. Both serine- (Kassam *et al.*, 2001) and metalloproteinases (Cornelius *et al.*, 1998) are inefficient at cleaving intact plasmin at physiological pH. It seems unlikely that the time frame for proteolysis of intact plasmin would be biologically relevant. We suggest that the disulfide cleavage in kringle 5 makes plasmin vulnerable to proteolytic attack and release of angiostatin.

The model indicates a sequential order of events beginning with the conversion of plasminogen to plasmin, followed by reduction of plasmin by PGK, then serine proteinase-dependent release of kringle 1–4$\frac{1}{2}$ (Gately *et al.*, 1996, 1997; Stathakis *et al.*, 1997, 1999;

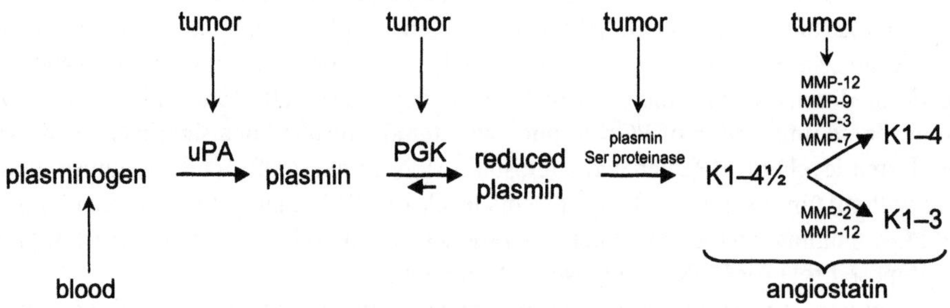

Figure 7.4. Model of the mechanism of angiostatin formation by tumors.

Kassam *et al.*, 2001) and finally matrix metalloproteinase-dependent trimming of kringle 1–4$\frac{1}{2}$ to either kringle 1–4 or 1–3 (Dong *et al.*, 1997; Cornelius *et al.*, 1998; O'Reilly *et al.*, 1999). An advantage of this mechanism is the opportunity for fine control of angiostatin formation at the level of both plasmin formation and plasmin reduction. Tumor production of angiostatin, therefore, will be controlled by the production and secretion of both urokinase plasminogen activator and PGK.

9. Future Directions

There is still much to learn about how PGK facilitates cleavage of plasmin disulfide-bonds. Efforts are underway to identify and map the sequence of cleavage events in kringle 5 by mass spectrometry. The formation of sulfenic acid is being assessed as well as the role of hydroxide ion in the cleavage. One intriguing possibility is that the disulfide-bond cleavage is not mediated by hydroxide ion. We are exploring a mechanism whereby a positive charge (Arg) close to the kringle 5 Cys512–Cys536 disulfide-bond and a negative charge (Asp/Glu) from PGK or plasmin polarizes and then breaks the disulfide-bond. Brandt *et al.* (1999) used semi-empirical quantum chemical calculations as well as ab initio studies to show that the interaction of a carboxylic acid and a primary amine with a disulfide-bond can polarize and cleave the bond.

How PGK is secreted from tumor cells and how this secretion is controlled by the oxygen tension is also unknown. Proteins destined for secretion via the classical Golgi pathway contain a recognizable N-terminal signal sequence. PGK does not contain this signal sequence and, indeed, inhibitors of the Golgi pathway do not affect secretion of PGK from tumor cells (unpublished observations).

Plasmin reduction is the first example of cleavage of a disulfide-bond in a secreted soluble protein. It has been generally considered that disulfide-bonds in secreted proteins are inert because of the oxidizing nature of the extracellular milieu. The findings presented in this chapter indicate that cleavage of disulfide-bonds in secreted proteins may be an important means of regulating certain cleavage function. A second example of this phenomenon is the disulfide-bond cleavage in the blood protein, von Willebrand Factor. Reduction of von Willebrand Factor by thrombospondin-1 regulates its haemostatic activity (Xie *et al.*, 2001).

References

Alloush, H.M., López-Ribot, J.L., Masten, B.J., and Chaffin, W.L. (1997). 3-phosphoglycerate kinase: A glycolytic enzyme protein present in the cell wall of Candida albicans. *Microbiology UK* **143**, 321–330.

Banks, R.D., Blake, C.C.F., Evans, P.R., Haser, R., Rice, D.W., Hardy, G.W., Merrett, M., Phillips, A.W. (1979). Sequence, structure and activity of phosphoglycerate kinase: A possible hinge-bending enzyme. *Nature* **279**, 773–777.

Bannai, S. and Tsukeda, H. (1979). The export of glutathione from human diploid cells in culture. *J. Biol. Chem.* **254**, 3444–3450.

Blake, C.C.F. and Evans, P.R. (1974). Structure of horse muscle phosphoglycerate kinase. Some results on the chain conformation, substrate binding and evolution of the molecule from a 3 angstrom Fourier map. *J. Mol. Biol.* **84**, 585–601.

Brandt, W., Golbraikh, A., Täger, M., and Lendeckel, U. (1999). A molecular mechanism for the cleavage of a disulfide-bond as the primary function of agonist binding to G-protein-coupled receptors based on theoretical calculations supported by experiments. *Eur. J. Biochem.* **261**, 89–97.

Chang, Y., Mochalkin, I., McCance, S.G., Cheng, B., Tulinsky, A., and Castellino, F.J. (1998). Structure and ligand binding determinants of the recombinant kringle 5 domain of human plasminogen. *Biochemistry* **37**, 3258–3271.

Claesson-Welsh, L., Welsh, M., Ito, N., Anand-Appte, B., Soker, S., Zetter, B., O'Reilly, M., and Folkman, J. (1998). Angiostatin induces endothelial cell apoptosis and activation of focal adhesion kinase independently of the integrin-binding motif RGD. *Proc. Natl. Acad. Sci. USA* **95**, 5579–5583.

Cornelius, L.A., Nehring, L.C., Harding, E., Bolanowski, M., Welgus, H.G., Kobayashi, D.K., Pierce, R.A., and Shapiro, S.D. (1998). Matrix metalloproteinases generate angiostatin: Effects on neovascularization. *J. Immunol.* **161**, 6845–6852.

Cserpán, I. and Vas, M. (1983). Effects of substrates on the heat stability and on the reactivities of thiol groups of 3-phosphoglycerate kinase. *Eur. J. Biochem.* **131**, 157–162.

Dékány, K. and Vas, M. (1984). Inactivation of pig muscle 3-phosphoglycerate kinase by thiol modification depends on reagent size. *Eur. J. Biochem.* **139**, 125–130.

Dethmers, J.K. and Meister, A. (1981). Glutathione export by human lymphoid cells: Depletion of glutathione by inhibition of its synthesis decreases export and increases sensitivity to irradiation. *Proc. Natl. Acad. Sci. USA* **78**, 7492–7496.

Dong, Z., Kumar, R., Yang, X., and Fidler, I.J. (1997). Macrophage-derived metalloelastase is responsible for the generation of angiostatin in Lewis lung carcinoma. *Cell* **88**, 801–810.

Folkman, J. (1995). Angiogenesis in cancer, vascular, rheumatoid and other disease. *Nature Med.* **1**, 27–31.

Gately, S., Twardowski, P., Stack, M.S., Cundiff, D.L., Grella, D., Castellino, F.J., Enghild, J., Kwaan, H.C., Lee, F., Kramer, R.A., Volpert, O., Bouck, N., and Soff, G.A. (1997). The mechanism of cancer-mediated conversion of plasminogen to the angiogenesis inhibitor angiostatin. *Proc. Natl. Acad. Sci. USA* **94**, 10868–10872.

Gately, S., Twardowski, P., Stack, M.S., Patrick, M., Boggio, L., Cundiff, D.L., Schnaper, H.W., Madison, L., Volpert, O., Bouck, N., Enghild, J., Kwaan, H.C., and Soff, G.A. (1996). Human prostate carcinoma cells express enzymatic activity that converts human plasminogen to the angiogenesis inhibitor, angiostatin. *Cancer Res.* **56**, 4887–4890.

Harlos, K., Vas, M., and Blake, C.C.F. (1992). Crystal structure of the binary complex of pig muscle phosphoglycerate kinase and its substrate 3-phospho-D-glycerate. *Proteins* **12**, 133–144.

Holmgren, A. (1989). Thioredoxin and glutaredoxin systems. *J. Biol. Chem.* **264**, 13963–13966.

Kassam, G., Kwon, M., Yoon, C.S., Graham, K.S., Young, M.K., Gluck, S., and Waisman, D.M. (2001). Purification and characterization of A61. An angiostatin-like plasminogen fragment produced by plasmin autodigestion in the absence of sulfhydryl donors. *J. Biol. Chem.* **276**, 8924–8933.

Katz, B.A. and Kossiakoff, A. (1986). The crystallographically determined structures of atypical strained disulfides engineered into subtilisin. *J. Biol. Chem.* **261**, 15480–15485.

Lannutti, B.J., Gately, S.T., Quevedo, M.E., Soff, G.A., and Paller, A.S. (1997). Human angiostatin inhibits murine hemangioendothelioma tumor growth in vivo. *Cancer Res.* **57**, 5277–5280.

Lay, A.J., Jiang, X.-M., Daly, E., Sun, L., and Hogg, P.J. (2002). Plasmin reduction by phosphoglycerate kinase is a thiol-independent process. *J. Biol. Chem.* **277**, 9062–9068.

Lay, A.J., Jiang, X.-M., Kisker, O., Flynn, E., Underwood, A., Condron, R., and Hogg, P.J. (2000). Phosphoglycerate kinase acts in tumour angiogenesis as a disulphide reductase. *Nature* **408**, 869–873.

May, A., Vas, M., Harlos, K., and Blake, C. (1996). 2.0 A resolution structure of a ternary complex of pig muscle phosphoglycerate kinase containing 3-phospho-D-glycerate and the nucleotide Mn adenylylimidodiphosphate. *Proteins* **24**, 292–303.

McHarg, J., Kelly, S.M., Price, N.C., Cooper, A., and Littlechild, J.A. (1999). Site-directed mutagenesis of proline 204 in the "hinge" region of yeast phosphoglycerate kinase. *Eur. J. Biochem.* **259**, 939–945.

Minard, P., Desmadril, M., Ballery, N., Perahia, D., and Mouawad, L. (1989). Study of the fast-reacting cysteines in phosphoglycerate kinase using chemical modification and site-directed mutagenesis. *Eur. J. Biochem.* **185**, 419–423.

Moser, T.L., Kenan, D.J., Ashley, T.A., Roy, J.A., Goodman, M.D., Misra, U.K., Cheek, D.J., and Pizzo, S.V. (2001). Endothelial cell surface F1-F0 ATP synthase is active in ATP synthesis and is inhibited by angiostatin. *Proc. Natl. Acad. Sci. USA* **98**, 6656–6661.

Mulligan-Kehoe, M.J., Wagner, R., Wieland, C., and Powell, R. (2001). A truncated plasminogen activator inhibitor-1 protein induces and inhibits angiostatin (kringles 1–3), a plasminogen cleavage product. *J. Biol. Chem.* **276**, 8588–8596.

Noiva, R. (1999). Protein disulfide isomerase: The multifunctional redox chaperone of the endoplasmic reticulum. *Cell Dev. Biol.* **10**, 481–493.

O'Reilly, M.S., Holmgren, L., Chen, C., and Folkman, J. (1996). Angiostatin induces and sustains dormancy of human primary tumors in mice. *Nature Med.* **2**, 689–692.

Ogino, T., Iwama, M., Kinouchi, J., Shibagaki, Y., Tsukamoto, T., and Mizumoto, K. (1999). Involvement of a cellular glycolytic enzyme, phosphoglycerate kinase, in Sendai virus transcription. *J. Biol. Chem.* **274**, 35999–36008.

O'Reilly, M.S., Holmgren, L., Shing, Y., Chen, C., Rosenthal, R.A., Moses, M., Lane, W.S., Cao, Y., Sage, E.H., and Folkman, J. (1994). Angiostatin: A novel angiogenesis inhibitor that mediates the suppression of metastases by a Lewis lung carcinoma. *Cell* **79**, 315–328.

O'Reilly, M.S., Wiederschain, D., Stetler-Stevenson, W.G., Folkman, J., and Moses, M.A. (1999). Regulation of angiostatin production by matrix metalloproteinase-2 in a model of concomitant resistance. *J. Biol. Chem.* **274**, 29568–29571.

Pjura, P.E., Matsumura, M., Wozniak, J.A., and Matthews, B.W. (1990). Structure of a thermostable disulfide-bridge mutant of phage T4 lysozyme shows that an engineered cross-link in a flexible region does not increase the rigidity of the folded protein. *Biochemistry* **29**, 2592–2598.

Popanda, O., Fox, G., and Thielmann, H.W. (1998). Modulation of DNA polymerases alpha, delta and epsilon by lactate dehydrogenase and 3-phosphoglycerate kinase. *Biochim. Biophys. Acta* **1397**, 102–117.

Sim, B.K.L., O'Reilly, M.S., Liang, H., Fortier, A.H., He, W., Madsen, J.W., Lapcevich, R., and Nacy, C.A. (1997). A recombinant human angiostatin protein inhibits experimental primary and metastatic cancer. *Cancer Res.* **57**, 1329–1334.

Stathakis, P., Fitzgerald, M., Matthias, L.J., Chesterman, C.N., and Hogg, P.J. (1997). Generation of angiostatin by reduction and proteolysis of plasmin. Catalysis by a plasmin reductase secreted by cultured cells. *J. Biol. Chem.* **272**, 20641–20645.

Stathakis, P., Lay, A.J., Fitzgerald, M., Schlieker, C., Matthias, L.J., and Hogg, P.J. (1999). Angiostatin formation involves disulfide-bond reduction and proteolysis in kringle 5 of plasmin. *J. Biol. Chem.* **274**, 8910–8916.

Szilágyi, A.N., Ghosh, M., Garman, E., and Vas, M. (2001). A 1.8 Å resolution structure of pig muscle 3-phosphoglycerate kinase with bound MgADP and 3-phosphoglycerate in open conformation: New insight into the role of the nucleotide in domain closure. *J. Mol. Biol.* **306**, 499–511.

Vas, M. and Csanády, G. (1987). The two fast-reacting thiols of 3-phosphoglycerate kinase are structurally juxtaposed. Chemical modification with bifunctional reagents. *Eur. J. Biochem.* **163**, 365–368.

Vishwanatha, J.K., Jindal, H.K., and Davis, R.G. (1992). The role of primer recognition proteins in DNA replication: Association with nuclear matrix in HeLa cells. *J. Cell Sci.* **101**, 25–34.

Watson, H.C., Walker, N.P.C., Shaw, P.J., Bryant, T.N., Wendell, P.L., Fothergill, L.A., Perkins, R.E., Conroy, S.C., Dobson, M.J., Tuite, M.F., Kingsman, A.J., and Kingsman, S.M. (1982). Sequence and structure of yeast phosphoglycerate kinase. *EMBO J.* **1**, 1635–1640.

Weiner, S.J., Kollman, P.A., Case, D.A., Singh, U.C., Ghio, C., Alagona, G., Profeta, S. Jr., and Weiner, P. (1984). A new force field for molecular mechanical simulation of nucleic acids and proteins. *J. Am. Chem. Soc.* **106**, 765–784.

Wells, J.A. and Power, D.B. (1986). In vivo formation and stability of engineered disulfide-bonds in subtilisin. *J. Biol. Chem.* **261**, 6564–6570.

Wetzel, R. (1987). Harnessing disulfide-bonds using protein engineering. *TIBS.* **12**, 478–482.

Wu, H.-L., Shi, G.-Y., and Bender, M.L. (1987a). Preparation and purification of microplasmin. *Proc. Natl. Acad. Sci. USA* **84**, 8292–8295.

Wu, H.-L., Shi, G.-Y., Wohl, R.C., and Bender, M.L. (1987b). Structure and formation of microplasmin. *Proc. Natl. Acad. Sci. USA* **84**, 8793–8795.

Xie, L., Chesterman, C.N., and Hogg, P.J. (2001). Control of von Willebrand factor multimer size by thrombospondin-1. *J. Exp. Med.* **193**, 1341–1349.

8

Mechanism of Angiostatin Formation from Plasminogen

Mijung Kwon and David M. Waisman

1. Introduction

An emerging concept in biology is that certain inactive precursor proteins are cleaved to produce biologically active fragments. This concept of bioregulatory proteolysis is dramatically illustrated for plasminogen where the proteolytic cleavage of the molecule produces the broad spectrum protease, plasmin. The further cleavage of plasmin produces anti-angiogenic plasminogen fragments typically consisting of the first three or four kringle domains. Since these fragments can inhibit primary tumor growth as well as angiogenesis-dependent growth of metastases, they have been referred to as angiostatins (O'Reilly *et al.*, 1994, 1996; Cao *et al.*, 1996).

Angiostatin was originally identified as a 38 kDa plasminogen fragment from mouse urine and serum (O'Reilly *et al.*, 1994). It is now apparent that angiostatin is a member of a large family of anti-angiogenic plasminogen fragments (AAPFs). These AAPFs can be produced *in vitro* either by plasmin autoproteolysis or by the cleavage of plasmin(ogen) by other proteinases. Physiologically relevant AAPFs include a 38 kDa AAPF isolated from the conditioned media of tumor-infiltrating macrophages (Dong *et al.*, 1997), and AAPFs of 48, 42, and 50 kDa produced from RAW 264.7 macrophage conditioned media (Falcone *et al.*, 1998). Other AAPFs include a 50 kDa AAPF isolated from the conditioned media of human prostrate carcinoma PC-3 cells (Gately *et al.*, 1996, 1997), and AAPFs of 66, 60, and 57 kDa detected in the conditioned media of HT1080 and Chinese hamster ovary cells (Stathakis *et al.*, 1999). Since the carboxyl-terminus of most of these AAPFs has not been determined, the exact primary sequence of most of the AAPFs is not known.

We have shown that the primary AAPF present in mouse and human blood as well as in cultured cells has a mol. wt. of 61 kDa (Kassam *et al.*, 2001). We have produced this AAPF, called A_{61}, in a cell-free system consisting of plasminogen and uPA. This has allowed the detailed biochemical characterization of these proteins. A_{61} was shown to be a novel four-kringle containing plasminogen fragment consisting of the amino acid

Mijung Kwon and David M. Waisman • Cancer Biology Research Group, Department of Biochemistry & Molecular Biology, University of Calgary, Calgary, Alberta, Canada T2N 4N1.

Plasminogen: Structure, Activation, and Regulation, edited by David M. Waisman. Kluwer Academic/Plenum Publishers, New York, 2003.

sequence, Lys^{78}–Lys^{468} (Kassam *et al.*, 2001). Plasmin autoproteolysis was suggested to be the primary mechanism for production of A_{61}. However, the release of A_{61} from plasmin not only requires cleavage of the Lys^{468}–Gly^{469} bond by plasmin autoproteolysis but also cleavage of the Cys^{462}–Cys^{541} disulfide bond. In fact, our recent observations have suggested that a plasminogen receptor, annexin II tetramer (AIIt), stimulates the formation of A_{61} (Kwon *et al.*, 2002).

This chapter will discuss the identification and characterization of two anti-angiogenic plasminogen fragments, A_{61} and p22. Furthermore, we will present recent data that establish the mechanism by which A_{61} and p22 are produced by cells.

2. Identification of Plasminogen Fragment, A_{61}, Produced by Plasmin Autoproteolysis

Plasminogen is synthesized in the liver and circulates in plasma at a concentration of about 2 μM. Upon binding to its cellular receptor, it is converted to the active serine protease, plasmin, by the cleavage of Arg^{561}–Val^{562} bond by tissue plasminogen activator (tPA) or urokinase-type plasminogen activator (uPA). The two polypeptide chains of plasmin are held together by disulfide bridges, and each polypeptide chain represents a distinct functional domain of plasmin (Figure 8.1). The larger polypeptide chain or *A* chain (Lys^{78}–Arg^{561}) contains five kringle domains (K), and the smaller polypeptide chain or *B* chain (Val^{562}–Asn^{791}) contains the serine protease domain. The kringles of the *A* chain are lysine-binding domains that mediate protein–protein interactions, such as those between plasminogen and the carboxyl-terminal lysines of its cell-surface receptors.

As mentioned above, the cleavage of plasmin produces AAPFs typically consisting of the first three or four kringle domains (Table 8.1). All of the AAPFs have been shown to inhibit microvascular endothelial cell proliferation and migration *in vitro* (O'Reilly *et al.*, 1994; Cao *et al.*, 1996). Many of the AAPFs also inhibit endothelial cell invasion and tube formation in three-dimensional lattices (Gately *et al.*, 1996; Claesson-Welsh *et al.*, 1998; Troyanovsky *et al.*, 2001). Many of the AAPFs have been shown to inhibit both the primary tumor growth as well as the angiogenesis-dependent growth of metastases in mice. As yet detectable toxicity or resistance to the AAPFs has not been reported (O'Reilly *et al.*, 1996).

The mechanism of the anti-angiogenic activity of the AAPFs is controversial. It has been suggested that the AAPFs induce endothelial cell apoptosis (Claesson-Welsh *et al.*, 1998; Lucas *et al.*, 1998). In contrast, it has also been suggested that AAPFs do not induce apoptosis, but actually trigger mitotic cell death of proliferating endothelial cells (Hari *et al.*, 2000). AAPFs have also been shown to bind to plasma membrane-associated ATP synthase, suggesting that AAPFs may disrupt the production of ATP and thus render the cell more susceptible to hypoxic damage (Moser *et al.*, 1999, 2001). Another laboratory has suggested that AAPFs bind to a cell motility regulatory protein called angiomotin, which blocks the function of angiomotin. Angiomotin is present at the lamellopodia of migrating endothelial cells and induces activation of focal adhesion kinase (Troyanovsky *et al.*, 2001). Most recently, it has been shown that endothelial cells bind to immobilized AAPFs, and this binding is mediated through $\alpha_V\beta_3$ integrin (Tarui *et al.*, 2001). The binding abrogates the induction of stress-fiber formation, suggesting that AAPF action is due to

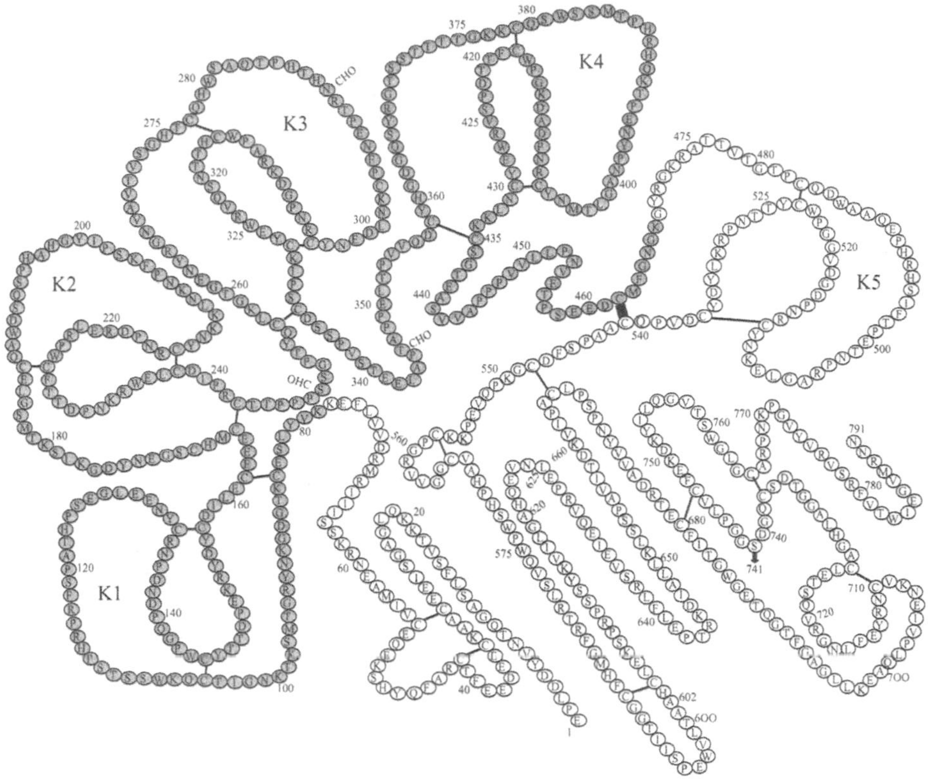

Figure 8.1. Diagrammatic illustration of the structure of plasminogen, A_{61}, and p22. The amino acids are indicated with the single letter code. Numbering is based on the sequence of human [Glu]-plasminogen (791 residues) excluding the 19 amino acid signal peptide that ends at Met^{19}. The shaded region represents the structure of A_{61} (Lys^{78}–Lys^{468}). The bold line indicates the Cys^{462}–Cys^{541} disulfide bond that must be reduced to generate A_{61}. p22 has the primary amino acid sequence of Lys^{78}–Lys^{180}.

perturbation of $\alpha_V\beta_3$ integrin-mediated signaling pathway. From the above discussion it is apparent that further studies will be necessary to establish the mechanism of action of the AAPFs.

Several laboratories have produced AAPFs in cell-free systems. Two mechanisms for the formation of the AAPFs have become well established. The first involves the cleavage of plasmin(ogen) by several distinct proteases. It has been shown that proteases such as elastase, matrilysin (matrix metalloprotease-7, MMP-7), gelatinase B (MMP-9), stromelysin-1 (MMP-3), metalloelastase (MMP-12), prostate-specific antigen (PSA), and cathepsin D can cleave plasminogen to AAPF (O'Reilly *et al.*, 1994; Patterson and Sang, 1997; Cornelius *et al.*, 1998; Lijnen *et al.*, 1998; Heidtmann *et al.*, 1999; Morikawa *et al.*, 2000).

Alternatively, a second mechanism of formation of AAPFs involves plasmin autoproteolysis. It has been observed that a K1–4 containing AAPF of 50 kDa (nonreduced SDS-PAGE) is generated in a cell-free system consisting of plasminogen, uPA, and *N*-acetyl-L-cysteine (Gately *et al.*, 1997). Interestingly, it was suggested that formation of the AAPF was absolutely dependent on the presence of a sulfhydryl donor such as

Table 8.1. Comparison of Various Anti-Angiogenic Plasminogen Fragments

Structure	Mol. wt. (kDa)	Kringles	IC_{50} (nM)	Method of Plasminogen Cleavage	Reference
Y^{80}–A^{440}	51–54	K1–4		Elastase	Takada *et al.* (1988)
Y^{80}–$V^{338/354}$	41–44	K1–3		Elastase	Takada *et al.* (1988)
K^{78}–P^{447}	58	K1–4		MMP-7	Patterson and Sang (1997)
K^{78}–P^{446}	58	K1–4		MMP-9	Patterson and Sang (1997)
N^{60}–P^{447}	55	K1–4		MMP-3	Lijnen *et al.* (1998)
K^{78}–R^{53}	55	K1–4 + 69rK5	0.05	autoproteolysis at pH 11	Cao *et al.* (1999)
unknown	**38**	**K1– (?)**		**Metalloelastase (MMP-12)**	**Dong *et al.* (1997)**
K^{78}–?	**50***	**K1–4 (?)**	**300**	**Reduction and autoproteolysis**	**Gately *et al.* (1997)**
K^{77}– ?	**42,48,50***	**K1–4 (?)**		**Autoproteolysis**	**Falcone *et al.* (1998)**
K^{78}–R^{530} (?)	**66,60,57**	**K1–4 + 69rK5**		**Plasmin reductase and serine protease**	**Stathakis *et al.* (1999)**
Y^{80}–?	**38**	**K1–3 (?)**	**5.2**	**MMP-2**	**O'Reilly *et al.* (1999)**
K^{78}– ?	**65,60,58**	**K1–5 (?)**	**100**	**13 kDa serine protease**	**Li *et al.* (2000)**
unknown	**42,45***	**K1–4 (?)**	**200**	**Procathepsin D**	**Morikawa *et al.* (2000)**
K^{78}–R^{530} (?)	**66,60,57**	**K1–4 + 69rK5**		**Phosphoglycerate kinase and serine protease**	**Lay *et al.* (2000)**
K^{78}–K^{468}	**61**	**K1–4 + 7rK5**	**35**	**Autoproteolysis**	**Kassam *et al.* (2001)**
unknown	**50,40,35***	**K1– (?)**		**PSA**	**Heidtmann *et al.* (1999)**
K^{78}–K^{180}	**22**	**K1+ 15rK2**	**14**	**unknown**	**Kwon *et al.* (2001)**
unknown	**32***	**K1–3 (?)**	**100**	**Neutrophil-secreted elastase**	**Scapini *et al.* (2002)**

Abbreviations: K, kringle; r, amino acid residues. Numbering is based on the sequence of human plasminogen (791 residues) excluding the 19 amino acid signal peptide that ends at Met^{19} and begins at Glu^{1}. The molecular weights are determined by reduced SDS-PAGE or nonreduced SDS-PAGE (*). The IC_{50} is determined as the concentration of AAPF required to half-maximally inhibit the proliferation of microvascular endothelial cells. The symbol (?) indicates proposed structures that are not confirmed by C-terminal sequencing. The regular font style indicates AAPFs produced in the cell-free system, where the bold font style indicates AAPFs generated by the cells.

N-acetyl-L-cysteine. In contrast, it has also been shown that plasmin autoproteolysis results in the generation of a similar AAPF in the absence of sulfhydryl donors (Fitzpatrick *et al.*, 2000). This AAPF was generated by incubating plasminogen with uPA at pH 9.0 in the absence of a sulfhydryl donor (Kassam *et al.*, 2001).

We have extensively characterized the plasminogen fragments produced in the absence of sulfhydryl donors. The plasminogen fragments produced in this reaction were purified by L-lysine–Sepharose affinity and gel permeation chromatography, and were found to be a doublet of apparent mass of 61 and 64 kDa, respectively, on reduced SDS-PAGE. Since the major plasminogen fragment produced by this procedure was 61 kDa on reduced SDS-PAGE, it was named as A_{61}. These two fragments also produced a diffuse band of an apparent mass of 50 kDa on non-reduced SDS-PAGE. These plasminogen fragments reacted with a monoclonal antibody against the K1–3 fragment of plasminogen, confirming that the plasminogen fragments isolated by this procedure were *A*-chain fragments.

The time course of generation of A_{61} indicated that A_{61} was generated after 2–4 hr of incubation of plasminogen and uPA (Kassam *et al.*, 2001). It was also shown that the uPA-dependent conversion of plasminogen to plasmin preceded the formation of A_{61}. In addition, a recombinant mutant human plasminogen (S741C) that is catalytically inactive when converted to plasmin cannot be converted to A_{61} (Kassam *et al.*, 2001). Based on these data, it was suggested that plasminogen was cleaved by uPA to form plasmin, which then underwent autoproteolysis resulting in the formation of A_{61}. The cleavage of disulfide bonds was necessary for the release of the plasminogen fragments from the parent molecule. However, the mechanism by which the disulfide bond(s) of plasmin was cleaved was uncertain.

Two autoproteolytic processes involving human plasmin suggested by Wu *et al.* (Wu *et al.*, 1987; Shi and Wu, 1988) may elucidate the mechanism by which plasmin disulfide bonds are cleaved. The cleavage of plasminogen by plasminogen activators, uPA or tPA results in the generation of disulfide linked heavy and light chains with the kringle-containing domains residing in the heavy chain. In a slightly acidic solution the plasmin light chain is cleaved faster than the heavy chain of plasmin. However, while both the heavy and light chain are cleaved at alkaline pH (between 6.5 and 11.0), the alkaline pH favours the cleavage of the heavy chain. The cleaved heavy chain fragment of mol. wt. 58 kDa was identified and the C-terminus of this was Arg^{530} (Figure 8.1). Collectively, it has been theorized that because of their accessibility to the solvent, the disulfide bonds between Cys^{512} and Cys^{536} and between Cys^{462} and Cys^{541} could be cleaved by hydroxyl ions (Wu *et al.*, 1987). Therefore, the cleavage of the disulfide bonds between Cys^{462} and Cys^{541} by hydroxyl ions could explain how plasmin autoproteolysis results in the release of A_{61} from plasmin at alkaline pH in the absence of sulfhydryl donors.

3. Characterization of A_{61}, an Anti-Angiogenic Plasminogen Fragment

The structure of A_{61} has been characterized (Kassam *et al.*, 2001). The N- and C-terminal sequencing established that the autoproteolytic cleavage sites of plasmin were Lys^{77}–Lys^{78}, Lys^{468}–Gly^{469}, and Arg^{471}–Gly^{472}. The major species of A_{61} has the primary amino acid sequence of Lys^{78}–Lys^{468}, while the minor species is Lys^{78}–Arg^{471} (Figure 8.1). Therefore, A_{61} contains the first four kringles of plasminogen plus seven residues of K5. Mass spectrometry of A_{61} revealed a mol. wt. of 46.6 kDa, whereas the average molecular mass measured from this amino acid sequence was 44.1 kDa. It is likely that this difference is due to the glycosylation of the molecule (Hayes and Castellino, 1979a,b).

The secondary structure of A_{61} has also been analyzed (Kassam *et al.*, 2001). The far-UV circular dichroism spectrum of A_{61} exhibited a strong negative band at 202 nm and a weak positive band at about 227 nm. This spectrum is similar to that reported for other kringle containing *A*-chain fragments such as K4 (Castellino *et al.*, 1986; Cleary *et al.*, 1989; Misselwitz *et al.*, 1994). Analysis of the secondary structure content from circular dichroism (CD) spectra yields about 21% β-strand, 14% β-turn, 18% 3_1-helix, 8% 3_{10}-helix, and 40% unordered (Kassam *et al.*, 2001). Similar to other kringle-containing structures, A_{61} does not contain any α-helix structure, but contains 3_1-helix (Marti *et al.*, 1999).

The kringles of plasminogen play a critical role in substrate recognition and in determining the conformation of plasminogen. The binding of lysine-type zwitterions such as ϵ-aminocaproic acid to the kringles of plasminogen results in an increase of about 7% in the fluorescence emission intensity (Violand *et al.*, 1978). The intrinsic fluorescence emission intensity of A_{61} increased in the presence of another lysine analogue, *trans*-4-(aminomethyl)-cyclohexanecarboxylic acid (AMCHA), by 27%, establishing that at least one of the kringles of A_{61} is functionally active.

The biological activity of A_{61} has also been examined (Kassam *et al.*, 2001). A_{61} inhibited the proliferation of bovine capillary endothelial (BCE) cells, but not cells of other lineages. The IC_{50} for inhibition of endothelial cell proliferation by A_{61}, prepared in the presence or absence of *N*-acetyl-L-cysteine, were identical. In comparison, the IC_{50} of 35 nM for A_{61} (Kassam *et al.*, 2001) was significantly lower than the IC_{50} of 300 nM for other AAPFs prepared in the presence of *N*-acetyl-L-cysteine (Gately *et al.*, 1996). In addition, A_{61} also blocked the growth of Lewis lung carcinoma metastatic foci in C57BL/6 mice. Interestingly, the dosage of A_{61}, 2.5 mg/kg/day, was lower than that of other AAPFs prepared in the presence of *N*-acetyl-L-cysteine, 12 mg/kg/day (Gately *et al.*, 1997). Therefore, A_{61} is as biologically active as the AAPFs produced in the presence of sulfhydryl donors.

4. Identification of Cell-Generated Plasminogen Fragments

The binding of plasminogen to the cell surface and its subsequent conversion to plasmin has been well established. While plasminogen can be converted to plasmin at the cell surface by the action of plasminogen activators such as uPA, it has recently been shown that in addition to plasmin, certain kringle-containing fragments of the plasmin(ogen) heavy chain are also produced at the cell surface (Gately *et al.*, 1996; O'Mahony *et al.*, 1998; Westphal *et al.*, 2000).

Several biologically active, kringle-containing fragments of plasminogen have been shown to occur physiologically. It has been suggested that either cancer cells or the tumor-infiltrating cells such as macrophages express proteinases that are capable of cleaving plasminogen into fragments that consist of various truncations of the kringle-containing *A*-chain. Typically, these kringle-containing truncations consist of the first three (K1–3) or four (K1–4) kringle-containing domains (Gately *et al.*, 1996, 1997; Dong *et al.*, 1997; Stathakis *et al.*, 1997, 1999; Falcone *et al.*, 1998; O'Reilly *et al.*, 1999). As described previously, different cancer cells have shown to produce different AAPFs (also reviewed in Table 8.1).

In order to examine whether A_{61} could be produced by cancer cells, HT1080 fibrosarcoma cells were incubated with plasminogen, and the cell-produced plasminogen fragments were analyzed by reduced or nonreduced SDS-PAGE followed by Western blot with the monoclonal anti-human plasminogen kringle 1–3 antibody. A plasminogen fragment of comparable molecular mass to A_{61} was easily detected in the conditioned media of HT1080 cells (Kassam *et al.*, 2001). The amount of cell-generated, A_{61}-like plasminogen fragment increased with increasing plasminogen concentrations. Based on these data, it has been suggested that the predominant cell-produced AAPF is A_{61}.

Recently, it has been reported that in addition to their production by cancer cells or cancer-associated cells, AAPFs can be produced in non-neoplastic conditions such as a consequence of the inflammatory response (Falcone *et al.*, 1998; Hoffman *et al.*, 1998). Furthermore, AAPFs have also been detected in the tears collected after overnight eye closure, suggesting that these fragments may also be produced under physiological conditions (Sack *et al.*, 1999). Consistent with this concept, A_{61}-like plasminogen fragments were shown to be produced by noncancerous cells such as Bovine Capillary Endothelial (BCE) cells (Kassam *et al.*, 2001).

The biological activity of cell-produced, A_{61}-like plasminogen fragments has also been examined and compared with the A_{61} produced in the cell-free system (Kassam *et al.*, 2001). Cultured cells were incubated with plasminogen and the conditioned media were collected. A_{61} was purified from conditioned media of HT1080 or BCE cells by L-lysine–Sepharose affinity chromatography. We found that A_{61} was released into the media by HT1080 or BCE cells. Furthermore, the cell-produced A_{61} inhibited the proliferation of BCE cells with similar potency to A_{61} produced in the cell-free system. The concentration of A_{61}-like plasminogen fragments produced by HT1080 and BCE cells necessary for half-maximal inhibition was about 45 nM. Therefore, these data suggest that normal cells as well as cancer cells produce A_{61} and the potency of the cancer cell-derived A_{61} is identical to the A_{61} produced by the normal cells.

5. Characterization of p22, the Smallest Anti-Angiogenic Plasminogen Fragment Produced by Cultured Cells

Interestingly, it has also been observed that incubation of plasminogen with HT1080 fibrosarcoma cells or HeLa cells also results in the generation of a novel plasminogen fragment of mol. wt. 22 kDa on reduced SDS-PAGE (Kwon *et al.*, 2001). Based on its mobility on reduced SDS-PAGE, the novel plasminogen fragment was named as p22. The amount of cell-generated p22 increased with increasing plasminogen concentrations. Importantly, the generation of p22 was observed at plasminogen concentration as low as 1 μM, well within the physiological concentration of circulating plasminogen. Like A_{61}, p22 was also generated from noncancerous cells such as BCE cells (Kwon *et al.*, 2001). That p22 originated from the plasminogen heavy chain was demonstrated by the reactivity of the p22 protein with a monoclonal antibody against the first three kringles of plasminogen. Based on its mobility on reduced SDS-PAGE, it was evident that p22 was slightly larger than recombinant kringle 1 (rK1), yet smaller than the mol. wt. of 38 or 25 kDa determined for recombinant kringle 1–3 or recombinant kringle 2–3, respectively. It was therefore likely that p22 was a single kringle-containing plasminogen fragment. The N-terminus of HT1080 cell-generated p22 was found to correspond to Lys^{78} of plasminogen. Based on these data, it was suggested that p22 consists of the kringle 1 of plasminogen and several additional N- and C-terminal residues (Kwon *et al.*, 2001). Thus, p22 was a novel plasminogen fragment.

In order to permit a detailed biochemical characterization of p22, we developed a cell-free system for its large-scale production. p22 was generated in this system by the action of plasmin on A_{61} (Kwon *et al.*, 2001). p22 was purified by L-lysine–Sepharose and octyl-Sepharose chromatography. The p22 produced by a cell-free system was indistinguishable on reduced SDS-PAGE from the p22 produced by incubation of plasminogen with HT1080 cells.

The N- and C-terminal sequencing of the p22 produced in the cell-free system established that p22 was formed by the cleavage of plasminogen at two sites, Lys^{77}–Lys^{78} and Lys^{180}–Thr^{181} (Kwon *et al.*, 2001). Therefore, p22 has the primary amino acid sequence of Lys^{78}–Lys^{180} (Figure 8.1). Thus, p22 encompasses four regions, which include six

N-terminal residues, plasminogen kringle 1, three residues that comprise the kringle 1–kringle 2 linker region, and 15 residues of kringle 2. The average mol. wt. measured from the amino acid sequence was 11,825, which compared favourably with the mol. wt. of 11,821 derived from mass spectrometry of p22. The similarity of the mol. wt. of p22 calculated from the amino acid sequence with that derived from mass spectrometry rules out the possibility of glycosylation or other posttranslational modifications of p22. By comparison, the mol. wt. of rK1 is 9926 as calculated from the amino acid composition, and 9815 by mass spectrometry.

Although p22 is larger than K1, it was unclear if these extended regions of p22 affected the structure or functional properties of the protein. We compared the secondary structure of p22 with that of rK1 (Kwon *et al.*, 2001). The CD spectrum of rK1 exhibited a positive band at 227.5 nm and a strong negative band at 197.4 nm. The overall shape of the CD spectrum of p22 was distinct from that of rK1. The spectrum of p22 showed a positive band at 227.5 nm and strong negative band at 202.7 nm, which is characteristic of a protein devoid of α-helical structure. Like A_{61}, the negative ellipticity near 200 nm is consistent with the presence of 3_1 helices (poly(pro)II helices) within the secondary structure of p22 or rK1. The major difference in the CD spectra between p22 and rK1 is the increased content of β turns in p22.

The intrinsic fluorescence excitation and emission maxima of p22 were 283 and 332 nm, respectively (Kwon *et al.*, 2001). In comparison, the intrinsic fluorescence excitation and emission maxima of rK1 were 283 and 339 nm, respectively (Kwon *et al.*, 2001). The binding of the lysine analog, AMCHA, to p22 caused a significant increase (22%) in the intrinsic fluorescence emission spectra. By contrast, although rK1 also showed a large increase in intrinsic fluorescence emission spectra (20%) upon binding to AMCHA, the emission maximum was blue-shifted from 339 to 331 nm. Therefore, these data established that the kringle domain of p22 was conformationally distinct from rK1. It was reasonable to suspect that the extended regions of p22 interacted with the kringle domain and modified its properties.

Although p22 and rK1 were shown to be conformationally distinct, it was unclear if these differences impacted on their biological activity. Therefore, the biological activity of p22 was compared with rK1 (Kwon *et al.*, 2001). The IC_{50} of p22 for the inhibition of BCE cell proliferation was 14 nM, while that of rK1 was 40 nM. It was also shown that concentrations of p22 as high as 2 μM did not inhibit the proliferation of several non-endothelial cell lines such as HT1080 fibrosarcoma cells, HeLa cells, and 293 cells. p22 was also found to prevent *de novo* growth of blood vessels on chick chorioallantoic membranes (CAMs), suggesting that p22 was an anti-angiogenic protein (Kwon *et al.*, 2001). p22 also blocked the growth of Lewis lung carcinoma metastatic foci in C57BL/6 mice (Kwon *et al.*, 2001). Using a similar dose of p22 to A_{61} (2.5 mg/kg/day), similar level of inhibition of metastases was observed. Considering p22 is a non-glycosylated protein, whereas A_{61} is heavily glycosylated, these results are thought to be interesting because the extent of glycosylation has been suggested to affect the biological activity and the half-life of AAPFs in circulation (Pirie-Shepherd, 1999). Therefore, these data demonstrate that p22 also has potent anti-metastatic activity, and indicate that glycosylation may not be required for the biological activity of the plasminogen fragments.

To date, p22 is the smallest plasminogen fragment generated under physiological conditions. Since p22 contains only a single kringle domain, future structure–function analysis of the anti-angiogenic activity of this protein will be simpler than similar studies of the multiple kringle-containing plasminogen fragments such as A_{61}.

6. Identification of Plasminogen Fragments Present in Sera

It was originally reported that an *A*-chain plasminogen fragment of 38 kDa was present in the urine and serum of mice bearing Lewis lung carcinoma (O'Reilly *et al.*, 1994). This *A*-chain fragment was named angiostatin. Based on N-terminal sequencing data it was claimed that mouse angiostatin consisted of the amino acid sequence Tyr^{78}–Ala^{440}. In addition, angiostatin was shown to be present in the serum of tumor-bearing mice but not in the serum of normal mice or tumor-resected mice (O'Reilly *et al.*, 1994). Therefore, it was proposed that the primary tumor produced angiostatin as a means of inhibiting its metastases. The loss of primary tumor resulted in a loss of angiostatin production, which resulted in a rapid growth of distant metastases (O'Reilly, 1997).

Recently, however, contradictory results have been found by our group (Kassam *et al.*, 2001). First, the major plasminogen *A*-chain fragments found in the serum of Lewis lung carcinoma-bearing mice was similar to A_{61}. A 38 kDa protein was not detected. Second, it was observed that the sera from normal, tumor-bearing, and tumor-resected mice have similar levels of the A_{61}-like plasminogen fragment. This mouse plasminogen fragment was of similar size to the lower mol. wt. band of purified human A_{61}. Although it has been reported that monoclonal anti-human plasminogen kringle 1–3 antibody shows limited cross-reactivity with mouse angiostatin (Gorrin-Rivas *et al.*, 2000), both polyclonal anti-mouse angiostatin antibody and monoclonal anti-human plasminogen kringle 1–3 antibody detected the mouse A_{61}-like plasminogen fragment. Furthermore, it was observed that tumor resection did not change the concentration of A_{61}-like plasminogen fragment in mouse serum (Kassam *et al.*, 2001). This result was intriguing since it was expected that release of angiostatin-like plasminogen fragments by the primary tumor was responsible for inhibition of metastases, thus reduction of the primary tumor should cause a dramatic loss in the serum concentration of the mouse A_{61}-like protein.

Supporting this observation, an A_{61}-like plasminogen fragment was also found to be present in the serum of healthy human volunteers and patients diagnosed with several forms of cancer (Kassam *et al.*, 2001). In all cases tested, the plasminogen fragment present in human serum was of similar size to the higher mol. wt. form of A_{61}. These data suggest that *A*-chain fragments are not exclusively produced in a pathophysiological setting. Therefore, it is reasonable to suspect that A_{61}-like *A*-chain fragments could be produced merely as a function of plasmin autoproteolysis.

7. Annexin II Tetramer, a Putative Plasminogen Receptor

To date, eight distinct plasminogen-binding proteins have been identified on the cell surface of various types of cells. These plasminogen-binding proteins include (a) histidine–proline-rich glycoprotein (HPRG) (Lijnen *et al.*, 1980; Jespersen and Kluft, 1982; Leung *et al.*, 1983; Borza and Morgan, 1997), (b) gp 330 (Kanalas and Makker, 1991), (c) α-enolase (Miles *et al.*, 1991; Nakajima *et al.*, 1994; Redlitz *et al.*, 1995;

Andronicos *et al.*, 1997), (d) a 41 kDa prokaryotic protein (Broder *et al.*, 1991), (e) cytokeratin 8 (Hembrough *et al.*, 1995, 1996a,b), (f) endothelial cell-surface actin (Dudani and Ganz, 1996), (g) annexin II (Cesarman *et al.*, 1994; Hajjar *et al.*, 1994), and (h) the p11 subunit of annexin II tetramer (AIIt) (Kassam *et al.*, 1998a; reviewed by Fogg *et al.* in this book).

AIIt is a Ca^{2+}- and phospholipid-binding protein that can exist intracellularly and on the extracellular surface of the plasma membrane (reviewed in Waisman, 1995; Siever and Erickson, 1997). It consists of two copies of a 36 kDa subunit (p36) called annexin II, and two copies of a 11 kDa subunit (p11) called S100A10 (Figure 8.2). The p36 subunit can exist as a subunit of AIIt or as a monomer. The p11 subunit can also exist intracellularly as a free dimer. However, the complex formed by the interaction of annexin II and p11, AIIt, is the predominant form in most cells (Waisman, 1995). For example, in MDCK cells, bovine intestinal epithelial cells, and calf pulmonary artery endothelial cells, 90–95% of the total cellular annexin II is present as a subunit of AIIt (Gerke and Weber, 1984; Waisman, 1995; Nilius *et al.*, 1996).

Several laboratories have shown that both annexin II and AIIt exist on the extracellular surface, although both proteins lack a signal peptide for extracellular sorting (Tressler *et al.*, 1993; Yeatman *et al.*, 1993; Cesarman *et al.*, 1994; Ma *et al.*, 1994; Kassam *et al.*, 1998a; Kang *et al.*, 1999; Mai *et al.*, 2000). Therefore, these proteins cannot go through the conventional endoplasmic reticulum and Golgi secretory pathway. It has been speculated that Golgi-independent secretory mechanisms might exist for the extracellular transport of a variety of proteins such as fibroblast growth factors, interleukin-1, and galectins, all of

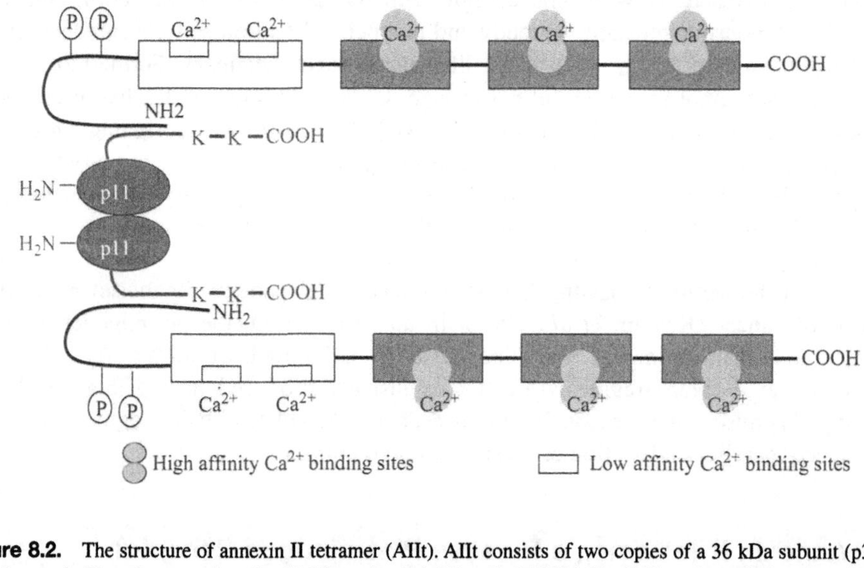

Figure 8.2. The structure of annexin II tetramer (AIIt). AIIt consists of two copies of a 36 kDa subunit (p36) called annexin II, and two copies of a 11 kDa subunit (p11) called S100A10. p36 has three distinct functional regions: the N-terminal region, the core region, and the C-terminal region. The N-terminal region contains the p11 binding sites, and the phosphorylation sites for $pp60^{src}$ and protein kinase C (PKC). The core region has three high-affinity Ca^{2+}-binding sites and two low-affinity Ca^{2+}-binding sites, and phospholipid-binding sites. The C-terminal region contains heparin and F-actin binding sites. p36 has two free thiols at Cys^8 in the N-terminal region and Cys^{334} in the C-terminal region. p11 has C-terminal lysines that play a key role in the stimulation of tPA-catalyzed conversion of plasminogen to plasmin. p11 also has a free thiol at Cys^{82}.

which are secretory proteins lacking signal peptide (Mignatti and Rifkin, 1991; Mignatti *et al.*, 1992; Hughes, 1999). Most recently, it has been reported that annexin II is secreted from chromaffin cells due to the membrane disruption during exocytosis (Faure *et al.*, 2002).

It was initially reported that annexin II bound tPA, plasminogen, and plasmin, and stimulated the tPA-catalyzed conversion of plasminogen to plasmin on the surface of endothelial cells (Cesarman *et al.*, 1994; Hajjar *et al.*, 1994). However, it has been shown by several other groups that the binding of plasminogen to cells is inhibited by lysine analogs such as ϵ-aminocaproic acid, and removal of C-terminal amino acids of extracellular membrane proteins by carboxypeptidase B treatment also blocks plasminogen binding (Felez *et al.*, 1996). These data indicate that the binding of plasminogen to its cellular receptor involves the interaction of plasminogen with the C-terminal lysines of the receptor (Felez *et al.*, 1996). In contrast to annexin II, which lacks C-terminal lysines, the last ten residues of the C-terminus of the p11 subunit of AIIt are [87]V–V–H–M–**K**–Q–**K**–G–**K**–**K**[96]. Therefore, p11 has the prerequisite C-terminal lysine that has been shown to be necessary for extracellular surface plasminogen activation.

As discussed, plasminogen is converted to plasmin by the action of the plasminogen activators, tPA or uPA. Typically the rate of cleavage of plasminogen by tPA is very slow. Although both annexin II and AIIt can stimulate the rate of plasminogen activation by the plasminogen activators, AIIt is a more potent stimulator. For example, AIIt stimulates the rate of tPA-catalyzed activation of [Glu]-plasminogen about 341-fold compared to an approximate 6-fold stimulation by annexin II (Kassam *et al.*, 1998a). AIIt possesses three properties that distinguish it from annexin II.

First, we have observed that while AIIt and its p11 subunit bind tPA and plasminogen, annexin II does not (MacLeod *et al.*, 2003). Using surface plasmon resonance to measure the interactions between AIIt and plasminogen in real time, we have shown that AIIt binds tPA (Kd=0.68 μM) and plasminogen (Kd=0.11 μM). Furthermore, the p11 subunit is the plasminogen binding subunit and binds both tPA (Kd=0.45 μM) and plasminogen (Kd=1.81 μM). In addition, the carboxyl terminal lysine of the p11 subunit is shown to be a common binding site for both tPA and plasminogen. Contrary to previous reports, we have shown that the annexin II subunit does not bind tPA or plasminogen.

Second, while both annexin II (Kd=0.78 μM) and AIIt (Kd=0.75 μM) bind plasmin (MacLeod *et al.*, 2003), only AIIt protects plasmin from inactivation by α_2-antiplasmin (Kassam *et al.*, 1998b). This is a critical difference between annexin II and AIIt because it suggests that the small amounts of plasmin produced by annexin II-dependent stimulation of the tPA-catalyzed cleavage of plasminogen will be rapidly inactivated by its physiological inhibitor.

Third, we have shown that AIIt, but not annexin II, stimulates the autoproteolysis of plasmin at the cell surface (Fitzpatrick *et al.*, 2000) or on the fibrin clot surface (Choi *et al.*, 2001). These data imply that the conformation of plasmin bound to AIIt is distinct from that bound to annexin II. These data have important implications. It suggests that AIIt regulates plasmin activity by stimulating its formation and destruction. Our interpretation of this observation is that AIIt stimulates the generation of a steady-state concentration of plasmin at the cell surface. Furthermore, these data help to understand the mechanism by which AIIt stimulates A_{61} formation from plasmin on the cell surface. The simplest mechanism consistent with these data is that AIIt supplies the precursors for the A_{61} reaction by promoting plasmin formation and its subsequent autoproteolysis.

The key question to be answered is why do AIIt and annexin II have so fundamentally different properties. The answer to this question lies in our observation that the p11 subunit of AIIt is the critical subunit for plasminogen binding and that this subunit interacts via its C-terminal lysine residues with plasminogen. Since annexin II does not possess C-terminal lysines, its interaction with plasminogen must be fundamentally different.

Therefore, an important question is—does the annexin II subunit play a role in plasmin generation or destruction? We have assessed the role of annexin II subunit of AIIt in plasminogen regulation by two methods. First, we examined the activity of a truncation mutant of AIIt, which consisted of the fifteen amino acids of the amino-terminus of annexin II (the entire p11 binding domain) and intact p11. We observed that the activity of this truncation mutant was similar to that of wild-type recombinant AIIt. This suggests that within AIIt, the amino-terminus of annexin II functions to stimulate the activity of the p11 subunit. Second, we have removed the C-terminal lysines of the p11 subunit by carboxypeptidaseB treatment of AIIt or by formation of a mutant recombinant AIIt consisting of wild-type p36 and $p11_{desKK}$ (a C-terminal double lysine deletion mutant of the p11 subunit). We found that this mutant AIIt possessed about 10–15% of the activity of the wild-type protein (Kassam *et al.*, 1998b). Therefore, these data establish that AIIt functions as a plasminogen receptor and that its p11 subunit stimulates the tPA-catalyzed formation of plasmin from plasminogen as well as the destruction of plasmin via stimulation of the plasmin autoproteolytic reaction.

8. Plasmin Reductase Activity of Annexin II Tetramer

As previously discussed, A_{61} is an internal fragment of plasminogen that encompasses the sequence Lys^{78}–Lys^{468} (Figure 8.1). In order for A_{61} to be released from plasmin, plasmin must be cleaved at the Lys^{77}–Lys^{78} and Lys^{468}–Gly^{469} peptide bonds, and the Cys^{462}–Cys^{541} disulfide bond of plasmin must be reduced. Since A_{61} is generated in a cell-free system from plasmin at alkaline pH in the absence of sulfhydryl donors, it is likely that cleavage of the Cys^{462}–Cys^{541} disulfide bond is catalyzed by hydroxyl ions *in vitro*. In contrast, it has been observed that at physiological pH the conversion of plasminogen to A_{61} is very slow in a cell-free system (unpublished observation). These data contrast with the observation that HT1080 fibrosarcoma and BCE cells stimulate the rapid formation of A_{61} from plasminogen at physiological pH (Kassam *et al.*, 2001). The mechanism by which these cells stimulate plasmin reduction and the release of A_{61} from plasmin was unclear.

We have reported that AIIt stimulates the generation of A_{61} in a cell-free system consisting of plasminogen, uPA, and AIIt at pH 7.5 (Kwon *et al.*, 2002). Consistent with the proposed mechanism, the A_{61} generated in these reactions contains a free sulfhydryl at Cys^{462}. The free sulfhydryl-reactive reagent, 3-(*N*-maleimidylpropionyl)biocytin (MPB), was used to label the generated free sulfhydryl-containing proteins in the reaction. The reaction of free sulfhydryl-containing proteins with MPB resulted in the biotinylation of the protein, which allowed easy detection with horseradish peroxidase-conjugated streptavidin in Western blot analysis. Since plasminogen and plasmin contain only disulfides, they do not react with MPB. In the absence of AIIt, plasminogen and uPA do not generate any free sulfhydryl-containing A_{61} in the same reaction conditions (Kwon *et al.*, 2002). Since AIIt stimulated the generation of A_{61} from plasminogen in the absence of sulfhydryl donors, we reported that AIIt facilitated the cleavage of a plasmin disulfide bond, presumably the Cys^{462}–Cys^{541} disulfide, resulting in the release of A_{61} from plasmin and the generation of a free sulfhydryl (Cys^{462}) in A_{61}. AIIt stimulated the dose- and time-dependent conversion of plasminogen to A_{61} (Kwon *et al.*, 2002). The maximal conversion of plasminogen to A_{61} occurred at approximately equimolar concentrations of AIIt and plasminogen. At equimolar concentrations, the half-maximal conversion of plasminogen to A_{61} occurred between 30 and 60 min.

The observation that AIIt caused the cleavage of plasmin(ogen) disulfide bonds was very exciting. Mechanistically, it was unclear how this reaction occurred. In this regard, AIIt could act as a plasmin reductase and utilize its cysteinyl residues to reduce the plasmin(ogen) disulfide bonds. Alternatively, the interaction of AIIt with plasmin(ogen) might induce a conformational contortion of plasmin(ogen) disulfide bonds, thus allowing the cleavage of the disulfide bonds by hydroxyl ions. However, we observed that AIIt itself lost reactivity to MPB during the reaction, which established that AIIt became oxidized during plasmin reduction reaction. Therefore, AIIt utilized its cysteinyl residues to reduce plasmin(ogen) disulfides.

The possible participation of the cysteines of each subunit of AIIt in the plasmin reductase reaction was examined, and it was shown that either the p36 or p11 subunit stimulated the formation of A_{61} (Kwon *et al.*, 2002). However, AIIt was a more potent plasmin reductase than either subunit, suggesting that the interaction of the subunits potentiates the plasmin reductase activity of either subunit. In order to determine which cysteinyl residue was involved in the plasmin reduction reaction, we mutated the individual cysteinyl residues of both subunits and examined the mutant AIIt for plasmin reductase activity. The p11 subunit contains two cysteinyl residues: Cys^{61} that plays a critical role in the binding to p36 and Cys^{82} that is a free thiol (Johnsson and Weber, 1990). The substitution of either of these cysteine residues resulted in a loss of plasmin reductase activity of p11 (Kwon *et al.*, 2002). This suggests that both cysteinyl residues of p11 are required to sustain the plasmin reductase activity of the p11 subunit. The p36 subunit (annexin II) consists of two thiol-containing cysteines, Cys^{8} and Cys^{334} (Burger *et al.*, 1996). The substitution of Cys^{334}, but not Cys^{8}, blocks the plasmin reductase activity of p36 (Kwon *et al.*, 2002). Most importantly, we found that the AIIt composed of p36Cys^{334}Ser and wild-type p11 showed a dramatic loss in plasmin reductase activity (unpublished observation). This result suggests that the Cys^{334} thiol of the p36 subunit is critical for plasmin reductase activity of AIIt. Our data provide a function for the plasminogen-binding property of annexin II. Although annexin II is not involved in the formation or destruction of plasmin, it utilizes its Cys^{334} thiol residues to reduce plasmin and thereby promote the generation of A_{61}.

Since the Cys^{334} residue is highly conserved among many of the annexins (Raynal and Pollard, 1994), we compared the plasmin reductase activity of seven different annexins. However, only annexin II (p36) and AIIt possessed plasmin reductase activity (Kwon *et al.*, 2002). This result establishes that plasmin reductase activity is not a common feature of the annexins.

Protein disulfide isomerase, thioredoxin, and phosphoglycerate kinase have been shown to be three major protein disulfide reductases that are secreted by cultured cells (Rubartelli *et al.*, 1992; Terada *et al.*, 1995; Lay *et al.*, 2000; Soderberg *et al.*, 2000). The three reductases have been shown to act as plasmin reductases (Stathakis *et al.*, 1997; Lay *et al.*, 2000). However, in recent observation AIIt has been shown to be a more potent plasmin reductase than the other reductases under the assay conditions (Kwon *et al.*, 2002). Protein disulfide reductases typically contain the Cys–X–X–Cys motif in their active sites which participates in the cleavage, formation, and reshuffling of disulfide bonds. Members of this family of proteins include thioredoxin, protein disulfide isomerase, fibronectin, von Willebrand factor, and platelet integrin $\alpha_{IIb}\beta_3$ (Mayadas and Wagner, 1992; Essex and Li, 1999; Langenbach and Sottile, 1999; Lahav *et al.*, 2000; O'Neill *et al.*, 2000; Soderberg *et al.*, 2000). Typically, these protein disulfide reductases share the general property of catalyzing the reduction of insulin disulfides (Heuck and Wolosiuk, 1997a,b). However, AIIt does not contain the Cys–X–X–Cys motif, and does not catalyze the reduction of insulin disulfides (Kwon *et al.*, 2002). This further suggests that the intrinsic plasmin reductase activity of AIIt is distinct from that of the traditional protein reductases. Collectively, these data establish that AIIt regulates the generation of the anti-angiogenic plasminogen fragment, A_{61} by catalyzing the reduction of plasmin disulfides, which allows the release of A_{61} from plasmin in a cell-free system.

9. Mechanism of Action of Annexin II Tetramer on A_{61} Formation

Many laboratories have established that, depending on the type of tumor, either cancer cells themselves or tumor-associated cells such as macrophages, express enzymatic activity that is capable of converting plasminogen into AAPF. Two distinct pathways have been identified for the formation of AAPF. First, macrophage-derived metalloelastase (MMP-12) was shown to cleave plasminogen into AAPF (Dong *et al.*, 1997). In addition, it was shown that gelatinase A (MMP-2), produced directly by the Lewis lung carcinoma cells, was responsible for the conversion of plasminogen to AAPF (O'Reilly *et al.*, 1999). Other studies have shown that PSA, a serine protease secreted by human prostate and human prostate carcinoma cells, and procathepsin D, secreted by human prostate carcinoma cells, can convert plasminogen to AAPF (Heidtmann *et al.*, 1999; Morikawa *et al.*, 2000). In most cases, the generation of these AAPFs involves cleavage in the K3–K4 linker region, which does not contain a disulfide. Therefore, plasmin reduction is not necessary to produce these AAPFs (Table 8.1).

Second, AAPFs are also generated by a three-step mechanism, which involves the conversion of plasminogen to plasmin by uPA, the autoproteolytic cleavage of plasmin, and the release of the resultant plasmin fragment by cleavage of the disulfide bonds. Typically, these cleavages are in the K4–K5 linker region or in the K5 region. Therefore, reduction of plasmin must occur for release of the fragment from the parent molecule (Table 8.1). The cleavage of the plasmin disulfide bonds could be accomplished by free sulfhydryl group donors such as reduced glutathione (Gately *et al.*, 1996, 1997) or by hydroxyl ions at alkaline pH (Wu *et al.*, 1987; Kassam *et al.*, 2001). Alternatively, the plasmin disulfide bonds could be cleaved by plasmin reductases such as phosphoglycerate kinase (Lay *et al.*, 2000; Hogg P.J. in this book) and AIIt (Kwon *et al.*, 2002).

In order to see whether AIIt could play an essential role in the generation of A_{61} at the cell surface, we developed a strategy to either increase or decrease the extracellular expression of p11 subunit. Specifically, we transduced HT1080 fibrosarcoma cells with the retrovirus expressing the p11 gene in the sense (pLin-p11S) or antisense (pLin-p11AS) orientation (Kwon *et al.*, 2002), and selected clones with increased (pLin-p11S) or decreased (pLin-p11AS) surface expression of p11. These stably transduced cells were incubated with plasminogen, and the formation of A_{61} was measured. We found that incubation of the pLin-p11S cells with plasminogen resulted in enhanced A_{61} formation

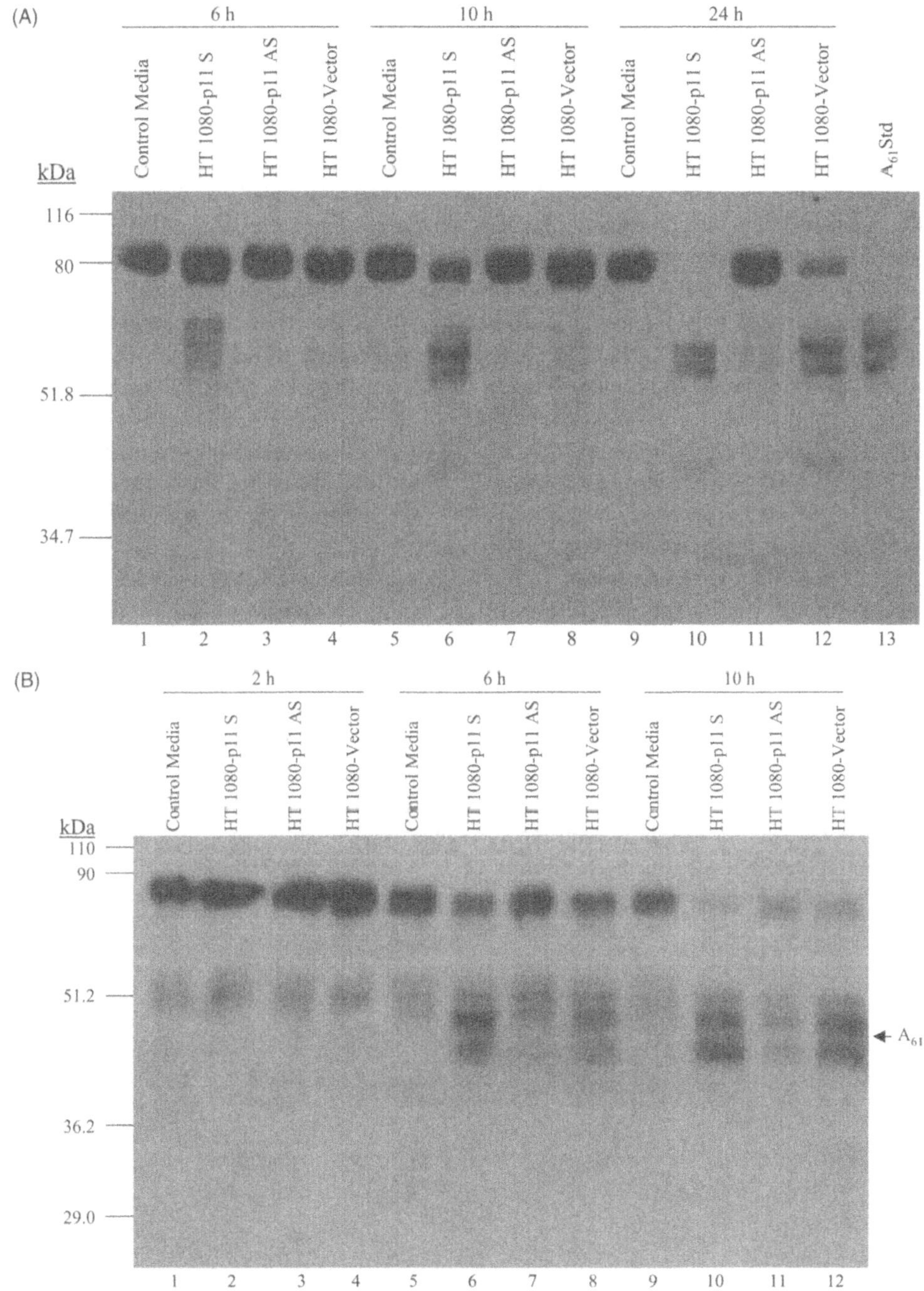

Figure 8.3. Downregulation of AIIt blocks A_{61} generation by HT1080 cells. Transduced HT1080 cells were incubated with DMEM containing 2 μM [Glu]-plasminogen (A) or plasmin (B). After the indicated time of incubation, the medium was analyzed by reduced SDS-PAGE (A) or nonreduced SDS-PAGE (B) followed by Western blot with monoclonal anti-human plasminogen kringle 1–3 antibody. As a control, each protein was incubated with DMEM in the absence of cells (*lane 1, 5, 9*). The following HT1080 clones were used: p11 sense-transduced cells (*lane 2, 6, 10*); p11 antisense-transduced cells (*lane 3, 7, 11*); control vector-transduced cells (*lane 4, 8, 12*). A_{61} standard is also shown (*lane 13*).

compared to the pLin-Vector control cells. In contrast, the pLin-p11AS cells failed to produce A_{61} (Figure 8.3A). It was also observed that A_{61} produced by HT1080 cells contained a free sulfhydryl, which reacted with MPB. The same results have been observed when these cells are incubated with plasmin instead of plasminogen (Figure 8.3B). However, when the diisopropylfluorophosphate-treated, catalytically inactive plasmin was incubated with these cells, A_{61} was not produced. Furthermore, the catalytically inactive plasmin that was incubated with the cells was not reduced, that is, it did not react with MPB. This result establishes that plasmin autoproteolysis is required before plasmin reduction could occur.

Our data form the basis for a model to define the mechanism by which HT1080 cells generate A_{61} from plasminogen. The model postulates that HT1080 fibrosarcoma cells convert plasminogen to A_{61} in a three-step process (Figure 8.4). First, uPA cleaves the Arg^{561}–Val^{562} peptide bond of plasminogen, resulting in the formation of plasmin. Second, plasmin autoproteolysis results in the cleavage of the Lys^{77}–Lys^{78} and Lys^{468}–Gly^{469} peptide bonds. However, the presence of a Cys^{462}–Cys^{541} disulfide bond prevents release of A_{61} (Lys^{78}–Lys^{468}) from the parent molecule. Third, AIIt catalyzes the reduction of the Cys^{462}–Cys^{541} disulfide bond, which then allows the release of A_{61}. It is of note that AIIt stimulates all of these three steps.

This model of A_{61} generation differs from earlier reports. It was reported that the mechanism of AAPF formation involved plasmin autoproteolysis and reduction by sulfhydryl donors such as reduced glutathione (Gately *et al.*, 1997). In contrast, another group proposed that AAPF was formed by the reduction of plasmin by phosphoglycerate kinase followed by the serine protease-dependent release of AAPF from plasmin (Stathakis *et al.*, 1999; Lay *et al.*, 2000). The proposed mechanism of A_{61} generation is unique in that plasmin autoproteolysis must precede plasmin reduction, and that plasmin reduction is performed by cysteinyl residues of AIIt. The AIIt present at the surface of HT1080 fibrosarcoma cells is required to convert plasminogen to A_{61}, and the loss of extracellular AIIt results in the inhibition of A_{61} generation (Kwon *et al.*, 2002). Since large amounts of plasminogen are converted to A_{61} by these cells, AIIt must act as a

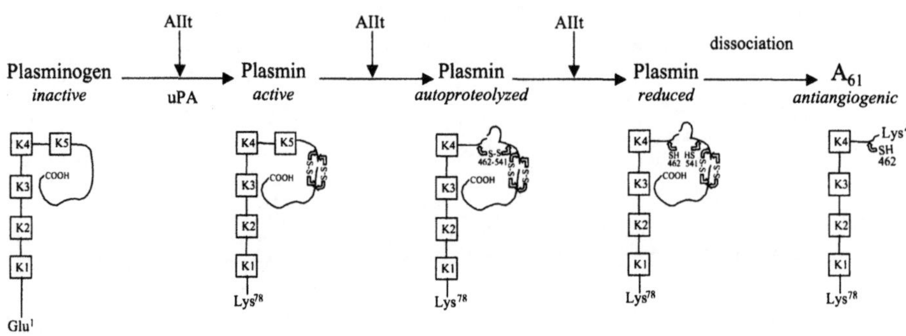

Figure 8.4. Diagrammatic illustration of the mechanism of A_{61} formation. AIIt stimulates the uPA-catalyzed cleavage of plasminogen to plasmin at Arg^{561}–Val^{562} peptide bond. This reaction is promoted by the C-terminal lysines of the p11 subunit on HT1080 cell surface. Plasmin then catalyzes the cleavage of the Lys^{77}–Lys^{78} and Lys^{468}–Gly^{469} peptide bonds of plasmin by autoproteolysis. AIIt then cleaves the Cys^{462}–Cys^{541} disulfide bond. This reaction is catalyzed by the Cys^{334} residue of the p36 subunit. The indications are K, kringle domain of plasminogen, S–S, disulfide bond, and SH, free thiols.

catalyst. This suggests that in order for AIIt to continually reduce plasmin, it must proceed through cycles of oxidation by plasmin and reduction by unknown reducing equivalents. Although the mechanism by which oxidized extracellular AIIt is reduced is currently unknown, it is likely that HT1080 cell-secreted reductase(s) could reduce the oxidized AIIt, thus regenerating the reductase activity of AIIt and allowing AIIt to function as a catalyst.

10. Concluding Remarks

Depending on the type of tumor, either cancer cells or tumor-associated cells express enzymatic activity that is capable of converting plasminogen into AAPF. Two distinct pathways have been suggested for the generation of AAPF. First, it can be direct enzymatic catalysis by proteases such as metalloelastase, PSA, and procathepsin D. Second, AAPF can be generated by a three-step mechanism, which involves the conversion of plasminogen to plasmin by uPA, the autoproteolytic cleavage of plasmin, and the release of the resultant plasmin fragment by cleavage of the disulfide bonds.

This chapter has discussed our data that identify the plasminogen receptor, AIIt, as a key regulatory molecule that stimulates all of three steps necessary for the formation and release of A_{61} from plasminogen. Most importantly, AIIt is identified as a plasmin reductase, which stimulates the reduction of plasmin disulfides, the final step in the generation of A_{61}, a novel AAPF. The AIIt present at the surface of HT1080 fibrosarcoma cells is required to convert plasminogen to A_{61}, and the loss of extracellular AIIt results in the inhibition of A_{61} generation.

Future challenges include identifying the extracellular protein(s) that maintain AIIt in a reduced state, while it participates in plasmin reduction. In addition, defining the interactions that regulate the plasmin reductase activity of AIIt, and determining how the subunits participate in this activity will also be focused on. For example, although the Cys^{334} residue is essential for plasmin reductase activity, it is not labeled with the thiol-specific reagent MPB or iodoacetic acid. It is therefore reasonable to suspect that Cys^{334} may be shielded and only accessible for reduction of plasmin upon binding of AIIt to autoproteolyzed plasmin.

Many anti-angiogenic proteins including AAPFs are currently under evaluation as anticancer drugs. Therefore, understanding the mechanism regulating the generation of these AAPFs will help in defining potential anticancer therapeutic approaches.

Acknowledgment

This work was supported by a grant from CIHR.

References

Andronicos, N.M., Ranson, M., Bognacki, J., and Baker, M.S. (1997). The human *ENO1* gene product (recombinant human alpha-enolase) displays characteristics required for a plasminogen binding protein. *Biochim. Biophys. Acta* **1337**, 27–39.

Borza, D.B. and Morgan, W.T. (1997). Acceleration of plasminogen activation by tissue plasminogen activator on surface-bound histidine–proline-rich glycoprotein. *J. Biol. Chem.* **272**, 5718–5726.

Broder, C.C., Lottenberg, R., von Mering, G.O., Johnston, K.H., and Boyle, M.D. (1991). Isolation of a prokaryotic plasmin receptor. Relationship to a plasminogen activator produced by the same micro-organism. *J. Biol. Chem.* **266**, 4922–4928.

Burger, A., Berendes, R., Liemann, S., Benz, J., Hofmann, A., Gottig, P., Huber, R., Gerke, V., Thiel, C., Romisch, J., and Weber, K. (1996). The crystal structure and ion channel activity of human annexin II, a peripheral membrane protein. *J. Mol. Biol.* **257**, 839–847.

Cao, Y., Ji, R.W., Davidson, D., Schaller, J., Marti, D., Sohndel, S., McCance, S.G., O'Reilly, M.S., Llinas, M., and Folkman, J. (1996). Kringle domains of human angiostatin. Characterization of the anti-proliferative activity on endothelial cells. *J. Biol. Chem.* **271**, 29461–29467.

Cao, R., Wu, H.L., Veitonmaki, N., Linden, P., Farnebo, J., Shi, G.Y., and Cao, Y. (1999). Suppression of angiogenesis and tumor growth by the inhibitor K1–5 generated by plasmin-mediated proteolysis. *Proc. Natl. Acad. Sci. USA* **96**, 5728–5733.

Castellino, F.J., de Serrano, V.S., Powell, J.R., Johnson, W.R., and Beals, J.M. (1986). Examination of the secondary structure of the kringle 4 domain of human plasminogen. *Arch. Biochem. Biophys.* **247**, 312–320.

Cesarman, G.M., Guevara, C.A., and Hajjar, K.A. (1994). An endothelial cell receptor for plasminogen/tissue plasminogen activator (t-PA). II. Annexin II-mediated enhancement of t-PA-dependent plasminogen activation. *J. Biol. Chem.* **269**, 21198–21203.

Choi, K.S., Fitzpatrick, S.L., Filipenko, N.R., Fogg, D.K., Kassam, G., Magliocco, A.M., and Waisman, D.M. (2001). Regulation of plasmin-dependent fibrin clot lysis by annexin II heterotetramer. *J. Biol. Chem.* **276**, 25212–25221.

Claesson-Welsh, L., Welsh, M., Ito, N., Anand-Apte, B., Soker, S., Zetter, B., O'Reilly, M., and Folkman, J. (1998). Angiostatin induces endothelial cell apoptosis and activation of focal adhesion kinase independently of the integrin-binding motif RGD. *Proc. Natl. Acad. Sci. USA* **95**, 5579–5583.

Cleary, S., Mulkerrin, M.G., and Kelley, R.F. (1989). Purification and characterization of tissue plasminogen activator kringle-2 domain expressed in *Escherichia coli. Biochemistry* **28**, 1884–1891.

Cornelius, L.A., Nehring, L.C., Harding, E., Bolanowski, M., Welgus, H.G., Kobayashi, D.K., Pierce, R.A., and Shapiro, S.D. (1998). Matrix metalloproteinases generate angiostatin: Effects on neovascularization. *J. Immunol.* **161**, 6845–6852.

Dong, Z., Kumar, R., Yang, X., and Fidler, I.J. (1997). Macrophage-derived metalloelastase is responsible for the generation of angiostatin in Lewis lung carcinoma. *Cell* **88**, 801–810.

Dudani, A.K., and Ganz, P.R. (1996). Endothelial cell surface actin serves as a binding site for plasminogen, tissue plasminogen activator and lipoprotein(a). *Br. J. Haematol.* **95**, 168–178.

Essex, D.W. and Li, M. (1999). Protein disulphide isomerase mediates platelet aggregation and secretion. *Br. J. Haematol.* **104**, 448–454.

Falcone, D.J., Khan, K.M., Layne, T., and Fernandes, L. (1998). Macrophage formation of angiostatin during inflammation. A byproduct of the activation of plasminogen. *J. Biol. Chem.* **273**, 31480–31485.

Faure, A.V., Migne, C., Devilliers, G., and Ayala-Sanmartin, J. (2002). Annexin 2 "secretion" accompanying exocytosis of chromaffin cells: Possible mechanisms of annexin release. *Exp. Cell Res.* **276**, 79–89.

Felez, J., Miles, L.A., Fabregas, P., Jardi, M., Plow, E.F., and Lijnen, R.H. (1996). Characterization of cellular binding sites and interactive regions within reactants required for enhancement of plasminogen activation by tPA on the surface of leukocytic cells. *Thromb. Haemost.* **76**, 577–584.

Fitzpatrick, S.L., Kassam, G., Choi, K.S., Kang, H.M., Fogg, D.K., and Waisman, D.M. (2000). Regulation of plasmin activity by annexin II tetramer. *Biochemistry* **39**, 1021–1028.

Gately, S., Twardowski, P., Stack, M.S., Patrick, M., Boggio, L., Cundiff, D.L., Schnaper, H.W., Madison, L., Volpert, O., Bouck, N., Enghild, J., Kwaan, H.C., and Soff, G.A. (1996). Human prostate carcinoma cells express enzymatic activity that converts human plasminogen to the angiogenesis inhibitor, angiostatin. *Cancer Res.* **56**, 4887–4890.

Gately, S., Twardowski, P., Stack, M.S., Cundiff, D.L., Grella, D., Castellino, F.J., Enghild, J., Kwaan, H.C., Lee, F., Kramer, R.A., Volpert, O., Bouck, N., and Soff, G.A. (1997). The mechanism of cancer-mediated conversion of plasminogen to the angiogenesis inhibitor angiostatin. *Proc. Natl. Acad. Sci. USA* **94**, 10868–10872.

Gerke, V. and Weber, K. (1984). Identity of p36K phosphorylated upon Rous sarcoma virus transformation with a protein purified from brush borders; calcium-dependent binding to non-erythroid spectrin and F-actin. *EMBO J.* **3**, 227–233.

Gorrin-Rivas, M.J., Arii, S., Furutani, M., Mizumoto, M., Mori, A., Hanaki, K., Maeda, M., Furuyama, H., Kondo, Y., and Imamura, M. (2000). Mouse macrophage metalloelastase gene transfer into a murine melanoma suppresses primary tumor growth by halting angiogenesis. *Clin. Cancer Res.* **6**, 1647–1654.

Hajjar, K.A., Jacovina, A.T., and Chacko, J. (1994). An endothelial cell receptor for plasminogen/tissue plasminogen activator. I. Identity with annexin II. *J. Biol. Chem.* **269**, 21191–21197.

Hari, D., Beckett, M.A., Sukhatme, V.P., Dhanabal, M., Nodzenski, E., Lu, H., Mauceri, H.J., Kufe, D.W., and Weichselbaum, R.R. (2000). Angiostatin induces mitotic cell death of proliferating endothelial cells. *Mol. Cell Biol. Res. Commun.* **3**, 277–282.

Hayes, M.L. and Castellino, F.J. (1979a). Carbohydrate of the human plasminogen variants. II. Structure of the asparagine-linked oligosaccharide unit. *J. Biol. Chem.* **254**, 8772–8776.

Hayes, M.L. and Castellino, F.J. (1979b). Carbohydrate of the human plasminogen variants. III. Structure of the O-glycosidically linked oligosaccharide unit. *J. Biol. Chem.* **254**, 8777–8780.

Heidtmann, H.H., Nettelbeck, D.M., Mingels, A., Jager, R., Welker, H.G., and Kontermann, R.E. (1999). Generation of angiostatin-like fragments from plasminogen by prostate-specific antigen. *Br. J. Cancer* **81**, 1269–1273.

Hembrough, T.A., Vasudevan, J., Allietta, M.M., Glass, W.F., and Gonias, S.L. (1995). A cytokeratin 8-like protein with plasminogen-binding activity is present on the external surfaces of hepatocytes, HepG2 cells and breast carcinoma cell lines. *J. Cell Sci.* **108**(Pt. 3), 1071–1082.

Hembrough, T.A., Kralovich, K.R., Li, L., and Gonias, S.L. (1996a). Cytokeratin 8 released by breast carcinoma cells *in vitro* binds plasminogen and tissue-type plasminogen activator and promotes plasminogen activation. *Biochem. J.* **317**(Pt. 3), 763–769.

Hembrough, T.A., Li, L., and Gonias, S.L. (1996b). Cell-surface cytokeratin 8 is the major plasminogen receptor on breast cancer cells and is required for the accelerated activation of cell-associated plasminogen by tissue-type plasminogen activator. *J. Biol. Chem.* **271**, 25684–25691.

Heuck, A.P. and Wolosiuk, R.A. (1997a). Di-fluoresceinthiocarbamyl-insulin: A fluorescent substrate for the assay of protein disulfide oxidoreductase activity. *Anal. Biochem.* **248**, 94–101.

Heuck, A.P. and Wolosiuk, R.A. (1997b). Fluoresceinthiocarbamyl-insulin: A potential analytical tool for the assay of disulfide bond reduction. *J. Biochem. Biophys. Methods* **34**, 213–225.

Hoffman, R., Starkey, S., and Coad, J. (1998). Wound fluid from venous leg ulcers degrades plasminogen and reduces plasmin generation by keratinocytes. *J. Invest. Dermatol.* **111**, 1140–1144.

Hughes, R.C. (1999). Secretion of the galectin family of mammalian carbohydrate-binding proteins. *Biochim. Biophys. Acta* **1473**, 172–185.

Jespersen, J. and Kluft, C. (1982). Decreased levels of histidine-rich glycoprotein (HRG) and increased levels of free plasminogen in women on oral contraceptives low in estrogen. *Thromb. Haemost.* **48**, 283–285.

Johnsson, N. and Weber, K. (1990). Alkylation of cysteine 82 of p11 abolishes the complex formation with the tyrosine-protein kinase substrate p36 (annexin 2, calpactin 1, lipocortin 2). *J. Biol. Chem.* **265**, 14464–14468.

Kanalas, J.J. and Makker, S.P. (1991). Identification of the rat Heymann nephritis autoantigen (GP330) as a receptor site for plasminogen. *J. Biol. Chem.* **266**, 10825–10829.

Kang, H.M., Choi, K.S., Kassam, G., Fitzpatrick, S.L., Kwon, M., and Waisman, D.M. (1999). Role of annexin II tetramer in plasminogen activation. *Trends Cardiovasc. Med.* **9**, 92–102.

Kassam, G., Choi, K.S., Ghuman, J., Kang, H.M., Fitzpatrick, S.L., Zackson, T., Zackson, S., Toba, M., Shinomiya, A., and Waisman, D.M. (1998a). The role of annexin II tetramer in the activation of plasminogen. *J. Biol. Chem.* **273**, 4790–4799.

Kassam, G., Le, B.H., Choi, K.S., Kang, H.M., Fitzpatrick, S.L., Louie, P., and Waisman, D.M. (1998b). The p11 subunit of the annexin II tetramer plays a key role in the stimulation of t-PA-dependent plasminogen activation. *Biochemistry* **37**, 16958–16966.

Kassam, G., Kwon, M., Yoon, C.S., Graham, K.S., Young, M.K., Gluck, S., and Waisman, D.M. (2001). Purification and characterization of A61. An angiostatin-like plasminogen fragment produced by plasmin autodigestion in the absence of sulfhydryl donors. *J. Biol. Chem.* **276**, 8924–8933.

Kwon, M., Yoon, C.S., Fitzpatrick, S., Kassam, G., Graham, K.S., Young, M.K., and Waisman, D.M. (2001). p22 is a novel plasminogen fragment with antiangiogenic activity. *Biochemistry* **40**, 13246–13253.

Kwon, M., Caplan, J.F., Filipenko, N.R., Choi, K.S., Fitzpatrick, S.L., Zhang, L., and Waisman, D.M. (2002). Identification of annexin II heterotetramer as a plasmin reductase. *J. Biol. Chem.* **277**, 10903–10911.

Lahav, J., Gofer-Dadosh, N., Luboshitz, J., Hess, O., and Shaklai, M. (2000). Protein disulfide isomerase mediates integrin-dependent adhesion. *FEBS Lett.* **475**, 89–92.

Langenbach, K.J. and Sottile, J. (1999). Identification of protein-disulfide isomerase activity in fibronectin. *J. Biol. Chem.* **274**, 7032–7038.

Lay, A.J., Jiang, X.M., Kisker, O., Flynn, E., Underwood, A., Condron, R., and Hogg, P.J. (2000). Phosphoglycerate kinase acts in tumour angiogenesis as a disulphide reductase. *Nature* **408**, 869–873.

Leung, L.L., Harpel, P.C., Nachman, R.L., and Rabellino, E.M. (1983). Histidine-rich glycoprotein is present in human platelets and is released following thrombin stimulation. *Blood* **62**, 1016–1021.

Li, F., Yang, J., Liu, X., He, P., Ji, S., Wang, J., Han, J., Chen, N., and Yao, L. (2000). Human glioma cell BT325 expresses a proteinase that converts human plasminogen to kringle 1–5-containing fragments. *Biochem. Biophys. Res. Commun.* **278**, 821–825.

Lijnen, H.R., Hoylaerts, M., and Collen, D. (1980). Isolation and characterization of a human plasma protein with affinity for the lysine binding sites in plasminogen. Role in the regulation of fibrinolysis and identification as histidine-rich glycoprotein. *J. Biol. Chem.* **255**, 10214–10222.

Lijnen, H.R., Ugwu, F., Bini, A., and Collen, D. (1998). Generation of an angiostatin-like fragment from plasminogen by stromelysin-1 (MMP-3). *Biochemistry* **37**, 4699–4702.

Lucas, R., Holmgren, L., Garcia, I., Jimenez, B., Mandriota, S.J., Borlat, F., Sim, B.K., Wu, Z., Grau, G.E., Shing, Y., Soff, G.A., Bouck, N., and Pepper, M.S. (1998). Multiple forms of angiostatin induce apoptosis in endothelial cells. *Blood* **92**, 4730–4741.

Ma, A.S., Bell, D.J., Mittal, A.A., and Harrison, H.H. (1994). Immunocytochemical detection of extracellular annexin II in cultured human skin keratinocytes and isolation of annexin II isoforms enriched in the extracellular pool. *J. Cell Sci.* **107**(Pt. 7), 1973–1984.

MacLeod, T.J., Kwon, M., Filipenko, N.R., and Waisman, D.M. (2003). Phospholipid-associated annexin A2-S100 A10 heterotetramer and its subunits: Characterization of the interaction with tissue plasminogen activator, plasminogen, and plasmin. *J. Biol. Chem.* in press.

Mai, J., Finley, R.L., Jr., Waisman, D.M., and Sloane, B.F. (2000). Human procathepsin B interacts with the annexin II tetramer on the surface of tumor cells. *J. Biol. Chem.* **275**, 12806–12812.

Marti, D.N., Schaller, J., and Llinas, M. (1999). Solution structure and dynamics of the plasminogen kringle 2-AMCHA complex: 3(1)-helix in homologous domains. *Biochemistry* **38**, 15741–15755.

Mayadas, T.N. and Wagner, D.D. (1992). Vicinal cysteines in the prosequence play a role in von Willebrand factor multimer assembly. *Proc. Natl. Acad. Sci. USA* **89**, 3531–3535.

Migita, T., Oda, Y., Naito, S., Morikawa, W., Kuwano, M., and Tsuneyoshi, M. (2001). The accumulation of angiostatin-like fragments in human prostate carcinoma. *Clin. Cancer Res.* **7**, 2750–2756.

Mignatti, P. and Rifkin, D.B. (1991). Release of basic fibroblast growth factor, an angiogenic factor devoid of secretory signal sequence: A trivial phenomenon or a novel secretion mechanism? *J. Cell Biochem.* **47**, 201–207.

Mignatti, P., Morimoto, T., and Rifkin, D.B. (1992). Basic fibroblast growth factor, a protein devoid of secretory signal sequence, is released by cells via a pathway independent of the endoplasmic reticulum–Golgi complex. *J. Cell Physiol.* **151**, 81–93.

Miles, L.A., Dahlberg, C.M., Plescia, J., Felez, J., Kato, K., and Plow, E.F. (1991). Role of cell-surface lysines in plasminogen binding to cells: Identification of alpha-enolase as a candidate plasminogen receptor. *Biochemistry* **30**, 1682–1691.

Misselwitz, R., Welfle, K., and Welfle, H. (1994). Conformations and stabilities of human Glu1- and Lys78-plasminogen and of the fragments mini- and microplasminogen, analysed by circular dichroism and differential scanning calorimetry. *Int. J. Biol. Macromol.* **16**, 187–194.

Morikawa, W., Yamamoto, K., Ishikawa, S., Takemoto, S., Ono, M., Fukushi, J., Naito, S., Nozaki, C., Iwanaga, S., and Kuwano, M. (2000). Angiostatin generation by cathepsin D secreted by human prostate carcinoma cells. *J. Biol. Chem.* **275**, 38912–38920.

Moser, T.L., Stack, M.S., Asplin, I., Enghild, J.J., Hojrup, P., Everitt, L., Hubchak, S., Schnaper, H.W., and Pizzo, S.V. (1999). Angiostatin binds ATP synthase on the surface of human endothelial cells. *Proc. Natl. Acad. Sci. USA* **96**, 2811–2816.

Moser, T.L., Kenan, D.J., Ashley, T.A., Roy, J.A., Goodman, M.D., Misra, U.K., Cheek, D.J., and Pizzo, S.V. (2001). Endothelial cell surface F1-F0 ATP synthase is active in ATP synthesis and is inhibited by angiostatin. *Proc. Natl. Acad. Sci. USA* **98**, 6656–6661.

Nakajima, K., Hamanoue, M., Takemoto, N., Hattori, T., Kato, K., and Kohsaka, S. (1994). Plasminogen binds specifically to alpha-enolase on rat neuronal plasma membrane. *J. Neurochem.* **63**, 2048–2057.

Nilius, B., Gerke, V., Prenen, J., Szucs, G., Heinke, S., Weber, K., and Droogmans, G. (1996). Annexin II modulates volume-activated chloride currents in vascular endothelial cells. *J. Biol. Chem.* **271**, 30631–30636.

O'Mahony, C.A., Seidel, A., Albo, D., Chang, H., Tuszynski, G.P., and Berger, D.H. (1998). Angiostatin generation by human pancreatic cancer. *J. Surg. Res.* **77**, 55–58.

O'Neill, S., Robinson, A., Deering, A., Ryan, M., Fitzgerald, D.J., and Moran, N. (2000). The platelet integrin alpha IIbbeta 3 has an endogenous thiol isomerase activity. *J. Biol. Chem.* **275**, 36984–36990.

O'Reilly, M.S., Holmgren, L., Shing, Y., Chen, C., Rosenthal, R.A., Moses, M., Lane, W.S., Cao, Y., Sage, E.H., and Folkman, J. (1994). Angiostatin: A novel angiogenesis inhibitor that mediates the suppression of metastases by a Lewis lung carcinoma. *Cell* **79**, 315–328.

O'Reilly, M.S., Holmgren, L., Chen, C., and Folkman, J. (1996). Angiostatin induces and sustains dormancy of human primary tumors in mice. *Nat. Med.* **2**, 689–692.

O'Reilly, M.S. (1997). Angiostatin: An endogenous inhibitor of angiogenesis and of tumor growth. *EXS* **79**, 273–294.

O'Reilly, M.S., Wiederschain, D., Stetler-Stevenson, W.G., Folkman, J., and Moses, M.A. (1999). Regulation of angiostatin production by matrix metalloproteinase-2 in a model of concomitant resistance. *J. Biol. Chem.* **274**, 29568–29571.

Patterson, B.C. and Sang, Q.A. (1997). Angiostatin-converting enzyme activities of human matrilysin (MMP-7) and gelatinase B/type IV collagenase (MMP-9). *J. Biol. Chem.* **272**, 28823–28825.

Pirie-Shepherd, S.R. (1999). Role of carbohydrate on angiostatin in the treatment of cancer. *J. Lab Clin. Med.* **134**, 553–560.

Raynal, P. and Pollard, H.B. (1994). Annexins: The problem of assessing the biological role for a gene family of multifunctional calcium- and phospholipid-binding proteins. *Biochim. Biophys. Acta* **1197**, 63–93.

Redlitz, A., Fowler, B.J., Plow, E.F., and Miles, L.A. (1995). The role of an enolase-related molecule in plasminogen binding to cells. *Eur. J. Biochem.* **227**, 407–415.

Rubartelli, A., Bajetto, A., Allavena, G., Wollman, E., and Sitia, R. (1992). Secretion of thioredoxin by normal and neoplastic cells through a leaderless secretory pathway. *J. Biol. Chem.* **267**, 24161–24164.

Sack, R.A., Beaton, A.R., and Sathe, S. (1999). Diurnal variations in angiostatin in human tear fluid: A possible role in prevention of corneal neovascularization. *Curr. Eye Res.* **18**, 186–193.

Scapini, P., Nesi, L., Morini, M., Tanghetti, E., Belleri, M., Noonan, D., Presta, M., Albini, A., and Cassatella, M.A. (2002). Generation of biologically active angiostatin kringle 1–3 by activated human neutrophils. *J. Immunol.* **168**, 5798–5804.

Shi, G.Y. and Wu, H.L. (1988). Differential autolysis of human plasmin at various pH levels. *Thromb. Res.* **51**, 355–364.

Siever, D.A. and Erickson, H.P. (1997). Extracellular annexin II. *Int. J. Biochem. Cell Biol.* **29**, 1219–1223.

Soderberg, A., Sahaf, B., and Rosen, A. (2000). Thioredoxin reductase, a redox-active selenoprotein, is secreted by normal and neoplastic cells: Presence in human plasma. *Cancer Res.* **60**, 2281–2289.

Stathakis, P., Fitzgerald, M., Matthias, L.J., Chesterman, C.N., and Hogg, P.J. (1997). Generation of angiostatin by reduction and proteolysis of plasmin. Catalysis by a plasmin reductase secreted by cultured cells. *J. Biol. Chem.* **272**, 20641–20645.

Stathakis, P., Lay, A.J., Fitzgerald, M., Schlieker, C., Matthias, L.J., and Hogg, P.J. (1999). Angiostatin formation involves disulfide bond reduction and proteolysis in kringle 5 of plasmin. *J. Biol. Chem.* **274**, 8910–8916.

Takada, A., Sugawara, Y., and Takada, Y. (1988). Degradation of Glu- and Lys-plasminogen by elastase in the presence or absence of tranexamic acid. *Thromb. Res.* **50**, 285–294.

Tarui, T., Miles, L.A., and Takada, Y. (2001). Specific interaction of angiostatin with integrin alpha(v)beta(3) in endothelial cells. *J. Biol. Chem.* **276**, 39562–39568.

Terada, K., Manchikalapudi, P., Noiva, R., Jauregui, H.O., Stockert, R.J., and Schilsky, M.L. (1995). Secretion, surface localization, turnover, and steady state expression of protein disulfide isomerase in rat hepatocytes. *J. Biol. Chem.* **270**, 20410–20416.

Tressler, R.J., Updyke, T.V., Yeatman, T., and Nicolson, G.L. (1993). Extracellular annexin II is associated with divalent cation-dependent tumor cell-endothelial cell adhesion of metastatic RAW117 large-cell lymphoma cells. *J. Cell Biochem.* **53**, 265–276.

Troyanovsky, B., Levchenko, T., Mansson, G., Matvijenko, O., and Holmgren, L. (2001). Angiomotin: An angiostatin binding protein that regulates endothelial cell migration and tube formation. *J. Cell Biol.* **152**, 1247–1254.

Violand, B.N., Byrne, R., and Castellino, F.J. (1978). The effect of alpha-, omega-amino acids on human plasminogen structure and activation. *J. Biol. Chem.* **253**, 5395–5401.

Waisman, D.M. (1995). Annexin II tetramer: Structure and function. *Mol. Cell Biochem.* **149–150**, 301–322.

Westphal, J.R., Van't Hullenaar, R., Geurts-Moespot, A., Sweep, F.C., Verheijen, J.H., Bussemakers, M.M., Askaa, J., Clemmensen, I., Eggermont, A.A., Ruiter, D.J., and De Waal, R.M. (2000). Angiostatin generation by human tumor cell lines: Involvement of plasminogen activators. *Int. J. Cancer* **86**, 760–767.

Wu, H.L., Shi, G.Y., Wohl, R.C., and Bender, M.L. (1987). Structure and formation of microplasmin. *Proc. Natl. Acad. Sci. USA* **84**, 8793–8795.

Yeatman, T.J., Updyke, T.V., Kaetzel, M.A., Dedman, J.R., and Nicolson, G.L. (1993). Expression of annexins on the surfaces of non-metastatic and metastatic human and rodent tumor cells. *Clin. Exp. Metastasis* **11**, 37–44.

Part II

Physiological Roles of Plasminogen

9

Lessons Learned from the Plg Deficient Mice (Plg−/−)

Jane L. Hoover-Plow

1. Introduction

The generation of the Plg−/− mice has led to important insights about critical and essential functions of the plasminogen system (Plg) in the whole animal. Prior to studies of the Plg−/− mice, the primary role of the Plg system was considered to be its role in fibrinolysis. Other proteolytic functions were observed for the Plg system with *in vitro* protein–protein interactions and cultured cells. In the Plg−/− mice, the role of the Plg system as the major enzymatic system for the degradation of the fibrin and clot lysis was confirmed (Bugge *et al.*, 1995; Ploplis *et al.*, 1995). In addition, the generation of the Plg−/− mice, has allowed the investigation of the functional importance of other roles of the Plg system, such as cell migration (Plow *et al.*, 1999), activation of metalloproteinases (Lijnen and Collen, 1999), and extracellular matrix degradation (Tsirka *et al.*, 1997; Pohl *et al.*, 2001). The Plg system plays a prominent role in physiological and pathological processes that require tissue remodeling. Some of the phenotypes observed in the Plg−/− mice were not anticipated. In this chapter, three roles of the Plg system in mice will be reviewed: (1) the lack of a critical role for Plg in physical and behavioral development; (2) a requirement of Plg for stressed induced behavior; and (3) an important role of the Plg system in adipose tissue development.

2. Development

Prior to the generation of the Plg deficient mice (Plg−/−), the Plg system had been implicated in a number of processes of development particularly in the nervous system. *In vitro* studies showed the production of Plg activators in several model systems related

Jane L. Hoover-Plow • Joseph J. Jacobs Center for Thrombosis and Vascular Biology, Department of Molecular Cardiology, Lerner Research Institute, Cleveland Clinic Foundation.

Plasminogen: Structure, Activation, and Regulation, edited by David M. Waisman. Kluwer Academic/Plenum Publishers, New York, 2003.

to nervous system development: migration of cephalic neural crest cells (Valinsky and Le Douarin, 1985); migration of neurons in rat cerebellum explants (Moonen *et al.*, 1982); extension of neuronal growth cones in cultured neuroblastoma cells (Krystosek and Seeds, 1981); migration of sympathetic neurons in primary cultures of cervical ganglia (Pittman, 1990); differentiation of sensory neurons (McGuire and Seeds, 1990); and regeneration of motor neurons (Nakajima *et al.*, 1996). Plasmin(ogen) and plasmin activity were implicated in models of neuronal migration, neuronal outgrowth, and differentiation of sensory neurons (Moonen *et al.*, 1982; McGuire and Seeds, 1990), as well as mediating the interaction between microglia and dopaminergic neurons (Nakajima *et al.*, 1994). In addition to a role in neuronal development, studies suggested a role of components of the Plg system in embryonic development (Strickland *et al.*, 1976; Sumi *et al.*, 1989; Kaaekuahiwi and Menino, 1990; Stern, 1996), testicular and sperm formation (Vihko *et al.*, 1986; Huarte *et al.*, 1987), ovulation (Beers *et al.*, 1975), placental development (Teesalu *et al.*, 1998), and kidney (Kanwar *et al.*, 1997), skeletal muscle (Hantai *et al.*, 1989) and lung development (Price and Stiles, 1996). Surprisingly, the Plg−/− mice (Hoover-Plow *et al.*, 1999) revealed no major impairment of development of the whole animal or of specific organs. These mice survived, were fertile and viable. A systematic study of early physical and behavioral development revealed only minor differences between the Plg−/− mice and wild-type animals.

2.1. Physical Development

Reproduction (Ny *et al.*, 1999), a complex process involving ovulation, fertilization, embryo implantation, and embryogenesis, is reduced in the Plg−/− mice. Although ovulation efficiency is similar to WT mice (Ny *et al.*, 1999), fertilization is impaired in the Plg−/− mice (Bugge *et al.*, 1995; Ploplis *et al.*, 1995; Ny *et al.*, 1999). A significant number of Plg−/− mice do not become pregnant, and although the number of mice per litter is similar in Plg−/− and WT mice, the Plg−/− mice rarely have more than one litter (Lund *et al.*, 2000). Compromised mammary gland involution and lactation in the Plg−/− mice, may also contribute to the impaired reproduction in the Plg−/− mice (Lund *et al.*, 2000). In tPA−/−:uPA−/− mice (Carmeliet *et al.*, 1995) fertility is also reduced with about 50% of the females producing small litters. While reproduction is reduced in the Plg−/− mice, the number of mice born per litter was similar for heterozygous (Plg+/−) and wild-type (WT, Plg+/+) parents. Overall the expected percentage of WT, Plg+/−, Plg−/− offspring was found (Ploplis *et al.*, 1995; Bugge *et al.*, 1996; Hoover-Plow *et al.*, 1999), indicating embryonic development was not compromised in the Plg−/− mice. No difference was found for milestones of development for ear detachment (4 days of age), lower teeth eruption (6 days of age), upper teeth eruption (11 days of age), or eye opening (14 days of age) among the male or female WT, Plg+/−, and Plg−/− mice (Hoover-Plow *et al.*, 1999). Testicular descent was similar in all three genotypes (3 weeks of age), however vaginal patency was delayed in Plg−/− mice by 4 days compared to the WT mice (4 weeks of age). The delayed vaginal patency may contribute to the reduced reproduction in the Plg−/− mice. Body weight for Plg−/− mice was reported as similar (Bugge *et al.*, 1996) or reduced at 2 months of age (Ploplis *et al.*, 1995; Hoover-Plow *et al.*, 1999). From 2 to 21 days of age, body weight gain was not different for WT, Plg+/−, and Plg−/− mice for both females and males (Figure 9.1). The growth rate in the Plg−/− mice began to

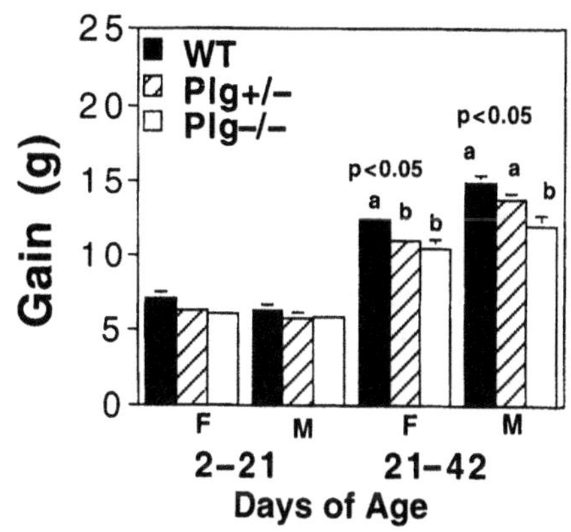

Figure 9.1. Body weight gain was determined for female (F) and male (M) mice for 2–21 days of age and for 21–42 days of age. Values are the mean ± SEM. Bars with a different letter indicate a significant difference ($p<0.05$) for that gender.

decline as early as 4 weeks of age and body weight gain from 21 to 42 days was higher by 2–3 g in the WT mice compared to the Plg−/− mice. Values for the Plg+/− mice were intermediate between the WT and Plg−/− mice, suggesting a dose effect of Plg concentration on body weight gain. At 6 weeks of age, liver, and adrenal weight was similar for Plg−/− and WT mice, and brain weight was larger in Plg−/− mice compared to the WT mice. Body weight, pituitary weight, and adipose depots (inguinal and gonadal) were smaller in the Plg−/− mice than WT mice. A reduced accumulation of adipose tissue in Plg−/− mice was consistent with a reduced body weight (Hoover-Plow *et al.*, 2002). The reduced adipose tissue in the Plg−/− mice will be discussed below. Thus, in the Plg−/− mice physical development was not impaired during embryogenesis and early development (2–21 days of age). Only after 4 weeks of age, were delays observed in physical development in the Plg−/− mice with a decrease in body weight gain and a delay in vaginal patency.

2.2. Behavioral Development

Behavioral development is a sensitive measure of brain and neurological function in the young animal (Altman and Sudarshan, 1975; Adams and Buelke-Sam, 1981). A number of developmental behavior patterns have been standardized in rodents. These measures are utilized to evaluate prenatal and postnatal drugs and environmental factors, as well as genetic mutations and manipulations (Jensh and Brent, 1988; Coscia and Fentress, 1993; Zhong *et al.*, 1996). The pattern and rate of development of negative geotaxis (response to placement on slope at a 45° angle), olfactory discrimination, and swimming were similar in the WT, Plg+/−, and Plg−/− mice (Hoover-Plow *et al.*, 1999). These behaviors were measured at early (negative geotaxis, swimming) and late (olfactory discrimination, swimming) periods of development, and evaluated neuromotor ability, reflex development, reaction to gravitational positioning, and sensory development. Two developmental behaviors, surface righting and pivoting, were slightly accelerated in the Plg−/− mice. Surface righting is the time required to return to all four paws from a supine position and measures reflex development and the integration of motor and vestibular systems (Jensh and Brent, 1988; Zhong *et al.*, 1996). The rate of increase in completing the behavior during 2–9 days of age was similar in the Plg−/− and WT mice, but the Plg−/− mice were generally faster at all ages than the WT mice. Pivoting is a behavior that occurs at about 2 days of age, reaches a maximum and then declines by 8 days of age as other forms of locomotion develop (Altman and Sudarshan, 1975; Adams and Buelke-Sam, 1981). The maximum number of pivots occurred sooner in the Plg−/− and Plg+/− than WT mice.

Grooming is a stereotypic behavior that measures neuromotor function (van Erp *et al.*, 1994), and performance of this sequence of behavior is impaired by excitotoxin lesions of the corpus striatum (Coscia and Fentress, 1993). The rate and pattern of development of the complete grooming sequence (after fur moistening) was similar in the Plg−/− and WT mice (Figure 9.2). The number of episodes and time spent grooming paralleled the increase in the score, which is minimal from days 7 to 13, then accelerates at 15 days of age and reaches a plateau by 18 days of age. The number of episodes of grooming was higher for Plg−/− mice than for WT mice. Thus, the Plg−/− mice developed this stereotypic behavior normally, but appeared to have a greater response than the WT mice.

These studies conducted in the developing mice indicate that Plg does not play a critical role in early physical or behavioral development. Despite the evidence that components of the Plg system, assessed by both *in situ* mRNA and enzyme activity, are localized in the brain during development (Soreq and Miskin, 1981; Ware *et al.*, 1995; Tsirka *et al.*, 1997; Levin *et al.*, 2000; Basham and Seeds, 2001), Plg deficiency does not delay neurological development. In addition to plasmin activity, other proteolytic enzymes, including the calcium-dependent thiol proteases (calpain) (Pittman, 1990) and metalloproteinases (Rao *et al.*, 1996) may predominate or substitute for Plg during development. In addition to replacement by other proteolytic enzyme systems during development, components of the Plg system, the Plg activators, uPA and tPA, may have alternative substrates which are their targets during neurological development. For example, hepatocyte-growth

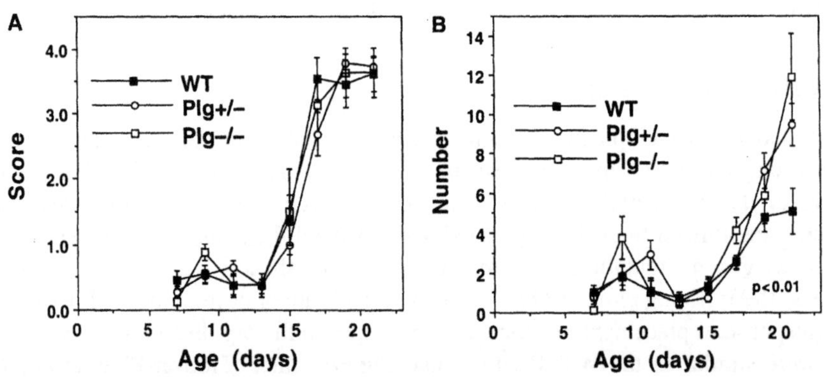

Figure 9.2. Grooming behavior. Panel A: grooming development. Mice were scored from 0–4 based on the completion of the stereotypic sequence. Panel B: grooming episodes. Mice were observed for 5 min after a 1 min session in water and the number of episodes recorded. Values are the mean ± SEM. Grooming episodes were higher in Plg+/− and Plg−/− mice compared to WT ($p < 0.05$) (Panel A from Hoover-Plow, J., Wang, N., and Ploplis, V. Growth and behavioral development in plasminogen gene-targeted mice. (*Growth, Dev. Aging*, **63**, 13–32, 1999, reprinted with permission).

factor (scatter factor), has kringle-containing structural domains similar to Plg (Liu *et al.*, 1994), has been localized in the brain (Jung *et al.*, 1994), and is activated by uPA and tPA (Mars *et al.*, 1993). Recently identified proteins or their homologues, containing at least one kringle domain, such as macrophage stimulating protein (Yoshimura *et al.*, 1993), brain-specific serine protease (Yamamura *et al.*, 1997), hepatocyte growth factor-like protein (Bezerra *et al.*, 1998), and the *N*-methyl-D-aspartate (NMDA) receptor (Nicole *et al.*, 2001) may be candidates for activation by the Plg activators. Thus, a lack of effect of a Plg deficiency on neurological development may reflect the involvement of alternative proteolytic enzyme systems and substrates for the Plg activators.

3. Role of the Plg System in Stress

In adolescence in animals and humans, there is considerable remodeling of the brain (Spear, 2000). While Plg does not appear to be critical in development (less than 3 weeks of age), studies (Hoover-Plow *et al.*, 2001) suggest an important role of the Plg system for brain function in the mature mice. The higher brain weight in Plg−/− mice suggests that brain remodeling may be reduced during the maturing stage of the mice (5–10 weeks of age). Sensitive measures of behavior were again used to examine brain function in the Plg−/− mice. As mentioned above, the level of response was altered in the Plg−/− mice compared to WT mice and suggested an altered response to stress. Behaviors which evaluated locomotor skills (hind-leg gait), learning ability (water maze), and fear (open field) were not altered in the Plg−/− mice. Only selected behaviors were altered in maturing mice, including stress-induced grooming to open field placement and fur moistening and the acoustic startle reflex (ASR). The Plg−/− mice provide a powerful model to examine the relationship of the Plg system and stress.

3.1. Alterations of the Plg System in Response to Stress

Alterations in the Plg system are induced by long- or short-term stress in both human and animal studies. Plasma levels of endothelium-derived tPA are increased during mental stress induced by arithmetic and color tests, similar to the increases found during exercise and epinephrine infusion (Jern *et al.*, 1989). Chronic stress is positively associated with increased plasma concentrations of PAI-1 in healthy middle-aged men (Raikkonen *et al.*, 1996). Workers with higher job demands have lower tPA and higher PAI-1 levels, suggesting that psychosocial job stress may be related to the development of cardiovascular disease via changes in plasma fibrinolytic activity (Ishizaki *et al.*, 1996). Earthquake-induced stress increased both plasma fibrinogen and tPA in human subjects, suggesting increased fibrin turnover (Kario *et al.*, 1997). Electric-foot-shock induced the suppression of fibrinolytic activity by decreasing tPA in rats (Malyszko *et al.*, 1994). A recent study reported that tissue and plasma PAI-1 were induced in mice after restraint stress and that thrombosis was induced less frequently in the PAI-1 deficient mice after restraint stress than in wild-type mice (Yamamoto *et al.*, 2002). It is well-established that the response to stress is mediated by the neuroendocrine hypothalamic–pituitary–adrenal axis (Vamrakapoulos and Chrousos, 1994) and sympathetic nervous system (Koob, 1999), thereby, linking the Plg system to the areas of these pathways.

3.2. Grooming, a Stress-Induced Behavior is Increased in the Plg−/− Mice

In adult animals, grooming is considered as an index of behavioral adaptation to stressful situations. In rodents, excessive grooming is induced by environmental factors such as fur moistening, the novel environment of the open field, and restraint. Grooming behavior has been used to evaluate drugs (Traber *et al.*, 1982; Moody *et al.*, 1988) and neuroendocrine regulation (Wimersma Greidanus *et al.*, 1985; Hitchcock *et al.*, 1994). Grooming, evaluated in an open field at 31 and 65 days of age, was increased in Plg−/− compared to WT mice (Hoover-Plow *et al.*, 2001). The number of grooming episodes was 2-fold higher in the Plg−/− than for WT mice. The Plg−/− mice responded to the novel environment of an open field with excessive grooming, while measures of locomotor activity and fear were unaffected in the Plg−/− mice in this evaluation. As for the younger mice (Figure 9.2), the number of grooming episodes was also significantly higher in response to fur moistening in 6 weeks old Plg−/− mice than WT mice. These results indicate that with two different stressors, an open field and fur moistening, the Plg−/− mice exhibit excessive grooming.

Grooming elicited by stress is modulated by adrenocorticotropic hormone (ACTH) and other neuropeptides (Gispen and Isaacson, 1981; Spruijt *et al.*, 1992; van Erp *et al.*, 1994). Intracerebroventricular (icv) injections of other peptide hormones, including corticotrophin releasing hormone (CRH), bombesin, β-endorphin (β-end), melanocyte stimulating hormones (α-MSH and β-MSH), may increase grooming as well (Wimersma Greidanus *et al.*, 1985). Animals after hypophysectomy and adrenalectomy still exhibit grooming when injected icv with ACTH, indicating a direct effect of ACTH on brain receptors (Gispen and Isaacson, 1981). Lewis strain (Sternberg *et al.*, 1992) rats are hyporesponsive to stress and exhibit decreased grooming as well as decreased secretion of the ACTH and corticosterone in an open field test. Dopamine pathways also modulate grooming, and to investigate whether this pathway was involved in excessive grooming of the Plg−/− mice, mice were tested following fur moistening after an injection of saline or arecoline. Arecoline is an agonist for the muscarinic receptors, which are located postsynaptically on the dopamine neurons, and has been shown to inhibit the grooming response to stress in mice (Bhattacharya and Sen, 1991). The Plg−/− deficient mice had a greater inhibition

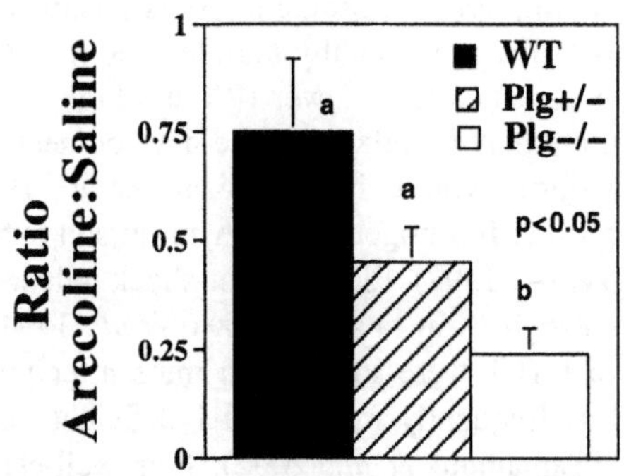

Figure 9.3. Grooming episodes. Mice were injected with either saline or arecoline (a muscarinic receptor agonist which inhibits grooming), and the number of grooming episodes was evaluated after fur moistening. Values are the mean ±SEM.

(expressed as the ratio of the time or number of bouts after arecoline to values after saline injection) of the grooming response than the WT mice (Figure 9.3). This suggests that Plg is involved in the modulation of the grooming response by dopamine pathways.

3.3. Acoustic Startle Reflex, a Stress-Induced Behavior is Decreased in Plg−/− Mice

ASR is an easily quantified reflexive musculature contraction due to an intense acoustic stimulus and is a widely used measure of neurological function (Davis, 1984). The ASR has been used as a model system to study how drugs alter stress and fear, and has been proposed as a model to isolate genetic variables associated with various components of the neurobiology of stress (Davis, 1984; Swerdlow *et al.*, 1986). In addition, abnormalities in the ASR in humans have been noted in clinical syndromes including schizophrenia (Geyer *et al.*, 2001), post-traumatic stress disorder (Shalev *et al.*, 2000), and Huntington's disease (Swerdlow *et al.*, 1995). ASR is enhanced by CRH (Koob *et al.*, 1993), an initiator of the hypothalamic pituitary axis, but in some paradigms, low concentrations of CRH have arousal actions, and at high doses CRH suppresses behavior (Koob *et al.*, 1993). The ASR in the Plg−/− mice was markedly reduced compared to WT mice (Hoover-Plow *et al.*, 2001). The average peak value is nearly 4-fold lower in the Plg−/− mice. Since this behavior is sound induced, a hearing impairment in the Plg−/− mice could cause the reduced ASR peak values. Using independent measures of hearing (ASR frequency testing and audiogenic seizure induction) no impairment has been detected in the Plg−/− mice. Either a mixed C57BL:129 or C57BL background did not affect the expression of the reduced ASR in the Plg−/− mice. In activator deficient mice (uPA−/−, tPA−/−), ASR was not significantly different compared to their WT control, which suggests that Plg is required for the normal ASR response.

Hormonal mediators and inhibitors of the ASR require direct infusion into the brain (Davis, 1984; Swerdlow *et al.*, 1986) to elicit their effect. Since Plg does not cross the blood brain barrier (Chen *et al.*, 1999), but is synthesized in the brain (Nakajima *et al.*, 1996; Tsirka *et al.*, 1997; Basham and Seeds, 2001) predictably, a peripheral bolus injection of Plg into the tail vein did not change the baseline ASR in WT or Plg−/− mice immediately or 24 hr after the injection. Infusion of Plg into the jugular vein over 24 and 72 hr was also ineffective in restoring the ASR in the Plg−/− mice. On the other hand, icv (intracerebroventricular) infusion of either mouse or human Plg reconstituted the ASR response in the Plg−/− mice (Hoover-Plow *et al.*, 2001). The icv infusion of Plg to WT did not increase the ASR. After infusion, human Plg was not detected in the plasma of either WT or Plg−/− mice, but similar amounts of human Plg were detected in brain homogenates from WT and Plg−/− mice infused for 72 hr. Saline or human serum albumin icv infusions were ineffective in restoring the ASR or increasing the ASR in WT mice. Thus, reconstitution of the ASR in Plg−/− requires Plg infused directly into the brain.

3.4. The Plg System in Neuroendocrine Tissues and Alterations in Plg−/− Mice

Pro-opiomelanocortin (POMC) is the precursor of many important bioactive peptide hormones, such as ACTH, β-lipotropin (β-LPH), α-MSH, corticotropin-like intermediate-lobe peptide, γ-lipotropin, and β-end (Rosa *et al.*, 1980). The POMC gene is expressed only in a limited number of cell types, such as pituitary corticotropic cells, melanotropic cells, and neuronal cells of the arcuate nucleus (Rosa *et al.*, 1980). Plg and Plg activators have been localized in neuroendocrine tissues suggesting a role for the Plg system in the neuroendocrine regulation. Immunohistochemistry and *in situ* hybridization have found protein and mRNA for both tPA and uPA found in endocrine cells. Isolated rat islets of Langerhans secrete uPA that is correlated with insulin production (Virji and Dimasi, 1987). Immunohistochemical staining of intact rat pancreas showed that tPA is found in the somatostatin cells of the islets (Kristensen *et al.*, 1987). In addition, nerve endings of the hypothalamic median eminence (rich in somatostatin) (Kristensen *et al.*, 1987) and norepinephrine-containing chromaffin cells in the adrenal medulla contain high amounts of tPA (Eriksen *et al.*, 1989). Rat pituitary maintained in organ culture produces both tPA and uPA. The pituitary cells maintained in these cultures are prolactin, luteinizing hormone, and ACTH producing cells (Granelli-Piperno and Reich, 1983). Parmer *et al.* (1997) reported that tPA is expressed in chromaffin cells from rat, bovine, or humans, sorted into the regulated pathway of secretion, and co-released with catecholamines by chromaffin cell stimulation. *In vivo* stimuli that increase tPA, such as exercise (Vind *et al.*, 1993) and mental stress (Jern *et al.*, 1994), also activate the sympathoadrenal system causing release of amines and proteins from storage vesicles of the adrenal medulla and sympathetic neurons. Bulens *et al.* (1997) have identified a multihormone responsive enhancer upstream from the human tPA gene. Studies (Parmer *et al.* 2000; Jiang *et al.*, 2001) report that Plg binds to pheochromocytoma cells, and plasmin can cleave chromogranin A to produce peptides that inhibit nicotine-induced cholinergic stimulation of catecholamine release from the pheochromocytoma cells.

ACTH, β-LPH, two bioactive peptide hormones of the pituitary generated from its precursor, POMC, are degraded by plasmin (Wang and Hoover-Plow, 2000). Plasmin is a broad-spectrum antibody and degradation of peptides *in vitro* may not be indicative of proteolytic action *in vivo*. In AtT-20/D16v-F2 cells, a mouse pituitary ACTH secreting cell line, incubated with Plg and tPA or uPA showed that POMC (30 kDa) and β-LPH (12 kDa) from medium was markedly degraded, but that POMC and β-LPH from the cell lysate did not change after incubation (Wang and Hoover-Plow, 2000). This suggested an extracellular role for plasmin processing of POMC hormones. In addition, pituitary and plasma concentrations of α-MSH and β-end were less in the Plg−/− mice than WT mice (Hoover-Plow *et al.*, 2000). Electrophoresis and immuodetection of pituitary extracts indicated a reduction in the POMC products in the Plg−/−mice compared to the WT mice. Thus, both functional behavior performance and biochemical evidence substantiate a role of the Plg system in neuroendocrine regulation *in vitro* and *in vivo*.

4. Role of the Plg System in Adipose Tissue Development

4.1. Role of the Plg System in Obesity

Obesity is a risk factor for cardiovascular disease, and numerous studies (Thompson *et al.*, 1991) show that in obesity endogenous fibrinolysis is impaired (Urano *et al.*, 1993; McGill *et al.*, 1994). The impaired fibrinolysis arises from abnormalities in levels of components of the Plg system and increased fibrinogen. The decreased fibrinolytic activity is associated with a consistent increase in PAI-1 activity and antigen. PAI-1 activity and levels correlate with body mass index, waist-to-hip ratio, and insulin and triglyceride levels. PAI-1 is elevated in patients with obesity and in an obese mouse model (Samad and Loskutoff, 1999). Studies in large populations (Juhan-Vague and Alessi, 1996) have found PAI-1 and t-PA antigen are strongly related to fasting insulin levels, body mass index, triglycerides, and HDL cholesterol. Exercise, blood pressure reduction, diet or drugs which reduce insulin, hypertriglyceridemia and cause weight loss in obesity also reduce PAI-1 and improve fibrinolysis (Juhan-Vague and Alessi, 1996). While these studies indicate impaired fibrinolysis with obesity, studies in the Plg−/− mice also indicate a role of the Plg system in the regulation of adipose tissue development.

4.2. Plg and Plg System Components in Adipose Tissue

PAI-1 mRNA (Salonen *et al.*, 1984) in mice was found to be expressed in high levels in adipose tissue and increased in response to injections of LPS, TNFα, and TGFβ (Samad *et al.*, 1996). Bastelica *et al.* (2002) recently reported that the stromal cells are the primary PAI-1 producing cells in adipose tissue. Differentiated 3T3 cells, a model of adipocytes, also have increased PAI-1 activity, protein, and mRNA after stimulation with TGFβ and TNFα (Lundgren *et al.*, 1996; Samad *et al.*, 1996). Seki *et al.* (2001) report a reciprocal regulation of tPA and uPA in the differentiation of 3T3 cells with a decrease of tPA and an increase of uPA. Both tPA and uPA mRNA has been detected in adipose tissue in relatively large amounts (Yamamoto and Loskutoff, 1996). Recent studies of PAI-1−/− mice have reported contradictory results on the role of PAI-1 in obesity. In one model PAI-1−/− mice (Morange *et al.*, 2000) fed a high fat diet accumulated adipose tissue faster than control mice with a marked increase in insulinemia and glycemia. This increase in body weight with the PAI-1−/− mice would be consistent with the reduced fat accumulation found in the Plg−/− mice (Hoover-Plow *et al.*, 2002). In another study (Samad and Loskutoff, 1999), PAI−/− mice were crossed with ob/ob mice, and body weight was reduced, as was hyperinsulinemia and hyperglycemia. Thc difference in the two PAI-1−/− models may relate to whether the obesity was in the active phase of rapid growth and adipose accumulation, as would be the case for the mice fed a high fat diet; or whether the mice initially had excess fat as would be the case with mice with the ob/ob mutation. In the tPA−/− and uPA−/− mice, weight gain and adipose tissue were reduced in the tPA−/− mice, but not the uPA−/− mice (Morange *et al.*, 2002).

Plg binding to differentiated 3T3 cells and mature mouse adipocytes (Figure 9.4) is substantially greater than for undifferentiated 3T3 cells and mouse stromal cells (Hoover-Plow and Yuen, 2001). While the k_d values for Plg binding to the undifferentiated and differentiated 3T3 cells was similar, the number of binding sites was nearly 10-fold higher for the differentiated cells than undifferentiated cells. In adipocytes freshly isolated from WT and Plg−/− mice, Plg binding was higher in mature cells than undifferentiated stromal cells for both WT and Plg−/− mice. However, the Plg binding to mature and differentiated stromal cells was 5-fold higher in the Plg−/− mice than WT mice. Activation of Plg occurs at the cell surface, and Plg binding to cells is an indicator of proteolytic capacity. The increase in binding during differentiation of the 3T3 cells and stromal mouse cells suggests that proteolytic capacity of plasmin activity during differentiation at least in part is regulated by receptor number. In addition, the increase in the binding in the cells

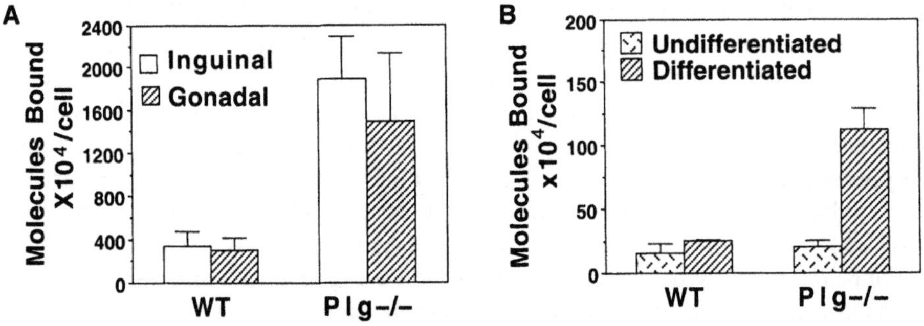

Figure 9.4. ^{125}I-Plg binding to mature fat cells (Panel A) and undifferentiated and differentiated stromal cells (Panel B). Cells were isolated from inguinal adipose depots from WT and Plg−/− mice and incubated with 100 nM ^{125}I-Plg for 1 hr at 37°C. The number of mature fat cells was 1×10^4 per tube and the number of stromal cells per well was 2×10^6 (from Hoover-Plow, J. and Yuen, L. Plasmingen binding is increased with adipocyte differentiation. (Reprinted from *Biochem. Biophys. Res. Comm.*, **284**, 389–394, 2001, with permission from Elsevier).

from deficient mice suggests regulation of the Plg system by receptor availability. The presence of components of the Plg system in adipose tissue and the increased Plg binding to differentiated fat cells compared to undifferentiated cells suggests an important role of the Plg system in adipose tissue development.

4.3. Adipose Tissue Development in Plg−/− Mice

Unlike other tissues, cell number expansion in adipose tissue does not reach a plateau until adolescence in humans and 6–10 weeks of age in rodents (Stern and Johnson, 1978). During 3–6 weeks of age, there is a rapid increase in adipose tissue accumulation in mice. As mentioned above, body weight was lower in both female and male Plg−/− mice (Figure 9.1) compared to WT mice (Hoover-Plow *et al.*, 2002). The reduction in weight was due to a reduced growth rate, rather than due to a weight loss. No difference was found in organ weights including liver, GI tract, or heart in the WT and Plg−/− mice. However, the weight of the adipose depots (gonadal, inguinal) was lower in the Plg−/− than for WT mice at 6 weeks of age. Depot weights at 3 weeks of age were not different in the Plg−/− and WT mice. Whole body fat content was reduced in the Plg−/− mice, but fat-free dry weight was not different in the Plg−/− mice. Since only the lipid weight was lower in the Plg−/− mice, reduced body weight in the Plg−/− mice can be attributed to a reduced fat accumulation, rather than a reduction in body size. During the rapid growth period (42 days of age) food intake was similar in the Plg−/− and WT mice. The lack of difference in the liver weight between the WT and Plg−/− mice supports the conclusion that the reduced body weight and fat accumulation in the Plg−/− mice are not due to a reduced food intake. Also, spontaneous activity in the open field or activity box was not different between WT and Plg−/− mice. An increased activity in the Plg−/− mice cannot account for the reduced body weight and fat accumulation. By 10–12 weeks of age, Plg−/− mice have fibrin deposits and ulcerations in the GI tract, which could impede absorption. In 6-week-old mice, fecal fat excretion was not different in Plg−/− and WT mice, and suggests absorption of dietary fat was equivalent in Plg−/− and WT mice at an age when body weight and adipose tissue are less in the Plg−/− mice.

There was no remarkable difference in mature fat cell size and number in Plg−/− and WT adipose tissue. The striking difference, however, between the adipose tissue of the WT and Plg−/− mice was the higher number of stromal cells from both gonadal and inguinal depots of the Plg−/− mice compared to the WT mice (Figure 9.5). The higher number of stromal cells suggests a reduced ability in the Plg−/− mice for differentiation of adipoblasts or preadipocytes, resulting in the accumulation of excess precursor cells.

Induction of increased adipose tissue accumulation and obesity in mice by a high fat diet is a well-established model of human obesity. Both Plg−/− and WT mice (Figure 9.6)

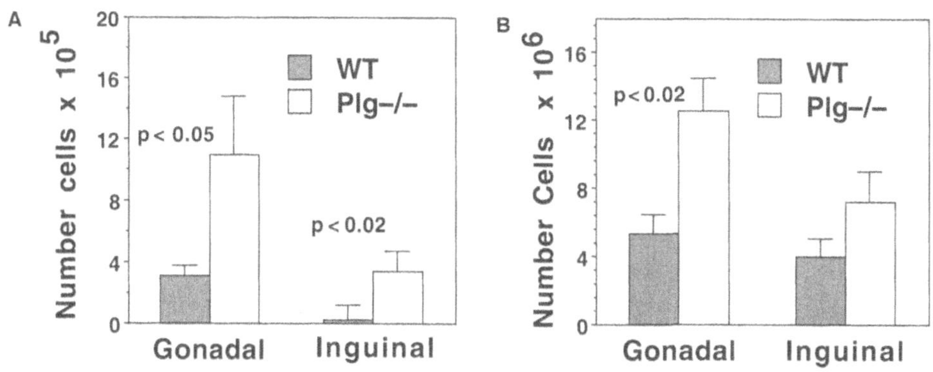

Figure 9.5. Stromal cell count from gonadal and inguinal depots from WT and Plg−/− mice fed a chow (Panel A) or high fat (Panel B) diet. Values are the mean ± SEM. Statistical differences between WT and Plg−/− mice are indicated above the bars (from Hoover-Plow, J., Ellis, J., and Yuen, L. *In vivo* plasminogen deficiency reduces fat accumulation. (*Thromb. Haemost.*, **87**, 1011–1019, 2002, reprinted with permission).

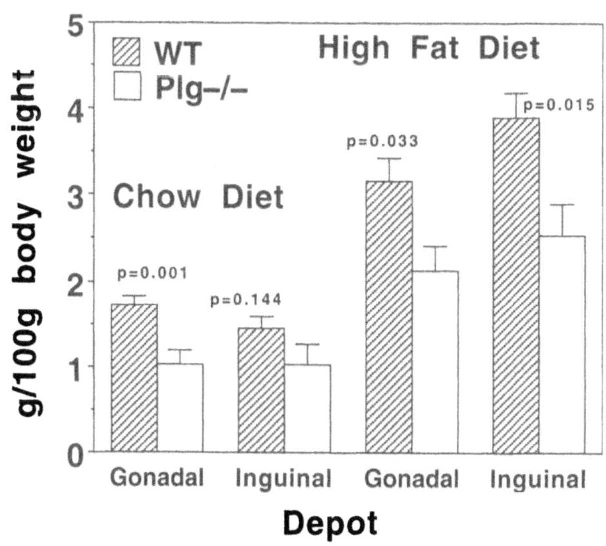

Figure 9.6. Weight of adipose depots from male WT and Plg−/− mice fed a chow diet or high fat diet. Values are the mean ±SEM. Statistical differences between WT and Plg−/− mice are indicated above the bars (from Hoover-Plow, J., Ellis, J., and Yuen, L. *In vivo* plasminogen deficiency reduces fat accumulation. (*Thromb. Haemost.*, **87**, 1011–1019, 2002, reprinted with permission).

gained weight and accumulated more adipose tissue when fed a high fat diet (Hoover-Plow *et al.*, 2002). Glycerol phosphate dehydrogenase, a late marker of adipocyte differentiation, was similar in the WT and Plg−/− mice. The Plg−/− mice have the capacity for lipid accumulation, but this may occur at a slower rate than in WT mice. This was the case in cultured stromal cells with a delay in the differentiation of the preadipocytes of nearly 4 weeks for the cells from Plg−/− mice compared to the cells from the WT mice. The increased number of stromal cells in the adipose tissue of the Plg−/− mice suggests that composition of this cell population may be different in the Plg−/− mice. Stromal cells isolated from adipose tissue contain, not only preadipocytes, but also endothelial cells, leukocytes, and fibroblasts (Springhorn *et al.*, 1995). Plg−/− mice exhibit reduced cell migration in models of inflammation, such the intraperitoneal injection of thioglycollate (Plow *et al.*, 1999) or intraperitoneal insertion of a biopolymer disk (Ploplis *et al.*, 1997), as well as reduced neointimal formation in models of vascular injury (Carmeliet *et al.*, 1997). Crandall *et al.* (2000) found that cell migration of human preadipocytes, was inhibited by PAI-1. Thus, in the Plg−/− mice, a reduced proliferation and differentiation in adipose tissue may relate to reduced recruitment of particular supporting cells required in this process.

4.4. Vascularization of Adipose Tissue

Histological sections of adipose tissue indicate a reduced number of vessels and capillaries in the Plg−/− mice compared to sections from WT mice (Hoover-Plow *et al.*, 2002). In studies of adipose tissue development, Hausman *et al.* (Hausman and Thomas, 1986; Hausman and Wright, 1996) found that in perirenal or visceral fat depots in fetal pigs, new adipocytes formed near large vessels and a capillary network was formed prior to lipid deposition. In mice, new mature fat cells are also found near blood vessels. Thus, the surrounding vasculature may have a profound effect on the ability of preadipocytes to replicate and differentiate. The interaction of capillary and aortic endothelial cells with adipocytes may be mediated by regulation of the subendothelial matrix (Aratani and Kitagawa, 1988; Richardson *et al.*, 1988; Fukai *et al.*, 1993). Cultured endothelial cells secrete fibronectin and laminin and factors that promote and regulate the differentiation of adipocytes (Varzaneh *et al.*, 1994). Some of these factors are regulated via the extracellular matrix and may involve components of the Plg system.

Recent studies (Murphy *et al.*, 1992; Lijnen *et al.*, 1998a,b) have demonstrated a role of the Plg system in activation of metalloproteinases, enzymes important for the degradation of extracellular matrices during tissue remodeling. MMP-2 is secreted by adipocytes and is thought to play a role in the clustering of the differentiated cells (Brown *et al.*, 1997). Lijnen *et al.* (2001) recently reported increased expression of MMP-2 in adipose tissue of mice fed a high fat diet or in mice which were genetically obese (db/db mice). Salvarajan *et al.* (2001) reported that adipocyte differentiation is impaired during involution of the mammary gland from Plg−/− mice and that in 3T3 cells differentiation is dependent on activation of plasmin by plasma kallikrein rather than uPA or tPA. This may not be the case for peripheral adipose tissue development, since tPA−/− mice are reported (Morange *et al.*, 2002) to accumulate less fat that WT mice. The rapid expansion of adipose tissue in growing animals and the requirements for cell migration and extracellular matrix remodeling would suggest a role for the Plg system in adipose development. The reduced adipose tissue accumulation in the Plg−/− mice provides evidence that the Plg system does play an important role in adipose tissue development *in vivo*.

5. Summary

While the Plg system is not critical for physical or behavioral development, this has led to the investigation of the role of tPA and uPA in the activation of other proteins other than Plg. An *in vivo* role for the Plg system in stress-induced behavior or in adipose tissue, had not been appreciated prior to the generation of the Plg−/− mice. Clearly, these studies of development, behavior, and adipose tissue lead to the conclusion that the Plg system is an enzyme system of tissue remodeling rather than tissue formation.

References

Adams, J. and Buelke-Sam, J. (1981). Behavioral assessment of the postnatal animal: Testing and methods development. In C. Kimmel and J. Buelke-Sam (eds) *Developmental Toxicology*. Plenum Press, New York. pp. 45–55.

Altman, J. and Sudarshan, K. (1975). Postnatal development of locomotion in the laboratory rat. *Anim. Behav.* **23**, 896–920.

Aratani, Y. and Kitagawa, Y. (1988). Enhanced synthesis and secretion of type IV collagen and entactin during adipose conversion of 3T3-L1 cells and production of unorthodox laminin complex. *J. Biol. Chem.*, **263**, 16163–16169.

Basham, M.E. and Seeds, N.W. (2001). Plasminogen expression in the neonatal and adult mouse brain. *J. Neurochem.*, **77**, 318–325.

Bastelica, D., Morange, P., Berthet, B., Borghi, H., Lacroix, O., Grino, M., Juhan-Vague, I., and Alessi, M.C. (2002). Stromal cells are the main plasminogen activator inhibitor-1-producing cells in human fat: Evidence of differences between visceral and subcutaneous deposits. *Arterioscler. Thromb. Vasc. Biol.*, **22**, 173–178.

Beers, W.H., Strickland, S., and Reich, E. (1975). Ovarian plasminogen activator: Relationship to ovulation and hormonal regulation. *Relationship to Ovulation and Hormonal Regulation*, **6**, 387–394.

Bezerra, J.A., Carrick, T.L., Degen, J.L., Witte, D., and Degen, S.J.F. (1998). Biological effects of targeted inactivation of hepatocyte growth factor-like protein in mice. *J. Clin. Invest.*, **101**, 1175–1183.

Bhattacharya, S.K. and Sen, A.P. (1991). Effects of muscarinic receptor agonists and antagonists on dopamine-mediated behavioural paradigms. *J. Neural Transm.*, **84**, 241–246.

Brown, L.M., Fox, H.L., Hazen, S.A., Lanoue, K.F., Rannels, S.R., and Lynch, C.J. (1997). Role of the matrix in MMP-2 in multicellular organization of adipocytes cultured in basement membrane components. *Am. J. Physiol. Cell Physiol.*, **272**, C937–C949.

Bugge, T.H., Flick, M.J., Daugherty, C.C., and Degen, J.L. (1995). Plasminogen deficiency causes severe thrombosis but is compatible with development and reproduction. *Genes Dev.*, **9**, 794–807.

Bugge, T.H., Kombrinck, K.W., Flick, J.J., Daugherty, C.C., Danton, M.J.S., and Degen, J.L. (1996). Loss of fibrinogen rescues mice from the pleiotropic effects of plasminogen deficiency. *Cell*, **87**, 709–719.

Bulens, F., Merchiers, P., Ibanez-Tallon, I., De Vriese, A., Nelles, L., Claessens, F., Belayew, A., and Collen, D. (1997). Identification of a multihormone responsive enhancer far upstream from the human tissue-type plasminogen activator gene. *J. Biol. Chem.*, **272**, 663–671.

Carmeliet, P., Bouche, A., De Clercq, C., Janssen, S., Pollefeyt, S., Wyns, S., Mulligan, R.C., and Collen, D. (1995). Biological effects of disruption of the tissue-type plasminogen activator, urokinase-type plasminogen activator, and plasminogen activator inhibitior-1 genes in mice. *Ann. NY Acad. Sci.*, **748**, 367–382.

Carmeliet, P., Moons, L., Ploplis, V., Plow, E., and Collen, D. (1997). Impaired arterial neointima formation in mice with disruption of the plasminogen gene. *J. Clin. Invest.*, **99**, 200–208.

Chen, Z.L., Indyk, J.A., Bugge, T.H., Kombrinck, K.W., Degen, J.L., and Strickland, S. (1999). Neuronal death and blood–brain barrier breakdown after excitotoxic injury are independent processes. *J. Neurosci.*, **19**, 9813–9820.

Coscia, E.M. and Fentress, J.C. (1993). Neurological dysfunction expressed in the grooming behavior of developing Weaver mutant mice. *Behav. Genet.*, **23**, 533–541.

Crandall, D.L., Busler, D.E., McHendry-Rinde, B., Groeling, T.M., and Kral, J.G. (2000). Autocrine regulation of human preadipocyte migration by plasminogen activator inhibitor-1. *J. Clin. Endocrinol. Metab.*, **85**, 2609–2614.

Davis M. (1984). The mammalian startle response. In R.C. Eaton (ed.) *Neural Mechanisms of Startle Behavior*. Plenum Publishing Corporation, New York. pp. 287–351.

Eriksen, P., Kristensen, P., Pyke, C., and Dano, K. (1989). Plasminogen activator inhibitor (type-1) in rat adrenal medulla. *Histochemistry*, **92**, 377–383.

Fukai, F., Iso, T., Sekiguchi, K., Miyatake, N., Tsugita, A., and Katayama, T. (1993). An amino-terminal fibronectin fragment stimulates the differentiation of ST-13 preadipocytes. *Biochemistry*, **32**, 5746–5751.

Geyer, M.A., Krebs-Thomson, K., Braff, D.L., and Swerdlow, N.R. (2001). Pharmacological studies of prepulse inhibition models of sensorimotor gating deficits in schizophrenia: A decade in review. *Psychopharmacology (Berl)*, **156**, 117–154.

Gispen, W.H. and Isaacson, R.L. (1981). ACTH-induced excessive grooming in the rat. *Pharmacol. Therap.*, **12**, 209–246.

Granelli-Piperno, A. and Reich, E. (1983). Plasminogen activators of the pituitary gland: Enzyme characterization and hormonal modulation. *J. Cell Biol.*, **97**, 1029–1037.

Hantai, D., Rao, J.S., Kahler, C., and Festoff, B.W. (1989). Decrease in plasminogen activator correlates with synapse elimination during neonatal development of mouse skeletal muscle. *Proc. Natl. Acad. Sci.*, **86**, 362–366.

Hausman, G.J. and Thomas, G.B. (1986). Structural and histochemical aspects of perirenal adipose tissue in fetal pigs: Relationships between stromal-vascular characteristics and fat cell concentration and enzyme activity. *J. Morphol.*, **190**, 271–283.

Hausman, G.J. and Wright, J.T. (1996). Cytochemical studies of adipose tissue-associated blood vessels in untreated and thyroxine-treated hypophysectomized pig fetuses. *J. Anim. Sci.*, **74**, 354–362.

Hitchcock, D.M., Bowers, M.B., Berridge, C.W., Melia, K.R., and Roth, R.H. (1994). Stress-induced activation of prefrontal cortex dopamine turnover: Blockage by lesions of the amygdala. *Brain Res.*, **664**, 207–210.

Hoover-Plow, J., Ellis, J., and Yuen, L. (2002). *In vivo* plasminogen deficiency reduces fat accumulation. *Thromb. Haemost.* **87**, 1011–1019.

Hoover-Plow, J., Skomorovska-Prokvolit, O., and Welsh, S. (2001). Selective behaviors altered in plasminogen-deficient mice are reconstituted with intracerebroventricular injection of plasminogen. *Brain Res.*, **898**, 256–264.

Hoover-Plow, J., Wang, N., and Ploplis, V. (1999). Growth and behavioral development in plasminogen gene-targeted mice. *Growth Dev. Aging*, **63**, 13–32.

Hoover-Plow, J., Wang, N., Welsh, S. (2000). Plasminogen deficiency alters basal β-endorphin and corticosterone hormones. *Fibrinolysis Proteolysis*, **14**, 29.

Hoover-Plow, J. and Yuen, L. (2001). Plasminogen binding is increased with adipocyte differentiation. *Biochem. Biophys. Res. Commun.*, **284**, 389–394.

Huarte, J., Belin, D., Bosco, D., Sappino, A.P., and Vassalli, J.D. (1987). Plasminogen activator and mouse spermatozoa: Urokinase synthesis in the male genital tract and binding of the enzyme to the sperm cell surface. *J. Cell. Biol.*, **104**, 1281–1289.

Ishizaki, M., Tsuritani, I., Noborisaka, Y., Yamada, Y., Tabata, M., and Nakagawa, H. (1996). Relationship between job stress and plasma fibrinolytic activity in male Japanese workers. *Int. Arch. Occup. Environ. Health*, **68**, 315–320.

Jensh, R.P. and Brent, R.L. (1988). Effect of prenatal X-irradiation on the 14th–18th days of gestation on postnatal growth and development in the rat. *Teratology*, **38**, 431–441.

Jern, C., Eriksson, E., Tengborn, L., Risberg, B., Wadenvik, H., and Jern, S. (1989). Changes of plasma coagulation and fibrinolysis in response to mental stress. *Thromb. Haemost.*, **62**, 767–771.

Jern, C., Selin, L., and Jern, S. (1994). *In vivo* release of tissue-type plasminogen activator across the human forearm during mental stress. *Thromb. Haemost.*, **72**, 285–291.

Jiang, Q.J., Taupenot, L., Mahata, S.K., Mahata, M., O'Connor, D.T., Miles, L.A., and Parmer, R.J. (2001). Proteolytic cleavage of chromogranin A (CgA) by plasmin—Selective liberation of a specific bioactive CgA fragment that regulates catecholamine release. *J. Biol. Chem.*, **276**, 25022–25029.

Juhan-Vague, I. and Alessi, M.C. (1996). Fibrinolysis and risk of coronary artery disease. *Fibrinolysis*, **10**, 127–136.

Jung, W., Castren, E., Odenthal, M., VandeWoude, G.F., Ishii, T., Dienes, H., Lindholm, D., and Schirmacher, P. (1994). Expression and functional interaction of hepatocyte growth factor and its receptor c-met in mammalian brain. *J. Cell Biol.*, **126**, 485–494.

Kaaekuahiwi, M.A. and Menino, A.R.J. (1990). Relationship between plasminogen activator production and bovine embryo development *in vitro*. *J. Anim. Sci.*, **68**, 2009–2014.

Kanwar, Y.S., Carone, F.A., Kumar, A., Wada, J., Ota, K., and Wallner, E.I. (1997). Role of extracellular matrix, growth factors and proto-oncogenes in metanephric development. *Kidney Int.*, **52**, 589–606.

Kario, K., Matsuo, T., Kobayashi, H., Yamamoto, K., and Shimada, K. (1997). Earthquake-induced potentiation of acute risk factors in hypertensive elderly patients: Possible triggering of cardiovascular events after a major earthquake. *J. Am. Coll. Cardiol.*, **29**, 926–933.

Koob, G.F. (1999). Corticotropin-releasing factor, norepinephrine, and stress. *Biol. Psychiatry*, **46**, 1167–1180.

Koob, G.F., Heinrichs, S.C., Pich, E.M., Menzaghi, F., Baldwin, H., Miczek, K., and Britton, K.T. (1993). The role of corticotropin-releasing factor in behavioural responses to stress. *Ciba Found. Symp.*, **172**, 277–289.

Kristensen, P., Nielsen, J.H., Larsson, L.I., and Dano, K. (1987). Tissue-type plasminogen activator in somatostatin cells of rat pancreas and hypothalamus. *Endocrinology*, **121**, 2238–2244.

Krystosek, A. and Seeds, N.W. (1981). Plasminogen activator release at the neuronal growth cone. *Science*, **213**, 1532–1534.

Levin, E.G., Banka, C.L., and Parry, G.C. (2000). Progressive and transient expression of tissue plasminogen activator during fetal development. *Arterioscler. Thromb. Vasc. Biol.*, **20**, 1668–1674.

Lijnen, H.R. and Collen, D. (1999). Matrix metalloproteinase system deficiencies and matrix degradation. *Thromb. Haemost.*, **82**, 837–845.

Lijnen, H.R., Maquoi, E., Holvoet, P., Mertens, A., Lupu, F., Morange, P., Alessi, M.C., and Juhan-Vague, I. (2001). Adipose tissue expression of gelatinases in mouse models of obesity. *Thromb. Haemost.*, **85**, 1111–1116.

Lijnen, H.R., Van Hoef, B., Lupu, F., Moons, L., Carmeliet, P., and Collen, D. (1998a). Function of the plasminogen/plasmin and matrix metalloproteinase systems after vascular injury in mice with targeted inactivation of fibrinolytic system genes. *Arterioscler. Thromb. Vasc. Biol.*, **18**, 1035–1045.

Lijnen, H.R., Van Hoef, B., Soloway, P., and Collen, D. (1998b). Plasminogen/plasmin system function in mice deficient in stromelysis-1 (MMP-3) or in tissue inhibitor of metalloproteinases type-1 (TIMP-1). *Fibrinolysis Proteolysis*, **12**, 1–8.

Liu, Y., Bell, A.W., Michalopoulos, G.K., and Zarnegar, R. (1994). The mouse hepatocyte growth factor-encoding gene: Structural organization and evolutionary conservation. *Gene*, **144**, 179–187.

Lund, L.R., Bjorn, S.F., Sternlicht, M.D., Nielsen, B.S., Solberg, H., Usher, P.A., Osterby, R., Christensen, I.J., Stephens, R.W., Bugge, T.H., Dano, K., and Werb, Z. (2000). Lactational competence and involution of the mouse mammary gland require plasminogen. *Development*, **127**, 4481–4492.

Lundgren, C.H., Brown, S.L., Nordt, T.K., Sobel, B.E., and Fujii, S. (1996). Elaboration of type-1 plasminogen activator inhibitor from adipocytes: A potential pathogenetic link between obesity and cardiovascular disease. *Circulation*, **93**, 106–110.

Malyszko, J., Urano, T., Takada, Y., and Takada, A. (1994). Stress-dependent changes in fibrinolysis, serotonin and platelet aggregation in rats. *Life Sci.*, **54**, 1275–1280.

Mars, W.M., Zarnegar, R., and Michalopoulos, G.K. (1993). Activation of hepatocyte growth factor by the plasminogen activators uPA and tPA. *Am. J. Pathol.*, **143**, 949–958.

McGill, J.B., Schneider, D.J., Arfken, C.L., Lucore, C.L., and Sobel, B.E. (1994). Factors responsible for impaired fibrinolysis in obese subjects and NIDDM patients. *Diabetes*, **43**, 104–109.

McGuire, P.G. and Seeds, N.W. (1990). Degradation of underlying extracellular matrix by sensory neurons during neurite outgrowth. *Neuron*, **4**, 633–642.

Moody, T.W., Merali, Z., and Crawley, J.N. (1988). The effects of anxiolytics and other agents on rat grooming behavior. *Ann. NY Acad. Sci.*, **515**, 281–290.

Moonen, G., Grau-Wagemans, M.P., and Selak, I. (1982). Plasminogen activator-plasmin system and neuronal migration. *Nature*, **298**, 753–755.

Morange, P.E., Bastelica, D., Bonzi, M.F., Van, H.B., Collen, D., Juhan-Vague, I., and Lijnen, H.R. (2002). Influence of t-pA and u-PA on adipose tissue development in a murine model of diet-induced obesity. *Thromb. Haemost.*, **87**, 306–310.

Morange, P.E., Lijnen, H.R., Alessi, M.C., Kopp, F., Collen, D., and Juhan-Vague, I. (2000). Influence of PAI-1 on adipose tissue growth and metabolic parameters in a murine model of diet-induced obesity. *Arterioscler. Thromb. Vasc. Biol.*, **20**, 1150–1154.

Murphy, G., Atkinson, S., Ward, R., Gavrilovic, J., and Reynolds, J.J. (1992). The role of plasminogen activators in the regulation of connective tissue metalloproteinases. *Ann. NY Acad. Sci.*, **667**, 1–12.

Nakajima, K., Nagata, K., and Kohsaka, S. (1994). Plasminogen mediates an interaction between microglia and dopaminergic neurons. *Eur. Neurol.*, 34(Suppl. 3), 10–16.

Nakajima, K., Reddington, M., Kohsaka, S., and Kreutzberg, G.W. (1996). Induction of urokinase-type plasminogen activator in rat facial nucleus by axotomy of the facial nerve. *J. Neurochem.*, **66**, 2500–2505.

Nicole, O., Docagne, F., Ali, C., Margaill, I., Carmeliet, P., MacKenzie, E.T., Vivien, D., and Buisson, A. (2001). The proteolytic activity of tissue-plasminogen activator enhances NMDA receptor-mediated signaling. *Nature Med.*, **7**, 59–64.

Ny, A., Leonardsson, G., Hägglund, A.C., Hägglöf, P., Ploplis, V.A., Carmeliet, P., and Ny, T. (1999). Ovulation in plasminogen-deficient mice. *Endocrinology*, **140**, 5030–5035.

Parmer, R.J., Mahata, M., Gong, Y., Mahata, S.K., Jiang, Q.J., O'Connor, D.T., Xi, X.P., and Miles, L.A. (2000). Processing of chromogranin A by plasmin provides a novel mechanism for regulating catecholamine secretion. *J. Clin. Invest.* **106**, 907–915.

Parmer, R.J., Mahata, M., Mahata, S., Sebald, M.T., O'Connor, D.T., and Miles, L.A. (1997). Tissue plasminogen activator (t-PA) is targeted to the regulated secretory pathway: Catecholamine storage vesicles as a reservoir for the rapid release of t-PA. *J. Biol. Chem.*, **272**, 1976–1982.

Pittman, R.N. (1990). Developmental roles of proteases and inhibitors. *Semin. Develop. Biol.*, **1**, 65–74.

Ploplis V., Bussutil, S., French, E., Tang, L., Carmeliet, P., Collen, D., and Plow, E. (1997). The inflammatory response to biopolymer implants in plasminogen deficient mice. *Thromb. Haemost.*, (Suppl.) abstract 2234.

Ploplis, V.A., Carmeliet, P., Vazirzadeh, S., Van Vlaenderen, I., Moons, L., Plow, E.F., and Collen, D. (1995). Effects of disruption of the plasminogen gene in mice on thrombosis, growth and health. *Circulation*, **92**, 2585–2593.

Plow, E.F., Ploplis, V.A., Carmeliet, P., and Collen, D. (1999). Plasminogen and cell migration *in vivo*. *Fibrinolysis Proteolysis* **13**, 49–53.

Pohl, J.F., Melin-Aldana, H., Sabla, G., Degen, J.L., and Bezerra, J.A. (2001). Plasminogen deficiency leads to impaired lobular reorganization and matrix accumulation after chronic liver injury. *Am. J Pathol.*, **159**, 2179–2186.

Price, W.A. and Stiles, A.D. (1996). New insights into lung growth and development. *Curr. Opin. Pediatr.*, **8**, 202–208.

Raikkonen, K., Lassila, R., Keltikangas-Jarvinen, L., and Hautanen, A. (1996). Association of chronic stress with plasminogen activator inhibitor-1 in healthy middle-aged men. *Arterioscler. Thromb. Vasc. Biol.*, **16**, 363–367.

Rao, J.S., Sawaya, R., Gokaslan, Z.L., Yung, W.K.A., Goldstein, G.W., and Laterra, J. (1996). Modulation of serine proteinases and metalloproteinases during morphogenic glial-endothelial interactions. *J. Neurochem.*, **66**, 1657–1664.

Richardson, R.L., Campion, D.R., and Hausman, G.J. (1988). Adhesion, proliferation, and adipogenesis in primary rat cell cultures: Effects of collagenous substrata, fibronectin, and serum. *Cell Tissue Res.*, **251**, 123–128.

Rosa, P., Policastro, P., and Herbert, E. (1980). A cellular basis for the differences in regulation of synthesis and secretion of ACTH/endorphin peptides in anterior and intermediate lobes of the pituitary. *J. Exp. Biol.*, **89**, 215–237.

Salonen, E.M., Zitting, A., and Vaheri, A. (1984). Laminin interacts with plasminogen and its tissue-type activator. *FEBS Lett.*, **172**, 29–32.

Salvarajan, S., Lund, L.R., Takeuchi, T., Craik, C.S., and Werb, Z. (2001). A plasma kallikrein-dependent plasminogen cascade required for adipocyte differentiation. *Nat. Cell Biol.*, **3**, 267–275.

Samad, F. and Loskutoff, D.J. (1999). Hemostatic gene expression and vascular disease in obesity: Insights from studies of genetically obese mice. *Thromb. Haemost.*, **82**, 742–747.

Samad, F., Yamamoto, K., and Loskutoff, D.J. (1996). Distribution and regulation of plasminogen activator inhibitor-1 in murine adipose tissue *in vivo*. *J. Clin. Invest.*, **97**, 37–46.

Seki, T., Miyasu, T., Noguchi, T., Hamasaki, A., Sasaki, R., Ozawa, Y., Okukita, K., Declerck, P.J., and Ariga, T. (2001). Reciprocal regulation of tissue-type and urokinase-type plasminogen activators in the differentiation of murine preadipocyte line 3T3-L1 and the hormonal regulation of fibrinolytic factors in the mature adipocytes. *J. Cell Physiol.*, **189**, 72–78.

Shalev, A.Y., Peri, T., Brandes, D., Freedman, S., Orr, S.P., and Pitman, R.K. (2000). Auditory startle response in trauma survivors with posttraumatic stress disorder: A prospective study. *Am. J. Psychiatry*, **157**, 255–261.

Soreq, H. and Miskin, R. (1981). Plasminogen activator in the rodent brain. *Brain Res.*, **216**, 361–374.
Spear, L.P. (2000). The adolescent brain and age-related behavioral manifestations. *Neurosci. Biobehav. Rev.*, **24**, 417–463.
Springhorn, J.P., Madri, J.A., and Squinto, S.P. (1995). Human capillary endothelial cells from abdominal wall adipose tissue: Isolation using an anti-PECAM antibody. *In Vitro. Cell. Dev. Biol. Anim.*, **31**, 473–481.
Spruijt, B.M., Van Hooff, J.A.R.A., and Gispen, W.H. (1992). Ethology and neurobiology of grooming behavior. *Physiol. Rev.*, **72**, 825–852.
Stern, J.S. and Johnson, P.R. (1978). Size and number of adipocytes and their implications. *Diabetes, Obes., Vasc. Dis.* **2**, 303–333.
Stern, T.C. (1996). Roles of kringle domain-containing serine proteases in epithelial-mesenchymal transitions during embryonic development. *Acta. Anat.*, **156**, 162–172.
Sternberg, E.M., Glowa, J.R., Smith, M.A., Calogero, A.E., Listwak, S.J., Aksentijevich, S., Chrousos, G.P., Wilder, R.L., and Gold, P.W. (1992). Corticotropin releasing hormone related behavioral and neuroendocrine responses to stress in Lewis and Fischer rats. *Brain Res.*, **570**, 54–60.
Strickland, S., Reich, E., and Sherman, M.I. (1976). Plasminogen activator in early embryogenesis: Enzyme production by trophoblast and parietal endoderm. *Cell*, **9**, 231–240.
Sumi, Y., Ichikawa, Y., Nakamura, Y., Miura, O., and Aoki, N. (1989). Expression and characterization of pro α2-plasmin inhibitor. *J. Biochem.*, **106**, 703–707.
Swerdlow, N.R., Geyer, M.A., Vale, W.W., and Koob, G.F. (1986). Corticotropin-releasing factor potentiates acoustic startle in rats: Blockade by chlordiazepoxide. *Psychopharmacology*, **88**, 147–152.
Swerdlow, N.R., Paulsen, J., Braff, D.L., Butters, N., Geyer, M.A., and Swenson, M.R. (1995). Impaired prepulse inhibition of acoustic and tactile startle response in patients with Huntington's disease. *J. Neurol. Neurosurg. Psychiatry*, **58**, 192–200.
Teesalu, T., Blasi, F., and Talarico, D. (1998). Expression and function of the urokinase type plasminogen activator during mouse hemochorial placental development. *Dev. Dyn.*, **13**, 27–38.
Thompson, C.J., Ryu, J.E., Craven, T.E., Kahl, F.R., and Crouse, J.R.I. (1991). Central adipose distribution is related to coronary atherosclerosis. *Arterioscler. Thromb.*, **11**, 327–333.
Traber, J., Klein, H.R., and Gispen, W.H. (1982). Actions of anti-depressant and neuroleptic drugs on ACTH- and novelty-induced behaviour in the rat. *Eur. J. Pharmacol.*, **80**, 407–414.
Tsirka, S.E., Rogove, A.D., Bugge, T.H., Degen, J.L., and Strickland, S. (1997). An extracellular proteolytic cascade promotes neuronal degeneration in the mouse hippocampus. *J. Neurosci.*, **17**, 543–552.
Urano, T., Kojima, Y., Takahashi, M., Serizawa, K., Sakakibara, K., Takada, Y., and Takada, A. (1993). Impaired fibrinolysis in hypertension and obesity due to high plasminogen activator inhibitor-1 level in plasma. *Jpn. J. Physiol.*, **43**(2), 221–228.
Valinsky, J.E. and Le Douarin, N.M. (1985). Production of plasminogen activator by migrating cephalic neural crest cells. *EMBO J.*, **4**, 1403–1406.
Vamrakapoulos, N.C. and Chrousos, G.P. (1994). Hormonal regulation of human corticotrophin releasing hormone gene expression: Implications for the stress response and immune/inflammatory reaction. *Endoc. Rev.*, **15**, 409–430.
van Erp, A.M., Kruk, M.R., Meelis, W., and Willekens-Bramer, D.C. (1994). Effect of environmental stressors on time course, variability and form of self-grooming in the rate: Handling, social contact, defeat, novelty, restraint and fur moistening. *Behav. Brain Res.*, **65**, 47–55.
Varzaneh, F.E., Shillabeer, G., Wong, K.L., and Lau, D.C.W. (1994). Extracellular matrix components secreted by microvascular endothelial cells stimulate preadipocyte differentiation *in vitro*. *Metabolism*, **43**, 906–912.
Vihko, K.K., Toppari, J., Saksela, O., Suominen, J.J., and Parvinen, M. (1986). Testicular plasminogen activators during postnatal development in the rat. *Acta Endocrinol.* (*Copenh*), **112**, 431–435.
Vind, J., Gleerup, G., Nielsen, P.T., and Winther, K. (1993). The impact of static work on fibrinolysis and platelet function. *Thromb. Res.*, **72**, 441–446.
Virji, M.A. and Dimasi, M.J. (1987). Selective secretion of only the urokinase plasminogen activator from rat islets of Langerhans. *Diabetes Res.*, **6**, 9–13.
Wang, N. and Hoover-Plow, J. (2000). The effect of the plasminogen system on pituitary peptide hormone processing. *FASEB J.*, **14**, A570.
Ware, J.H., Dibenedetto, A.J., and Pittman, R.N. (1995). Localization of tissue plasminogen activator mRNA in adult rat brain. *Brain Res. Bull.*, **37**, 275–281.

Wimersma Greidanus, T., Donker, D.K., Zinnicq Bergmann, F.F.M., Bekenkamp, R., Maigret, C., and Spruyt, B.M. (1985). Comparison between excessive grooming induced by bombesin or by ACTH: The differential elements of grooming and development of tolerance. *Peptides*, **6**, 369–372.

Yamamoto, K. and Loskutoff, D.J. (1996). Fibrin deposition in tissues from endotoxin-treated mice correlates with decreases in the expression of urokinase-type but not tissue-type plasminogen activator. *J. Clin. Invest.*, **97**, 2440–2451.

Yamamoto, K., Takeshita, K., Shimokawa, T., Yi, H., Isobe, K.K., Loskutoff, D.J., and Saito, H. (2002). Plasminogen activator inhibitor-1 is a major stress-regulated gene: Implications for stress-induced thrombosis in aged individuals. *Proc. Natl. Acad. Sci. USA*, **99**, 890–895.

Yamamura, Y., Yamashiro, K., Tsuruoka, N., Nakazato, H., Tsujimura, A., and Yamaguchi, N. (1997). Molecular cloning of a novel brain-specific serine protease with a kringle-like structure and three scavenger receptor cysteine-rich motifs. *Biochem. Biophys. Res. Commun.*, **239**, 386–392.

Yoshimura, T., Yuhki, N., Wang, M.H., Skeel, A., and Leonard, E.J. (1993). Cloning, sequencing, and expression of human macrophage stimulating protein (MSP, MST1) confirms MSP as a member of the family of kringle proteins and locates and MSP gene on chromosome 3. *J. Biol. Chem.* **268**, 15461–15468.

Zhong, D.Z., Pei, C., and Xiu-Qin, L. (1996). Neurobehavioral study of prenatal exposure to hyperthermia combined with irradiation in mice. *Neurotoxicol. Teratol.* **18**, 703–709.

10

Plasminogen-Directed Phenotypes in Mice

Victoria A. Ploplis and Francis J. Castellino

1. Introduction

While the major physiological function of the fibrinolytic system is maintaining vascular patency, a number of *in vitro* and *in vivo* observations have implicated this proteolytic pathway in diverse nonhemostatic functions from embryogenesis to cancer. With the development of mice with specific gene deletions of the fibrinolytic system, more direct approaches to further delineate its regulatory capacity in physiological and pathophysiological events have evolved.

This chapter presents a brief review of studies performed in plasminogen-deficient ($PG^{-/-}$) mice investigating both spontaneous and challenge-induced phenotypes utilizing models that mimic human clinical conditions.

2. Generation and Initial Characterization of Plasminogen-Deficient Mice ($PG^{-/-}$)

Plasminogen-deficient mice were generated utilizing two different approaches. One involved replacing the 3′ end of intron 14 to intron 16 with the *NEO* gene. This resulted in eliminating exons 15 and 16 which contain codons for 2 of the 3 amino acids that comprise the catalytic triad (Ploplis *et al.*, 1995). Another approach replaced a 9 kb region of the gene that encompassed 5′ flanking sequences (proximal promoter sequences) and the first two exons (signal peptide and 43 residues of the activation peptide) with the phosphoglycerate kinase promoter (PGK)-HPRT gene (Bugge *et al.*, 1995).

Both approaches resulted in viable mice with a complete Pg deficiency ($PG^{-/-}$) that survived well into adulthood. However, $PG^{-/-}$ mice exhibited diminished growth rates shortly after weaning with some evidence of apathic behavior. While there appeared to be no hematological abnormalities, spontaneous fibrin deposits in a number of organs, with

---

Victoria A. Ploplis and Francis J. Castellino • W. M. Keck Center for Transgene Research, Department of Chemistry and Biochemistry, University of Notre Dame, Notre Dame, Indiana.

Plasminogen: Structure, Activation, and Regulation, edited by David M. Waisman. Kluwer Academic/Plenum Publishers, New York, 2003.

some evidence of an induced pathology, was a common phenotype of this deficiency. For example, gastric and colonic ulcerations were commonly found, most likely precipitated by small blood vessel thrombosis. Additionally, rectal prolapse, another frequently observed pathology in these mice, was most likely due to vasocclusion, which results in ulcerations of the rectum, necrotic tissue formation, inflammation, and bacterial contamination. Another spontaneous pathology observed in these mice that appears to be fibrin-mediated is ligneous conjunctivitis. Prevalence for this pathology in $PG^{-/-}$ mice increases with age and the genetic strain of the mice. Clinically, this pathology has been observed in patients with Type-1 Pg deficiency (Mingers *et al.*, 1999; Kraft *et al.*, 2000). Other pathologies clinically observed with heterozygous and homozygous deficiencies of Pg are occlusive hydrocephalus and gingival overgrowth with ulcerations (Schuster *et al.*, 1997, 1999; Scully *et al.*, 2001).

3. Challenge-Induced Phenotypes in Mice Deficient for Plasminogen

3.1. Pulmonary Clot Lysis

The relative fibrinolytic capacity of $PG^{-/-}$ mice when challenged was determined by measuring, *in vivo*, the spontaneous lysis of a radiolabeled plasma clot after injection into the jugular vein. Clot lysis was determined by comparing the radioactivity of clots lodged in the lungs and/or heart, at various timepoints, to the total activity injected. Relative to WT mice, clot lysis was severely compromised in $PG^{-/-}$ mice (Figure 10.1). An intermediate reduction in fibrinolytic capacity was observed in $PG^{+/-}$ mice, indicative of a dose dependency on this activity (Ploplis *et al.*, 1995). Evidence that *in vivo* clot lysis is critically dependent on the presence of Pg was further demonstrated by bolus i.v. administration of

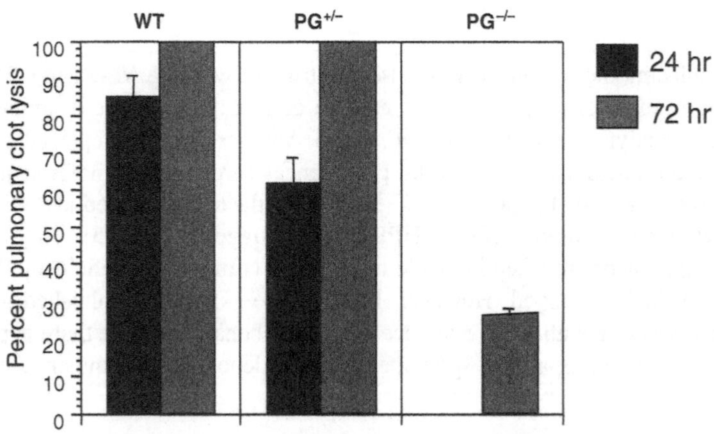

Figure 10.1. Spontaneous pulmonary clot lysis 24 (black bars) and 72 (grey bars) hours after injection of an ^{125}I fibrin-labeled plasma clot into the jugular veins of WT, $PG^{+/-}$, and $PG^{-/-}$ mice. Clot lysis expressed as the percent of total labeled clot injected.

Pg into deficient mice which resulted in normal clot lysis and dissolution of spontaneous fibrin deposits in the liver (Figure 10.2, see color plate facing p. 180) (Lijnen *et al.*, 1996).

3.2. Responses to Vascular Injury

Studies determining critical players in events involved in vascular remodeling have implicated components of the fibrinolytic system. The generation of mice deficient for components of this system have led to the development of a number of vascular injury challenges in these gene deficient mice (Carmeliet *et al.*, 1997a; Lijnen *et al.*, 2000; Ploplis *et al.*, 2001). Arterial injuries involving mechanical and electrical inducers in $PG^{-/-}$ demonstrated a delay in the removal of necrotic tissue or fibrin, impaired wound healing, and reduced inflammatory cell and smooth muscle cell accumulation relative to WT mice (Carmeliet *et al.*, 1997b; Busuttil *et al.*, 2000). However, in the electrical injury model, no effect was observed on smooth muscle cell proliferation or re-endothelialization of the injured site (Carmeliet *et al.*, 1997b). Other vascular remodeling studies that are clinically more relevant have been performed in $PG^{-/-}$ mice. For example, a carotid artery transplant model (arterial graft-versus-host disease) demonstrated a similar phenotype in $PG^{-/-}$ mice with reduced inflammatory cell infiltration and diminished media necrosis and smooth muscle proliferation and migration into the intimal layer (Figure 10.3, see color plate facing p. 180) (Moons *et al.*, 1998). In contrast, utilizing a model of venous graft stenosis, no differences were observed in neointima formation between WT and $PG^{-/-}$ mice (Shi *et al.*, 1999). Finally, in a diet-induced hypercholesterolemia model, $APOE^{-/-}/PG^{-/-}$ mice demonstrated accelerated lesion formation in the proximal and distal aorta relative to that observed in mice with a single deficiency of apo E (Xiao *et al.*, 1997). While these studies appear to result in contradictory outcomes, the method and location of injury, as well as the local environment at the time of injury, may play very diverse roles in the remodeling events that follow.

3.3. Pathogen Susceptibility and Inflammation

Components of the fibrinolytic system appear to be associated with the inflammatory process based primarily on *in vitro* observations. The plasminogen activator, urokinase (uPA) and its receptor (uPAR) have been identified on the surface of monocytes at the migratory edges (Estreicher *et al.*, 1990) and stimulated macrophages produce and secrete uPA (Vassalli *et al.*, 1984). The fibrinolytic system also affects the expression and/or activation of cytokines. For example, plasmin (Pm) has been shown to release interleukin-1 from macrophages and to activate transforming growth factor-β (TGF-β) (Matsushima *et al.*, 1986; Keski-Oja *et al.*, 1987). *In vivo* studies confirming these observations were generated in $PG^{-/-}$ mice where it was demonstrated that 72 hr after peritoneal stimulation with thioglycollate a significant reduction in macrophage recruitment was observed in $PG^{-/-}$ mice relative to WT mice (Figure 10.4) (Ploplis *et al.*, 1998).

Plasmin has also been implicated in pathogen invasion through surface association which then facilitates in host tissue barrier degradation and ultimately dissemination of the invading organism. For example, a number of gram-positive *streptococci* and *Yersinia* are able to activate Pg through their endogenous activators and at least one of the activators from *Yersinia pestis* plays multifunctional roles in the invasion process some of which are

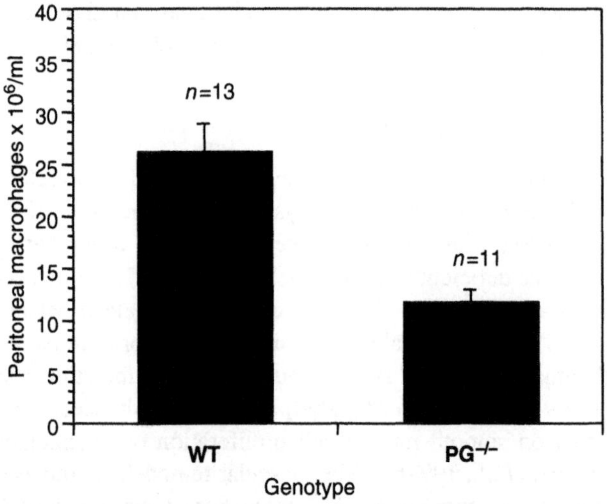

Figure 10.4. Peritoneal macrophage levels 96 hr after stimulation with thioglycollate in WT and $PG^{-/-}$ mice demonstrating diminished recruitment in $PG^{-/-}$ mice.

nonproteolytic in nature (Lottenberg *et al.*, 1992; Sodeinde *et al.*, 1992; Lahteenmaki *et al.*, 2001). Other studies have identified an interaction of plasminogen with prion resulting in spongiform encephalopathies (Fischer *et al.*, 2000).

Since the development of $PG^{-/-}$ mice, several studies have utilized these mice in pathogen-challenged models and have compared the responses to WT counterparts. As indicated, *Y. pestis*, which is the causative pathogen for plague, express plasminogen activators on their cell surface and could therefore contribute to systemic infection through surface-associated plasmin activity (Sodeinde *et al.*, 1992). This was confirmed in studies utilizing $PG^{-/-}$ mice in which it was demonstrated that there was an enhanced resistance to this organism relative to WT mice (Goguen *et al.*, 2000). Additional studies assessing pathogen infectivity in $PG^{-/-}$ mice have demonstrated that a lack of host Pg during relapsing *Borrelia* fever and infection by *Borrelia crocidurae* resulted in decreased spirochetal burden in the hearts, brains, and kidneys of these mice compared to WT mice (Gebbia *et al.*, 1999; Nordstrand *et al.*, 2001). Other studies have looked at the dependence of plasminogen on the erythrocytic life cycle of the malaria parasites, *Plasmodium falciparum* and *Plasmodium chabaudi*, and have found no differences between WT and $PG^{-/-}$ mice (Rosenthal *et al.*, 1998). These results were surprising since plasminogen activator activity has been identified in erythrocytes that have been infected with these two strains of parasites (Roggwiller *et al.*, 1997). Additionally, inhibitors and antibodies to urokinase attenuated the parasite-mediated rupture of erythrocytes (Roggwiller *et al.*, 1997). Contrary to these observations, $PG^{-/-}$ mice infected with an isolate of Group A *streptococcus* (CS101) demonstrated a reduced virulence compared to WT mice (Li *et al.*, 1999). This demonstrates a role of plasminogen in the virulence of this pathogen and supports *in vitro* observations demonstrating that human plasma incubated with Group A streptococcus (CS101) acquires plasmin activity on its surface (D'Costa and Boyle, 1998). Further characterization of this interaction demonstrated that the a1a2 repeats of PAM,

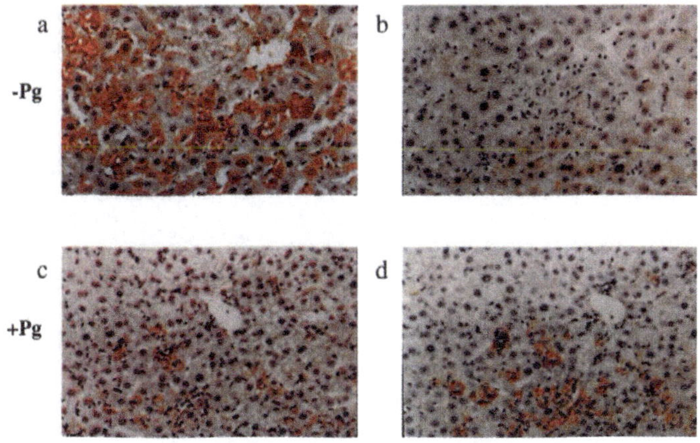

Figure 10.2. Liver sections (200×) from $PG^{-/-}$ mice without (a,b) or with (c,d) injection of plasminogen immunostained for murine fibrin(ogen) (a,c) or for murine plasminogen (b,d). From: Lijnen, H.R., Carmeliet, P., Bouche, A., Moons, L., Ploplis, V.A., Plow, E.F., and Collen, D. (1996). Restoration of thrombolytic potential in plasminogen-deficient mice by bolus administration of plasminogen. Blood 88:870–876. Copyright American Society of Hematology, used by permission.

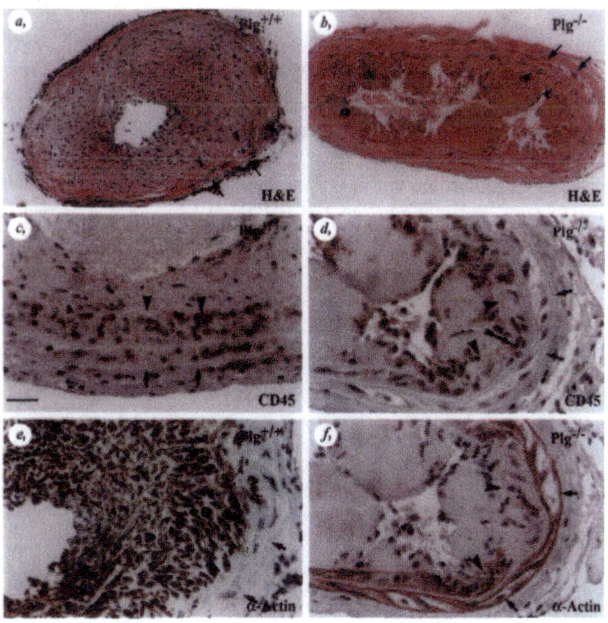

Figure 10.3. Carotid arterial grafts from WT (a, c, e) and $PG^{-/-}$ (b, d, f) recipient mice 45 days post-transplantation. Hematoxylin & eosin stains of grafts from WT (a) and $PG^{-/-}$ (b) recipient mice demonstrated a multilayered neointima in grafts from WT mice and only a few cells, intermingled with thrombi, in $PG^{-/-}$ mice. CD45 immunostained grafts from WT (c) and $PG^{-/-}$ (d) mice indicated the presence of leukocytes in the intimal and medial layers of the graft in WT mice but only in intimal layer in grafts from $PG^{-/-}$ mice. Smooth muscle alpha actin immunostains indicated an abundance of smooth muscle cells in the intima from WT mice (e) but primarily confined to the medial compartment in $PG^{-/-}$ mice (f). Arrowheads indicate internal elastica laminae and arrows the external elastic lamina. From: Moons, L., Shi, C., Ploplis, V., Plow, E., Haber, E., and Collen, D., Reduced transplant arteriosclerosis in plasminogen-deficient mice, The Journal of Clinical Investigation, 102:1788–1797 (1998), reproduced with permission.

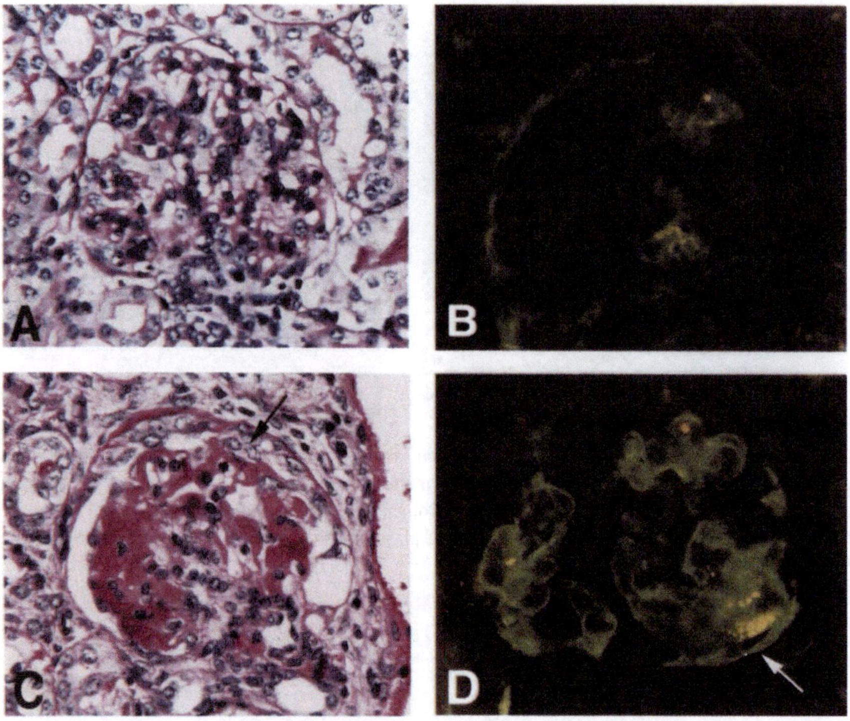

Figure 10.5. Photomicrographs of glomeruli from WT mice (A,B) and $PG^{-/-}$ mice (C,D) with glomerulonephritis (GN). PAS+ staining (350×) in glomeruli from WT (A) and $PG^{-/-}$ (B) mice demonstrated proliferative GN in WT mice at day 10 and hypocellularity with extensive deposition of PAS+ material in the glomerular tuft and cellular crescent formation (arrow) in $PG^{-/-}$ mice at day 8. Fibrin deposition (immunofluorescence, 400×) was segmental dispersed in the glomerular tuft in WT mice but extensive in the glomerular tuft and Bowman's capsule (arrow) in $PG^{-/-}$ mice. Reproduced from The Journal of Experimental Medicine, vol:185, 963–968, 1997—fig. 1 by copyright permission of The Rockefeller University Press.

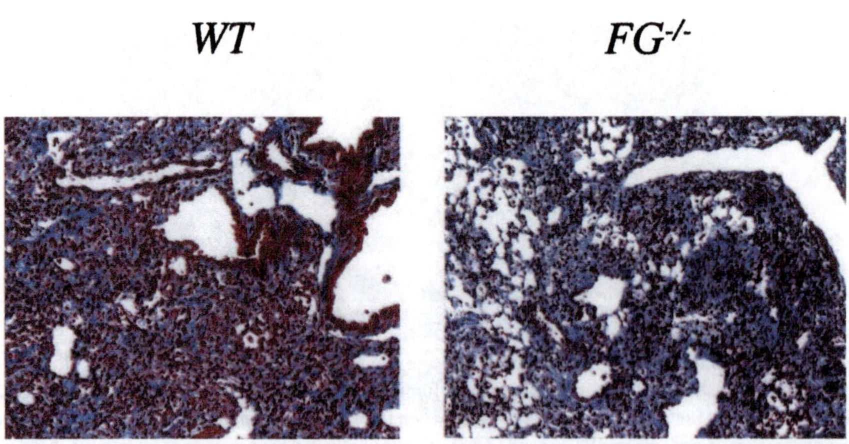

Figure 10.6. Masson's trichrome staining of lung tissue from WT and $FG^{-/-}$ mice, 14 days after intratracheal administration of bleomycin, demonstrating qualitatively similar amounts of collagen deposition (blue) within fibrotic lesions (200×). From: Ploplis, V.A., Wilberding, J., McLennan, L., Liang, Z., Cornelissen, I., DeFord, M.E., Rosen, E.D., and Castellino, F.J. (2000). A total fibrinogen deficiency is compatible with the development of pulmonary fibrosis in mice, The American Journal of Pathology, 157:703–708 (2000), reproduced with permission from the American Society for Investigative Pathology.

a M-like surface protein present in Group A *streptococci*, is the interactive component for human Pg and is mediated through the kringle 2 domain of Pg (Wistedt *et al.*, 1998). In association with streptokinase, a potent bacterial-derived Pg activator, plasmin can be generated on the bacterial surface and facilitate in the invasion into host tissue (Ringdahl *et al.*, 1998). Therefore, a number of these *in vivo* studies support *in vitro* observations related to the role of plasminogen in pathogen invasion and dissemination.

3.4. Glomerulonephritis

Fibrin deposition in the glomerulus is a hallmark feature of severe proliferative and crescentic glomerulonephritis (GN) which can result in renal failure (Kincaid-Smith, 1972). The crescentic forms of GN are the result of an abnormal accumulation of fibrin/macrophage aggregates in the Bowman's space. (Naish *et al.*, 1972; Holdsworth *et al.*, 1979). Additionally, alterations in the fibrinolytic system have been demonstrated in both experimental and human forms of this disease (Rondeau *et al.*, 1990; Malliaros *et al.*, 1993; Lee *et al.*, 2001). For example, patients with focal segmental glomerulosclerosis (FSG) and membranous nephropathy (MN) demonstrated significantly greater levels of glomerular PAI-1 mRNA relative to normal controls (Hamano *et al.*, 2002). In order to more directly determine a protective role for plasminogen in GN, proliferative GN was induced in WT and $PG^{-/-}$ mice (Kitching *et al.*, 1997). Progressive GN with renal impairment developed in both WT and $PG^{-/-}$ mice but was severely exacerbated in the deficient mice demonstrating enhance extracellular matrix (PAS+) and fibrin deposition in the glomerulus (Figure 10.5, see color plate facing p. 181). Results from this study corroborate indirect clinical and experimental investigations and support a protective function of plasminogen from acute inflammatory injury to the glomerulus.

3.5. Rheumatoid Arthritis

An autoimmune disease that involves chronic inflammation of the synovia and ultimately results in destruction of cartilage and bone is rheumatoid arthritis (Firestein, 1997). By regulating inflammatory responses and fibrin deposition in the joint (Clemmensen *et al.*, 1983; Weinberg *et al.*, 1991); the infiltration of synovial cells into the surrounding cartilage; and remodeling events of the cartilage; the fibrinolytic system could play an important role in this and other inflammatory joint diseases. Murine models of rheumatoid arthritis have identified increased expression of both coagulant and fibrinolytic proteins namely tissue factor, tissue factor pathway inhibitor, thrombin receptor, uPA and PAI-1 in affected joints (Salvi *et al.*, 2000). Additionally, studies in $TPA^{-/-}$ and $UPA^{-/-}$ mice indicated that the severity of the disease was greater in $TPA^{-/-}$ mice relative to $UPA^{-/-}$ and WT mice and the extent of fibrin deposition paralleled the severity of the disease (Yang *et al.*, 2001; Cook *et al.*, 2002). Thus, the fibrinolytic system could be protective by removing fibrin deposits that accumulate in the joints but it could also be detrimental by degrading cartilage and bone matrix protein. Therefore, in order to more directly assess the relative importance of fibrinolysis in inflammatory joint disease, $PG^{-/-}$ and WT mice were challenged utilizing a murine model of antigen-induced arthritis (Busso *et al.*, 1998). Inflammation in the joint was comparable between these mice during the early stages of the disease process but remained significantly elevated in $PG^{-/-}$ mice at a much later time

(30 days). While synovial thickness subsided in WT mice, it remained elevated up to 2 months in $PG^{-/-}$ mice. Thirty days after initiation of the disease, $PG^{-/-}$ mice developed bone erosion and fibrin deposits in the arthritic joints. Results from this study indicated that the fibrinolytic system plays primarily a protective role in arthritis by regulating fibrin accumulation in the joints and subsequent bone degradation.

3.6. Pulmonary Fibrosis

One of the underlying features in the development of pulmonary diseases is impaired fibrinolytic activity (Idell *et al.*, 1989a,b,c). For example, patients with adult respiratory distress syndrome (ADRS) present with extravascular fibrin deposits in the aveolar compartment (Idell, 1994). A well-established model of pulmonary fibrosis in mice involves administering the chemotherapeutic agent, bleomycin (Thrall and Scaliso, 1995), and a number of studies have been performed utilizing this model to investigate the role of the fibrinolytic system in the regulation of this disease. For example, in bleomycin-treated mice that are overexpressing PAI-1, enhanced collagen deposition was observed relative to WT mice while collagen deposition in PAI-$1^{-/-}$ mice was similar to WT mice (Eitzmann *et al.*, 1996). Additional studies in bleomycin-treated rats and mice indicated a beneficial effect of administrating uPA resulting in diminished pulmonary collagen accumulation (Hart *et al.*, 1994; Sisson *et al.*, 1999). In support of these observations, enhanced collagen deposition was observed in bleomycin-treated $PG^{-/-}$ mice relative to control bleomycin-treated WT mice (Swaisgood *et al.*, 2000). While it has clinically been demonstrated that fibrin deposition is a component of pulmonary disease and experimental studies have identified colocalization of fibrin with collagen-rich fibrotic lesions, similar studies in $FG^{-/-}$ mice indicated that its presence was not critical for fibrotic lesion formation (Figure 10.6, see color plate facing p. 181) (Hattori *et al.*, 2000; Ploplis *et al.*, 2000).

3.7. Wound Healing

Skin wound healing is a complex physiological process that involves a number of biochemical and cellular events to occur in a regulated spatial and temporal manner. Some of the key events associated with this process are the formation of a temporary fibrin-rich matrix, recruitment of inflammatory cells, proliferation and migration of keratinocytes at the edges of the wound, and the formation of granulation tissue. Components of the fibrinolytic system have been implicated in a number of these events. Additional studies have demonstrated a functional overlap of this proteolytic system and the matrix metalloprotease system in skin wound healing processes and indicated that inhibition of both pathways leads to a complete arrest of wound healing and closure (Lund *et al.*, 1999). Studies in $PG^{-/-}$ mice indicated a delay in skin wound healing relative to WT mice due to impaired migration of keratinocytes (Romer *et al.*, 1996). Altered migration of these cells from the wound edges was associated with an inability to degrade the provisional fibrin matrix associated with the wound as determined from studies in $PG^{-/-}/FG^{-/-}$ mice (Bugge *et al.*, 1996). Other wound healing studies in $PG^{-/-}$ mice utilized two corneal models, an excimer laser photorefractive keratectomy injury model (Drew *et al.*, 2000) and a blade to create corneal epithelial defects (Kao *et al.*, 1998). Results from these studies supported observations made in the skin wound healing model demonstrating unresolved fibrin deposition which led to corneal opacity and scarring. These phenotypes were also resolved in $PG^{-/-}/FG^{-/-}$ mice. $PG^{-/-}$ mice also demonstrated inability to efficiently remove necrotic tissue and heal wounds in models of liver injury/repair (Bezerra *et al.*, 1999) and myocardial infarction (Creemers *et al.*, 2000). While the phenotype observed in $PG^{-/-}$ mice during the wound healing process was not always dependent on Fg (liver injury model), an inability to efficiently eliminate provisional matrix barriers was a common observation in these studies and underscores the importance of a balanced fibrinolytic system in wound healing events.

3.8. Neuronal and Axonal Degeneration and Demyelination

The hippocampus is a region of the brain that is particularly sensitive to neuronal death and while Pg is primarily produced in the liver, it has also been shown to be produced in other organs namely adrenal, kidney, testis, heart, lung, uterus, spleen, thymus, gut, and brain (cerebral, hippocampal, and cerebellar neurons) (Sappino *et al.*, 1993; Zhang *et al.*, 2002). Recent studies have identified a co-localization pattern of plasminogen and its activator, tPA, in the hippocampus and, therefore, these proteins could potentially mediate neuronal degeneration (Salles and Strickland, 2002). As a result of these observations, an excitoxin-induced neuronal injury model was used in $PG^{-/-}$ mice to determine if these mice were protected from neuronal death (Tsirka *et al.*, 1997a). For this study, neuronal survival was determined 5 days after injecting a glutamate analog, kainate, into the hippocampus of WT and $PG^{-/-}$ mice and it was found that $PG^{-/-}$ mice were resistant to excitoxin injury compared to WT controls. Studies in which kainate-treated WT mice were injected with α_2-anti-plasmin demonstrated a similar resistance to excitotoxin-induced injury, therefore, confirming plasmin's role in neuronal survival (Tsirka *et al.*, 1997a). Results from additional studies utilizing this model ($PG^{-/-}/FG^{-/-}$) indicated that laminin and not fibrin is the substrate involved in excitoxin-induced neuronal death and that tPA mediated activation of Pg results in disruption of critical neuron-laminin interactions (Tsirka *et al.*, 1997b; Chen and Strickland, 1997). Contrary to these studies in brain, another neurological injury model utilized in $PG^{-/-}$ mice involved injury to the sciatic nerve with resultant axonal demyelination (Akassoglou *et al.*, 2000). Axonal demyelination was found to be enhanced in $PG^{-/-}$ mice, compared to WT control mice, paralleled by an increase in fibrin deposition. Additional studies with this model confirmed that fibrin was the plasmin substrate that affected the axonal degenerative process (Akassoglou *et al.*, 2000).

3.9. Tumor Development, Metastasis, and Angiogenesis

Clinical studies have indicated that elevated plasma levels of proteins associated with the fibrinolytic system in patients with cancer correlated with a poor prognosis for survival (Swiercz *et al.*, 1998). In order for primary tumors to grow and to disseminate to secondary sites, surrounding barrier protein matrices need to be compromised. Since studies have identified uPA/uPAR complexes on a number of tumors, the fibrinolytic system has been implicated in tumor growth-promoting activities (Dano *et al.*, 1985). Indeed, one study demonstrated that inhibition of this interaction can lead to dormancy of some tumors (Yu *et al.*, 1997). Only two tumor model studies have been performed in $PG^{-/-}$ to date.

The first study, which utilized the Lewis Lung Carcinoma model, indicated that primary tumors were smaller, less hemorrhagic, and there was reduced skin ulceration in $PG^{-/-}$ mice compared to that observed in WT mice but the rate of appearance of the primary tumor was similar for both genotypes (Bugge *et al.*, 1997). While dissemination to the lymph nodes was delayed in $PG^{-/-}$ mice, relative to WT mice, there were no quantitative differences in lung metastases between the two genotypes. The other model utilized $PG^{-/-}$ mice that were cross-bred with transgenic mice expressing Polyoma middle T antigen under the control of the mouse mammary tumor virus long terminal repeat (Bugge *et al.*, 1998). Utilizing this model, well-differentiated adenocarcinomas developed in both WT and $PG^{-/-}$ mice with no obvious histological differences, but unlike the Lewis Lung Carcinoma study, lung metastases were significantly reduced in $PG^{-/-}$ mice.

Results from these studies, would indicate that primary tumor growth and dissemination can be affected by tumor microenvironment, dominance of other proteolytic pathways, that is, matrix metalloproteases, and the genetic composition of both host and tumor tissue. Deficiency of plasminogen or other components of the fibrinolytic system may alter these and other regulatory factors and thus determine the pathology of the disease.

4. Conclusion

While earlier indirect studies have indicated that a plasminogen deficiency would not be compatible with life, the development and survival of these mice into adulthood implicate other proteases in contributing to a functional redundancy that would support embryonic development and postnatal survival in a compromised fibrinolytic state. Studies in $PG^{-/-}$ mice have confirmed, in a direct *in vivo* fashion, its major role in regulating fibrin turnover. However, aside from its hemostatic role, a number of other plasmin(ogen)-mediated physiological and pathophysiological functions have been confirmed utilizing these mice. Not surprising, a number of the phenotypes observed in a plasminogen deficiency were resolved with an additional fibrinogen deficiency but a few were not, implicating an effect of plasminogen on regulating turnover of other matrix proteins, whether through direct degradation of matrix or activation of other proteolytic pathways. These and other ongoing studies in $PG^{-/-}$ mice, and mice with related gene deletions of the fibrinolytic system, should further contribute to an appreciation of the diverse functions of this proteolytic pathway.

References

Akassoglou, K., Kombrinck, K.W., Degen, J.L., and Strickland, S. (2000). Tissue plasminogen activator-mediated fibrinolysis protects against axonal degeneration and demyelination after sciatic nerve injury. *J. Cell Biol.* **149**, 1157–1166.

Bezerra, J.A., Bugge, T.H., Melin-Aldana, H., Sabla, G., Kombrinck, K.W., Witte, D.P., and Degen, J.L. (1999). Plasminogen deficiency leads to impaired remodeling after toxic injury to the liver. *Proc. Natl. Acad. Sci. USA* **96**, 15143–15148.

Bugge, T.H., Flick, M.J., Daugherty, C.C., and Degen, J.L. (1995). Plasminogen deficiency causes severe thrombosis but is compatible with development and reproduction. *Genes Dev.* **9**, 794–807.

Bugge, T.H., Kombrinck, K.W., Flick, M.J., Daugherty, C.C., Danton, M.J., and Degen, J.L. (1996). Loss of fibrinogen rescues the pleiotropic effects of plasminogen deficiency. *Cell* **87**, 709–719.

Bugge, T.H., Kombrinck, K.W., Xiao, Q., Holmbäck, K., Daugherty, C.C., Witte, D.P., and Degen, J.L. (1997). Growth and dissemination of Lewis Lung Carcinoma in plasminogen-deficient mice. *Blood* **90**, 4522–4531.

Bugge, T.H., Lund, L.R., Kombrinck, K.W., Nielsen, B.S., Holmbäck, K., Drew, A.F., Flick, M.J., Witte, M.D.P., Dano, K., and Degen, J.L. (1998). Reduced metastasis of polyoma virus middle T antigen-induced mammary cancer in plasminogen-deficient mice. *Oncogene* **16**, 3097–3104.

Busso, N., Péclat, V., Van Ness, K., Kolodziesczyk, E., Degen, J., Bugge, T., and So, A. (1998). Exacerbation of antigen-induced arthritis in urokinase-deficient mice. *J. Clin. Invest.* **102**, 41–50.

Busuttil, S.J., Drumm, C., Ploplis, V.A., and Plow, E.F. (2000.) Endoluminal arterial injury in plasminogen deficient mice. *J. Surg. Res.* **91**, 159–164.

Carmeliet, P., Moons, L., Herbert, J.M., Crawley, J., Lupu, F., Lijnen, R., and Collen, D. (1997a). Urokinase but not tissue plasminogen activator mediates arterial neointima formation in mice. *Circ. Res.* **81**, 829–839.

Carmeliet, P., Moons, L., Ploplis, V., Plow, E., and Collen, D. (1997b). Impaired arterial neointima formation in mice with disruption of the plasminogen gene. *J. Clin. Invest.* **99**, 200–208.

Chen, Z.L. and Strickland, S. (1997). Neuronal death in the hippocampus is promoted by plasmin-catalyzed degradation of laminin. *Cell* **26**, 917–925.

Clemmensen, I., Holund, D., and Andersen, R.B. (1983). Fibrin and fibronectin in rheumatoid synovial membrane and rheumatoid synovial fluid. *Arthritis Rheum.* **26**, 479–485.

Cook, A.D., Braine, E.L., Campbell, I.K., and Hamilton, J.A. (2002). Differing roles for urokinase and tissue-type plasminogen activator in collagen-induced arthritis. *Am. J. Pathol.* **160**, 917–926.

Creemers, E., Cleutjens, J., Smits, J., Heymans, S., Moons, L., Collen, D., Daemen, M., and Carmeliet, P. (2000). Disruption of plasminogen gene in mice abolishes wound healing after myocardial infarction. *Am. J. Pathol.* **156**, 1865–1873.

Dano, K., Andreasen, P.A., Grondahl-Hansen, J., Kristensen, P., Nielsen, L.S., and Skriver, L. (1985). Plasminogen activators, tissue degradation, and cancer. *Adv. Cancer Res.* **44**, 139–267.

D'Costa, S.S. and Boyle, M.D. (1998). Interaction of a group A streptococcus within human plasma results in assembly of a surface plasminogen activator that contributes to occupancy of surface plasmin binding structures. *Microbial. Pathogen* **24**, 341–349.

Drew, A.F., Schiman, H.L., Kombrinck, K.W., Bugge, T.H., Degen, J.L., and Kaufman, A.H. (2000). Persistent corneal haze after excimer laser photokeratectomy in plasminogen-deficient mice. *Invest. Ophthalmol. Vis. Sci.* **41**, 67–72.

Eitzmann, D.T., McCoy, R.D., Zheng, X., Fay, W.P., Shen, T., Ginsburg, D., and Simon, R.H. (1996). Bleomycin-induced pulmonary fibrosis in transgenic mice. *J. Clin. Invest.* **97**, 232–237.

Estreicher, A., Mulhauser, J., Carpentier, J.L., Orci, L., and Vassalli, J.D. (1990). The receptor for urokinase type plasminogen activator polarizes expression of the protease to the leading edge of migrating monocytes and promotes degradation of enzyme inhibitor complexes. *J. Cell Biol.* **111**, 783–792.

Firestein, G.S. (1997). Etiology and pathogenesis of rheumatoid arthritis. In W.N. Kelley, E.D. Harris, S. Ruddy, and C.B. Sledge (eds) *Textbook of Rheumatology*. W.B. Saunders, Philadelphia, pp. 851–897.

Fischer, M.B., Roeckl, C., Parizek, P., Schwarz, H.P., and Aguzzi, A. (2000). Binding of disease-associated prion protein to plasminogen. *Nature* **408**, 479–483.

Gebbia, J.A., Monco, J.C., Degen, J.L., Bugge, T.H., and Benach, J.L. (1999). The plasminogen activation system enhances brain and heart invasion in murine relapsing fever borreliosis. *J. Clin. Invest.* **103**, 81–87.

Goguen, J.D., Bugge, T., and Degen, J.L. (2000). Role of pleiotropic effects of plasminogen deficiency in infection experiments with plasminogen-deficient mice. *Methods* **21**, 179–183.

Hamano, K., Iwano, M., Akai, Y., Sato, H., Kubo, a., Nishitani, Y., Uyama, H., Yoshida, Y., Miyazaki, M., Shiiki, H., Kohno, S., and Dohi, K. (2002). Expression of glomerular plasminogen activator inhibitor type 1 in glomerulonephritis. *Am. J. Kidney Dis.* **39**, 695–705.

Hart, D.A., Whidden, P., Green, F., Henkin, J., and Woods, D.E. (1994). Partial reversal of established bleomycin-induced pulmonary fibrosis by rh-urokinase in a rat model. *Clin. Invest. Med.* **17**, 69–76.

Hattori, N., Degen, J.L., Sisson, T.H., Liu, H., Moore, B.B., Pandrangi, R.G., Simon, R.H., and Drew, A.F. (2000). Bleomycin-induced fibrosis in fibrinogen-null mice. *J. Clin. Invest.* **106**, 1341–1350.

Holdsworth, S.R., Thomson, N.M., Glasgow, E.F., and Atkins, R.C. (1979). The effect of defibrination on macrophage participation in rabbit nephrotoxic nephritis: Studies using glomerular culture and electron microscopy. *Clin. Exp. Immunol.* **37**, 38–43.

Idell, S. (1994). Extravascular coagulation and fibrin deposition in acute lung injury. *New Horiz.* **2**, 566–574.

Idell, S., James, K.K., Gilles, C., Fair, D.S., and Thrall, R.S. (1989a). Abnormalities of pathways of fibrin turnover in lung lavage of rats with oleic acid and bleomycin-induced lung injury support alveolar fibrin deposition. *Am. J. Pathol.* **135**, 387–399.

Idell, S., James, K.K., Levin, E.G., Schwartz, B.S., Manchanda, N., Maunder, R.J., Martin, T.R., McLarty, J., and Fair, D.S. (1989b). Local abnormalities of coagulation and fibrinolytic pathways predispose to alveolar fibrin deposition in the adult respiratory distress syndrome. *J. Clin. Invest.* **84**, 695–705.

Idell, S., Peters, J., James, K.K., Fair, D.S., and Coalson, J.J. (1989c). Local abnormalities of coagulation and fibrinolytic pathways that promote alveolar fibrin deposition in the lungs of baboons with diffuse alveolar damage. *J. Clin. Invest.* **84**, 181–193.

Kao, W.W., Kao, C.W., Kaufman, A.H., Kombrinck, K.W., Converse, R.L., Good, W.V., Bugge, T.H., and Degen, J.L. (1998). Healing of corneal epithelial defects in plasminogen- and fibrinogen-deficient mice. *Invest. Ophthalmol. Vis. Sci.* **39**, 502–508.

Keski-Oja, J., Lyons, R.M., and Moses, H.L. (1987). Inactive secreted forms of transforming growth factor β: Activation by proteolysis. *J. Cell Biochem.* **11**A, 60.

Kincaid-Smith, P. (1972). Coagulation and renal disease. *Kidney Int.* **2**, 183–190.

Kitching, R., Holdsworth, S.R., Ploplis, V.A., Plow, E.F., Collen, D., Carmeliet, P., and Tipping, P.G. (1997). Plasminogen and plasminogen activators protect against renal injury in crescentic glomerulonephritis. *J. Exp. Med.* **185**, 963–968.

Kraft, J., Lieb, W., Zeitler, P., and Schuster, V. (2000). Ligneous conjunctivitus in a girl with severe type I plasminogen deficiency. *Graefeís Arch. Clin. Exp. Ophthalmol.* **238**, 797–800.

Lahteenmaki, K., Kukkonen, M., and Korhonen, T.K. (2001). The Pla surface protease/adhesin of *Yersinia pestis* mediates bacterial invasion into human endothelial cells. *FEBS Lett.* **504**, 69–72.

Lee, H.S., Park, S.Y., Moon, K.C., Hong, H.K., Song, C.Y., and Hong, S.Y. (2001). mRNA expression of urokinase and plasminogen activator inhibitor-1 in human crescentic glomerulonephritis. *Histopathology* **39**, 203–209.

Li, Z., Ploplis, V.A., French, E.L., and Boyle, M.D. (1999). Interaction between group A streptococci and the plasmin(ogen) system promotes virulence in a mouse skin infection model. *J. Infect. Dis.* **179**, 907–914.

Lijnen, H.R., Carmeliet, P., Bouche, A., Moons, L., Ploplis, V.A., Plow, E.F., and Collen, D. (1996). Restoration of thrombolytic potential in plasminogen-deficient mice by bolus administration of plasminogen. *Blood* **88**, 870–876.

Lijnen, H.R., Van Hoef, B., Dewerchin, M., and Collen, D. (2000). Alpha (2)-antiplasmin gene deficiency in mice does not affect neointima formation after vascular injury. *Arterioscler. Thromb. Vasc. Biol.* **20**, 1488–1492.

Lottenberg, R., des Jardin, L.E., Wang, H., and Boyle, M.D.P. (1992). Streptokinase-producing strep-tococci grown in human plasmin acquire unregulated cell-associated plasmin activity. *J. Infect. Dis.* **166**, 436–440.

Lund, L., Romer, J., Bugge, T.H., Nielsen, B.S., Frandsen, T.L., Degen, J.L., Stephens, R.W., and Dano, K. (1999). Functional overlap between classes of matrix-degrading proteases in wound healing. *EMBO J.* **18**, 4645–4656.

Malliaros, J., Holdsworth, S.R., Wojta, J., Erlich, J., and Tipping, P.G. (1993). Glomerular fibrinolytic activity in antiGBM glomerulonephritis in rabbits. *Kidney Int.* **44**, 557–564.

Matsushima, K., Taguchi, M., Kovacs, E.J., Yuong, H.A., and Oppenheim, J.J. (1986). Intracellular localization of human monocyte associated-interleukin (IL-1) activity and release of biologically active IL-1 from monocytes by trypsin and plasmin. *J. Immunol.* **136**, 2883–2891.

Mingers, A.M., Philapitsch, A., Zeitler, P., Schuster, V., Schwarz, H.P., and Kreith, H.W. (1999). Human homozygous type I plasminogen deficiency and ligneous conjunctivitus. *APMIS* **107**, 62–67.

Moons, L., Shi, C., Ploplis, V., Plow, E., Haber, E., and Collen, D. (1998). Reduced transplant arteriosclerosis in plasminogen-deficient mice. *J. Clin. Invest.* **102**, 1788–1797.

Naish, P., Penn, G.B., Evans, D.J., and Peters, D.K. (1972). The effect of defibrination on nephrotoxic serum nephritis in rabbits. *Clin. Sci.* **42**, 643–646.

Nordstrand, A., Shamaei-Tousi, A., Ny, A., and Bergstrom, S. (2001). Delayed invasion of the kidney and brain by *Borrelia crocidurae* in plasminogen-deficient mice. *Infect. Immun.* **69**, 5832–5839.

Ploplis, V.A., Carmeliet, P., Vazirzadeh, S., Van Vlaenderen, I., Moons, L., Plow, E.F., and Collen, D. (1995). Effects of disruption of the plasminogen gene on thrombosis, growth, and health in mice. *Circulation* **92**, 2585–2593.

Ploplis, V.A., Cornelissen, I., Sandoval-Cooper, M.J., Weeks, L., Noria, F.A., and Castellino, F.J. (2001). Remodeling of the vessel wall after copper-induced injury is highly attenuated in mice with a total deficiency of plasminogen activator inhibitor-1. *Am. J. Pathol.* **158**, 107–117.

Ploplis, V.A., French, E.L., Carmeliet, P., Collen, D., and Plow, E.F. (1998). Plasminogen deficiency differentially affects recruitment of inflammatory cell populations in mice. *Blood* **91**, 2005–2009.

Ploplis, V.A., Wilberding, J., McLennan, L., Liang, Z., Cornelissen, I., DeFord, M.E., Rosen, E.D., and Castellino, F.J. (2000) A total fibrinogen deficiency is compatible with the development of pulmonary fibrosis in mice. *Am. J. Pathol.* **157**, 703–708.

Ringdahl, U., Svensson, M., Wistedt, A.C., Renn, T., Kellner, R., Muller-Esterl, W., and Sjobring, U. (1998). Molecular cooperation between protein PAM and streptokinase for plasmin acquisition by *Streptococcus pyogenes*. *J. Biol. Chem.* **273**, 6424–6430.

Roggwiller, E., Fricaud, A.-C., Blisnick, T., and Braun-Breton, B. (1997). Host urokinase-type plasminogen activator participates in the release of malaria merozoites from infected erythrocytes. *Mol. Biochem. Parasitol.* **86**, 49–59.

Romer, J., Bugge, T.H., Pyke, C., Lund, L.R., Flick, M.J., Degen, J.L., and Dano, K. (1996) Impaired wound healing in mice with a disrupted plasminogen gene. *Nat. Med.* **2**, 287–292.

Rondeau, E., Mougenot, B., Lacave, R., Peraldi, M.N., Kruithof, E.K.O., and Sraer, J.D. (1990). Plasminogen activator inhibitor 1 in renal deposits of human nephropathies. *Clin. Nephrol.* **33**, 55–60.

Rosenthal, P.J., Semenov, A., Ploplis, V.A., and Plow, E.F. (1998). Plasminogen activators are not required in the erythrocytic life cycle of malaria parasites. *Mol. Biochem. Parasitol.* **97**, 253–257.

Salles, F.J. and Strickland, S. (2002). Localization and regulation of the tissue plasminogen activator-plasmin system in the hippocampus. *J. Neurosci.* **22**, 2125–2134.

Salvi, R., Péclat, V., So, A., and Busso, N. (2000). Enhanced expression of genes involved in coagulation and fibrinolysis in murine arthritis. *Arthritis Res.* **2**, 504–512.

Sappino, A., Madani, R., Huarte, J., Belin, D., Kiss, J., Wohlwend, A., and Vassali, J.-D. (1993). Extracellular proteolysis in the adult murine brain. *J. Clin. Invest.* **92**, 679–685.

Schuster, V., Mingers, A.-M., Seidenspinner, S., Nüssgens, Z., Pukrop, T., and Kreth, H.W. (1997). Homozygous mutations in the plasminogen gene of two unrelated girls with ligneous conjunctivitis. *Blood* **90**, 958–966.

Schuster, V., Seidenspinner, S., Zeitler, P., Escher, C., Pleyer, U., Bernauer, W., Stiehm, E.R., Isenberg, S., Seregard, S., Olsson, T., Mingers, A.-M., Schambeck, C., and Kreth, H.W. (1999). Compound-heterozygous mutations in the plasminogen gene predispose to the development of ligneous conjunctivitis. *Blood* **93**, 3457–3466.

Scully, C., Gokbuget, A.Y., Allen, C., Bagan, J.V., Efeoglu, A., Erseven, G., Flaitz, C., Cintan, S., Hodgson, T., Porter, S.R., and Speight, P. (2001). Oral lesions indicative of plasminogen deficiency (hypoplasminogenemia). *Oral Surg. Oral Med. Oral Pathol. Oral Radiol. Endod.* **91**, 334–337.

Shi, C., Patel, A., Zhang, D., Wang, H., Carmeliet, P., Reed, G.L., Lee, M.-E., Haber, E., and Sibinga, N.E.S. (1999). Plasminogen is not required for neointima formation in a mouse model of vein graft stenosis. *Circ. Res.* **84**, 883–890.

Sisson, T.H., Hattori, N., Xu, Y., and Simon, R.H. (1999). Treatment of bleomycin-induced pulmonary fibrosis by transfer of urokinase-type plasminogen activator genes. *Hum. Gene Ther.* **10**, 2315–2323.

Sodeinde, O.A., Subrahmanyam, Y.V., Stark, K., Quan, T., Bao, Y., and Goguen, J.D. (1992). A surface protease and the invasive character of plaque. *Science* **258**, 1004–1007.

Swaisgood, C.M., French, E.L., Noga, C., Simon, R.H., and Ploplis, V.A. (2000). The development of bleomycin-induced pulmonary fibrosis in mice deficient for components of the fibrinolytic system. *Am. J. Pathol.* **157**, 177–187.

Swiercz, R., Wolfe, J.D., Zaher, A., and Jankun, J. (1998). Expression of the plasminogen activation system in kidney cancer correlates with its aggressive phenotype. *Clin. Cancer Res.* **4**, 869–877.

Thrall, R.S. and Scaliso, P.J. (1995) Bleomycin. In S.H. Phan and R.S. Thrall (eds) *Pulmonary Fibrosis*. Marcel Dekker, Inc., New York, pp. 231–292.

Tsirka, S.E., Bugge, T.H., Degen, J.L., and Strickland, S. (1997a). Neuronal death in the central nervous system demonstrates a non-fibrin substrate for plasmin. *Proc. Natl. Acad. Sci. USA* **94**, 9779–9781.

Tsirka, S.E., Rogove, A.D., Bugge, T.H., Degen, J., and Strickland, S. (1997b). An extracellular proteolytic cascade promotes neuronal degeneration in the mouse hippocampus. *J. Neurosci.* **17**, 543–552.

Vassalli, J.D., Dayer, J.-M., Wohlwend, A., and Belin, D. (1984). Concomitant secretion of prourokinase and a plasminogen activator-specific inhibitor by cultured human monocytes-macrophages. *J. Exp. Med.* **159**, 1653–1668.

Weinberg, J.B, Pippen, A.M., and Greenberg, C.S. (1991). Extravascular fibrin formation and dissolution in synovial tissue of patients with osteoarthritis and rheumatoid arthritis. *Arthritis Rheum.* **34**, 996–1005.

Wistedt, A.C., Kotarsky, H., Marti, D., Ringdahl, U., Castellino, F.J., Schaller, J., and Sjobring, U. (1998). Kringle 2 mediates high affinity binding of plasminogen to an internal sequence in streptococcal protein PAM. *J. Biol. Chem.* **273**, 24420–24424.

Xiao, Q., Danton, M.J.S., Witte, D.P., Kowala, M.C., Valentine, M.T., Bugge, T.H., and Degen, J.L. (1997). Plasminogen deficiency accelerates vessel wall disease in mice predisposed to atherosclerosis. *Proc. Natl. Acad. Sci. USA* **94**, 10335–10340.

Yang, Y.H., Carmeliet, P., and Hamilton, J.A. (2001). Tissue-type plasminogen activator deficiency exacerbates arthritis. *J. Immunol.* **167**, 1047–1052.

Yu, W., Kim, J., and Ossowski, L. (1997). Reduction in surface urokinase receptor forces malignant cells into a protracted state of dormancy. *J. Cell Biol.* **137**, 767–777.

Zhang, L., Seiffert, D., Fowler, B.J., Jenkins, G.R., Thinnes, T.C., Loskutoff, D.J., Parmer, R.J., and Miles, L.A. (2002). Plasminogen has a broad extrahepatic distribution. *Thromb. Haemost.* **87**, 493–501.

11

Role of the Plasminogen and MMP Systems in Wound Healing

H. Roger Lijnen and Désiré Collen

1. Introduction

Wound healing involves cell migration and tissue remodeling, which require degradation of extracellular matrix (ECM). Two proteolytic systems, the fibrinolytic (plasminogen/plasmin) and matrix metalloproteinase (MMP) systems can degrade most ECM components. Plasmin can only degrade some components of the ECM directly, such as laminin and fibronectin, whereas other components such as elastin and collagen are degraded by MMPs. The plasminogen/plasmin system can, however, play a role in the activation of several proMMPs. Thus, in concert, both systems can degrade the ECM. In this chapter we will review the contribution of both proteolytic systems to neointima formation and arterial restenosis after vascular injury, allograft transplant stenosis, skin wound healing, and myocardial ischemia.

2. Plasminogen/Plasmin and MMP Systems

The main components of the plasminogen/plasmin and MMP systems and their potential molecular interactions are discussed below and schematically represented in Figure 11.1.

2.1. Plasminogen/Plasmin System

The blood fibrinolytic (plasminogen/plasmin) system comprises an inactive proenzyme, plasminogen, that can be converted to the active enzyme, plasmin, that degrades fibrin into soluble fibrin degradation products. Two immunologically distinct physiologic plasminogen activators (PA) have been identified: the tissue-type PA (t-PA) and the urokinase-type PA (u-PA). t-PA-mediated plasminogen activation is mainly

---

H. Roger Lijnen and Désiré Collen • Center for Molecular and Vascular Biology, University of Leuven, Campus Gasthuisberg, O & N, Herestraat 49, B-3000 Leuven, Belgium

Plasminogen: Structure, Activation, and Regulation, edited by David M. Waisman. Kluwer Academic/Plenum Publishers, New York, 2003.

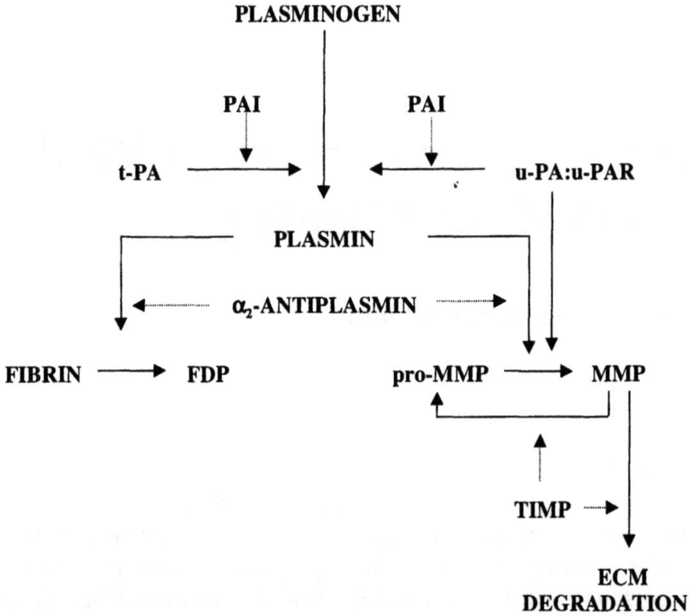

Figure 11.1. The zymogen plasminogen is converted to the active enzyme plasmin by tissue-type plasminogen activator (t-PA) or by urokinase-type plasminogen activator (u-PA) bound to its cellular receptor (u-PAR). Plasmin can convert the fibrin component of a blood clot into soluble degradation products (FDP) and can also convert several proenzyme forms of matrix metalloproteinases (pro-MMP) into active MMPs. Several active MMPs can activate other proMMPs, and collectively these are able to degrade the extracellular matrix (ECM). The plasminogen activators t-PA and u-PA are inhibited by plasminogen activator inhibitors (PAI), plasmin is inhibited mainly by α_2-antiplasmin, and MMPs by tissue inhibitors of metalloproteinases (TIMP). (→ activation or converstion; ·····> inhibition).

involved in the dissolution of fibrin in the circulation. u-PA binds to a specific cellular receptor (u-PAR) resulting in enhanced activation of cell-bound plasminogen. The main role of u-PA appears to be in the induction of pericellular proteolysis via the degradation of matrix components or via activation of latent proteinases or growth factors. Inhibition of the fibrinolytic system may occur either at the level of the PA, by specific plasminogen activator inhibitors (PAI-1 and PAI-2), or at the level of plasmin, mainly by α_2-antiplasmin (Lijnen and Collen, 1995). Physiological fibrinolysis is regulated by specific molecular interactions between its main components as well as by controlled synthesis and release, primarily from endothelial cells, of PAs and PAIs.

The plasminogen/plasmin system may play a role in a variety of biological phenomena, including blood clot dissolution (thrombolysis), reproduction, embryogenesis, cell invasion, angiogenesis, and brain function. In addition, the fibrinolytic system might participate in thrombosis, restenosis, atherosclerosis, neoplasia, metastasis, and chronic lung or kidney inflammatory disorders (Collen and Lijnen, 1991; Vassalli *et al.*, 1991; Blasi, 1993). The physiological importance of the fibrinolytic system is demonstrated by the association between abnormal fibrinolysis and a tendency toward bleeding or thrombosis (Declerck *et al.*, 1994).

2.2. MMP System

MMPs constitute a tightly regulated family of zinc dependent endopeptidases that degrade most components of the extracellular matrix and basement membrane, and can also function in many other proteolytic processes.

MMPs comprise several domain motifs. The signal peptide (17–29 amino acids) is followed by the propeptide domain (77–87 amino acids) that constitutes the amino-terminus of the enzyme. It contains a unique conserved sequence with a cysteine residue that ligates the catalytic zinc to maintain the latency of the proenzymes. Some MMPs, for example, MMP-11, MMP-14, MMP-15, MMP-16, and MMP-17 have a furin recognition site, and *Xenopus* MMP has an additional vitronectin-like domain. The catalytic domain (about 170 amino acids) contains the catalytic machinery including the zinc-binding site and a conserved methionine. This domain contains additional zinc and calcium ions, which maintain the three dimensional structure of MMPs, required for their stability and enzymatic activities. The two gelatinases (MMP-2 and MMP-9) contain additional repeats of fibronectin-like domains, which interact with collagen and gelatins. The carboxy-terminal hemopexin-like domain (about 210 amino acids) has an ellipsoidal disc shape and plays a key role in substrate specificity. The hemopexin domain of proMMP-2 is also required for its cell surface activation by MT1-MMP. MMP-7 and MMP-26 lack this domain entirely and in some members of the MMP family, for example, MMP-23, the hemopexin domain is substituted by a cysteine-rich, proline-rich, and interleukin-1 receptor-like motif. The catalytic and the hemopexin domains are linked by a proline-rich linker domain (the hinge region). A transmembrane domain occurs in MT-MMPs, which anchors them to the cell surface (Birkedal-Hansen *et al.*, 1993; Nagase and Woessner, 1999; Lijnen, 2001).

The members of the MMP family are divided into five classes based on their structure and substrate specificities. These are collagenases, gelatinases, stromelysins, membrane type MMPs (MT-MMPs) and others. Although each MMP is encoded by a different gene, there is a high degree of sequence homology and structural domain conservation. MMPs are synthesized, and most also secreted as zymogens that are extracellularly activated by several proteinases (including plasmin, trypsin, chymotrypsin, kallikrein, cathepsin, or neutrophil elastase and other MMPs) (Nagase, 1997).

Active MMPs are inhibited by tissue inhibitors (TIMPs); four homologous TIMPs (TIMP-1 to -4) have been identified. They form noncovalent complexes with the active forms of MMPs as well as with some proMMPs, thereby inhibiting active MMPs or impairing the activation of proMMPs (Brew *et al.*, 2000).

2.3. Molecular Interactions Between the Plasminogen/Plasmin and MMP Systems

In vitro, plasmin directly activates proMMP-1, proMMP-3, proMMP-9, proMMP-10, and proMMP-13 (Eeckhout and Vaes, 1977; He *et al.*, 1989; Suzuki *et al.*, 1990; Okada *et al.*, 1992). Activation of proMMP-2 involves hydrolysis by MT1-MMP, yielding an intermediate that is activated by plasmin (Baramova *et al.*, 1997). Activation of proMMP-9 may occur via plasmin-dependent mechanisms, or via plasmin-independent mechanisms (Lijnen *et al.*, 1998a). Thus, plasmin may play an important role in the *in vivo* activation of proMMPs. Several active MMPs can further activate other proMMPs, thus constituting positive feedback mechanisms. MMP-3 can activate proMMP-9 (Ogata *et al.*, 1992), and MMP-3 and MMP-10 can superactivate procollagenase, generating collagenase with higher specific activity (He *et al.*, 1989; Suzuki *et al.*, 1990). MMP-3 (Ugwu *et al.*, 1998) and MMP-7 (Marcotte *et al.*, 1992) convert scu-PA to a lower molecular weight form and remove the u-PAR binding domains. MMP-3 (Lijnen *et al.*, 1998b), MMP-7 and MMP-9 (Patterson and Sang, 1997), and MMP-12 (Dong *et al.*, 1997) cleave plasminogen generating an angiostatin-like fragment containing the receptor-binding domains. MMP-3 may thus downregulate cellular fibrinolysis by reducing the amount of activatable cell-bound plasminogen (Ugwu *et al.*, 1999). MMP-3 and MMP-7 also degrade fibrinogen and cross-linked fibrin (Bini *et al.*, 1996, 1999), whereas the catalytic domain of MMP-11 cleaves fibrinogen but not fibrin (Lijnen *et al.*, 1998c).

3. Arterial Restenosis

Vascular interventions for the treatment of atherothrombosis induce restenosis of the vessel within 3–6 months in 30–50% of treated patients. Arterial stenosis may result from remodeling of the vessel wall (such as occurs predominantly after balloon angioplasty) and/or from accumulation of cells and ECM in the intimal layer (such as occurs predominantly after intraluminal stent application).

Proteinases from the plasminogen/plasmin and MMP systems participate in the proliferation and migration of smooth muscle cells (SMC), and in matrix remodeling during arterial wound healing (Dollery *et al.*, 1995; Celentano and Frishman, 1997; Carmeliet and Collen, 1998). Of the MMPs, only MMP-2 appears to be expressed in the quiescent SMC, whereas expression of MMP-3, MMP-7, MMP-9, MMP-12, and MMP-13 is induced in injured, transplanted, or atherosclerotic arteries (Irizarry *et al.*, 1993; Newman *et al.*, 1994; Galis *et al.*, 1995; Halpert *et al.*, 1996; Sakalihasan *et al.*, 1996; Moons *et al.*, 1998). In a balloon-injured rat carotid artery model, a correlation was observed between changes in plasminogen activators (Clowes *et al.*, 1990; Jackson and Reidy, 1992; Reidy *et al.*, 1996) or MMPs (Bendeck *et al.*, 1994; Zempo *et al.*, 1994) and SMC migration. Increased levels of PAI-1 and TIMP-2 in this model suggested that changes in the proteolytic balance play a role in SMC migration after arterial injury (Hasenstab *et al.*, 1997). This is supported by the findings that *in vivo* SMC migration in the rat was inhibited by administration of the plasmin inhibitor tranexamic acid (Hasenstab *et al.*, 1997) or of MMP inhibitors (Bendeck *et al.*, 1994; Zempo *et al.*, 1996), and by overexpression of TIMP-1 (Forough *et al.*, 1996). Adenovirus-mediated gene transfer of the human TIMP-1 gene was also shown to inhibit SMC migration and neointimal formation in the human saphenous vein (George *et al.*, 1998). Furthermore, it was demonstrated that SCM migration in baboon aortic explants requires plasminogen activators, MMP-2, MMP-9, basic fibroblast growth factor (bFGF), and platelet-derived growth factor (PDGF) (Kenagy *et al.*, 1996, 1997). PDGF-induced migration of SMC was found to depend on MMP-2, whereas bFGF-induced migration depended on both MMP-2 and MMP-9 (Kenagy *et al.*, 1997). Studies in balloon-injured rabbit carotid arteries suggested that MMP-2 facilitates migration of SMC from the media to the intima and does not affect replication (Aoyagi *et al.*, 1998).

To assess the role of the plasminogen/plasmin and MMP systems in SMC migration and neointima formation, a perivascular electric injury model in the mouse has been extensively used (Carmeliet *et al.*, 1997a). In this model, surgically exposed femoral arteries are injured perivascularly via delivery of an electric current, which destroys all medial smooth muscle cells, denudes the injured segment of intact endothelium, and transiently induces platelet-rich mural thrombosis. A vascular wound-healing response results that is characterized by degradation of the mural thrombus, transient infiltration of the vessel wall by inflammatory cells, and progressive removal of the necrotic debris. Topographic analysis revealed repopulation of the media and accumulation in the neointima of smooth muscle cells originating from the uninjured borders and progressing into the necrotic center. Within 3 weeks after injury, a neointima was formed that contained up to 12 layers of smooth muscle α-actin-immunoreactive cells. Evans blue staining in injured arteries revealed progressive re-endothelialization from the uninjured borders.

3.1. Role of the Plasminogen/Plasmin System in Neointima Formation

The electric injury model described above has been applied to wild-type mice and to mice with deficiency of the main components of the plasminogen/plasmin system.

At 1 week after vascular injury in wild-type mice, t-PA activity in arterial sections or extracts was not significantly altered, whereas u-PA activity levels were 2- to 3-fold higher than control at 2 days after injury of the femoral artery. Prolonged fibrin overlay with femoral or carotid artery sections from t-PA deficient mice revealed that the fibrinolytic activity in injured versus control segments was markedly enhanced. This activity was reduced by approximately 50% upon inhibition of u-PA; it was also markedly inhibited upon addition of anti-MMP-2 IgG, which may be explained by a weak direct fibrinolytic activity of MMP-2 (Lijnen *et al.*, 1998d).

In plasminogen deficient mice, wound healing was significantly impaired with delayed removal of necrotic debris, reduced leucocyte infiltration and SMC accumulation, and decreased neointima formation. SMC accumulated at the uninjured borders but failed to migrate into the necrotic center (Carmeliet *et al.*, 1997b). Neointima formation and neointimal cell accumulation were also reduced in u-PA deficient and in combined t-PA and u-PA deficient arteries but not in t-PA deficient arteries. Similar to the plasminogen-deficient arteries, SMC accumulated at the uninjured borders but failed to migrate into the necrotic center in u-PA deficient and in the double deficient arteries (Carmeliet *et al.*, 1997c). Proliferation of SMC and re-endothelialization were not affected by either plasminogen, u-PA or t-PA deficiency. Thus, u-PA and plasminogen play a significant role in vascular wound healing and arterial neointima formation after injury, most likely by promoting cellular migration.

Deficiency of PAI-1, the main inhibitor of both u-PA and t-PA, in contrast improved vascular wound healing. SMC migrated more rapidly from the uninjured borders into the necrotic center of the arterial wound than wild-type SMC. When PAI-1 deficient mice were intravenously injected with replication-defective adenovirus expressing human PAI-1, plasma PAI-1 antigen levels increased in a dose-dependent fashion and luminal stenosis was significantly suppressed. By impairing cellular migration, PAI-1 thus plays an inhibitory role in vascular wound healing and arterial neointima formation after injury (Carmeliet *et al.*, 1997d).

In wild-type mice and in mice deficient in α_2-antiplasmin, the main physiological plasmin inhibitor, the neointimal and medial areas at 1–3 weeks after electric injury of the femoral artery were similar, resulting in comparable intima/media ratios. Nuclear cell counts in cross-sectional areas of the intima of the injured region were also comparable. Fibrin deposition was not significantly different in arteries of both genotypes at 1 day after injury, and no mural thrombosis was detected at 1 week after injury. Thus, α_2-antiplasmin does not seem to play a major role in SMC migration and neointima formation in mice (Lijnen *et al.*, 2000).

3.2. Role of the MMP System in Neointima Formation

The temporal and topographic expression pattern of MMPs (MMP-2, MMP-3, MMP-9, MMP-12, and MMP-13) after vascular injury is compatible with a role in ECM degradation and SMC migration. Evaluation of the expression after injury of these MMPs by electron microscopic immunogold labeling and immunofluorescence confocal microscopy indeed revealed gradually enhanced expression during vessel remodeling, with the highest levels obtained 1–2 weeks after injury (Lijnen *et al.*, 1999a). In noninjured femoral arteries and in injured arteries after 1 week, MMP-2 and MMP-3 were produced by adventitial cells, which were not stained for α-actin (fibroblasts and/or noncontractile SMC), whereas 2 weeks after injury, when the SMC have regained their contractile phenotype, MMP-2 and MMP-3 were also detected in some α-actin stained cells. MMP-9, MMP-12, and MMP-13 were found in macrophages located mainly in the adventitia and only occasionally in subintimal areas. A high amount of MMP-2 remains associated with the cell surface suggesting the presence of a specific binding site, possibly one of the membrane-type metalloproteinases (MT-MMPs), which are known to bind and activate MMP-2. Interestingly, MMP-2 was found in high concentration extracellularly, associated with the internal elastic lamina and with elastin bundles in the neointima, whereas virtually no extracellular MMP-9 was detected. Total MMP-9 levels (active plus latent) were also markedly increased after injury. MMP-3 expression after injury gradually increased to reach a maximum at about 2 weeks. It was produced mainly by fibroblasts (and SMC) in the adventitia, where it was found in typical storage granules and was extracellularly associated with the basement membrane and occasionally with collagen fibers. These data are compatible with findings in the rat carotid artery model, showing constitutive expression of latent MMP-2 and markedly enhanced levels of active MMP-2 up to 7 days after balloon injury, whereas also active and latent MMP-9 were induced after injury (Bendeck *et al.*, 1994; Webb *et al.*, 1997). In this rat model, MMP-3 was detectable as early as 2 hr after balloon injury and reached a maximum (2-fold increase) at 7 days (Webb *et al.*, 1997).

Arterial neointima formation after vascular injury was studied in mice with deficiency of TIMP-1 (Lijnen *et al.*, 1999b). At 1–3 weeks after injury, the intimal areas were significantly larger in TIMP-1 deficient mice as compared to wild-type mice, and contained abundant SMC, whereas the medial areas were comparable, resulting in significantly higher intima/media ratio's in the TIMP-1 deficient mice. These data thus confirm a physiological role of TIMP-1 in vascular remodeling, most likely via monitoring of MMP activity. This is further supported by the finding that stromelysin mRNA antisense oligonucleotides inhibited phenotypic modulation of rat arterial SMC and thereby caused a decrease in migration and proliferation and in neointima formation after vessel wall injury (Lovdahl *et al.*, 1999).

Taken together, these data indicate a role of MMPs in extracellular matrix degradation and SMC migration after vascular injury.

4. Allograft Transplant Stenosis

Obstruction of the lumen due to intimal thickening limits the success of organ transplantation in many patients and constitutes a major limitation to long-term survival. Transplant arteriosclerosis is characterized by concentric lesions, which progress more rapidly than common atherosclerosis. The molecular mechanisms remain poorly understood. It may be caused by an immune reaction of the recipient to donor graft antigens (Libby, 1995), as indeed inflammatory cells infiltrate the blood vessels of the transplanted organ and produce cytokines, growth factors, and chemotactic agents that induce vascular SMC proliferation and migration (Libby, 1995; Russel *et al.*, 1995; Nagano *et al.*, 1997). To migrate across anatomical borders such as the elastic lamina, cells require proteolytic activity. The role of the plasminogen/plasmin system was studied in a mouse model of transplant arteriosclerosis, in which carotid arteries from (B.10A,2R) wild-type mice were transplanted in (C57Bl6:129) plasminogen-deficient mice. Within 45 days, the neointima was much smaller and less SMC were present in plasminogen-deficient than in wild-type mice. Media necrosis was less pronounced and degradation of the elastic laminae and formation of a neoadventitia were more severe in wild-type mice. At 15 days after transplantation, when cells actively migrate, expression of u-PA, t-PA, MMP-2, MMP-3, MMP-9, MMP-12, and MMP-13 was significantly enhanced. Destruction of the medial stroma occurred only after prior degradation of the elastic lamina by macrophages (Moons *et al.*, 1998). Since plasmin does not directly degrade elastin or collagen it may act through activation of proMMPs. Subsequently, leukocytes in the media activate SMC to proliferate and migrate into the intima, a process that is also mediated by plasmin.

5. Skin Wound Healing

Skin wound healing has been studied in a mouse model in which incisional skin wounds are made middorsally; a new epidermal layer is formed by keratinocytes migrating from the wound edges. During skin wound healing, migrating keratinocytes express u-PA and PAI-1 (Rømer *et al.*, 1991, 1994), as well as several MMP system components (Hewitt and Danø, 1996). With this model impaired skin wound healing was observed in mice with plasminogen deficiency. These showed a decreased migration rate of keratinocytes from the wound edge during re-epithelialization (Rømer *et al.*, 1996). The finding that skin wound healing was normal in mice with combined deficiency of plasminogen and fibrin(ogen) suggests that the defect in the plasminogen deficient mice is due to a diminished ability of the cells to migrate through the fibrin-rich ECM beneath the wound crust (Bugge *et al.*, 1996).

Although wound healing in plasminogen-deficient mice was delayed, it eventually did occur. However, complete inhibition of wound healing was observed in plasminogen-deficient mice treated with the broad spectrum MMP inhibitor galardin. This suggests that there is a functional overlap between the plasminogen/plasmin and MMP systems. Each system alone can maintain sufficient keratinocyte migration to regenerate the epidermal surface (Lund *et al.*, 1999).

6. Myocardial Ischemia

Cardiac rupture accounts for significant in-hospital mortality after acute myocardial infarction (MI) (AMI). The plasminogen/plasmin and MMP systems may be involved in cardiac remodeling and in growth and remodeling of collateral vessels (Knoepfler *et al.*, 1995; Heymans *et al.*, 1999; Rohde *et al.*, 1999). Increased MMP activity was reported in experimental MI (Cleutjens *et al.*, 1995) and in clinical dilated cardiomyopathy (Thomas *et al.*, 1998). Furthermore, animal models of congestive heart failure demonstrated that the activity of MMP-1, MMP-2, and MMP-3 increased together with left ventricular end-diastolic dimension (Coker *et al.*, 1998). Increased proteolysis may facilitate infarct expansion and ventricular dilation, whereas impaired proteolysis may be involved in cardiac fibrosis (Robert *et al.*, 1997) and in coronary or myocardial remodeling after AMI (Tyagi *et al.*, 1996). Proteinase inhibitors may improve cardiac function by protecting against vessel wall thinning or dilatation. Indeed, administration of a broad spectrum MMP inhibitor to mice attenuated early left ventricular dilation after experimental MI (Rohde *et al.*, 1999). This effect was most prominent in animals with initial baseline dilatation, suggesting a beneficial effect of MMP inhibition particularly in mice with more extensive damage. MMP inhibition was also reported to attenuate left ventricular enlargement in pacing-induced models of congestive heart failure (Spinale *et al.*, 1997) and in spontaneous heart failure in rats (Peterson *et al.*, 1997).

The role of the plasminogen/plasmin and MMP systems was recently studied using a combination of gene-inactivation and gene-transfer techniques in mice. In these studies, AMI resulted in rupture in wild-type mice and in mice deficient in t-PA, u-PAR, MMP-3, or MMP-12, whereas u-PA deficient mice were completely protected against rupture, and MMP-9 deficient mice partially. However, u-PA deficient mice showed impaired scar formation and infarct revascularization and died of cardiac failure. Temporary administration of PAI-1 or TIMP-1 completely protected the wild-type mice against rupture but did not abort infarct healing. Thus, transient administration of proteinase inhibitors may provide a new approach to prevent fatal cardiac rupture, but prolonged inhibition may cause cardiac failure after AMI (Heymans *et al.*, 1999).

7. Summary

The plasminogen/plasmin and MMP systems contribute to cell migration and tissue remodeling, which require degradation of extracellular matrix. Proteolytic activity generated via either system can degrade some components of the ECM. Several proMMPs can be activated by the plasminogen/plasmin system or by other active MMPs. In concert, both proteolytic systems can degrade all components of the ECM and thus may play a role in wound healing.

References

Aoyagi, M., Yamamoto, M., Azuma, H., Nagashima, G., Niimi, Y., Tamaki, M., Hirakawa, K., and Yamamoto, K. (1998). Immunolocalization of matrix metalloproteinases in rabbit carotid arteries after balloon denudation. *Histochem. Cell Biol.* **109**, 97–102.

Baramova, E.N., Bajou, K., Remacle, A., L'Hoir, C., Krell, H.W., Weidle, U.H., Noel, A., and Foidart, J.M. (1997). Involvement of PA/plasmin system in the processing of proMMP-9 and in the second step of proMMP-2 activation. *FEBS Lett.* **405**, 157–162.

Bendeck, M.P., Zempo, N., Clowes, A.W., Galardy, R.E., and Reidy, M.A. (1994). Smooth muscle cell migration and matrix metalloproteinase expression after arterial injury in the rat. *Circ. Res.* **75**, 539–545.

Bini, A., Itoh, Y., Kudryk, B.J., and Nagase, H. (1996). Degradation of cross-linked fibrin by matrix metalloproteinase 3 (stromelysin 1): Hydrolysis of the gamma Gly 404–Ala 405 peptide bond. *Biochemistry* **35**, 13056–13063.

Bini, A., Wu, D., Schnuer, J., and Kudryk, B.J. (1999). Characterization of stromelysin-1 (MMP-3), matrilysin (MMP-7), and membrane type 1 matrix metalloproteinase (MT1-MMP) derived fibrin(ogen) fragments D-dimer and D-like monomer: NH_2-terminal sequences of late-stage digest fragments. *Biochemistry* **38**, 13928–13936.

Birkedal-Hansen, H., Moore, W.G., Bodden, M.K., Windsor, L.J., Birkedal-Hansen, B., DeCarlo, A., and Engler, J.A. (1993). Matrix metalloproteinases: A review. *Crit. Rev. Oral Biol. Med.* **4**, 197–250.

Blasi, F. (1993). Urokinase and urokinase receptor: A paracrine/autocrine system regulating cell migration and invasiveness. *BioEssays* **15**, 105–111.

Brew, K., Dinakarpandian, D., and Nagase, H. (2000). Tissue inhibitors of metalloproteinases: Evolution, structure and function. *Biochim. Biophys. Acta* **1477**, 267–283.

Bugge, T.H., Kombrinck, K.W., Flick, M.J.,, Daugherty, C.C., Danton, M.J., and Degen, J.L. (1996). Loss of fibrinogen rescues mice from the pleiotropic effects of plasminogen deficiency. *Cell* **87**, 709–719.

Carmeliet, P. and Collen, D. (1998). Development and disease in proteinase-deficient mice: Role of the plasminogen, matrix metalloproteinase and coagulation system. *Thromb. Res.* **91**, 255–285.

Carmeliet, P., Moons, L., Stassen, J.-M., De Mol, M., Bouché, A., van den Oord, J.J., Kockx, M., and Collen, D. (1997a). Vascular wound healing and neointima formation induced by perivascular electric injury in mice. *Am. J. Pathol.* **150**, 761–776.

Carmeliet, P., Moons, L., Ploplis, V., Plow, E., and Collen, D. (1997b). Impaired arterial neointima formation in mice with disruption of the plasminogen gene. *J. Clin. Invest.* **99**, 200–208.

Carmeliet, P., Moons, L., Herbert, J.-M., Crawley, J., Lupu, F., Lijnen, R., and Collen, D. (1997c). Urokinase but not tissue plasminogen activator mediates arterial neointima formation in mice. *Circ. Res.* **81**, 829–939.

Carmeliet, P., Moons, L., Lijnen, R., Janssens, S., Lupu, F., Collen, D., and Gerard, R.D. (1997d). Inhibitory role of plasminogen activator inhibitor-1 in arterial wound healing and neointima formation. A gene targeting and gene transfer study in mice. *Circulation* **96**, 3180–3191.

Celentano, D.C. and Frishman, W.H. (1997). Matrix metalloproteinases and coronary artery disease: A novel therapeutic target. *J. Clin. Pharmacol.* **150**, 761–776.

Cleutjens, J.P.M., Kandala, J.C., Guarda, E., Guntaka, R.V., and Weber, K.T. (1995). Regulation of collagen degradation in the rat myocardium after infarction. *J. Mol. Cell Cardiol.* **27**, 1281–1292.

Clowes, A.W., Clowes, M.M., Au, Y.P., Reidy, M.A., and Belin, D. (1990). Smooth muscle cells express urokinase during mitogenesis and tissue-type plasminogen activator during migration in injured rat carotid artery. *Circ. Res.* **67**, 61–67.

Coker, M.L., Thomas, C.V., Clair, M.J., Hendrick, J.W., Kombrach, R.S., Galis, Z.S., and Spinale, F.G. (1998). Myocardial matrix metalloproteinase activity and abundance with congestive heart failure. *Am. J. Physiol.* **274**, H1516–H1523.

Collen, D. and Lijnen, H.R. (1991). Basic and clinical aspects of fibrinolysis and thrombolysis. *Blood* **78**, 3114–3124.

Declerck, P.J., Juhan-Vague, I., Felez, J., and Wiman, B. (1994). Pathophysiology of fibrinolysis. *J. Int. Med.* **236**, 425–432.

Dollery, C.M., McEwan, J.R., and Henney, A.M. (1995). Matrix metalloproteinases and cardiovascular disease. *Circ. Res.* **77**, 863–868.

Dong, Z., Kumar, R., Yang, X., and Fidler, I.J. (1997). Macrophage-derived metalloelastase is responsible for the generation of angiostatin in Lewis lung carcinoma. *Cell* **88**, 801–810.

Eeckhout, Y. and Vaes, G. (1977). Further studies on the activation of procollagenase, the latent precursor of bone collagenase. Effects of lysosomal cathepsin B, plasmin and kallikrein, and spontaneous activation. *Biochem. J.* **166**, 21–31.

Forough, R., Koyama, N., Hasenstab, D., Lea, H., Clowes, M., Nikkari, S.T., and Clowes, A.W. (1996). Overexpression of tissue inhibitor of matrix metalloproteinase-1 inhibits vascular smooth muscle cell functions *in vitro* and *in vivo*. *Circ. Res.* **79**, 812–820.

Galis, Z.S., Sukhova, G.K., Kranzhöfer, R., Clark, S., and Libby, P. (1995). Macrophage foam cells from experimental atheroma constitutively produce matrix-degrading proteinases. *Proc. Natl. Acad. Sci. USA* **92**, 402–406.

George, S.J., Johnson, J.L., Angelini, G.D., Newby, A.C., and Baker, A.H. (1998). Adenovirus-mediated gene transfer of the human TIMP-1 gene inhibits smooth muscle cell migration and neointimal formation in human saphenous vein. *Hum. Gene Ther.* **9**, 867–877.

Halpert, I., Sires, U.I., Roby, J.D., Potter-Perigo, S., Wight, T.N., Shapiro, S.D., Welgus, H.G., Wickline, S.A., and Parks, W.C. (1996). Matrilysin is expressed by lipid-laden macrophages at sites of potential rupture in atherosclerotic lesions and localizes to areas of versican deposition, a proteoglycan substrate for the enzyme. *Proc. Natl. Acad. Sci. USA* **93**, 9748–9753.

Hasenstab, D., Forough, R., and Clowes, A.W. (1997). Plasminogen activator inhibitor type 1 and tissue inhibitor of metalloproteinases-2 increase after arterial injury in rats. *Circ. Res.* **80**, 490–496.

He, C.S., Wilhelm, S.M., Pentland, A.P., Marmer, B.L., Grant, G.A., Eisen, A.Z., and Goldberg, G.I. (1989). Tissue cooperation in a proteolytic cascade activating human interstitial collagenase. *Proc. Natl. Acad. Sci. USA* **86**, 2632–2636.

Hewitt, R. and Danø, K. (1996). Stromal cell expression of components of matrix-degrading protease systems in human cancer. *Enzyme Protein* **49**, 163–173.

Heymans, S., Luttun, A., Nuyens, D., Theilmeier, G., Creemers, E., Moons, L., Dyspersin, G.D., Cleutjens, J.P.M., Shipley, M., Angellillo, A., Levi, M., Nübe, O., Baker, A., Keshet, E., Lupu, F., Herbert, J.-M., Smiths, F.J.M., Shapiro, S.D., Baes, M., Borgers, M., Collen, D., Daemen, M.J.A.P., and Carmeliet, P. (1999). Inhibition of plasminogen activators or matrix metalloproteinases prevents cardiac rupture but impairs therapeutic angiogenesis and causes cardiac failure. *Nature Med.* **5**, 1135–1142.

Irizarry, E., Newman, K.M., Gandhi, R.H., Nackman, G.B., Halpern, V., Wishner, S., Scholes, J.V., and Tilson, M.D. (1993). Demonstration of interstitial collagenase in abdominal aortic aneurysm disease. *J. Surg. Res.* **54**, 571–574.

Jackson, C.L. and Reidy, M.A. (1992). The role of plasminogen activation in smooth muscle cell migration after arterial injury. *Ann. N. Y. Acad. Sci.* **667**, 141–150.

Kenagy, R.D., Vergel, S., Mattsson, E., Bendeck, M., Reidy, M.A., and Clowes, A.W. (1996). The role of plasminogen, plasminogen activators, and matrix metalloproteinases in primate arterial smooth muscle cell migration. *Arterioscler. Thromb. Vasc. Biol.* **16**, 1373–1382.

Kenagy, R.D., Hart, C.E., Stetler-Stevenson, W.G., and Clowes, A.W. (1997). Primate smooth muscle cell migration from aortic explants is mediated by endogenous platelet-derived growth factor and basic fibroblast growth factor acting through matrix metalloproteinases 2 and 9. *Circulation* **96**, 3555–3560.

Knoepfler, P.S., Bloor, C.M., and Carroll, S.M. (1995). Urokinase plasminogen activator activity is increased in the myocardium during coronary artery occlusion. *J. Mol. Cell. Cardiol.* **27**, 1317–1324.

Libby, P. (1995). Molecular bases of the acute coronary syndromes. *Circulation* **91**, 2844–2850.

Lijnen, H.R. (2001). Plasmin and matrix metalloproteinases in vascular remodeling. *Thromb. Haemost.* **86**, 324–333.

Lijnen, H.R. and Collen, D. (1995). Mechanisms of physiological fibrinolysis. *Baillière's Clin. Haematol.* **8**, 277–290.

Lijnen, H.R., Silence, J., Van Hoef, B., and Collen, D. (1998a). Stromelysin-1 (MMP-3)-independent gelatinase expression and activation in mice. *Blood* **91**, 2045–2053.

Lijnen, H.R., Ugwu, F., Bini, A., and Collen, D. (1998b). Generation of an angiostatin-like fragment from plasminogen by stromelysin-1 (MMP-3). *Biochemistry* **37**, 4699–4702.

Lijnen, H.R., Ugwu, F., Rio, M.C., and Collen, D. (1998c). Plasminogen/plasmin and matrix metalloproteinase system function in mice with targeted inactivation of stromelysin-3 (MMP-11). *Fibrinol. Proteol.* **12**, 155–164.

Lijnen, H.R., Van Hoef, B., Lupu, F., Moons, L., Carmeliet, P., and Collen, D. (1998d). Function of the plasminogen/plasmin and matrix metalloproteinase systems after vascular injury in mice with targeted inactivation of fibrinolytic system genes. *Arterioscler. Thromb. Vasc. Biol.* **18**, 1035–1045.

Lijnen, H.R., Lupu, F., Moons, L., Carmeliet, P., Goulding, D., and Collen, D. (1999a). Temporal and topographic matrix metalloproteinase expression after vascular injury in mice. *Thromb. Haemost.* **81**, 799–807.

Lijnen, H.R., Van Hoef, B., Soloway, P., and Collen, D. (1999b). Tissue inhibitor type 1 of matrix metalloproteinases (TIMP-1) impairs arterial neointima formation after vascular injury in mice. *Circ. Res.* **85**, 1186–1191.

Lijnen, H.R., Van Hoef, B., Dewerchin, M., and Collen, D. (2000). α_2-Antiplasmin gene deficiency in mice does not affect neointima formation after vascular injury. *Arterioscler. Thromb. Vasc. Biol.* **20**, 1488–1492.

Lovdahl, C., Thyberg, J., Cercek, B., Blomgren, K., Dimayuga, P., Kallin, B., and Hultgardh-Nilsson, A. (1999). Antisense oligonucleotides to stromelysin mRNA inhibit injury-induced proliferation of arterial smooth muscle cells. *Histol. Histopathol.* **14**, 1101–1112.

Lund, L.R., Rømer, J., Bugge, T.H., Nielsen, B.S., Frandsen, T.L., Degen, J.L., Stephens, R.W., and Danø, K. (1999). Functional overlap between two classes of matrix-degrading proteases in wound healing. *EMBO J.* **18**, 4645–4656.

Marcotte, P.A., Kozan, I.M., Dorwin, S.A., and Ryan, J.M. (1992). The matrix metalloproteinase pump-1 catalyzes formation of low molecular weight (pro)urokinase in cultures of normal human kidney cells. *J. Biol. Chem.* **267**, 13803–13806.

Moons, L., Shi, V., Ploplis, V., Plow, E., Haber, E., Collen,, D., and Carmeliet, P. (1998). Reduced transplant arteriosclerosis in plasminogen deficient mice. *J. Clin. Invest.* **102**, 1788–1797.

Nagano, H., Mitchell, R.N., Taylor, M.K., Hasegawa, S., Tilney, N.L., and Libby, P. (1997). Interferon-gamma deficiency prevents coronary arteriosclerosis but not myocardial rejection in transplanted mouse hearts. *J. Clin. Invest.* **100**, 550–557.

Nagase, H. (1997). Activation mechanisms of matrix metalloproteinases. *Biol. Chem.* **378**, 151–160.

Nagase, H. and Woessner, J.F. Jr. (1999). Matrix metalloproteinases. *J. Biol. Chem.* **274**, 21491–21494.

Newman, K.M., Jean Claude, J., Li, H., Scholes, J.V., Ogata, Y., Nagase, H., and Tilson, M.D. (1994). Cellular localization of matrix metalloproteinase in the abdominal aortic aneurysm wall. *J. Vasc. Surg.* **20**, 814–820.

Ogata, Y., Enghild, J.J., and Nagase, H. (1992). Matrix metalloproteinase 3 (stromelysin) activates the precursor of the human matrix metalloproteinase 9. *J. Biol. Chem.* **267**, 3581–3584.

Okada, Y., Gonoji, Y., Naka, K., Tomita, K., Nakanishi, I., Iwata, K., Yamashita, K., and Hayakawa, T. (1992). Matrix metalloproteinase 9 (92-kDa gelatinase/type IV collagenase) from HT1080 human fibrosarcoma cells. Purification and activation of the precursor and enzymic properties. *J. Biol. Chem.* **267**, 21712–21719.

Patterson, B.C. and Sang, Q.A. (1997). Angiostatin-converting enzyme activities of human matrilysin (MMP-7) and gelatinase B/type IV collagenase (MMP-9). *J. Biol. Chem.* **272**, 28823–28825.

Peterson, J.T., Rosebury, W.S., Robertson, A.W., Washington, R.A., Li, H., O'Brien, P.M., Sliskovic, D.R., Hallak, H., Uprichard, A.C.G., and Bocan, T.M.A. (1997). Matrix metalloproteinase inhibition blocks progression of heart failure. *Circulation* **96** (Suppl. I), I-520 (Abstract).

Reidy, M.A., Irvin, C., and Lindner, V. (1996). Migration of arterial wall cells. Expression of plasminogen activators and inhibitors in injured rat arteries. *Circ. Res.* **78**, 405–414.

Robert, V., Besse, S., Sabri, A., Silvestre, J.S., Assayag, P., Nguyen, V.T., Swynghedauw, B., and Delcayre, C. (1997). Differential regulation of matrix metalloproteinases associated with aging and hypertension in the rat heart. *Lab. Invest.* **76**, 729–738.

Rohde, L.E., Ducharme, A., Arroyo, L.H., Aikawa, M., Sukhova, G.H., Lopez-Anaya, A., McClure, K.F., Mitchell, P.G., Libby, P., and Lee, R.T. (1999). Matrix metalloproteinase inhibition attenuates early left ventricular enlargement after experimental myocardial infarction in mice. *Circulation* **99**, 3063–3070.

Rømer, J., Lund, L.R., Eriksen, J., Ralfkiær, E., Zeheb, R., Gelehrter, T.D., Danø, K., Kristensen, P. (1991). Differential expression of urokinase-type plasminogen activator and its type-1 inhibitor during healing of mouse skin wounds. *J. Invest. Dermatol.* **97**, 803–811.

Rømer, J., Lund, L.R., Eriksen, J., Pyke, C., Kristensen, P., and Danø, K. (1994). The receptor for urokinase-type plasminogen activator is expressed by keratinocytes at the leading edge during re-epithelialization of mouse skin wounds. *J. Invest. Dermatol.* **102**, 519–522.

Rømer, J., Bugge, T.H., Pyke, C., Lund, L.R., Flick, M.J., Degen, J.L., and Danø, K. (1996). Impaired wound healing in mice with a disrupted plasminogen gene. *Nature Med.* **2**, 287–292.

Russel, M.E., Wallace, A.F., Hancock, W.W., Sayegh, M.H., Adams, D.H., Sibinga, N.E., Wyner, L.R., and Karnovsky, M.J. (1995). Upregulation of cytokines associated with macrophage activation in the Lewis-to-F344 rat transplantation model of chronic cardiac rejection. *Transplantation* **59**, 572–578.

Sakalihasan, N., Delvenne, P., Nusgens, B.V., Limet, R., and Lapiere, C.M. (1996). Activated forms of MMP2 and MMP9 in abdominal aortic aneurysms. *J. Vasc. Surg.* **24**, 127–133.

Spinale, F.G., Krombach, R.S., Coker, M.L., Mukherjee, R., Houck, W.V., Clair, M.J., Kribbs, S.B., Hebbar, L., and Peterson, J.T. (1997). Matrix metalloproteinase inhibition with congestive heart failure improves left ventricular geometry and pump function. *Circulation* **96** (Suppl. I), I-520 (Abstract).

Suzuki, K., Enghild, J.J., Morodomi, T., Salvesen, G., and Nagase, H. (1990). Mechanisms of activation of tissue procollagenase by matrix metalloproteinase 3 (stromelysin). *Biochemistry* **29**, 10261–10270.

Thomas, C.V., Coker, M..L, Zellner, J.L., Handy, J.R., Crumbley, A.J. III, and Spinale, F.G. (1998). Increased matrix metalloproteinase activity and selective upregulation in LV myocardium from patients with end-stage dilated cardiomyopathy. *Circulation* **97**, 1708–1715.

Tyagi, S.C., Kumar, S., Cassatt, S., and Parker, J.L. (1996). Temporal expression of extracellular matrix metalloproteinases and tissue plasminogen activator in the development of collateral vessels in the canine model of coronary occlusion. *Can. J. Physiol. Pharmacol.* **74**, 983–995.

Ugwu, F., Van Hoef, B., Bini, A., Collen, D., and Lijnen, H.R. (1998). Proteolytic cleavage of urokinase-type plasminogen activator by stromelysin-1 (MMP-3). *Biochemistry* **37**, 7231–7236.

Ugwu, F., Lemmens, G., Collen, D., and Lijnen, H.R. (1999). Modulation of cell-associated plasminogen activation by stromelysin-1 (MMP-3). *Thromb. Haemost.* **82**, 1127–1131.

Vassalli, J.D., Sappino, A.P., and Belin, D. (1991). The plasminogen activator/plasmin system. *J. Clin. Invest.* **88**, 1067–1072.

Webb, K.E., Henney, A.M., Anglin, S., Humphries, S.E., and McEwan, J.R. (1997). Expression of matrix metalloproteinases and their inhibitor TIMP-1 in the rat carotid artery after balloon injury. *Arterioscler. Thromb. Vasc. Biol.* **17**, 1837–1844.

Zempo, N., Kenagy, R.D., Au, Y.P.T., Bendeck, M., Clowes, M.M., Reidy, M.A., and Clowes, A.W. (1994). Matrix metalloproteinases of vascular cells are increased in balloon-injured rat carotid artery. *J. Vasc. Surg.* **20**, 209–217.

Zempo, N., Koyama, N., Kenagy, R.D., Lea, H.J., and Clowes, AW. (1996). Regulation of vascular smooth muscle cell migration and proliferation *in vitro* and in injured rat arteries by a synthetic matrix metalloproteinase inhibitor. *Arterioscler. Thromb. Vasc. Biol.* **16**, 28–33.

12

Matrix Metalloproteinases and the Plasminogen System in Tumor Progression

Yves A. DeClerck and Walter E. Laug

1. Introduction

Extracellular proteases play an important contributory role in cancer progression. Proteases are expressed by tumor and stromal cells in the tumor microenvironment, where they proteolytically modify a large variety of extracellular proteins. The mechanisms by which they contribute to tumor progression are multiple and becoming increasingly complex. By degrading proteins that constitute the extracellular matrix (ECM), proteases promote the invasion of tumor cells into surrounding tissues and blood vessels and contribute to the development of distant metastases. They also play a positive role in angiogenesis by promoting the penetration of malignant tissues by endothelial cells. The proteolytic spectrum of these proteases is, however, not restricted to ECM proteins as many proteins that control critical cellular functions such as cell adhesion, growth and survival, like growth factors, cell surface receptors and growth factor binding proteins, are also the target of these proteases. Among the multiple proteases that play a key role in cancer progression are the matrix metalloproteinases (MMPs) and the plasminogen activator-plasmin (PA) system referred here as the plasminogen system. The contribution of these proteases to cancer has been the subject of an abundant scientific literature. However, these two families of proteases have often been studied independently of each other and it is uncertain whether their function in cancer overlaps in a redundant or complementary manner. In this article, interactive aspects between MMPs and the plasminogen system will be discussed first at a biochemical level. How these interactions could play a role in cancer biology will then be reviewed. Because the biochemistry of the plasminogen system is discussed in detail in other chapters, it will not be reviewed here. In contrast, a brief overview of the biochemistry of MMPs will be presented for the reader less familiar with this family of proteases.

Yves A. DeClerck and Walter E. Laug • Division of Hematology-Oncology and Departments of Pediatrics and Biochemistry and Molecular Biology, Childrens Hospital Los Angeles and USC Keck School of Medicine, Los Angeles, California, USA.

Plasminogen: Structure, Activation, and Regulation, edited by David M. Waisman. Kluwer Academic/Plenum Publishers, New York, 2003.

2. Matrix Metalloproteinases

2.1. MMP Classification and their Expression in Cancer

MMPs, also known as matrixins, consist of a large family of more than 20 proteases (Nagase and Woessner, 1999; Nelson *et al.*, 2000; Chang and Werb, 2001). On the basis of their primary structure, these neutral endopeptidases can be classified in six major subgroups (Figure 12.1). Matrilysins (MMP-7 and MMP-26) constitute the simplest and first group of MMPs. The primary structure of these proteins is composed of three domains. The first domain consists of a predomain (pre) or signal peptide that is necessary for the extracellular secretion of these proteases. The second domain is a prodomain (pro) that contains a PRCGVP peptide sequence preserved among most members of the MMP family. This domain prevents the activity of the enzyme (see below). The third domain consists of a catalytic domain responsible for the enzyme activity. This domain contains a highly preserved HEXXHXXGXXH motif in which a Zn atom is stabilized by three His residues forming a three-dimensional Zn pocket that includes a catalytic Glu residue. This domain is characteristic of all the Metzincins, the larger family of proteases of which the MMPs are members. A unique feature of matrilysins is their specific expression by epithelial cells and by tumor cells of epithelial origin such as adenocarcinoma (McDonnell *et al.*, 1991; Adachi *et al.*, 1999; Tanimoto *et al.*, 1999; Marchenko *et al.*, 2001). Interstitial collagenases (MMP-1, MMP-8, MMP-13) constitute a second subgroup of MMPs. These proteases have an additional C-terminal domain homologous to hemopexin that plays a critical role in substrate recognition. These MMPs are primarily involved in the degradation of fibrillar interstitial collagens (type I, II, III, and VII). MMP-1 and MMP-13 are present in many cancers including breast, prostate, colon, melanoma, and squamous cell carcinoma (Gray *et al.*, 1992, 1993; Freije *et al.*, 1994; Uria *et al.*, 1997). MMP-1 is produced by a large variety of malignant tumors of mesenchymal and epithelial origin and its expression in stromal cells in tumor tissues is enhanced by specific stimulatory factors derived from tumor cells such as EMMPRIN (Guo *et al.*, 1998). MMP-8, in contrast, is a neutrophil MMP and is not expressed by malignant cells. Stromelysins form a third subgroup of 3 MMPs, designated MMP-3, MMP-10 and MMP-11 (stromelysin 1, 2, and 3, respectively). Their primary structure is similar to the structure of interstitial collagenases. Stromelysin-1 (MMP-3) was initially described as transin, a protein expressed by phorbol ester transformed fibroblasts and is expressed by many malignant cells and stromal cells in tumor tissues (Matrisian and Bowden, 1990). It is also a promoter of mammary carcinogenesis (Sternlicht *et al.*, 1999). MMP-10 (stromelysin-2) is expressed in both tumor cells and stromal cells of head and neck carcinoma and its expression in keratinocytes is increased by phorbol ester (Windsor *et al.*, 1993). The role of this MMP in cancer progression has not been extensively investigated (Powell and Matrisian, 1996). MMP-11 (stromelysin-3) is expressed by stromal cells in breast tumors (Basset *et al.*, 1993). The physiological substrate of this enzyme still needs to be identified but overexpression of this MMP enhances tumor take in nude mice (Noel *et al.*, 1996). Gelatinases A and B (MMP-2 and MMP-9, respectively) are a fourth group of MMPs. These MMPs have a unique structural feature characterized by the presence, in the catalytic region, of three repeats of the type II module of fibronectin that binds to denatured collagen type I, IV, and V (gelatin binding domain), and a proline rich hinge domain between the catalytic and the hemopexin domains. Before being cloned and fully characterized, these two MMPs were described as type IV collagenases and were considered to play a major contributory role in tumor invasion and metastasis because of their ability to degrade basement membrane type IV collagen (Stetler-Stevenson, 1990). MMP-2 is expressed by most normal and malignant cells, and high levels of expression have been associated with most cancers (Poulsom *et al.*, 1992; Stetler Stevenson *et al.*, 1992; Sawaya *et al.*, 1996; Väisänen *et al.*, 1998). MMP-9, in contrast, is more selectively expressed by monocytes and macrophages and plays an important role in bone metabolism. In tumors, it is primarily produced by stromal cells, in particular inflammatory cells, and rarely by tumor cells (Rao *et al.*, 1993; Nielsen *et al.*, 1996). Membrane-type (MT) MMPs form a fifth subgroup that comprises six MMPs (MMP-14, -15, -16, -17, -24, -25). These proteases are bound to the plasma membrane either via a sequence rich in hydrophobic residues in their C-terminal end that functions as a transmembrane domain, or via a glycosyl phosphatidylinositol bond. These MMPs are expressed by many normal and malignant cells (Seiki, 1996). MMP-14 (MT-1 MMP) is the best characterized member of the subgroup and plays a key role in the activation of proMMP-2 (see below). The sixth subgroup of MMPs consists of a more heterogeneous group of MMPs that includes MMP-12 (macrophage metallo-elastase), MMP-18, MMP-19, MMP-20 (enamelysin), MMP-21, MMP-22, MMP-23A/B, MMP-27, and MMP-28. Until now the expression of these MMPs in cancer and their potential role in malignant progression has not been extensively studied (Grant *et al.*, 1999). In the extracellular milieu, the activity of all MMPs is controlled by a group of four specific natural inhibitors designated tissue inhibitors of MMPs (TIMPs). The balance between MMPs and TIMPs is a critical factor in the control of tissue homeostasis in cancer (DeClerck *et al.*, 1997).

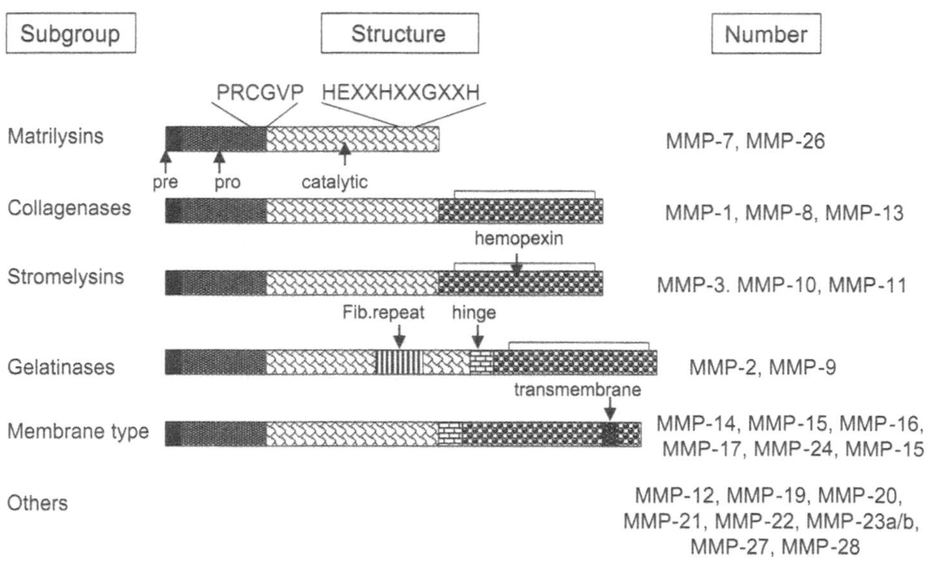

Figure 12.1. Classification and structure of matrix metalloproteinases. The figure shows the six subgroups of MMPs with the structure and the assignment of each MMP to these subgroups.

2.2. Substrate Specificity

An important aspect of these MMPs and the subject of very recent interest, is their substrate specificity. It was initially considered that these proteases were essentially targeted toward proteins of the extracellular matrix, a reason they received the "Matrix" descriptive. However it has now become apparent that they can also process a number of non-ECM proteins (McCawley and Matrisian, 2001). These include, to name a few, adhesion molecules such as E-cadherin (Crawford *et al.*, 1999) and CD-44 (Kajita *et al.*, 2001), growth factor binding proteins such as insulin-like growth factor binding protein (Manes *et al.*, 1999) and perlecan that sequesters bFGF (Whitelock *et al.*, 1996), Fas ligand (Powell *et al.*, 1999), and even proteins involved in bacterial control such as defensin (Wilson *et al.*, 1999). This latter proteolytic activity can affect essential cellular functions such as cell–cell and cell–ECM adhesion, growth factor bioavailability, cell growth, and apoptosis. Thus, the exact mechanisms by which MMPs affect malignant progression are more complex than initially anticipated and go beyond a permissive role in cancer invasion as initially proposed (Chang and Werb, 2001). There is also a significant degree of redundancy among MMPs in terms of substrate specificity and one single ECM protein is generally the target of multiple MMPs. This degree of redundancy has become apparent with the generation and analysis of transgenic mice deficient in the expression of a specific MMP. In most cases, there has been an absence of strong phenotype, with a few exceptions. Mice deficient in MMP-9 have a mild phenotype characterized by short femoral bones and a transient delay in the ossification and vascularization of the growth plate (Vu *et al.*, 1998). MT1-MMP knock-out mice have a stronger phenotype with craniofacial dysmorphism, arthritis, osteopenia, dwarfism, and fibrosis of soft tissues (Holmbeck *et al.*, 1999). None of the MMP knock-out mice reported so far have an embryonic lethal phenotype. Tumorigenesis is suppressed in MMP deficient mice (Wilson *et al.*, 1997; Coussens *et al.*, 2000) and mice overexpressing certain MMPs are more prone to develop cancer (Sternlicht *et al.*, 1999). These observations provide an argument for a role of MMP in early stages of malignancy.

2.3. Activation

Another important and relevant aspect of the biochemistry of MMPs is the fact that they are expressed and secreted in the extracellular milieu as inactive proteases (zymogens). Activation of proMMPs is the result of complex mechanisms that differ among each MMP, but they have a common theme which is the requirement for proteolytic removal of the prodomain and elimination of the Cys residue in the PRCGVP sequence that covers the Zn pocket. Removal of this Cys containing peptide sequence allows access of a molecule of water into the Zn pocket and the cleavage of a peptide bond (Nagase, 1997). Cleavage of the prodomain often involves a two-step process. A first proteolytic cleavage occurs up-stream of the Cys residue containing sequence and destabilizes the enzyme. This cleavage can be initiated by plasmin (see below), furins or another MMP. It is then followed by a second cleavage, this time down-stream of the Cys residue that results in the formation of a stable activated MMP. The contribution of the plasminogen system to MMP activation will be discussed in more detail in the following section.

3. Interactions Between MMPs and the Plasminogen System

Interactions between MMPs and the plasminogen system have been extensively examined at the biochemical level and these *in vitro* studies, usually done in the presence of purified enzymes, have pointed to four interactive aspects that are discussed here (Figure 12.2).

3.1. Activation of proMMPs by Plasmin

In vitro, plasmin contributes to the activation of several MMPs including proMMP-1, -3, -9, and -13. These proteases have in their prodomain peptide sequences that contain Arg or Lys residues that are targeted by plasmin. These sequences are often located up-stream of the Cys residue and therefore further proteolytic cleavage is required for the generation of a stably activated enzyme. For example, plasmin cleaves the Leu86–Arg87 peptide bonds in the prodomain of procollagenase 1 (MMP-1). Autoproteolysis then results in the release of the remaining prodomain generating a Phe100 N-terminal active collagenase (Santala *et al.*, 1999). Similarly, prostromelysin 1 (MMP-3) can be cleaved at Lys38–Asp39 by plasmin and by MMP-3 at His82–Phe83 to generate a fully active enzyme (Murphy *et al.*, 1992). Plasmin cleaves proMMP-13 at Lys38–Glu39 and Arg76–Cys77 bonds to activate the enzyme, however it is unlikely that it plays a physiological role in the activation of this MMP because it also cleaves the C-terminal domain of MMP-13, which results in loss of collagenolytic activity (Knauper *et al.*, 1996). Sometimes the role of plasmin in MMP activation is only indirect and requires another MMP, like MMP-9, which is activated by stromelysin which itself is activated by plasmin (Hahn-Dantona *et al.*, 1999). For other MMPs, the exact role of plasmin in the activation process is still unclear. This is in particular the case with MMP-2. This protease is

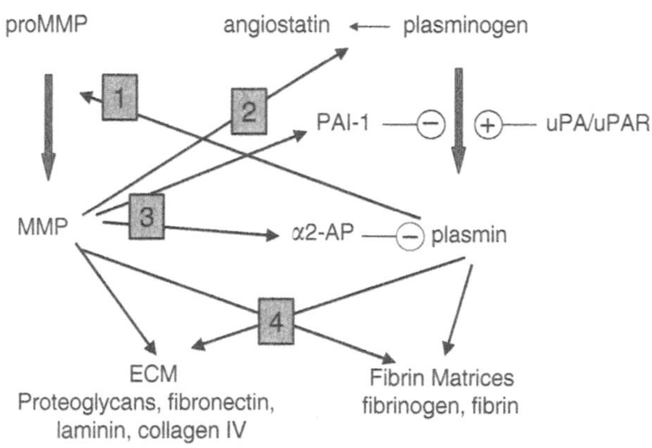

Figure 12.2. Interactions between MMPs and the plasminogen system. The figure represents four major interactions. 1. Plasmin activates several MMPs. 2. Several MMPs can cleave plasminogen to generate angiostatin-like peptides. 3. MMP-3 degrades and inactivates PAI-1 and α2-AP. 4. MMPs and plasmin are proteolytically active against several proteins of the ECM and against fibrin.

typically activated at the cell surface by MT1-MMP interacting with TIMP-2, which acts as an adaptor molecule bringing a TIMP-2/MMP-2 complex in contact with MT1-MMP (Shofuda *et al.*, 1998). Although activation of proMMP-2 from membrane vesicles in HT1080 cells by plasmin has been shown (Ginestra *et al.*, 1997; Mazzieri *et al.*, 1997), other investigators have failed to demonstrate the ability of plasmin to activate proMMP-2 *in vivo* (Okada *et al.*, 1990; Lijnen *et al.*, 1998a). Very recently, a cooperative effect of plasmin on MT1-MMP activation of proMMP-2 has been proposed. In this model, while MT1-MMP initiates the first step of activation, plasmin generated from plasminogen by uPA fully activates MMP-2 by being responsible for the second cleavage of the prodomain (Sounni *et al.*, 2002).

3.2. Cleavage of Plasmin by MMPs

Several MMPs including MMP-2, -3, -7, -9, and -12 can cleave plasminogen. The proteolytic cleavage of plasminogen by these proteases generates a series of biologically active peptides that inhibit endothelial cell proliferation (angiostatin). Cleavage of plasminogen by these MMPs occurs at several positions within the kringles region that is characteristic of plasminogen. For example, MMP-3 generates an angiostatin-like fragment from plasminogen containing kringles 1–4 (Lijnen *et al.*, 1998b). Both MMP-12 (metallo-elastase) and MMP-2 have been reported to generate angiostatin in Lewis Lung carcinoma bearing mice (Dong *et al.*, 1997; O'Reilly *et al.*, 1999). MMP-7 generates a 42 and 38 kDa fragment, which both have the N-terminal sequence VVLLPNVETP, corresponding to the amino acid sequence 467–476 between kringle domain 4 and 5. MMP-9 cleaves plasminogen to generate a 42 kDa fragment with the N-terminal sequence PVVLLPNVE, 1 residue upstream of the MMP-7 cleavage site (Patterson and Sang, 1997). Some MMPs can also interact with uPAR. For example, recombinant MMP-12 cleaves the uPA binding domain D1 of uPAR and, therefore, decreases pericellular plasmin generation (Koolwijk *et al.*, 2001). The relevance of the interactions between MMPs and the plasminogen system in angiogenesis will be discussed in a later section.

3.3. MMPs Degrade Plasmin Inhibitors

MMP-3 can proteolytically inactivate several physiological inhibitors of plasmin. It specifically hydrolyzes human α2-antiplasmin (α2-AP), which when cleaved, can no longer form a stable complex with plasmin and interact with plasminogen (Lijnen, 2002). MMP-3 also cleaves the Ser337–Ser338 and Val341–Ile342 peptide bonds in soluble human plasminogen activator inhibitor-1 (PAI-1). Stable PAI-1 bound to vitronectin is also cleaved and inactivated by MMP-3 in a manner comparable with that of free PAI-1 but the cleaved protein cannot bind to vitronectin anymore. Cleavage and inactivation of α2-AP and PAI-1 by MMP-3 constitute a mechanism that increases plasmin dependent ECM degradation. Inactivation of vitronectin bound PAI-1 by MMP-3, may also interfere with the inhibitory activity of PAI-1 on cell adhesion and migration on vitronectin (Lijnen *et al.*, 2000).

3.4. MMPs and Plasmin Degrade Several Common Proteins

Plasmin degrades several ECM proteins that are also targeted by MMPs offering, therefore, an alternate pathway for the degradation of ECM proteins during tissue remodeling.

These include in particular proteoglycans (Emonds-Alt *et al.*, 1980), glycoproteins such as fibronectin (Jilek and Hormann, 1977), laminin (Schlechte *et al.*, 1989), and type IV collagen (Mackay *et al.*, 1990). And vice versa, MMPs can proteolytically process fibrinogen and fibrin, the physiological substrates of plasmin. MMPs such as MMP-8, -12, -13, and -14 can proteolytically process fibrinogen and cross-linked fibrin and also inactivate Hageman factor, thus impairing clotting (Hiller *et al.*, 2000; Lijnen, 2002). MMP-3 specifically cleaves at gamma Gly404–Ala405 resulting in a D-like monomer fragment. Similarly MMP-7 and MT1-MMP can solubilize fibrin (Bini *et al.*, 1999). Thus, MMPs could offer an alternate fibrinolytic pathway. In cancer progression and angiogenesis as well as in wounds, fibrin deposition in tissues is an important feature, and the fibrinolytic function of MMPs may play an important role.

In summary, MMPs and the plasminogen system interact at multiple levels including proMMP activation, generation of angiostatin, inactivation of plasmin inhibitors and protein degradation in the ECM and in fibrin rich tissues. How these interactions influence cancer progression will be discussed now.

4. Interaction between MMPs and the Plasminogen System in Tumor Invasion and Metastasis

4.1. Expression of MMPs and PA in Cancer

MMPs and PA are co-expressed by many tumor cell lines and are abundantly found in most human tumor tissues where they are produced by malignant cells as well as stromal cells (Nelson *et al.*, 2000; Chang and Werb, 2001). In general, the levels of these proteases is higher in more advanced stages consistent with a positive role in cancer invasion and metastasis and high protease levels have a negative prognostic value in several cancers (Duffy, 1996). More recently the role of these proteases in earlier stages of tumor progression has also been documented (Koblinski *et al.*, 2000; Crawford *et al.*, 2002). An extensive review of this literature is, however, beyond the scope of this article. In our laboratory, we have more specifically examined the expression of MMPs and PAs in neuroblastoma, a cancer of neural crest origin that is the second most common solid malignancy in children. In an analysis of the expression of MMPs, TIMPs, PA, and PAI in established cell lines and primary tumor specimens we documented the expression of MMP-2, MMP-9, tPA, and uPA in the vast majority of the specimens tested (Sugiura *et al.*, 1998, 1999). While the amount of MMP-2 and MMP-9 present in these tumor specimens was significantly higher in those specimens obtained from patients with metastatic disease (stage IV), in comparison with specimens derived from patients with noninvasive disease (stage I and II), there was no correlation between tPA levels and clinical stage. MMP-2 was mainly present in an inactive proform because neuroblastoma cells do not express MT1-MMP and MMP-9 was exclusively expressed by stromal cells. This observation emphasizes the important contribution that stromal cells make to the production of proteases in cancer.

4.2. MMPs and Plasminogen Interactions *in vitro*

It has generally been assumed that PA and MMPs are interacting to optimize the penetration of tissues and blood vessels by invasive tumor cells. The proposed model is one where proMMPs secreted by tumor cells or stromal cells are specifically activated at the cell surface by plasmin generated upon activation by uPA bound to the uPA receptor (uPAR) (Vassalli and Pepper, 1994). A central question that this model raises is whether tumor cells require both proteolytic systems to invade or whether these two proteolytic systems act independently of each other. A more definitive answer to this question is still lacking. Several experiments *in vitro* have suggested that MMPs and PA act independently of each other. We have previously reported in M24met human melanoma cells, the degradation of complex ECM by MMPs in the presence of plasmin inhibitors, indicating that these cells have the ability to activate MMPs such as MMP-1 in the absence of plasmin (Montgomery *et al.*, 1993). In EGF and HGF/SF stimulated human squamous carcinoma cells that express uPA, uPAR, and several MMPs, invasion of type I collagen is not affected by the presence of inhibitors directed against the plasminogen activator-plasmin axis. In contrast, both recombinant and synthetic inhibitors of MMPs completely suppress invasion (Rosenthal *et al.*, 1998). In neuroblastoma we observed the opposite, namely an inhibition of Matrigel® invasion in the presence of a tPA inhibitor but not an MMP inhibitor. Thus, despite evidence supporting an interaction between MMPs and the plasmin system in test tube assays, experiments done *in vitro* on cell lines indicate that there are situations where tumor cells can invade ECM using either MMPs or the PA-plasmin system and that efficient invasion does not necessarily require the contribution of both families of proteases.

4.3. MMPs and Plasminogen Interaction *in vivo*

Experiments *in vitro* do not always allow prediction of tumor cell behavior *in vivo*. In an attempt to address this question, we have used a human xenograft melanoma model (M24met cells) in scid mice. These cells express several MMPs such as MMP-1, MMP-3, and MMP-2 as well as uPA and uPAR (Montgomery *et al.*, 1993). The relative contribution of these two families of proteases to melanoma progression in mice was then studied by overexpressing specific inhibitors of these proteases (PAI-2 and TIMP-2) in melanoma cells and examining their behavior when transplanted in immunodeficient mice. Overexpression of PAI-2 inhibited the spread of distant metastases indicating a role for uPA/plasmin in melanoma invasion and metastasis (Mueller *et al.*, 1995). Overexpression of TIMP-2 in contrast, reduced the growth rate of subcutaneous tumors but did not inhibit metastasis, indicating that MMP activities promote melanoma growth in the skin and may not be required for metastatic dissemination (Montgomery *et al.*, 1994). The data, thus, indicate that in this cancer, both uPA and MMP activities are involved in different processes but both contribute to melanoma malignancy (Mueller, 1996). More recently we have examined the relative contribution of MMPs and PA in neuroblastoma and made a similar observation on an absence of a role of MMPs in metastasis. We have determined that MMPs play a contributory role to angiogenesis and hematogeneous spread but not to metastasis. In other models, however, plasmin seems to play a predominant role on the development of the primary tumor and no role on metastatic spread. In plasminogen-deficient mice, lewis lung carcinoma cells form smaller and less hemorrhagic tumors when injected subcutaneously but are not impaired in their ability to form distant lung metastases, suggesting that in these mice, there is sufficient plasminogen-independent proteolytic activity for efficient tumor development and metastasis (Bugge *et al.*, 1997).

4.4. Conclusion and Directions

Although there is convincing evidence indicating that several MMPs can be activated by plasmin, inhibition of plasmin activity does not seem to necessarily eliminate MMP activity. It appears thus that in many cancers, MMPs and plasmin have overlapping functions in tumor progression as they have in wound healing (Lund *et al.*, 1999). How this functional overlap affects the fate of many MMPs and PA in human cancer will require further *in vivo* experiments in mice deficient in both classes of proteases.

5. Interaction between MMPs and the Plasminogen System in Angiogenesis

5.1. *In vitro* and *in vivo* Studies

The relative contribution of MMPs and plasmin in angiogenesis has been the subject of much investigation (Figure 12.3). The importance of the plasminogen system and the MMPs in endothelial cell proliferation and invasion has been well demonstrated in many *in vitro* systems (Mignatti and Rifkin, 1996; Pepper, 1997, 2001). Although these *in vitro* studies suggest that both plasmin and MMPs play a critical role in angiogenesis, experiments *in vivo* have not always supported a role for plasmin. During cutaneous and corneal wound healing, for example, angiogenesis proceeds normally in plasminogen-deficient animals. Similarly, the growth of most neoplasms is unimpaired in the absence of plasminogen (Hajjar and Deora, 2000). In contrast neovessel formation, *in vitro* and *in vivo*, can be achieved by an MMP-dependent fibrinolytic activity from endothelial cell-derived MT1-MMP. This MMP, when transfected in invasion-incompetent endothelial cells, allows

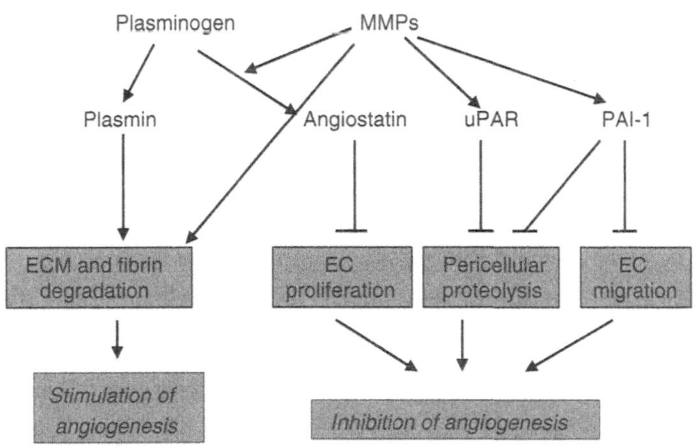

Figure 12.3. MMPs and the plasminogen system in angiogenesis. MMPs and the plasminogen system can have positive (Left) and negative (Right) effects on angiogenesis. By degrading ECM proteins and fibrin, MMPs and plasmin promote EC migration and stimulate angiogenesis. In contrast, the proteolytic activity of MMP on plasminogen, uPAR and PAI-1 generates angiostatin, decreases cellular proteolysis and inhibits EC migration toward fibronectin which inhibits angiogenesis.

the penetration of fibrin barriers (Hiraoka *et al.*, 1998). In MMP-9 deficient mice there is a delayed angiogenic response associated with the formation of autochthonous squamous cell tumors (Coussens *et al.*, 2000). On the other hand, hypoxia-driven vascular proliferation may require plasmin-like activity, and angiogenesis within atherosclerotic plaques seems to be associated with increased expression of fibrinolytic proteins (Levi *et al.*, 2001). Inhibition of angiogenesis by several natural and synthetic inhibitors of MMPs has also been demonstrated *in vitro* (Anand-Apte *et al.*, 1997; Takigawa *et al.*, 1990). In our laboratory we have observed a significant inhibition of angiogenesis in Matrigel® plugs implanted under the skin of MMP-9 -/- mice. These observations have supported the hypothesis that in order to invade the tumor microevironment, endothelial cells use the proteolytic activity of MMPs and plasmin to degrade fibrin and collagen rich matrices, but that MMP activity alone is sufficient.

5.2. Paradoxical Aspects

On the basis of these observations, it has been assumed that inhibitors of MMPs (TIMPs and synthetic compounds) and PA (PAIs) will have a strong anti-angiogenic activity. Although this concept is supported by *in vitro* data, experiments in mice and also in human have indicated that these inhibitors may have a paradoxically positive effect on angiogenesis. For example, the PA inhibitor PAI-1 has been shown to play a positive rather than a negative role in angiogenesis. PAI-1 deficient mice lack the angiogenic response that follows the implantation of squamous carcinoma cells. This activity of PAI-1 seems dependent on its anti-proteolytic activity (Bajou *et al.*, 1998, 2001). In our laboratory we observed that PAI-1 acts as a positive switch in angiogenesis by promoting the migration of human microvascular endothelial cells from vitronectin toward fibronectin. This activity is dependent on the ability of PAI-1 to bind to vitronectin and to interfere with endothelial cell adhesion to vitronectin (Isogai *et al.*, 2001). Because MMPs such as MMP-3 can proteolytically cleave PAI-1 to generate fragments that do not bind to vitronectin (Lijnen *et al.*, 2000), MMP-3 may also have an indirect negative effect on angiogenesis by suppressing the stimulatory effect of PAI-1 on endothelial cell migration. As a consequence, inhibition of MMP activity may potentiate rather than inhibit angiogenesis. A second mechanism by which MMP/plasminogen system interaction has an anti-angiogenic effect is linked to the ability of MMPs to cleave plasminogen and to generate angiostatin-like peptides. This mechanism has been well illustrated recently in mice deficient in the integrin $\alpha 1$. In these mice there is a marked increase in the expression of MMP-7 and MMP-9. When implanted with tumors, integrin $\alpha 1$ deficient mice develop slower rather than faster growing tumors with a markedly decreased vascularization and reduction in capillary number and size. This is the result of an increase in plasma levels of angiostatin due to the action of MMP-7 and MMP-9 on circulating plasminogen. *In vitro* integrin $\alpha 1$-deficient endothelial cells show a marked reduction of their proliferation on both integrin $\alpha 1$-dependent (collagenous) and independent (noncollagenous) substrates. This reduction is prevented by culturing integrin $\alpha 1$-deficient cells with plasma derived from plasminogen-null animals, thus omitting the source from which to generate angiostatin (Pozzi *et al.*, 2000). A third mechanism by which MMPs interacting with the plasminogen system could inhibit rather enhance angiogenesis is provided by the ability of MMP-12 to proteolytically cleave uPAR. Endothelial cell derived MMP-12 can release the

D1 (uPA binding) domain of cellularly expressed uPAR, therefore promoting uPA solubilization and decreasing cell associated uPA-dependent proteolysis. This activity results in an inhibition of outgrowth of capillary-like structures by endothelial cells in a fibrin matrix. Interestingly, the addition of BB94, an inhibitor of MMPs to the culture system increases the outgrowth of these capillary-like structures (Koolwijk *et al.*, 2001). These experiments suggest that inhibition of MMPs may have a paradoxical effect on cancer progression by stimulating rather than inhibiting tumor angiogenesis.

6. Relevance of these Studies in Human Cancer

To better understand the interaction between MMPs and the plasminogen system and to obtain more definitive evidence that plasmin can substitute for MMPs in tumor invasion represents more that an academic exercise and has significant implications in cancer therapy. Our recent experience in clinical trials with synthetic inhibitors of MMPs illustrates clearly this point. Over the last decade, several synthetic MMP inhibitors have been developed by pharmaceutical companies and have been tested in clinical trials. Most of these clinical trials have shown little or no benefit to patients and have therefore been discontinued (Coussens *et al.*, 2002). One problem with these clinical trials has been that they were generally performed on patients with highly advanced and drug-resistant tumors for which a treatment with an MMP inhibitor alone or even in combination with chemotherapy was unlikely to be of dramatic benefit, in particular considering the increased amount of evidence suggesting a role for MMPs in earlier aspects of tumor progression (Zucker *et al.*, 2000). Interestingly enough, in one trial with the BAY 12-9566 compound, administered to patients with non-small cell lung cancer, an acceleration of the cancer was observed. It is conceivable that administration of this compound may have prevented the generation of angiostatin by MMPs and that as a result, a stimulatory rather an inhibitory effect on tumor growth was achieved. This possibility is supported by the recent observation that reduction of plasma levels of MMP-9 in either normal or integrin $\alpha 1$-null mice leads to decreased synthesis of angiostatin and consequently to an increase in tumor growth and vascularization. In contrast, specifically enhancing MMP-9 expression in these mice causes a reduction in tumor vascularization (Pozzi *et al.*, 2002). This example clearly illustrates the need to better understand the complex interactions between MMPs and the plasminogen system in malignant progression and to realize how these interactions can differ from one type of cancer to another. This information will be critical if new clinical trials with protease inhibitors are to be considered in cancer therapies. A more definite answer to these important question will require models that take into consideration the tumor micro-environment that actively contributes to tumor progression and angiogenesis. In this aspect, genetically defined mice suitable for tumor models will be invaluable.

Acknowledgments

This work has been supported by grants CA 42919, CA 84103, and CA 82989 from the National Institutes of Health.

References

Adachi, Y., Yamamoto, H., Itoh, F., Hinoda, Y., Okada, Y., and Imai, K. (1999). Contribution of matrilysin (MMP-7) to the metastatic pathway of human colorectal cancers. *Gut* **45**, 252–258.

Anand-Apte, B., Pepper, M.S., Voest, E., Montesano, R., Olsen, B., Murphy, G., Apte, S.S., and Zetter, B. (1997). Inhibition of angiogenesis by tissue inhibitor of metalloproteinase-3. *Invest. Ophthalmol. Vis. Sci.* **38**, 817–823.

Bajou, K., Masson, V., Gerard, R.D., Schmitt, P.M., Albert, V., Praus, M., Lund, L.R., Frandsen, T.L., Brunner, N., Dano, K., Fusenig, N.E., Weidle, U., Carmeliet, G., Loskutoff, D., Collen, D., Carmeliet, P., Foidart, J.M., and Noel, A. (2001). The plasminogen activator inhibitor PAI-1 controls *in vivo* tumor vascularization by interaction with proteases, not vitronectin. Implications for antiangiogenic strategies. *J. Cell. Biol.* **152**, 777–784.

Bajou, K., Noel, A., Gerard, R.D., Masson, V., Brunner, N., Holst-Hansen, C., Skobe, M., Fusenig, N.E., Carmeliet, P., Collen, D., and Foidart, J.M. (1998). Absence of host plasminogen activator inhibitor 1 prevents cancer invasion and vascularization. *Nat. Med.* **4**, 923–928.

Basset, P., Wolf, C., and Chambon, P. (1993). Expression of the stromelysin-3 gene in fibroblastic cells of invasive carcinomas of the breast and other human tissues: A review. *Breast Cancer Res. Treat.* **24**, 185–193.

Bini, A., Wu, D., Schnuer, J., and Kudryk, B.J. (1999). Characterization of stromelysin 1 (MMP-3), matrilysin (MMP-7), and membrane type 1 matrix metalloproteinase (MT1-MMP) derived fibrin(ogen) fragments D-dimer and D-like monomer: NH2-terminal sequences of late- stage digest fragments. *Biochemistry* **38**, 13928–13936.

Bugge, T.H., Kombrinck, K.W., Xiao, Q., Holmbäck, K., Daugherty, C.C., Witte, D.P., and Degen, J.L. (1997). Growth and dissemination of Lewis lung carcinoma in plasminogen-deficient mice. *Blood* **90**, 4522–4531.

Chang, C. and Werb, Z. (2001). The many faces of metalloproteases: Cell growth, invasion, angiogenesis and metastasis. *Trends Cell. Biol.* **11**, S37–S43.

Coussens, L.M., Fingleton, B., and Matrisian, L.M. (2002). Matrix metalloproteinase inhibitors and cancer: Trials and tribulations. *Science* **295**, 2387–2392.

Coussens, L.M., Tinkle, C.L., Hanahan, D., and Werb, Z. (2000). MMP-9 supplied by bone marrow-derived cells contributes to skin carcinogenesis. *Cell* **103**, 481–490.

Crawford, H.C., Fingleton, B.M., Rudolph-Owen, L.A., Goss, K.J.H., Rubinfeld, B., Polakis, P., and Matrisian, L.M. (1999). The metalloproteinase matrilysin is a target of b-catenin transactivation in intestinal tumors. *Oncogene* **18**, 2883–2891.

Crawford, H.C., Scoggins, C.R., Washington, M.K., Matrisian, L.M., and Leach, S.D. (2002). Matrix metalloproteinase-7 is expressed by pancreatic cancer precursors and regulates acinar-to-ductal metaplasia in exocrine pancreas. *J. Clin. Invest.* **109**, 1437–1444.

DeClerck, Y.A., Imren, S., Montgomery, A.M P., Mueller, B.M., Reisfeld, R.A., and Laug, W.E. (1997). Proteases and protease inhibitors in tumor progression. *Adv. Exp. Med. Biol.* **425**, 89–97.

Dong, Z., Kumar, R., Yang, X., and Fidler, I.J. (1997). Macrophage-derived metalloelastase is responsible for the generation of angiostatin in Lewis lung carcinoma. *Cell* **88**, 801–810.

Duffy, M.J. (1996). Proteases as prognostic markers in cancer. *Clin. Cancer Res.* **2**, 613–618.

Emonds-Alt, X., Quisquater, E., and Vaes, G. (1980). Proteoglycan- and fibrin-degrading neutral proteinase activities of Lewis lung carcinoma cells. *Eur. J. Cancer* **16**, 1257–1261.

Freije, J.M., Diez Itza, I., Balbin, M., Sanchez, L.M., Blasco, R., Tolivia, J., and Lopez Otin, C. (1994). Molecular cloning and expression of collagenase-3,a novel human matrix metalloproteinase produced by breast carcinomas. *J. Biol. Chem.* **269**, 16766–16773.

Ginestra, A., Monea, S., Seghezzi, G., Dolo, V., Nagase, H., Mignatti, P., and Vittorelli, M.L. (1997). Urokinase plasminogen activator and gelatinases are associated with membrane vesicles shed by human HT1080 fibrosarcoma cells. *J. Biol. Chem.* **272**, 17216–17222.

Grant, G.M., Giambernardi, T.A., Grant, A.M., and Klebe, R.J. (1999). Overview of expression of matrix metalloproteinases (MMP-17, MMP- 18, and MMP-20) in cultured human cells. *Matrix Biol.* **18**, 145–148.

Gray, S.T., Wilkins, R.J., and Yun, K. (1992). Interstitial collagenase gene expression in oral squamous cell carcinoma. *Am. J. Pathol.* **141**, 301–306.

Gray, S.T., Yun, K., Motoori, T., and Kuys, Y.M. (1993). Interstitial collagenase gene expression in colonic neoplasia. *Am. J. Pathol.* **143**, 663–671.

Guo, H.M., Majmudar, G., Jensen, T.C., Biswas, C., Toole, B.P., and Gordon, M.K. (1998). Characterization of the gene for human EMMPRIN, a tumor cell surface inducer of matrix metalloproteinases. *Gene* **220**, 99–108.

Hahn-Dantona, E., Ramos-DeSimone, N., Sipley, J., Nagase, H., French, D.L., and Quigley, J.P. (1999). Activation of ProMMP-9 by a plasmin/MMP-3 cascade in a tumor cell model. Regulation by tissue inhibitors of metalloproteinases. *Ann. NY Acad. Sci.* **878**, 372–387.

Hajjar, K.A. and Deora, A. (2000). New concepts in fibrinolysis and angiogenesis. *Curr Atheroscler. Rep.* **2**, 417–421.

Hiller, O., Lichte, A., Oberpichler, A., Kocourek, A., and Tschesche, H. (2000). Matrix metalloproteinases collagenase-2, macrophage elastase, collagenase-3, and membrane type 1-matrix metalloproteinase impair clotting by degradation of fibrinogen and factor XII. *J. Biol. Chem.* **275**, 33008–33013.

Hiraoka, N., Allen, E., Apel, I.J., Gyetko, M.R., and Weiss, S.J. (1998). Matrix metalloproteinases regulate neovascularization by acting as pericellular fibrinolysins. *Cell* **95**, 365–377.

Holmbeck, K., Bianco, P., Caterina, J., Yamada, S., Kromer, M., Kuznetsov, S.A., Mankani, M., Robey, P.G., Poole, A.R., Pidoux, I., Ward, J.M., and Birkedal-Hansen, H. (1999). MT1-MMP-deficient mice develop dwarfism, osteopenia, arthritis, and connective tissue disease due to inadequate collagen turnover. *Cell* **99**, 81–92.

Isogai, C., Laug, W.E., Shimada, H., Declerck, P.J., Stins, M.F., Durden, D.L., Erdreich-Epstein, A., and DeClerck, Y.A. (2001). Plasminogen activator inhibitor-1 promotes angiogenesis by stimulating endothelial cell migration toward fibronectin. *Cancer Res.* **61**, 5587–5594.

Jilek, F. and Hormann, H. (1977). Cold-insoluble globulin, II[1,2]. Cyanogen bromide and plasminolysis fragments containing a label introduced by transamidation. *Hoppe Seylers. Z. Physiol. Chem.* **358**, 1165–1168.

Kajita, M., Itoh, Y., Chiba, T., Mori, H., Okada, A., Kinoh, H., and Seiki, M. (2001). Membrane-type 1 matrix metalloproteinase cleaves CD44 and promotes cell migration. *J. Cell. Biol.* **153**, 893–904.

Knauper, V., Will, H., Lopez-Otin, C., Smith, B., Atkinson, S.J., Stanton, H., Hembry, R.M., and Murphy, G. (1996). Cellular mechanisms for human procollagenase-3 (MMP-13) activation. Evidence that MT1-MMP (MMP-14) and gelatinase a (MMP-2) are able to generate active enzyme. *J. Biol. Chem.* **271**, 17124–17131.

Koblinski, J.E., Ahram, M., and Sloane, B.F. (2000). Unraveling the role of proteases in cancer. *Clin. Chim. Acta* **291**, 113–135.

Koolwijk, P., Sidenius, N., Peters, E., Sier, C.F., Hanemaaijer, R., Blasi, F., and van Hinsbergh, V.W. (2001). Proteolysis of the urokinase-type plasminogen activator receptor by metalloproteinase-12: Implication for angiogenesis in fibrin matrices. *Blood* **97**, 3123–3131.

Levi, M., Moons, L., Bouche, A., Shapiro, S.D., Collen, D., and Carmeliet, P. (2001). Deficiency of urokinase-type plasminogen activator-mediated plasmin generation impairs vascular remodeling during hypoxia-induced pulmonary hypertension in mice. *Circulation* **103**, 2014–2020.

Lijnen, H.R. (2002). Matrix metalloproteinases and cellular fibrinolytic activity. *Biochemistry (Mosc.)* **67**, 92–98.

Lijnen, H.R., Arza, B., Van Hoef, B., Collen, D., and Declerck, P.J. (2000). Inactivation of plasminogen activator inhibitor-1 by specific proteolysis with stromelysin-1 (MMP-3). *J. Biol. Chem.* **275**, 37645–37650.

Lijnen, H.R., Silence, J., Lemmens, G., Frederix, L., and Collen, D. (1998a). Regulation of gelatinase activity in mice with targeted inactivation of components of the plasminogen/plasmin system. *Thromb. Haemost.* **79**, 1171–1176.

Lijnen, H.R., Ugwu, F., Bini, A., and Collen, D. (1998b). Generation of an angiostatin-like fragment from plasminogen by stromelysin-1 (MMP-3). *Biochemistry* **37**, 4699–4702.

Lund, L.R., Romer, J., Bugge, T.H., Nielsen, B.S., Frandsen, T.L., Degen, J.L., Stephens, R.W., and Dano, K. (1999). Functional overlap between two classes of matrix-degrading proteases in wound healing. *EMBO J.* **18**, 4645–4656.

Mackay, A.R., Corbitt, R.H., Hartzler, J.L., and Thorgeirsson, U.P. (1990). Basement membrane type IV collagen degradation: Evidence for the involvement of a proteolytic cascade independent of metalloproteinases. *Cancer Res.* **50**, 5997–6001.

Manes, S., Llorente, M., Lacalle, R.A., Gomez-Mouton, C., Kremer, L., Mira, E., and Martinez, A. (1999). The matrix metalloproteinase-9 regulates the insulin-like growth factor-triggered autocrine response in DU-145 carcinoma cells. *J. Biol. Chem.* **274**, 6935–6945.

Marchenko, G.N., Ratnikov, B.I., Rozanov, D.V., Godzik, A., Deryugina, E.I., and Strongin, A.Y. (2001). Characterization of matrix metalloproteinase-26, a novel metalloproteinase widely expressed in cancer cells of epithelial origin. *Biochem. J.* **356**, 705–718.

Matrisian, L.M. and Bowden, G.T. (1990). Stromelysin/transin and tumor progression. *Semin. Cancer Biol.* **1**, 107–115.

Mazzieri, R., Masiero, L., Zanetta, L., Monea, S., Onisto, M., Garbisa, S., and Mignatti, P. (1997). Control of type IV collagenase activity by components of the urokinase-plasmin system: A regulatory mechanism with cell-bound reactants. *EMBO J.* **16**, 2319–2332.

McCawley, L.J. and Matrisian, L.M. (2001). Matrix Metalloproteinases: They're not just for matrix anymore. *Curr. Opin. Cell. Biol.* **13**, 534–540.

McDonnell, S., Navre, M., Coffey, R., Jr., and Matrisian, L.M. (1991). Expression and localization of the matrix metalloproteinase pump- 1 (MMP-7) in human gastric and colon carcinomas. *Mol. Carcinog.* **4**, 527–533.

Mignatti, P. and Rifkin, D.B. (1996). Plasminogen activators and matrix metalloproteinases in angiogenesis. *Enzyme Protein* **49**, 117–137.

Montgomery, A.M., DeClerck, Y.A., Langley, K.E., Reisfeld, R.A., and Mueller, B.M. (1993). Melanoma-mediated dissolution of extracellular matrix: Contribution of urokinase-dependent and metalloproteinase-dependent proteolytic pathways. *Cancer Res.* **53**, 693–700.

Montgomery, A.M., Mueller, B.M., Reisfeld, R.A., Taylor, S.M., and DeClerck, Y.A. (1994). Effect of tissue inhibitor of the matrix metalloproteinases-2 expression on the growth and spontaneous metastasis of a human melanoma cell line. *Cancer Res.* **54**, 5467–5473.

Mueller, B.M. (1996). Different roles for plasminogen activators and metalloproteinases in melanoma metastasis. *Curr. Top. Microbiol. Immunol.* **213**, 65–80.

Mueller, B.M., Yu, Y.B., and Laug, W.E. (1995). Overexpression of plasminogen activator inhibitor 2 in human melanoma cells inhibits spontaneous metastasis in scid mice. *Proc. Natl. Acad. Sci. USA* **92**, 205–209.

Murphy, G., Ward, R., Gavrilovic, J., and Atkinson, S. (1992). Physiological mechanisms for metalloproteinase activation. *Matrix Suppl.* **1**, 224–230.

Nagase, H. (1997). Activation mechanisms of matrix metalloproteinases. *Biol. Chem. Hoppe Seyler* **378**, 151–160.

Nagase, H. and Woessner, J.F., Jr. (1999). Matrix metalloproteinases. *J. Biol. Chem.* **274**, 21491–21494.

Nelson, A.R., Fingleton, B., Rothenberg, M.L., and Matrisian, L.M. (2000). Matrix metalloproteinases: Biologic activity and clinical implications. *J. Clin. Oncol.* **18**, 1135–1149.

Nielsen, B.S., Timshel, S., Kjeldsen, L., Sehested, M., Pyke, C., Borregaard, N., and Dano, K. (1996). 92 kDa type IV collagenase (MMP-9) is expressed in neutrophils and macrophages but not in malignant epithelial cells in human colon cancer. *Int. J. Cancer.* **65**, 57–62.

Noel, A., Lefebvre, O., Maquoi, E., VanHoorde, L., Chenard, M.P., Mareel, M., and Foidart, J.M. (1996). Stromelysin-3 expression promotes tumor take in nude mice. *J. Clin. Invest.* **97**, 1924–1930.

O'Reilly, M.S., Wiederschain, D., Stetler-Stevenson, W.G., Folkman, J., and Moses, M.A. (1999). Regulation of angiostatin production by matrix metalloproteinase-2 in a model of concomitant resistance. *J. Biol. Chem.* **274**, 29568–29571.

Okada, Y., Morodomi, T., Enghild, J.J., Suzuki, K., Yasui, A., Nakanishi, I., Salvesen, G., and Nagase, H. (1990). Matrix metalloproteinase 2 from human rheumatoid synovial fibroblasts. Purification and activation of the precursor and enzymic properties. *Eur. J. Biochem.* **194**, 721–730.

Patterson, B.C. and Sang, Q.A. (1997). Angiostatin-converting enzyme activities of human matrilysin (MMP-7) and gelatinase B/type IV collagenase (MMP-9). *J. Biol. Chem.* **272**, 28823–28825.

Pepper, M.S. (1997). Manipulating angiogenesis—from basic science to the bedside. *Arterioscler. Thromb. Vasc. Biol.* **17**, 605–619.

Pepper, M.S. (2001). Role of the matrix metalloproteinase and plasminogen activator-plasmin systems in angiogenesis. *Arterioscler. Thromb. Vasc. Biol* **21**, 1104–1117.

Poulsom, R., Pignatelli, M., Stetler-Stevenson, W.G., Liotta, L.A., Wright, P.A., Jeffery, R.E., Longcroft, J.M., Rogers, L., and Stamp, G.W. (1992). Stromal expression of 72 kda type IV collagenase (MMP-2) and TIMP-2 mRNAs in colorectal neoplasia. *Am. J. Pathol.* **141**, 389–396.

Powell, W.C., Fingleton, B., Wilson, C.L., Boothby, M., and Matrisian, L.M. (1999). The metalloproteinase matrilysin proteolytically generates active soluble Fas ligand and potentiates epithelial cell apoptosis. *Curr. Biol.* **9**, 1441–1447.

Powell, W.C. and Matrisian, L.M. (1996). Complex roles of matrix metalloproteinases in tumor progression. *Curr. Top. Microbiol. Immunol.* **213**, 1–21.

Pozzi, A., LeVine, W.F., and Gardner, H.A. (2002). Low plasma levels of matrix metalloproteinase 9 permit increased tumor angiogenesis. *Oncogene* **21**, 272–281.

Pozzi, A., Moberg, P.E., Miles, L.A., Wagner, S., Soloway, P., and Gardner, H.A. (2000). Elevated matrix metalloprotease and angiostatin levels in integrin alpha 1 knockout mice cause reduced tumor vascularization. *Proc. Natl. Acad. Sci. USA* **97**, 2202–2207.

Rao, J.S., Steck, P.A., Mohanam, S., Stetler-Stevenson, W.G., Liotta, L.A., and Sawaya, R. (1993). Elevated levels of M(r) 92,000 type IV collagenase in human brain tumors. *Cancer Res.* **53**, 2208–2211.

Rosenthal, E.L., Johnson, T.M., Allen, E.D., Apel, I.J., Punturieri, A., and Weiss, S.J. (1998). Role of the plasminogen activator and matrix metalloproteinase systems in epidermal growth factor- and scatter factor-stimulated invasion of carcinoma cells. *Cancer Res.* **58**, 5221–5230.

Santala, A., Saarinen, J., Kovanen, P., and Kuusela, P. (1999). Activation of interstitial collagenase, MMP-1, by *Staphylococcus aureus* cells having surface-bound plasmin: A novel role of plasminogen receptors of bacteria. *FEBS Lett.* **461**, 153–156.

Sawaya, R.E., Yamamoto, M., Gokaslan, Z.L., Wang, S.W., Mohanam, S., Fuller, G.N., McCutcheon, I.E., Stetler-Stevenson, W.G., Nicolson, G.L., and Rao, J.S. (1996). Expression and localization of 72 kDa type IV collagenase (*MMP-2*) in human malignant gliomas *in vivo. Clin. Exp. Metastasis* **14**, 35–42.

Schlechte, W., Murano, G., and Boyd, D. (1989). Examination of the role of the urokinase receptor in human colon cancer mediated laminin degradation. *Cancer Res.* **49**, 6064–6069.

Seiki, M. (1996). Membrane type matrix metalloproteinase and tumor invasion. *Curr. Top. Microbiol. Immunol.* **213**, 23–32.

Shofuda, K., Moriyama, K., Nishihashi, A., Higashi, S., Mizushima, H., Yasumitsu, H., Miki, K., Sato, H., Seiki, M., and Miyazaki, K. (1998). Role of tissue inhibitor of metalloproteinases-2 (TIMP-2) in regulation of pro-gelatinase A activation catalyzed by membrane- type matrix metalloproteinase-1 (MT1-MRP) in human cancer cells. *J. Biochem. (Tokyo)* **124**, 462–470.

Sounni, N.E., Baramova, E.N., Munaut, C., Maquoi, E., Frankenne, F., Foidart, J.M., and Noel, A. (2002). Expression of membrane type 1 matrix metalloproteinase (MT1-MMP) in A2058 melanoma cells is associated with MMP-2 activation and increased tumor growth and vascularization. *Int. J. Cancer* **98**, 23–28.

Sternlicht, M.D., Lochter, A., Sympson, C.J., Huey, B., Rougier, J.P., Gray, J.W., Pinkel, D., Bissell, M.J., and Werb, Z. (1999). The stromal proteinase MMP3/stromelysin-1 promotes mammary carcinogenesis. *Cell* **98**, 137–146.

Stetler-Stevenson, W.G. (1990). Type IV collagenases in tumor invasion and metastasis. *Cancer Metastasis Rev.* **9**, 289–303.

Stetler Stevenson, W.G., Liotta, L.A., and Brown, P.D. (1992). Role of type IV collagenases in human breast cancer. *Cancer Treat. Res.* **61**, 21–41.

Sugiura, Y., Ma, L.Q., Sun, B., Shimada, H., Laug, W.E., Seeger, R.C., and DeClerck, Y.A. (1999). The plasminogen–plasminogen activator (PA) system in neuroblastoma: Role of PA inhibitor-1 in metastasis. *Cancer Res.* **59**, 1327–1336.

Sugiura, Y., Shimada, H., Seeger, R.C., Laug, W.E., and DeClerck, Y.A. (1998). Matrix metalloproteinases-2 and -9 are expressed in human neuroblastoma: Contribution of stromal cells to their production and correlation with metastasis. *Cancer Res.* **58**, 2209–2216.

Takigawa, M., Nishida, Y., Suzuki, F., Kishi, J., Yamashita, K., and Hayakawa, T. (1990). Induction of angiogenesis in chick yolk-sac membrane by polyamines and its inhibition by tissue inhibitors of metalloproteinases (TIMP and TIMP-2). *Biochem. Biophys. Res. Commun.* **171**, 1264–1271.

Tanimoto, H., Underwood, L.J., Shigemasa, K., Parmley, T.H., Wang, Y.X., Yan, Y., Clarke, J., and O'Brien, T.J. (1999). The matrix metalloprotease pump-1 (MMP-7, matrilysin): A candidate marker/target for ovarian cancer detection and treatment. *Tumor Biol.* **20**, 88–98.

Uria, J.A., Stahle-Backdahl, M., Seiki, M., Fueyo, A., and Lopez-Otin, C. (1997). Regulation of *collagenase-3* expression in human breast carcinomas is mediated by stromal-epithelial cell interactions. *Cancer Res.* **57**, 4882–4888.

Vassalli, J.D. and Pepper, M.S. (1994). Tumour biology. Membrane proteases in focus. *Nature* **370**, 14–15.

Väisänen, A., Kallioinen, M., Taskinen, P.J., and Turpeenniemi-Hujanen, T. (1998). Prognostic value of MMP-2 immunoreactive protein (72 kD type IV collagenase) in primary skin melanoma. *J. Pathol.* **186**, 51–58.

Vu, T.H., Shipley, J.M., Bergers, G., Berger, J.E., Helms, J.A., Hanahan, D., Shapiro, S.D., Senior, R.M., and Werb, Z. (1998). MMP-9/gelatinase B is a key regulator of growth plate angiogenesis and apoptosis of hypertrophic chondrocytes. *Cell* **93**, 411–422.

Whitelock, J.M., Murdoch, A.D., Iozzo, R.V., and Underwood, P.A. (1996). The degradation of human endothelial cell-derived perlecan and release of bound basic fibroblast growth factor by stromelysin, collagenase, plasmin, and heparanases. *J. Biol. Chem.* **271**, 10079–10086.

Wilson, C.L., Heppner, K.J., Labosky, P.A., Hogan, B.L.M., and Matrisian, L.M. (1997). Intestinal tumorigenesis is suppressed in mice lacking the metalloproteinase matrilysin. *Proc. Natl. Acad. Sci. USA* **94**, 1402–1407.

Wilson, C.L., Ouellette, A.J., Satchell, D.P., Ayabe, T., López-Boado, Y.S., Stratman, J.L., Hultgren, S.J., Matrisian, L.M., and Parks, W.C. (1999). Regulation of intestinal α-defensin activation by the metalloproteinase matrilysin in innate host defense. *Science* **286**, 113–117.

Windsor, L.J., Grenett, H., Birkedal-Hansen, B., Bodden, M.K., Engler, J.A., and Birkedal-Hansen, H. (1993). Cell type-specific regulation of SL-1 and SL-2 genes. Induction of the SL-2 gene but not the SL-1 gene by human keratinocytes in response to cytokines and phorbolesters. *J. Biol. Chem.* **268**, 17341–17347.

Zucker, S., Cao, J., and Chen, W.T. (2000). Critical appraisal of the use of matrix metalloproteinase inhibitors in cancer treatment. *Oncogene* **19**, 6642–6650.

13

Role of Plasminogen Activation in Hematopoietic Malignancies and in Normal Hematopoiesis

Satu Mustjoki, Riitta Alitalo, and Antti Vaheri

1. Introduction

Adhesion and invasion are important processes for both normal and malignant cells. However, malignant cells often differ from normal cells in their capability to produce large amounts of proteolytic enzymes, which they use to degrade extracellular matrix proteins during cell invasion. The plasminogen activation cascade is known to be one of the most important extracellular proteolytic mechanisms for solid tumors, whereas its role in leukemia has remained elusive. It has been thought that the traditional pericellular proteolytic roles of plasminogen activators are unimportant in leukemia, since leukemic cells grow in "suspension" in the circulation, and are not thought to invade in the manner adherent tumor cells do. It is very likely, however, that the plasminogen activation cascade plays an important role in leukemia as well, because in some circumstances leukemia cells invade extramedullary organs and even the central nervous system. In addition, novel non-proteolytic roles for the plasminogen activation system have recently been discovered. In this review we will focus on the possible roles of the plasminogen activation (PA) system in human leukemia and in normal hematopoiesis and review the existing literature from this area.

2. Normal Hematopoiesis and its Control

The life-span of normal blood cells is limited, and these cells are constantly produced in the bone marrow. It has been estimated that approximately $0.5–1 \times 10^{12}$ new blood cells are formed every day in a healthy adult (Gale and Butturini, 1992).

Satu Mustjoki and Riitta Alitalo • Department of Clinical Chemistry, Hematology and Stem Cell Laboratories, Helsinki University Central Hospital, Finland **Satu Mustjoki** • Biomedicum Helsinki, Hematopoietic Stem Cell Research, Helsinki University Central Hospital, Finland **Antti Vaheri** • Haartman Institute, University of Helsinki, Finland

Plasminogen: Structure, Activation, and Regulation, edited by David M. Waisman. Kluwer Academic/Plenum Publishers, New York, 2003.

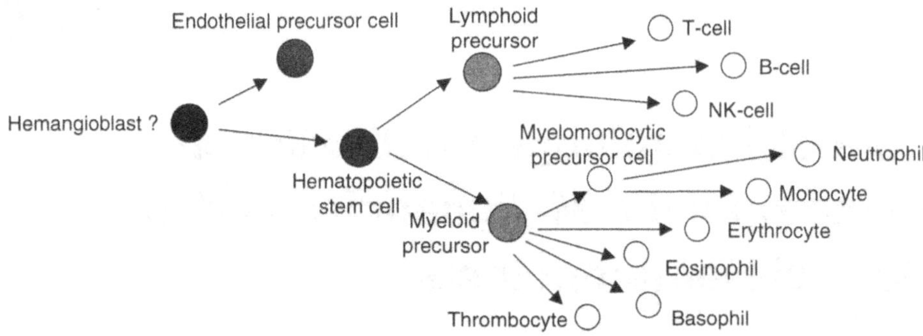

Figure 13.1. Hematopoiesis. A schematic representation of normal hematopoiesis

Furthermore, in response to various stimuli (hypoxia, infection, bleeding) the production of certain blood cell types is multiplied. The real hematopoietic stem cell has not been characterized, but recent reports suggest that blood cells and endothelial cells may have a common progenitor cell, called the hemangioblast (Choi *et al.*, 1998). In clinical practice, CD34-positive cells are regarded as hematopoietic stem cells, although it is likely that only a small fraction of these cells are true stem cells. However, these cells are able to differentiate into mature blood cells and to repopulate the bone marrow. They are used as stem-cell support in connection with high-dose chemotherapy of various cancers. Pluripotent stem cells give rise to myeloid and lymphoid precursor cells, which in turn can differentiate into mature granulocytes, erythrocytes, monocytes, platelets, and T- and B-cells (see Figure 13.1). Several humoral factors have been characterized which regulate the growth and differentiation of hematopoietic precursor cells. Various interleukins, erythropoietin, trombopoietin, granulocyte–macrophage colony-stimulating factor (GM-CSF), macrophage colony-stimulating factor (M-CSF), and granulocyte colony-stimulating factor (G-CSF) can induce the proliferation of hematopoietic cells (Groopman *et al.*, 1989; Metcalf, 1991; Ogawa, 1993). In contrast, interferons, transforming growth factor-β (TGF-β), and macrophage inflammatory protein-1 reduce the production of hematopoietic cells (Fortunel *et al.*, 2000). In addition, retinoic acid and interferons can induce differentiation of various leukemia cell types (*in vitro*) (Warrell *et al.*, 1991).

3. Leukemia

3.1. Classification

In leukemia, the growth control and differentiation of the neoplastic cells is disrupted, and gradually, these cells overgrow normal differentiating blood cells. These immature cells subsequently burst into the circulation and also into other organs. Leukemias can be roughly divided into acute and chronic leukemias based on the degree of cellular-maturation, and myeloid and lymphoid leukemias based on the lineage of differentiation of the malignant cells. In chronic leukemias, malignant cells usually resemble mature blood cells and/or their precursors, but in acute leukemias the malignant cells are frozen at a much earlier blast stage of differentiation. The French–American–British (FAB) classification is used widely (Bennett *et al.*, 1976) to classify acute leukemias based on morphological and cytochemical criteria, but a new classification system of hematological malignancies has also been introduced (the WHO classification) (Harris *et al.*, 1999) taking into account also the recurrent chromosomal abnormalities in leukemia. According to the FAB classification, acute myeloid leukemias (AML) are divided into eight subclasses (M0–M7), whereas in acute lymphoid leukemias (ALL) there are three subclasses (L1–L3). However, in ALL, the FAB classification has very limited clinical value; instead, the immunophenotypic classification is usually employed. In adults, AML is the most common (80%) acute leukemia. In contrast, in children, most acute leukemias are of lymphoid origin. Yet, acute leukemia is, overall, quite a rare cancer compared to many solid tumors. Thus, for example, in Finland, the annual incidence of adult acute leukemia is about 150 new cases ($3/10^5$ inhabitants), compared to about 3,000 breast cancers ($60/10^5$ inhabitants) (Dickman *et al.*, 1999).

The etiology of acute leukemia is largely unknown. No solid evidence exists of its inheritability, although it seems that in some families the risk of developing leukemia is higher than in others (Cartwright and Staines, 1992). Some chemicals (such as benzene and ethylene oxide), irradiation, and cytotoxic agents are known risk factors (Cartwright and Staines, 1992), because these can cause gene alterations and chromosomal instability blocking normal blood cell differentiation. Several characteristic nonrandom chromosomal abnormalities occur in various types of leukemias, and some of these abnormalities can be used as markers or even as hallmarks for diagnosis, such as the 15;17 translocation in acute promyelocytic leukemia (Grignani *et al.*, 1994; Larson *et al.*, 1984). Furthermore, some chromosomal abnormalities provide prognostic information, such as the Philadelphia chromosome (t9;22) in ALL.

Treatment of acute leukemia consists of several cycles of high-dose combination chemotherapy using various combinations of cytotoxic drugs active against leukemia. After the intensive induction and consolidation phases, patients with ALL usually receive low-dose maintenance therapy. Stem cell transplantation is an alternative consolidation therapy for younger patients with suitable donors. Treatment results have improved considerably during the last two decades, and at present, approximately 30–40% of adult patients under the age of 65 years become long-term survivors. In childhood acute leukemia the prognosis is more optimistic: of every four children with acute leukemia, three are cured. Without treatment, the median survival time of acute leukemia patients is only 2–3 months.

3.2. Clinical Findings

Most of the signs and symptoms of acute leukemia relate to cytopenias caused by impaired production of normal blood cells. The shortage of thrombocytes can cause petechiae and bleeding diathesis, and the shortage of normal leukocytes can predispose patients to infections. Fatigue and fever are relatively common. Leukemia cells infiltrating various organs, such as skin and gums, cause some symptoms. Infiltration of blast cells into the central nervous system is a feared and often fatal complication. In addition to these symptoms, the intensive treatment of acute leukemia causes numerous other symptoms. On the other hand, patients with chronic leukemia are often asymptomatic for prolonged periods of time, and the abnormal blood cell counts are found accidentally during routine laboratory tests. At more advanced stages, problems caused by the large mass of abnormal cells, infections, and also immunological problems may appear. But overall, at presentation the general condition of chronic leukemia patients is often better than that of patients with acute leukemia.

4. Plasminogen Activation System in General

The PA system comprises several activators, inhibitors, and receptors, all of which have strictly controlled roles (see Figure 13.2). Under physiological conditions, activation of plasminogen to proteolytically active enzyme plasmin takes part in many processes, such as fibrinolysis, cell migration, wound healing, trophoblast invasion, angiogenesis, tissue destruction, and activation of growth factors and of other proteolytic enzymes. Many of these roles of the PA system in normal physiology are equally important in cancer. Cancer cells need to adhere to interstitial connective tissue matrices and basement membranes and subsequently detach from them to be able to move on in the extracellular matrix (ECM). They need to degrade various proteins and maintain their own growth by activating growth factors and enzymes.

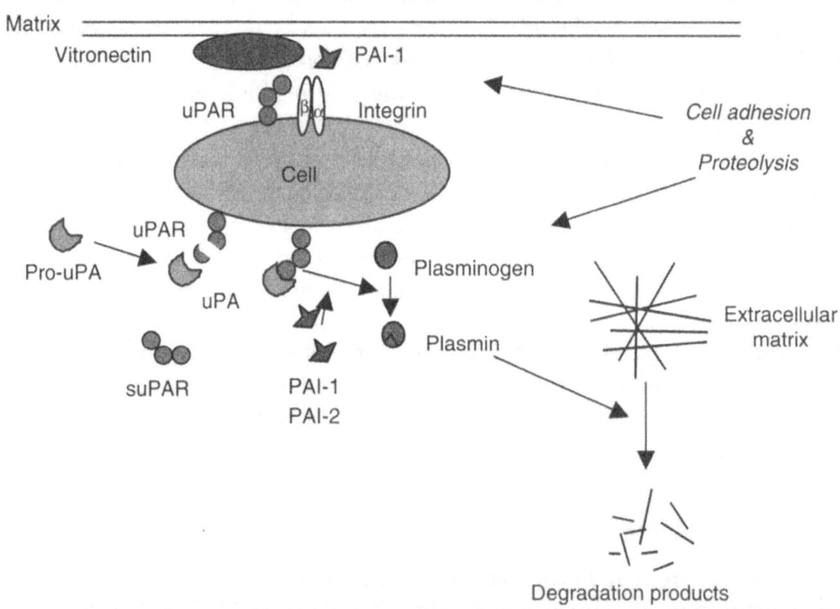

Figure 13.2. Overview of the uPA-mediated plasminogen activation system. Plasminogen is a proenzyme that needs to be activated to be able to degrade extracellular matrix proteins. In humans, there are two main plasminogen activators: urokinase-type plasminogen activator (uPA) and tissue-type plasminogen activator (tPA). In addition to the activators, two inhibitors, PAI-1 and -2 play major roles. uPAR is a specific high-affinity cell-surface receptor for uPA. Binding of uPA to uPAR localizes plasminogen activation at the cell surface. In addition to its function in proteolytic cascades, uPAR is also associated with several important adhesion proteins such as vitronectin and integrins and plays an important role in cell migration and adhesion.

The PA system will be covered in more detail in other chapters of this book, but in the next section we will briefly describe the urokinase plasminogen activator receptor, uPAR, in view of its importance in the following parts of this chapter.

4.1. Structure and General Function of uPAR

A high-affinity specific cell-surface receptor for uPA, uPAR, was first discovered in monocytes (Vassalli *et al.*, 1985) and monocytoid U937 cells (Stoppelli *et al.*, 1985; Vassalli *et al.*, 1985), and since then it has been characterized in a variety of cell types such as granulocytes, endothelial cells, keratinocytes, fibroblasts, and smooth-muscle cells (Blasi *et al.*, 1994; Danø *et al.*, 1994; Behrendt and Stephens, 1998). uPAR binds to the cell surface via a glycosyl-phosphatidylinositol (GPI) anchor (Ploug *et al.*, 1991) (Figure 13.3) and thus has no cytoplasmic tail. The heavily glycosylated protein consists of three homologous domains (D1–D3), which have different binding properties for various ligands. The amino-terminal domain D1 is responsible for the binding of uPA (Behrendt *et al.*, 1990, 1991), whereas the other two, D2 + the carboxyterminal D3 domain, are important for vitronectin binding (Wei *et al.*, 1994). However, the full-size receptor is required for high-affinity interactions with both of these ligands (Ploug *et al.*, 1994; Wei *et al.*, 1994; Behrendt *et al.*, 1996; Riittinen *et al.*, 1996; Høyer-Hansen *et al.*, 1997a; Sidenius and Blasi, 2000) (see Figures 13.2 and 13.3).

Initially, uPAR was thought to occur only in its full-size form on the cell surface, but later, also a cleaved form (D2D3) lacking the amino-terminal uPA-binding domain D1 was discovered in cultured U937 cells (Høyer-Hansen *et al.*, 1992) and also *in vivo* in transplanted murine tumors (Solberg *et al.*, 1994). Several enzymes (uPA, plasmin, and chymotrypsin) can cleave the receptor at the linker region between domains 1 and 2 (Behrendt *et al.*, 1991; Høyer-Hansen *et al.*, 1992, 1997b). The receptor cleavage by uPA occurs quite slowly in solution, whereas on the cell surface the cleavage is highly accelerated (Høyer-Hansen *et al.*, 1997b). Furthermore, intact GPI-anchor is needed for uPA-induced uPAR cleavage, and the soluble form of the uPA receptor (suPAR) devoid of

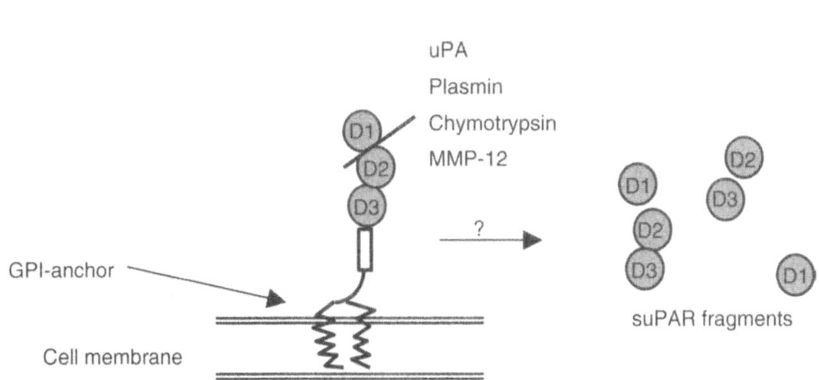

Figure 13.3. Cleavage of urokinase receptor from the cell surface. uPA, plasmin, MMP-12, and chymotrypsin can cleave uPAR between domains D1 and D2, but the enzyme responsible for cleavage of the whole receptor from the cell surface is unknown.

the GPI-anchor is resistant to this cleavage (Høyer-Hansen *et al.*, 2001). Recently, it has also been shown that metalloproteases may be involved in this cleavage process, since MMP-12 is able to cleave D1 from the full-length receptor (Koolwijk *et al.*, 2001). All this suggests that the receptor cleavage is a negative feedback mechanism for plasminogen activation on the cell surface. Both phosphatidylinositol-specific phospholipase C and D are able to catalyze the cleavage of the whole receptor from the cell surface *in vitro* (Ploug *et al.*, 1991; Wilhelm *et al.*, 1999), but the physiological enzyme functioning *in vivo* remains unknown (see Figure 13.3).

There have been problems in understanding how uPAR can activate intracellular signal transduction pathways, because it has no transmembrane or cytoplasmic domain. One possibility is that the activation of uPA, when bound to the receptor, induces the activation of latent growth factors (like TGF-β and pro-HGF), which in turn can activate signal transduction pathways (Besser *et al.*, 1996; Koshelnick *et al.*, 1999). However, current results suggest that the proteolytic activity of uPA is not needed to induce the activation of intracellular kinases (Resnati *et al.*, 1996). Furthermore, synthetic uPAR peptides, which carry a specific epitope, are able to induce chemotaxis in THP-1 monocytic leukemia cells and activate tyrosine kinases (Fazioli *et al.*, 1997). It has therefore been suggested that a special transmembrane adaptor molecule exists, which can mediate the signals between extracellular uPAR and intracellular kinase pathways (Resnati *et al.*, 1996; Blasi, 1999b). However, such an adaptor molecule has not been identified as yet, and it is also possible that several of them exist, since not all signals induced by uPA–uPAR interaction are uniform.

The primary function of uPAR originally appeared to be the localization of uPA and the whole PA system onto specific areas on the cell surface (Myöhänen *et al.*, 1993). Through this function, uPAR participates in localized proteolysis in wound healing, ovulation, angiogenesis, and cancer invasion (Danø *et al.*, 1994). Unexpectedly, the first knockout studies found no abnormalities in uPAR-deficient mice: the mice were viable, healthy, and fertile (Dewerchin *et al.*, 1996). Later on, new, non-proteolytic functions for uPAR have been discovered, implying that the function of uPAR is not limited to the localization of plasminogen activation on the cell surface (Danø *et al.*, 1994; Blasi, 1997, 1999a, b; Chapman, 1997; Preissner *et al.*, 2000; Waltz *et al.*, 2000). Especially, the results from chemotaxis and cell adhesion experiments have shed more light on the roles of uPAR and uPAR fragments (Gyetko *et al.*, 1996; Resnati *et al.*, 1996; Fazioli *et al.*, 1997; Degryse *et al.*, 1999, 2001).

4.2. Soluble uPAR (suPAR)

The suPAR was first discovered in plasma of patients with paroxysmal nocturnal hemoglobinuria (Ploug *et al.*, 1992) and ascitic fluid of ovarian cancer patients (Pedersen *et al.*, 1993). No uPA-ligand was bound to it, and it was also devoid of the GPI anchor. Later, suPAR was also found in plasma of healthy individuals (Rønne *et al.*, 1995; Stephens *et al.*, 1997), and the levels were quite stable at a median concentration of 1.2 ng/ml (range 0.8–1.7 ng/ml), with no correlation to age (Stephens *et al.*, 1997). In addition to plasma/serum, suPAR has been demonstrated in carcinomatous cystic fluid (Wahlberg *et al.*, 1998) and more recently in urine (Sier *et al.*, 1999). Urine samples have been shown to contain uPAR fragments D2D3 and D1 in addition to the full-length receptor (Sier *et al.*,

1999; Sidenius *et al.*, 2000a), whereas only the full-length form of uPAR is present in plasma (Sidenius *et al.*, 2000a). However, the physiological mechanisms involved in the shedding of fragments or the entire receptor from the cell surface have remained unknown (see Figure 13.3).

In several different types of malignancies, elevated levels of serum/plasma suPAR have been detected. For example, patients with non-small-cell lung cancer (Pappot *et al.*, 1997), and cancer of the breast (Stephens *et al.*, 1997), colon (Stephens *et al.*, 1999), ovary (Sier *et al.*, 1998; Riisbro *et al.*, 2001), rectum (Fernebro *et al.*, 2001), prostate (Miyake *et al.*, 1999), and head and neck squamous cell carcinoma (Schmidt and Hoppe, 1999) have significantly increased plasma suPAR levels. Furthermore, it has recently been shown that high suPAR levels correlate with shorter median survival in patients with ovarian, colon, and rectal cancer (Sier *et al.*, 1998; Stephens *et al.*, 1999; Fernebro *et al.*, 2001). The origin of suPAR in cancer patients' plasma has been unknown (Brünner *et al.*, 1999), and only recently it was demonstrated that mice carrying transplanted human xenograft tumors have human suPAR in their plasma (Holst-Hansen *et al.*, 1999), suggesting that suPAR may be derived from tumor cells. The reason for the low level of plasma suPAR in healthy individuals has also been unclear, but recent studies suggest that normal plasma suPAR is produced by endothelial (Mustjoki *et al.*, 2000b; Chavakis *et al.*, 2001) and blood mononuclear cells and that its release is enhanced when cells are in physical contact with each other, implying that receptor shedding is a possible regulatory mechanism in cell adhesion (Mustjoki *et al.*, 2000b). The functions and importance of the soluble receptors *in vivo* in body fluids are also largely unknown, although *in vitro* experiments have shown that suPAR fragments may be involved in chemotaxis (Resnati *et al.*, 1996; Fazioli *et al.*, 1997).

5. Plasminogen Activation in Leukemia

5.1. Components of the PA System in Normal Blood Cells

Several types of leukocytes express various components of the PA system. For example, uPAR and uPA are expressed by monocytes, macrophages, neutrophils, and myeloid precursors, but not usually by lymphocytes (Wilson *et al.*, 1983, 1992; Miles and Plow, 1987; Plesner *et al.*, 1994, 1997; Jardi *et al.*, 1996; Mustjoki *et al.*, 1999b). However, T-cells from HIV-positive patients express uPAR (Nykjaer *et al.*, 1994), and interleukins or PMA stimulation can also induce uPAR expression in normal lymphocytes (Nykjaer *et al.*, 1994). Increased serum suPAR levels have also been observed in HIV-positive patients (Sidenius *et al.*, 2000b). In addition, it has been recently reported that antibodies against uPAR inhibit dendritic cell-induced T-cell activation, suggesting that uPAR may also play a role in antigen-presenting cell/T-cell interaction (Woodhead *et al.*, 2000). No uPAR has been found in platelets and erythrocytes, but our previous studies demonstrated that megakaryocytes in the bone marrow express uPAR on the cell surface (Mustjoki *et al.*, 1999b). Both tPA and uPA are expressed by various types of leukocytes. Wilson and Francis (1987) observed that the type of plasminogen activator produced by leukocytes is a differentiation-linked property. In normal bone-marrow cells, the more differentiated cells secreted uPA, whereas the more primitive cells released tPA. tPA protein and mRNA have also been characterized in bone marrow megakaryocytes (Jeanneau and Sultan, 1988; Brisson-Jeanneau *et al.*, 1990). Plasminogen activator inhibitors, PAI-1 and PAI-2, are also produced by blood cells. PAI-2 is mostly expressed by monocytes and not by other leukocytes, whereas PAI-1 is released by platelets, and in addition, in normal bone marrow PAI-1 mRNA has been found in megakaryocytes (Alessi *et al.*, 1994).

An active fibrinolytic system exists in normal human bone marrow (McWilliam *et al.*, 1996). Compared to plasma, normal bone marrow has higher levels of active tPA, despite the inhibitors PAI-1 and PAI-2. In addition, the levels of uPA, α_2-antiplasmin, and tPA–PAI-1 complexes are low, favoring fibrinolysis in the bone marrow (McWilliam *et al.*, 1996). This tPA activity may be necessary for removing fibrin deposits and ensuring sufficient blood flow through the marrow.

5.2. Components of the PA System in Leukemia Cells

Similarly to normal leukocytes, leukemic cells were also found to produce plasminogen activators into their growth medium (Wilson *et al.*, 1983; Tapiovaara *et al.*, 1996). Leukemic cells from AML patients produced varying amounts of plasminogen activators into the growth medium, but the amount produced had no correlation with the clinical severity of the disease (Wilson *et al.*, 1983). Leukemic cells are able to produce both uPA and tPA, whereas normal granulocytes produce only uPA. Another interesting finding was that patients whose cells produced only tPA failed to respond to combination chemotherapy (Wilson *et al.*, 1983). As later experiments have shown that in healthy individuals more primitive leukocytes produce tPA, and mature leukocytes produce uPA, the authors postulated that this may be the reason for the tPA producers' poor response, as poor responses are more often seen in patients whose cells have early-progenitor-type features (Wilson and Francis, 1987). However, in patients with acute promyelocytic leukemia, tPA levels have been found to be significantly lower in relapse than at initial onset (Nakasaki *et al.*, 2000). Moreover, both plasminogen activators and their inhibitors occur in higher amounts in leukemic cell homogenates from AML patients than from ALL patients (Wada *et al.*, 1993), and plasma uPA levels are also increased in AML patients (Garcia Frade *et al.*, 1992).

Studies describing the expression of uPAR on leukemia cell surface have given somewhat discrepant results. Plesner *et al.* (1994) found uPAR to be expressed in bone marrow leukemic cells in 12 of 21 AML patients with FAB classes M2, M4, and M5, but not with FAB class M1. No uPAR was found in lymphoblastic leukemia cells. A few years later Jardi *et al.* (1996) found uPAR only on the cell surfaces of the most mature acute myeloid leukemias, M3 and M5 leukemia, but in these groups the pattern was also heterogeneous. In our studies, we found uPAR on the leukemic cell surface in 28 of 29 patients with AML, whereas most patients with ALL were negative for uPAR (7/9) (Mustjoki *et al.*, 1999a). In addition, two of three patients with a hybrid phenotype had uPAR on the cell surface. In AML patients, uPAR was expressed in all FAB subtypes, with the highest expression in subclasses M3, M4, and M5. The expression pattern of uPA resembled the pattern seen with uPAR: it was expressed by myeloid blasts and not by lymphoid blasts. However, usually a smaller percentage of cells was uPA- than uPAR-positive. Simultaneously with our studies, Lanza *et al.* (1998) showed that leukemia cells from all common AML FAB groups (M0–M5) can have uPAR on the cell surfaces. Furthermore, they described an AML subgroup (high uPAR positivity, chromosome 11 abnormalities, and M5 FAB morphology) characterized by a more aggressive clinical course. uPAR mRNA expression in leukemic blast cells has been shown to be limited to FAB classes M4 to M5 (Scherrer *et al.*, 1999). In summary, all these studies agree that leukemic cells in lymphoid malignancies do not express uPAR. Moreover, uPAR expression is associated with malignancies of myeloid origin, but among them the pattern is heterogeneous.

In vitro experiments with leukemia cell lines have revealed that leukemia cells have an enhanced capacity to activate the pro-uPA they produce, and mainly the active form of uPA is found on the cell surface and in the growth medium; this differs from adherent tumor cells, which produce only inactive pro-uPA in regular culture conditions (Stephens *et al.*, 1988, 1992; Tapiovaara *et al.*, 1991). uPA activity has also been demonstrated to play a role in extracellular-matrix invasion by lymphoblastic T-cell lines (Reiter *et al.*, 1997). This suggests that leukemic cells have an active proteolytic potential on the cell surface. These results are also applicable *in vivo*. Tapiovaara *et al.* (1993) found active uPA in the mononuclear cells from peripheral blood and bone marrow in patients with acute leukemia, but not in control subjects or patients with chronic leukemia. No correlation was found between the lineage of leukemia and PA activity. However, Scherrer *et al.* (1999) found recently that only leukemic cells from AML patients have uPA mRNA and produce active uPA protein, whereas cells from ALL patients lack this ability. These results are in good accordance with the recent report by Robbie *et al.* (2000), who found that only myeloid blast cells can lyse fibrin clots directly. Mature mononuclear and polymorphonuclear cells from healthy individuals and lymphoid blasts failed to lyse fibrin clots. This fibrinolytic activity was inhibited with antibodies against uPA, but not with tPA antibodies. Active uPA has also been found in the cerebrospinal fluid of patients with acute leukemia (Akenami *et al.*, 1996). In addition, in AML marrow, the active form of uPA is found, in contrast to normal bone marrow, which shows mostly tPA activity (McWilliam *et al.*, 1998). This uPA activity was suggested to contribute to the severe hemorrhages occurring in AML patients.

5.3. Soluble uPA and uPAR in Leukemia

As measurements of PA components in tumor tissue are often quite laborious and also difficult to standardize (Stephens *et al.*, 1998), attempts have been made to correlate the levels of uPA, uPAR, and PAI-1 in the circulation with prognosis. High circulating uPAR levels in cancer patients have already been shown to associate with poor survival prognosis in colon and ovarian carcinoma patients (Sier *et al.*, 1998; Stephens *et al.*, 1999). In leukemia only a few such studies have been published. When studying plasma samples from acute leukemia patients, we found that patients with AML and ALL had significantly higher levels of suPAR in their plasma than did healthy controls (Mustjoki *et al.*, 1999a). There was no correlation between suPAR levels and serum C-reactive protein levels, D-dimer values, or WBC count. However, suPAR levels correlated significantly with the tumor-cell count in the circulation and with the content of uPAR in tumor cell lysates, suggesting that excess plasma suPAR in leukemia patients is produced by the tumor cells (Mustjoki *et al.*, 2000a). Interestingly, the results also showed that in AML high suPAR plasma levels correlated with poor response to chemotherapy (Mustjoki *et al.*, 1999a). However, a low level of suPAR did not predict a rapid response to chemotherapy, as after several cycles of chemotherapy some patients had a normalized plasma suPAR level and residual disease in the bone marrow.

In solid tumors, there are no longitudinal studies describing the behavior of suPAR during treatment. In leukemia, the evaluation of the short-term effects of chemotherapy can already be done after one month of chemotherapy, and in most cases the tumor cells have disappeared from the circulation within a few weeks. Therefore, it was interesting to see that in acute leukemia patients, plasma suPAR levels normalized within 2 weeks of chemotherapy, in concert with the disappearance of tumor cells from the circulation (Mustjoki *et al.*, 2000a). The simultaneous decrease in plasma suPAR and in the number of circulating tumor cells strengthens the concept that the excess plasma suPAR is derived from tumor cells. Patients with normal plasma suPAR levels and no tumor cells in the circulation at diagnosis showed no marked changes in suPAR levels during chemotherapy.

Sier *et al.* (1999) were the first to show that urinary suPAR correlates with plasma suPAR levels. This proved to be the situation in leukemia as well, since urinary suPAR levels followed closely the plasma suPAR levels at diagnosis and during chemotherapy (Mustjoki *et al.*, 2000a). However, in the urine of leukemia patients there was a peak at day 5 after the start of chemotherapy. Such a peak was not observed in blood samples. This increase in urinary suPAR levels correlated with decreasing tumor-cell counts, and thus seems to be a reflection of the massive tumor cell death occuring in response to therapy. uPAR released from dying tumor cells then accumulates in the urine.

In addition to plasma suPAR levels, plasma uPA levels have also been shown to be significantly increased in acute leukemia patients (Garcia Frade *et al.*, 1992; Mustjoki *et al.*, 1999a). Most patients with high plasma uPA levels also had high plasma suPAR levels (Mustjoki *et al.*, 1999a). uPA and tPA activities have generally been found to be quite low in plasma samples (McWilliam *et al.*, 1996, 1998), which is in good agreement with the results of our studies (Mustjoki *et al.*, 1999a). However, in our material two patients with hybrid leukemia and one AML patient showed markedly elevated plasma uPA activity. Both of these hybrid leukemia patients suffered from symptomatic retinal bleeding with loss of visual field at the beginning of therapy. The AML patient with high uPA activity died on the third day after diagnosis, and a potential bleeding diathesis could not be properly evaluated.

5.4. Expression Patterns of uPAR Fragments in Leukemia Patients and in Healthy Controls

Since proteolytic cleavage of uPAR induces a potent chemotactic response *in vitro*, it is of importance to study whether these fragments exist *in vivo* in blood and urine samples of leukemia patients. In leukemia cell lysates of AML patients, both full-length receptor and a fragment corresponding to domains D2 + D3 were found (Mustjoki *et al.*, 2000a). In contrast, peripheral blood mononuclear cells from healthy controls contained only the full-length receptor, and leukemia cell lysates of ALL patients usually had no uPAR. The same pattern was observed in plasma from both peripheral blood and bone marrow aspirates. In AML patients, fragmented uPAR was more visible in plasma from bone marrow aspirates than from peripheral blood. This may have been due to the higher local concentration of tumor cells in the bone marrow. In bone marrow of healthy controls or patients with chronic leukemia, only full-length uPAR was seen. Bone marrow plasma samples were also analyzed with zymography, and samples with clearly visible fragmented receptor were found to have more active uPA than did samples containing only full-length receptor (unpublished observations). The fragment pattern changed towards normal during chemotherapy.

6. Clinical Impact of Plasminogen Activation in Leukemia

Over the last 20 years, acute leukemias have been grouped according to the FAB classification, which is based on morphological and cytochemical features. However, in recent years, knowledge in cytogenetics has markedly increased, and also much more is known of its significance for the outcome of the disease. Furthermore, other important features have been recognized regarding the clinical behavior of the leukemia. Based on all this, a new WHO classification of neoplastic diseases of the hematopoietic and lymphoid tissues was recently launched (Harris *et al.*, 1999). It tries to integrate genetic and clinical features with the morphology, cytochemistry, and immunophenotype of the neoplastic cells, and in that manner, to define disease entities which have clinical relevance. In the future, leukemias may be even more strictly divided into subclasses according to the biological characteristics of the malignant cells. This could assist in the design of individual therapy. It is therefore of great importance to search for biological factors which influence disease progression. Some components of the plasminogen activation cascade may turn out to be such biological factors and potential therapeutic targets in the future. In the following sections their clinical value will be discussed.

6.1. Plasminogen Activation System and Hemostasis

Leukemia patients have an increased bleeding tendency mostly due to the thrombocytopenia caused by defective production in the leukemic bone marrow. However, this bleeding tendency is not always readily attributable to thrombocytopenia.

Plasmin formation plays an important role in normal blood coagulation and fibrinolysis. Plasmin inhibitors in plasma, α_2-macroglobulin and α_2-antiplasmin strictly control plasmin formation. In normal hemostasis, a fibrin clot is formed at the site of tissue injury. When the clot is no longer needed, plasminogen is activated into plasmin which starts to degrade the clot. In hemostatic disorders, plasminogen can be unnecessarily activated, and plasmin as a wide-spectrum protease can degrade almost any plasma protein. In addition, if the plasmin inhibitors are not working properly, uncontrolled fibrinolysis will occur with a fatal outcome. This kind of a situation has been described in acute promyelocytic leukemia (APL) patients with disseminated intravascular coagulation (DIC). Plasma samples from these patients have been shown to contain reduced levels of α_2-antiplasmin and PAI-1 and increased levels of plasmin-α_2-antiplasmin complexes in addition to hypofibrinogenemia (Sakata *et al.*, 1991; Fukao *et al.*, 1992; Dombret *et al.*, 1993, 1995). Therefore, it has been suggested that the actual problem in DIC is excessive fibrinolysis rather than excessive coagulation. In our *in vitro* model for acute promyelocytic leukemia, we found that on the APL cell surface plasminogen can be activated and plasmin formed (Tapiovaara *et al.*, 1994; Mustjoki *et al.*, 1998). These results suggest that because plasmin has potent protease activity, it can interfere with the normal fibrinolytic balance and be part of a cascade leading to the excessive bleeding problems encountered in APL patients. Furthermore, we noticed that plasmin formation is transiently induced by retinoic acid (RA) treatment, but that the overall effect of RA treatment on plasmin formation is suppressive. This fits well with the clinical observation that in APL patients RA treatment diminishes the bleeding tendency. Our results on plasmin formation in APL were obtained using the promyelocytic NB4 cell line, and we have not studied the effect of RA on plasmin formation in fresh APL cells. However, we were able to examine promyelocytic cells from some APL patients with FACS analysis, and showed that these cells have increased amounts of both uPAR and uPA on the cell surface. Recently, Menell *et al.* (1999) have also shown that the expression of annexin II, a proposed receptor for tPA, is higher on the cell surface in APL than in other acute leukemias and that after RA treatment, its expression decreases.

Only a few studies have described the correlation between plasma plasminogen activator levels and *in vivo* bleeding events (Bennett *et al.*, 1989; Mustjoki *et al.*, 1999a). In 1989, Bennet *et al.* demonstrated that leukemia patients with clinical bleeding have free uPA in their plasma. In our study, three patients had markedly high plasma uPA activity (Mustjoki *et al.*, 1999a). All these three also had high uPA antigen levels. Notably, although receiving prophylactic platelet transfusions, during the first weeks of treatment two of these three patients suffered from symptomatic retinal bleeding with loss of visual field. The third patient died on the third day before proper evaluation and follow-up for bleeding diathesis. Although this bleeding diathesis episode with high plasma uPA activity may have been coincidental, it is noteworthy that in the other 53 patients no clinical bleeding was observed, and although some of these patients had increased plasma uPA protein levels, they still had normal uPA activity levels, indicating that uPA activity was controlled by complex-formation with inhibitors. However, more patient data are needed to clarify the real connection between bleeding events and uPA activity, and the possible value of uPA inhibitors in therapeutic approaches.

6.2. Cell Surface Plasminogen Activators and Urokinase Receptor: Diagnostic and Prognostic Tools in Leukemia?

The lineage of leukemic blast cells has therapeutic implications; for example, in the choice of the most effective chemotherapeutic drugs. For example, ALL and AML FAB M4–M5 patients routinely receive prophylactic intrathecal chemotherapeutic agents, because neuroleukemia is seen more commonly in these patient groups. However, the definition of blast cell lineage is not always self-evident, and additional phenotypic markers would be very useful. uPAR (CD87) could serve as such an additional marker, because its expression seems to be restricted to blast cells expressing myeloid markers (Plesner *et al.*, 1994; Jardi *et al.*, 1996; Lanza *et al.*, 1998; Mustjoki *et al.*, 1999a; Scherrer *et al.*, 1999). In addition to AML, also hybrid leukemias express uPAR (Lanza *et al.*, 1998; Mustjoki *et al.*, 1999a). Although the published studies have given somewhat discrepant results regarding the uPAR expression in different FAB subclasses, they all have agreed that lymphoid cells are negative for uPAR. This discrepancy may be due to the different antibodies and techniques used, and to the source of the blast cells (peripheral blood vs. bone marrow). By standardizing these measurements with larger size patient material, analysis of uPAR expression may have diagnostic value in acute leukemia in the future.

uPAR expression could also serve as a prognostic marker in AML, because in combination with chromosome 11 abnormalities and M5 FAB class, it predicted a more aggressive clinical course (Lanza *et al.*, 1998). Further studies are needed to establish whether cell-surface uPA or tPA expression could be used as prognostic markers. The initial observation by Wilson *et al.* (1983) that AML patients whose blast cells produced only tPA responded poorly to chemotherapy, has not been confirmed by more recent studies.

6.3. suPAR as a Marker for Prognosis in Leukemia

Quantitative and reliable methods for the follow-up of minimal residual disease and for the prediction of relapse in acute leukemia are badly needed. At present practical clinical tools for this purpose are only available for patients whose leukemic cells express an aberrant immunophenotype or carry a specific chromosomal or molecular marker. The methods required to detect these changes in a low percentage of cells in a given sample are quite laborious and time-consuming and have to be carried out in a sample-specific, nonautomated manner. A method amenable to automation would therefore be of great practical importance. Could measurement of suPAR in plasma or in urine serve as such a marker? In a large series of colon carcinoma patients (n = 591), Stephens *et al.* (1999) found an increasing risk of mortality with an increasing plasma suPAR level and the suPAR level to be an independent prognostic marker. Similarly, in AML, it has been shown that high suPAR levels at diagnosis correlate with poor response to chemotherapy and with increased risk for relapse within a short time (5 months) (Mustjoki *et al.*, 1999a). Although these results suggest that the suPAR level can be related to the aggressiveness of the disease, the small number of patients in this study does not allow definitive conclusions. It would therefore be of interest to study in a larger patient population, whether the plasma suPAR level is an independent prognostic marker, and whether it correlates with the long-term survival of leukemia patients.

In a longitudinal study, the first results suggested that suPAR could be used as a follow-up marker, because the levels correlated with the number of circulating blast cells, and because suPAR levels decreased rapidly during treatment (Mustjoki *et al.*, 2000a). Furthermore, there were patients whose suPAR levels decreased transiently during treatment and increased again in relapse, or if no remission was achieved. However, disappointingly it seemed that the levels did not increase before blast cells appeared in the circulation, that is, they failed to show the presence of blast cells in the bone marrow. In addition, there were also patients who had a normal plasma suPAR at diagnosis.

Thus far, it has been possible to measure only full-length suPAR and uPAR fragments together but not individual uPAR fragments (D1 and D2D3) separately. However, *in vitro* experiments have shown that especially uPAR fragments may have important roles, for example, in chemotaxis (Resnati *et al.*, 1996; Fazioli *et al.*, 1997; Degryse *et al.*, 1999). Full-length uPAR is required for efficient vitronectin binding (Wei *et al.*, 1994; Høyer-Hansen *et al.*, 1997a; Sidenius and Blasi, 2000), and it also associates with β_1-integrins, whereas the cleaved form does not (Montuori *et al.*, 1999). Different forms of uPAR seem to be required and be important for different functions and it would be important to analyze these fragments separately. Notably, in leukemia patients uPAR fragments were found in tumor cells, as well as in plasma and urine samples, whereas samples from healthy individuals usually contained only full-length uPAR.

During chemotherapy the proportion of fragments diminished and the uPAR fragmentation pattern started to resemble that seen in healthy individuals. Therefore, measurement of separate uPAR fragments could have clinical importance in leukemia as follow-up markers.

The difference in uPAR fragment patterns between cancer patients and healthy individuals is not likely to be accidental. On the contrary, it suggests that these fragments may have some unknown function in cancer pathophysiology. In the future, it would be of interest to isolate these fragments from various samples from cancer patients and to test their effects *in vitro*, to see whether they induce such processes as cytoskeletal reorganization, adherence of cells, cell migration, or chemotaxis. Such studies could shed more light on the possible role of uPAR and uPAR fragments in the pathophysiology of cancer in general, and on their significance in acute leukemia as well.

7. Perspectives

Because cancer cells resemble the body's nonmalignant cells in many ways, they are not effectively rejected by the immune system. However, cancer cells also possess advantages which assist them to overgrow normal cells. The ideal way to treat cancer would be to kill the cancer cells without affecting normal cells. For example in leukemia, this is very difficult to accomplish by means of conventional chemotherapy, since both normal blood cells and leukemia cells are rapidly dividing cells. In addition, both these cell types have similar phenotypic markers on the surface. Therefore, it is important to characterize any specific features of leukemia cells and to find possible differences between them and normal blood cells which could be exploited in the design of future therapy.

During several decades, strong evidence has accumulated that the plasminogen activation system plays an important role in tumor invasion and metastasis. The newly discovered non-proteolytic functions of the plasminogen activation system have reinforced interest in the plasminogen activation field and also explained some phenomena not understandable by the mechanisms of the proteolytic functions of the plasminogen activation system. Results from leukemia studies show that the components of the plasminogen activation system are expressed in increased amounts in leukemia cells *in vitro* and *in vivo* and that therapy with various drugs used in the treatment of leukemia results in changes in the expression level and the activity of these components. Although there are results suggesting that some of these components, such as uPAR, could be used as diagnostic or prognostic markers in leukemia, further studies are needed to clarify their clinical impact. However, as knowledge of the functions of uPAR increases, and when our understanding of the biological role of soluble uPAR becomes clearer, these results may also contribute to the understanding of the pathophysiology of leukemia and to more precise approaches in the treatment of leukemia patients.

References

Akenami, F.O.T., Sirén, V., Koskiniemi, M., Siimes, M.A., Teräväinen, H., and Vaheri, A. (1996). Cerebrospinal fluid activity of tissue plasminogen activator in patients with neurological diseases. *J. Clin. Pathol.* **49**, 577–580.

Alessi, M.C., Chomiki, N., Berthier, R., Schweitzer, A., Fossat, C., and Juhan-Vague, I. (1994). Detection of plasminogen activator inhibitor-1 (PAI-1) mRNA in human megakaryocytes by in situ hybridization. *Thromb. Haemost.* **72**, 931–936.

Behrendt, N., Rønne, E., Ploug, M., Petri, T., Løber, D., Nielsen, L.S., Schleuning, W.-D., Blasi, F., Appella, E., and Danø, K. (1990). The human receptor for urokinase plasminogen acivator. NH_2-terminal amino acid sequence and glycosylation variants. *J. Biol. Chem.* **265**, 6453–6460.

Behrendt, N., Ploug, M., Patthy, L., Houen, G., Blasi, F., and Danø, K. (1991). The ligand-binding domain of the cell surface receptor for urokinase-type plasminogen activator. *J. Biol. Chem.* **266**, 7842–7847.

Behrendt, N., Rønne, E., and Danø, K. (1996). Domain interplay in the urokinase receptor. Requirement for the third domain in high affinity ligand binding and demonstration of ligand contact sites in distinct receptor domains. *J. Biol. Chem.* **271**, 22885–22894.

Behrendt, N. and Stephens, R.W. (1998). The urokinase receptor. *Fibrinolysis Proteol.* **12**, 191–204.

Bennett, B., Booth, N.A., Droll, A., and Dawson, A.A. (1989). The bleeding disorder in APL: Fibrinolysis due to uPA rather than defibrination. *Br. J. Haematol.* **71**, 511–516.

Bennett, J.M., Catovsky, D., Daniel, M.T., Flandrin, G., Galton, D.A.G., Gralnick, H.R., and Sultan, C. (1976). Proposals for the classification of the acute leukaemias. The French–American–British (FAB) co-operative Group. *Br. J. Haematol.* **33**, 451–458.

Besser, D., Verde, P., Nagamine, Y., and Blasi, F. (1996). Signal transduction and the uPA/uPAR system. *Fibrinolysis* **10**, 215–237.

Blasi, F., Conese, M., Møller, L.B., Pedersen, N., Cavallaro, U., Cubellis, M.V., Hernandez-Marrero, F., Limongi, P., Munoz-Canoves, P., Resnati, M., Riittinen, L., Sidenius, N., Soravia, E., Soria, M.R., Stoppelli, M.P., Talarico, D., Teesalu, T., and Valcamonica, S. (1994). The urokinase receptor: structure, regulation and inhibitor-mediated internalization. *Fibrinolysis* **8**, 182–188.

Blasi, F. (1997). uPA, uPAR, PAI-1: Key intersection of proteolytic, adhesive and chemotactic highways? *Immunol. Today* **18**, 415–417.

Blasi, F. (1999a). Proteolysis, cell adhesion, chemotaxis, and invasiveness are regulated by the u-PA-u-PAR-PAI-1 system. *Thromb. Haemost.* **82**, 298–304.

Blasi, F. (1999b). The urokinase receptor. A cell surface, regulated chemokine. *APMIS* **107**, 96–101.

Brisson-Jeanneau, C., Nelles, L., Rouer, E., Sultan, Y., and Benarous, R. (1990). Tissue-plasminogen activator RNA detected in megakaryocytes by in situ hybridization and biotinylated probe. *Histochemistry* **95**, 23–26.

Brünner, N., Nielsen, H.J., Hamers, M., Christensen, I.J., Thorlacius-Ussing, O., and Stephens, R.W. (1999). The urokinase plasminogen activator receptor in blood from healthy individuals and patients with cancer. *APMIS* **107**, 160–167.

Cartwright, R.A. and Staines, A. (1992). Acute leukaemias. *Bailliere Clin. Haem.* **5**, 1–26.

Chapman, H.A. (1997). Plasminogen activators, integrins, and the coordinated regulation of cell adhesion and migration. *Curr. Opin. Cell Biol.* **9**, 714–724.

Chavakis, T., Willuweit, A.K., Lupu, F., Preissner, K.T., and Kanse, S.M. (2001). Release of soluble urokinase receptor from vascular cells. *Thromb. Haemost.* **86**, 686–693.

Choi, K., Kennedy, M., Kazarov, A., Papadimitriou, J.C., and Keller, G. (1998). A common precursor for hematopoietic and endothelial cells. *Development* **125**, 725–732.

Danø, K., Behrendt, N., Brunner, N., Ellis, V., Ploug, M., and Pyke, C. (1994). The urokinase receptor: Protein structure and role in plasminogen activation and cancer invasion. *Fibrinolysis* **8**, 189–203.

Degryse, B., Resnati, M., Rabbani, S.A., Villa, A., Fazioli, F., and Blasi, F. (1999). Src-dependence and pertussis-toxin sensitivity of urokinase receptor-dependent chemotaxis and cytoskeleton reorganization in rat smooth muscle cells. *Blood* **94**, 649–662.

Degryse, B., Orlando, S., Resnati, M., Rabbani, S.A., and Blasi, F. (2001). Urokinase/ urokinase receptor and vitronectin/ $\alpha v\beta 3$ integrin induce chemotaxis and cytoskeleton reorganization through different signaling pathways. *Oncogene* **20**, 2032–2043.

Dewerchin, M., Van Nuffelen, A., Wallays, G., Bouché, A., Moons, L., Carmeliet, P., Mulligan, R.C., and Collen, D. (1996). Generation and characterization of urokinase receptor-deficient mice. *J. Clin. Invest.* **97**, 870–878.

Dickman, P.W., Hakulinen, T., Luostarinen, T., Pukkala, E., Sankila, R., Soderman, B., and Teppo, L. (1999). Survival of cancer patients in Finland 1955–1994. *Acta Oncol.* **38**, 1–103.

Dombret, H., Scrobohaci, M.L., Ghorra, P., Zini, J.M., Daniel, M.T., Castaigne, S., and Degos, L. (1993). Coagulation disorders associated with acute promyelocytic leukemia: Corrective effect of all-trans retinoic acid treatment. *Leukemia* **7**, 2–9.

Dombret, H., Scrobohaci, M.-L., Daniel, M.-T., Micléa, J.-M., Castaigne, S., Chomienne, C., Fenaux, P., and Degos, L. (1995). In vivo thrombin and plasmin activities in patients with acute promyelocytic leukemia (APL): Effect of all-trans retinoic acid (ATRA) therapy. *Leukemia* **9**, 19–24.

Fazioli, F., Resnati, M., Sidenius, N., Higashimoto, Y., Appella, E., and Blasi, F. (1997). A urokinase-sensitive region of the human urokinase receptor is responsible for its chemotactic activity. *EMBO J.* **16**, 7279–7286.

Fernebro, E., Madsen, R.R., Fernö, M., Brunner, N., Bendahl, P.-O., Christensen, I.J., Johnson, A., and Nilbert, M. (2001). Prognostic importance of the soluble plasminogen activator receptor, suPAR, in plasma from rectal cancer patients. *Eur. J. Cancer* **37**, 486–491.

Fortunel, N.O., Hartzfield, A., and Hartzfield, J.A. (2000). Transforming growth factor-β: Pleiotropic role in regulation of hematopoiesis. *Blood* **96**, 2022–2036.

Fukao, H., Ueshima, S., Okada, K., Yamamoto, K., Matsuo, T., and Matsuo, O. (1992). Tissue-type plasminogen activator, type 1 plasminogen activator inhibitor and their complex in plasma with disseminated intravascular coagulation. *Thromb. Res.* **68**, 57–65.

Gale, R.P. and Butturini, A. (1992). Leukemia: Stem cells, preleukemia and cure. *Leukemia* **6**, 80–85.

Garcia Frade, L.J., Sureda, A., Torrado, M.C., and Garcia Avello, A. (1992). High plasma urokinase-type plasminogen activator levels are present in patients with acute nonlymphoblastic leukemia. *Acta Haematol.* **88**, 7–10.

Grignani, F., Fagioli, M., Alcalay, M., Longo, L., Pandolfi, P.P., Donti, E., Biondi, A., Lo Coco, F., Grignani, F., and Pelicci, G.P. (1994). Acute promyelocytic leukemia: From genetics to treatment. *Blood* **83**, 10–25.

Groopman, J.E., Molina, J.-M., and Scadden, D.T. (1989). Hematopoietic growth factors. Biology and clinical applications. *N. Eng. J. Med.* **321**, 1449–1459.

Gyetko, M.R., Chen, G.-H., McDonald, R.A., Goodman, R., Huffnagle, G.B., Wilkinson, C.C., Fuller, J.A., and Toews, G.B. (1996). Urokinase is required for the pulmonary inflammatory response to *Cryptococcus neoformans*. *J. Clin. Invest.* **97**, 1818–1826.

Harris, N.L., Jaffe, E.S., Diesbold, J., Flandrin, G., Muller-Hermelink, H.K., Vardiman, J., Lister, T.A., and Bloomfield, C.D. (1999). World Health Organization classification of neoplastic diseases of the hematopoietic and lymphoid tissues: Report of the Clinical Advisory Committee meeting-Airlie House, Virginia, November 1997. *J. Clin. Oncol.* **17**, 3835–3849.

Holst-Hansen, C., Hamers, M.J.A.G., Johannessen, B.E., Brünner, N., and Stephens, R.W. (1999). Soluble urokinase receptor released from human carcinoma cells; A plasma parameter for xenograft tumour studies. *Br. J. Cancer* **81**, 203–211.

Høyer-Hansen, G., Rønne, E., Solberg, H., Behrendt, N., Ploug, M., and Lund, L.R. (1992). Urokinase plasminogen activator cleaves its cell surface receptor releasing the ligand-binding domain. *J. Biol. Chem.* **267**, 18224–18229.

Høyer-Hansen, G., Behrendt, N., Ploug, M., Danø, K., and Preissner, K.T. (1997a). The intact urokinase receptor is required for efficient vitronectin binding: Receptor cleavage prevents ligand interaction. *FEBS Lett.* **420**, 79–85.

Høyer-Hansen, G., Ploug, M., Behrendt, N., Rønne, E., and Danø, K. (1997b). Cell-surface acceleration of urokinase-catalyzed receptor cleavage. *Eur. J. Biochem.* **243**, 21–26.

Høyer-Hansen, G., Pessera, U., Holm, A., Pass, J., Weidle, U., Danø, K., and Behrendt, N. (2001). Urokinase-catalysed cleavage of the urokinase receptor requires an intact glycolipid anchor. *Biochem. J.* **358**, 673–679.

Jardi, M., Ingles-Esteve, J., Burgal, M., Azqueta, C., Velasco, F., Lopez-Pedrera, C., Miles, L.A., and Felez, J. (1996). Distinct patterns of urokinase receptor (uPAR) expression by leukemic cells and peripheral blood cells. *Thromb. Haemost.* **76**, 1009–1019.

Jeanneau, C. and Sultan, Y. (1988). Tissue plasminogen activator in human megakaryocytes and platelets: immunocytochemical localization, immunoblotting and zymographic analysis. *Thromb. Haemost.* **59**, 529–534.

Koolwijk, P., Sidenius, N., Peters, E., Sier, C.F.M., Hanemaaijer, R., Blasi, F., and van Hindsbergh, V.W.M. (2001). Proteolysis of the urokinase-type plasminogen activator receptor by metalloproteinase-12: Implication for angiogenesis in fibrin matrices. *Blood* **97**, 3123–3131.

Koshelnick, Y., Ehart, M., Stockinger, H., and Binder, B.R. (1999). Mechanisms of signaling through urokinase receptor and the cellular response. *Thromb. Haemost.* **82**, 305–311.

Lanza, F., Castoldi, G.L., Castagnari, B., Todd R.F., III, Moretti, S., Spisani, S., Latorraca, A., Focarile, E., Roberti, M.G., and Traniello, S. (1998). Expression and functional role of urokinase-type plasminogen activator receptor in normal and acute leukaemic cells. *Br. J. Haematol.* **103**, 110–123.

Larson, R.A., Kondo, K., Vardiman, J.W., Butler, A.E., Golob, H.M., and Rowley, J.D. (1984). Evidence for a 15:17 translocation in every patient with acute promyelocytic leukemia. *Am. J. Med.* **76**, 827–841.

McWilliam, N.A., Robbie, L.A., Barelle, C.J., Adey, G., Prasad, S., Bennett, B., and Booth, N.A. (1996). Evidence for an active fibrinolytic system in normal human bone marrow. *Br. J. Haematol.* **93**, 170–176.

McWilliam, N., Robbie, L., Booth, N., and Bennet, B. (1998). Plasminogen activator in acute myeloid leukaemic marrows: u-PA in contrast to t-PA in normal marrow. *Br. J. Haematol.* **101**, 626–631.

Menell, J.S., Cesarman, G.M., Jacovina, A.T., McLaughlin, M.A., Lev, E.A., and Hajjar, K.A. (1999). Annexin II and bleeding in acute promyelocytic leukemia. *N. Engl. J. Med.* **340**, 994–1004.

Metcalf, D. (1991). Control of granulocytes and macrophages: Molecular, cellular, and clinical aspects. *Science* **254**, 529–533.

Miles, L.A. and Plow, E.F. (1987). Receptor mediated binding of the fibrinolytic components, plasminogen and urokinase, to peripheral blood cells. *Thromb. Haemost.* **58**, 936–942.

Miyake, H., Hara, I., Yamanaka, K., Gohji, K., Arakawa, S., and Kamidono, S. (1999). Elevation of serum levels of urokinase-type plasminogen activator and its receptor is associated with disease progression and prognosis in patients with prostate cancer. *Prostate* **39**, 123–129.

Montuori, N., Rossi, G., and Ragno, P. (1999). Cleavage of urokinase receptor regulates its interaction with integrins in thyroid cells. *FEBS Lett.* **460**, 32–36.

Mustjoki, S., Tapiovaara, H., Sirén, V., and Vaheri, A. (1998). Interferons and retinoids enhance and dexamethasone suppresses urokinase-mediated plasminogen activation in promyelocytic leukemia cells. *Leukemia* **12**, 164–174.

Mustjoki, S., Alitalo, R., Stephens, R.W., and Vaheri, A. (1999a). Blast cell-surface and plasma soluble urokinase receptor in acute leukemia patients: Relationship to classification and response to therapy. *Thromb. Haemost.* **81**, 705–710.

Mustjoki, S., Alitalo, R., Stephens, R.W., and Vaheri, A. (1999b). Plasminogen activation in human leukemia and in normal hematopoietic cells. *APMIS* **107**, 144–149.

Mustjoki, S., Sidenius, N., Sier, C.F.M., Blasi, F., Elonen, E., Alitalo, R., and Vaheri, A. (2000a). Soluble urokinase receptor levels correlate with number of circulating tumor cells in acute myeloid leukemia and decrease rapidly during chemotherapy. *Cancer Res.* **60**, 7126–7132.

Mustjoki, S., Sidenius, N., and Vaheri, A. (2000b). Enhanced release of soluble urokinase receptor by endothelial cells in contact with peripheral blood cells. *FEBS Lett.* **486**, 237–242.

Myöhänen, H., Stephens, R.W., Hedman, K., Tapiovaara, H., Rønne, E., Høyer-Hansen, G., Danø, K., and Vaheri, A. (1993). Distribution and lateral mobility of the urokinase-receptor complex at the cell surface. *J. Histochem. Cytochem.* **41**, 1291–1301.

Nakasaki, T., Wada, H., Mori, Y., Okugawa, Y., Watanabe, R., Nishikawa, M., Gabazza, E.C., Masuya, M., Kageyama, S., Kumeda, K., Kato, H., and Shiku, H. (2000). Decreased tissue factor and tissue-plasminogen activator antigen in relapsed acute promyelocytic leukemia. *Am. J. Hematol.* **64**, 145–150.

Nykjaer, A., Møller, B., Todd III, R.F., Christensen, T., Andreasen, P.A., Gliemann, J., and Petersen, C.M. (1994). Urokinase receptor. An activation antigen in human T lymphocytes. *J. Immunol.* **152**, 505–516.

Ogawa, M. (1993). Differentiation and proliferation of hematopoietic stem cells. *Blood* **81**, 2844–2853.

Pappot, H., Høyer-Hansen, G., Rønne, E., Hoi-Hansen, H., Brunner, N., Danø, K., and Grøndahl-Hansen, J. (1997). Elevated levels of receptor for urokinase plasminogen activator in patients with non-small cell lung cancer. *Eur. J. Cancer* **33**, 867–872.

Pedersen, N., Schmitt, M., Rønne, E., Nicoletti, M., Høyer-Hansen, G., Conese, M., Giavazzi, R., Danø, K., Kuhn, W., Jänicke, F., and Blasi, F. (1993). A ligand-free, soluble urokinase receptor is present in the ascitic fluid from patients with ovarian cancer. *J. Clin. Invest.* **92**, 2160–2167.

Plesner, T., Rafkier, E., Wittrup, M., Johnsen, H., Pyke, C., Pedersen, T.L., Hansen, N.E., and Danø, K. (1994). Expression of the receptor for urokinase-type plasminogen activator in normal and neoplastic blood cells and hematopoietic tissue. *Am. J. Clin. Pathol.* **102**, 835–841.

Plesner, T., Behrendt, N., and Ploug, M. (1997). Structure, function and expression on blood and bone marrow cells of the urokinase-type plasminogen activator receptor, uPAR. *Stem Cells* **15**, 398–408.

Ploug, M., Rønne, E., Behrendt, N., Jensen, A.L., Blasi, F., and Danø, K. (1991). Cellular receptor for urokinase plasminogen activator. Carboxylterminal processing and membrane anchoring by glycosyl-phosphatidylinositol. *J. Biol. Chem.* **266**, 1026–1033.

Ploug, M., Eriksen, J., Plesner, T., Hansen, N.E., and Danø, K. (1992). A soluble form of the glycolipid-anchored receptor for urokinase-type plasminogen activator is secreted from peripheral blood leukocytes from patients with paroxysmal nocturnal hemoglobinuria. *Eur. J. Biochem.* **208**, 397–404.

Ploug, M., Ellis, V., and Danø, K. (1994). Ligand interaction between urokinase-type plasminogen activator and its receptor probed with 8-anilino-1-naphtalenesulfonate. Evidence for a hydrophobic binding site exposed only on the intact receptor. *Biochemistry* **33**, 8991–8997.

Preissner, K.T., Kanse, S.M., and May, A.E. (2000). Urokinase receptor: A molecular organizer in cellular communication. *Curr. Opin. Cell Biol.* **12**, 621–628.

Reiter, L.S., Spertini, O., and Kruithof, E. (1997). Plasminogen activators play an essential role in extracellular-matrix invasion by lymphoblastic T cells. *Int. J. Cancer* **70**, 461–466.

Resnati, M., Guttinger, M., Valcamonica, S., Sidenius, N., Blasi, F., and Fazioli, F. (1996). Proteolytic cleavage of the urokinase receptor substitutes for the agonist-induced chemotactic effect. *EMBO J.* **15**, 1572–1582.

Riisbro, R., Stephens, R.W., Brunner, N., Christensen, I.J., Nielsen, H.J., Heilman, L., and von Templehoff, G.-F. (2001). Soluble urokinase plasminogen activator receptor in preoperatively obtained plasma from patients with gynecological cancer or benign gynecological diseases. *Gynecol. Oncol.* **82**, 523–531.

Riittinen, L., Limongi, P., Crippa, M.P., Conese, M., Hernandez-Marrero, L., Fazioli, F., and Blasi, F. (1996). Removal of domain D2 or D3 of the human urokinase receptor does not affect ligand affinity. *FEBS Lett.* **381**, 1–6.

Robbie, L., Berry, S., Moir, E., Booth, N.A., Culligan, D., Tighe, J., Watson, H., King, D., and Bennett, B. (2000). Myeloid leukaemic cells can lyse fibrin directly. *Br. J. Haem.* **111**, 524–529.

Rønne, E., Pappot, H., Grøndahl-Hansen, J., Høyer-Hansen, G., Plesner, T., Hansen, N.E., and Danø, K. (1995). The receptor for urokinase plasminogen activator is present in plasma from healthy donors and elevated in patients with paroxysmal nocturnal haemoglobinuria. *Br. J. Haematol.* **89**, 576–581.

Sakata, Y., Murakami, T., Noro, A., Mori, K., and Matsuda, M. (1991). The specific activity of plasminogen activator inhibitor-1 in disseminated intravascular coagulation with acute promyelocytic leukemia. *Blood* **77**, 1949–1957.

Scherrer, A., Wohlwend, A., Kruithof, E.K.O., Vassalli, J.-D., and Sappino, A.-P. (1999). Plasminogen activation in human acute leukaemias. *Br. J. Haematol.* **105**, 920–927.

Schmidt, M. and Hoppe, F. (1999). Increased levels of urokinase receptor in plasma of head and neck squamous cell carcinoma patients. *Acta Otolaryngol.* **119**, 949–953.

Sidenius, N. and Blasi, F. (2000). Domain 1 of the urokinase receptor (uPAR) is required for uPAR-mediated cell binding to vitronectin. *FEBS Lett.* **470**, 40–46.

Sidenius, N., Sier, C.F.M., and Blasi, F. (2000a). Shedding and cleavage of the urokinase receptor (uPAR): Identification and characterisation of uPAR fragments in vitro and in vivo. *FEBS Lett.* **475**, 52–56.

Sidenius, N., Sier, C.F.M., Ullum, H., Pedersen, B.K., Lepri, A.C., Blasi, F., and Eugen-Olsen, J. (2000b). Serum level of soluble urokinase-type plasminogen activator receptor is a strong and independent predictor of survival in human immunodeficiency virus infection. *Blood* **96**, 4091–4095.

Sier, C.F.M., Stephens, R.W., Bizik, J., Mariani, A., Bassan, M., Pedersen, N., Frigerio, L., Ferrari, A., Danø, K., Brünner, N., and Blasi, F. (1998). The level of urokinase type plasminogen activator receptor is increased in serum of ovarian cancer patients. *Cancer Res.* **58**, 1843–1849.

Sier, C.F.M., Sidenius, N., Mariani, A., Aletti, G., Agape, V., Ferrari, A., Casetta, G., Stephens, R.W., Brünner, N., and Blasi, F. (1999). Presence of urokinase-type plasminogen activator receptor in urine of cancer patients and its possible clinical relevance. *Lab. Invest.* **79**, 717–722.

Solberg, H., Rømer, J., Brünner, N., Holm, A., Sidenius, N., Danø, K., and Høyer-Hansen, G. (1994). A cleaved form of the receptor for urokinase-type plasminogen activator in invasive transplanted human and murine tumors. *Int. J. Cancer* **58**, 877–881.

Stephens, R., Alitalo, R., Tapiovaara, H., and Vaheri, A. (1988). Production of an active urokinase by leukemia cells: A novel distinction from cell lines of solid tumors. *Leukemia Res.* **12**, 419–422.

Stephens, R.W., Tapiovaara, H., Alitalo, R., Reisberg, T., Lymboussakis, A., Myöhänen, H., and Vaheri, A. (1992). Cell-surface plasminogen activation and leukemia. *Fibrinolysis* **6**, 35–39.

Stephens, R.W., Pedersen, A., Nielsen, H.J., Hamers, M., Høyer-Hansen, G., Rønne, E., Dybkjaer, E., Danø, K., and Brünner, N. (1997). ELISA determination of soluble urokinase receptor in blood from healthy donors and cancer patients. *Clin. Chem.* **43**, 1868–1876.

Stephens, R.W., Brünner, N., Jänicke, F., and Scmitt, M. (1998). The urokinase plasminogen activator system as a target for prognostic studies in breast cancer. *Breast Cancer Res. Tr.* **52**, 99–111.

Stephens, R.W., Nielsen, H.J., Christensen, I.J., Thorlacius-Ussing, O., Sørensen, S., Danø, K., and Brünner, N. (1999). Plasma urokinase receptor levels in patients with colorectal cancer: relationship to prognosis. *J. Natl. Cancer I.* **91**, 869–874.

Stoppelli, M.P., Corti, A., Sofficientini, A., Cassani, G., Blasi, F., and Assoian, R. (1985). Differentiation-enhanced binding of the amino-terminal fragment of human urokinase plasminogen activator to a specific receptor on U937 monocytes. *Proc. Natl. Acad. Sci. USA* **82**, 4939–4943.

Tapiovaara, H., Vaheri, A., Alitalo, R., Pheng-Siew, S., Woodrow, G., and Stephens, R.W. (1991). Enhanced activation of prourokinase by human RC2A leukemia cells. *Semin. Thromb. Hemostas.* **17**, 251–261.

Tapiovaara, H., Alitalo, R., Stephens, R.W., Myöhänen, H., Ruutu, T., and Vaheri, A. (1993). Abundant urokinase activity on the surface of mononuclear cells from blood and bone marrow of acute leukemia patients. *Blood* **82**, 914–919.

Tapiovaara, H., Matikainen, S., Hurme, M., and Vaheri, A. (1994). Induction of differentiation of promyelocytic NB4 cells by retinoic acid is associated with rapid increase in urokinase activity subsequently downregulated by production of inhibitors. *Blood* **83**, 1883–1891.

Tapiovaara, H., Alitalo, R., and Vaheri, A. (1996). Plasminogen acvtivation on tumor cell surface and its involvement in human leukemia. *Adv. Cancer Res.* **69**, 101–133.

Vassalli, J.-D., Baccino, D., and Belin, D. (1985). A cellular binding site for the M_r 55,000 form of the human plasminogen activator, urokinase. *J. Cell Biol.* **100**, 86–92.

Wada, H., Kumeda, Y., Ogasawara, Z., Ohiwa, M., Toshihiro, K., Tamaki, S., Ohno, T., Kageyama, S., Kobayashi, T., Deguchi, K., and Shirakawa, S. (1993). Plasminogen activators and their inhibitors in leukemic cell homogenates. *Am. J. Hematol.* **42**, 166–170.

Wahlberg, K., Hoyer-Hansen, G., and Casslen, B. (1998). Soluble receptor for urokinase plasminogen activator in both full-length and a cleaved form is present in high concentration in cystic fluid from ovarian cancer. *Cancer Res.* **58**, 3294–3298.

Waltz, D.A., Fujita, R.M., Yang, X., Natkin, L., Zhuo, S., Gerard, C.J., Rosenberg, S., and Chapman, H.A. (2000). Nonproteolytic role for the urokinase receptor in cellular migration in vivo. *Am. J. Respir. Cell Mol. Biol.* **22**, 316–322.

Warrell, R.P.J., Frankel, S.R., Miller, W.H., Scheinberg, D.A., Itri, L.M., Hittelmann, W.N., Vyas, R., Andreeff, M., Tafuri, A., Jakubowski, A., Gabrilove, J., Gordon, M.S., and Dmitrovsky, E. (1991). Differentiation therapy of acute promyelocytic leukemia with tretinoin (all-trans-retinoic acid). *N. Engl. J. Med.* **324**, 1385–1393.

Wei, Y., Waltz, D.A., Rao, N., Drummond, R.J., Rosenberg, S., and Chapman, H.A. (1994). Identification of the urokinase receptor as an adhesion receptor for vitronectin. *J. Biol. Chem.* **269**, 32380–32388.

Wilhelm, O.G., Wilhelm, S., Escott, G.M., Lutz, V., Magdolen, V., Schmitt, M., Rifkin, D.B., Wilson, E.L., Graeff, H., and Brunner, G. (1999). Cellular glycosylphosphatidylinositol-specific phospholipase D regulates urokinase receptor shedding and cell surface expression. *J. Cell Physiol.* **180**, 225–235.

Wilson, L. and Francis, G.E. (1987). Differentiation-linked secretion of urokinase and tissue plasminogen activator by normal human hemopoietic cells. *J. Exp. Med.* **165**, 1609–1623.

Wilson, E.L., Jacobs, P., and Dowdle, E.B. (1983). The secretion of plasminogen activators by human myeloid leukemic cells in vitro. *Blood* **61**, 568–574.

Wilson, E.L., Jacobs, P., Francis, G.E., Oliver, L., Burger, P., and Dowdle, E.B. (1992). Secretion of plasminogen activators by normal bone marrow cells and leukaemic myeloid cells. *Fibrinolysis* **6**, 77–79.

Woodhead, V.E., Stonehouse, T.J., Binks, M.H., Speidel, K., Fox, D.A., Gaya, A., Hardie, D., Henniker, A.J., Horejsi, V., Sagawa, K., Skubitz, K.M., Taskov, H., Todd III, R.F., van Agthoven, A., Katz, D.R., and Chain, B.M. (2000). Novel molecular mechanisms of dendritic cell-induced T cell activation. *Int. Immunol.* **12**, 1051–1061.

14

The Role of Plasminogen in Bone Remodeling

Evis Daci, Elizabeth H. Allan, and T. John Martin

The processes of bone formation and resorption are continuous throughout life and are necessary for the maintenance of a normal functional skeleton. Bone formation by cells of the osteoblast lineage results from the synthesis of matrix followed by mineralization, while resorption by osteoclasts involves matrix degradation and dissolution of the mineral phase. The two processes are tightly coupled. The activity of both osteoblasts and osteoclasts can be stimulated by a number of bone-resorbing hormones and local factors, although the formation and activity of osteoclasts *in vitro* requires the presence of cells of the osteoblast lineage (Suda *et al.*, 1992). This and other evidence led to the discovery of a system of control of osteoclast formation and activity by members of the tumor necrosis factor (TNF) ligand and receptor families. Receptor activator of nuclear factor kappa B (RANK) ligand (RANKL) is produced by cells of the osteoblastic lineage under the influence of several hormones and cytokines, and acts upon its receptor, RANK, in hemopoietic cells to promote osteoclast formation and maintain their activity. Osteoprotegerin (OPG), a secreted member of the TNF receptor family, is a decoy receptor for RANKL/RANK, produced by the same cells that produce RANKL, and acts as a powerful inhibitor of osteoclast formation and activity (Suda *et al.*, 1999).

It is likely that certain proteinases and their inhibitors, produced locally and regulated by osteotropic hormones and local factors, are involved in the processes of tissue degradation, cell motility, mitogenesis, and growth factor activation, which are essential to the processes of bone formation and resorption. The plasminogen activator (PA)/plasmin pathway is one such proteolytic pathway that is tightly regulated in bone, and has been extensively investigated.

1. Osteoblasts and Bone Formation

Bone is a specialized connective tissue, consisting of a complex mixture of cells integrated in a highly organized extracellular matrix, which is mostly mineralized.

Evis Daci • Onderwijs en Navorsing, Gasthuisberg, 3000 Leuven, Belgium **Elizabeth H. Allan and T. John Martin** • St Vincent's Institute of Medical Research, 9 Princes Street, Fitzroy Vic 3065, Australia

Plasminogen: Structure, Activation, and Regulation, edited by David M. Waisman. Kluwer Academic/Plenum Publishers, 2003.

Osteoblasts are the bone lining cells responsible for the production of the mineralized bone matrix. They originate from the local mesenchymal stem cells, which undergo proliferation and differentiation into mature osteoblasts. As proliferation ceases, the relatively undifferentiated preosteoblasts become progressively more cuboidal and acquire an osteoblastic phenotype as they start to produce an initially unmineralized extracellular matrix, the major component of which is type I collagen (MacDonald and Gowen, 1993), with the cells maximally expressing the alkaline phosphatase gene. As the unmineralized osteoid layer becomes thicker, crystals of mineral are deposited at discrete sites in the collagenous matrix. This onset of mineralization is associated with increased expression of markers of the mature osteoblast, such as osteocalcin, an osteoblast-specific marker, and bone sialoprotein (BSP). The mineral crystals become progressively larger as bone matures, due to the addition of ions and/or aggregation of the crystals (Boskey, 1999).

Osteoblasts are derived from the mesenchymal stem cells, which also give rise to chondrocytes and other mesenchymal cells (adipocytes, myoblasts, and fibroblasts). Osteoblast differentiation and function of the mature osteoblasts are activated transcriptionally by Runx 2 (formerly known as Cbfa1), a member of the runt family of transcription factors (Karsenty, 1999). Runx 2 was identified as the factor binding to an osteoblast-specific *cis*-acting element present in the osteocalcin promoter (Ducy and Karsenty, 1995; Ducy *et al.*, 1997), which binds to and regulates the expression of multiple genes expressed in osteoblasts. Ectopic expression of Runx 2 in non-osteoblastic cells induces the expression in these cells of osteoblast-specific genes such as osteocalcin (Ducy *et al.*, 1997). Deletion of Runx 2 in mice leads to mutant animals in which the skeleton is made of chondrocytes that produces a typical cartilaginous matrix, but is devoid of osteoblasts and lacks ossification (Komori *et al.*, 1997; Otto *et al.*, 1997). The skeletal phenotype of mice with a heterozygously mutated Runx 2 locus resembles that of cleidocranial dysplasia, a heritable human disorder characterized by hypoplastic clavicles, open fontanelles, supernumerary teeth and short stature (Otto *et al.*, 1997). Moreover, mutations of the Runx 2 locus are identified in patients with cleidocranial dysplasia (Mundlos *et al.*, 1997). Studies in genetically manipulated mice indicate that Runx 2 not only controls osteoblast differentiation during embryonic development, but also acts as a transcriptional activator of genes required for the maintenance of the differentiated osteoblast (Ducy *et al.*, 1999). An interesting new control mechanism has emerged with the discovery of a transcription factor, osterix (osx), which appears to act downstream of Runx 2 in the osteoblast differentiation pathway. Mice rendered null for the osterix gene also are born with an entirely cartilaginous skeleton, the subtle difference from Runx 2−/− mice being that the osx−/− mice are able to generate osteoclasts (Nakashima *et al.*, 2002).

As mineralization proceeds during skeletal development, some of the osteoblasts become trapped behind the advancing mineralization front, becoming connected to each other and to the surface cells by a complex system of canaliculae. These osteocytes retain some of the characteristics of osteoblasts, such as the ability to produce osteocalcin. They have numerous and long cell processes rich in microfilaments, which through gap junctions communicate with cell processes from other osteocytes or from the surface osteoblasts (Burger and Klein-Nulend, 1999). This network of thin canaliculi is believed to facilitate transmission of information between osteoblasts on the bone surface and the osteocytes within the structure of bone itself. Finally, the third major phenotypic expression of the "mature" osteoblast lineage is represented by the bone lining cells. After completing their tasks of synthesizing the bone matrix components, many of the osteoblasts lose their synthetic capacity, gradually becoming flattened, and are seen lining trabecular and endosteal surfaces.

The two PAs, urokinase-type PA (uPA), and tissue-type PA (tPA), are produced by osteoblasts, as is the specific inhibitor, PA inhibitor-1 (PAI-1), and a cellular receptor for uPA, uPAR. These components of the pathway are differentially regulated by various osteotropic hormones and local factors, and this will be discussed.

2. Osteoclasts and Bone Resorption

Osteoclasts are multinucleated giant cells, formed by fusion of mononuclear precursors and they are highly specialized to resorb mineralized matrices. Their formation is under the control of M-CSF and RANKL (Suda *et al.*, 1999; *vide supra*). Osteoclasts are usually found in contact with calcified bone surfaces or within Howship's lacunae, which are the result of their own resorptive activity (Baron, 1999). Their unique properties include a ruffled border, comprised of a series of fine finger-like cytoplasmic projections of the plasma membrane adjacent to bone, and the sealing zone, which is the site of attachment of osteoclasts to the bone surface (Roodman, 1996).

Ultrastructural studies have shown that the resorbing osteoclasts are highly polarized cells (Vaananen *et al.*, 1996). Resorption of mineralized bone matrix takes place in the specialized extracellular compartment under the ruffled border and involves the release of both acid and proteolytic enzymes. Acid is produced by carbonic anhydrase-II, an enzyme highly expressed in the osteoclasts, which catalyzes the hydration of CO_2 to H_2CO_3. A vacuolar-type ATP-ase pumps the protons through the ruffled border into the extracellular compartment, where they create the acid environment necessary for the dissolution of the mineral phase. Inhibition of the carbonic anhydrase-II or the vacuolar-type ATP-ase by specific inhibitors or antisense oligonucleotides markedly decreases bone resorption both *in vivo* and *in vitro* (Vaananen, 1996).

After the mineral phase is dissolved, proteinases degrade the organic matrix. Several groups of enzymes are involved in this proteolytic process, such as cysteine proteinases, matrix metalloproteinases (MMPs) and possibly also neutral serine proteinases. Osteoclasts express MMP-9 (Tezuka *et al.*, 1994) and MT1-MMP (Sato *et al.*, 1997), as well as a number of cysteine proteinases including cathepsins B, D, K, L (Goto *et al.*, 1994; Inaoka *et al.*, 1995). In addition, Yang *et al.* (1997), used RT-PCR with microdissected mouse osteoclasts to show the expression of mRNAs for tPA, uPA, PAI-1, PAI-2, protease nexin (PN), and urokinase receptor isoform 1 (uPAR1). The expression of uPAR mRNA in osteoclasts was confirmed by *in situ* hybridization with a uPAR1 probe. Whereas RNA encoding the uPAR isoform 2 was not detected in mouse osteoclasts, a novel unspliced uPAR RNA variant was detected in these cells. The novel uPAR variant and uPAR1 RNA were also detected in mouse calvarial osteoblasts, kidney, muscle, and the mouse macrophage cell line J774A.1 by RT-PCR. The presence of RNAs for most of the components of the PA system in osteoclasts suggests that it may have a functional role in these cells. When osteoclasts and osteoblasts were cultured together in fibrin overlay experiments the plasminogen dependence of fibrinolytic activity was clearly seen (Grills *et al.*, 1990). In the case of the osteoclasts, fibrin clearing was noted only on one side of the cell, whereas fibrinolysis was equally distributed around the perimeter of the osteoblasts. This difference reflects the motility behaviour of osteoclasts and could imply involvement of the PA pathway in the process. In another study osteoclasts in coculture with osteoblasts were observed to move underneath osteoblasts and maintain clear spaces around them. Inhibitors of plasmin inhibited this response (Gail I. Anderson, unpublished observations).

3. Bone Remodeling

Bone structure and integrity are maintained throughout life by bone remodeling, a continuous process of bone resorption and deposition, which requires coordination of the activities of osteoclasts and osteoblasts. Bone remodeling occurs in focal and discrete packets in both trabecular and cortical bone. It serves two main functions: (a) maintenance of mechanical strength, by replacing fatigued bone with new bone, and (b) mineral homeostasis, by providing access to skeletal stores of calcium and phosphorus (Dempster, 1999). While the turnover rate of cortical bone (2–3% per year) is consistent with maintenance of mechanical properties, the turnover rate of trabecular bone is much higher, suggesting its involvement in mineral metabolism. Moreover, the different turnover rates suggest that activation of the sequence of cellular events responsible for remodeling is probably locally controlled. The bone cells on the trabecular bone surface are in intimate contact with cells of the marrow cavity, which produce a variety of potent osteotropic cytokines and are, therefore, more prone to be influenced, compared to the cells in cortical bone (Mundy, 1999). Despite these differences, the sequence of events in each remodeling cycle is the same: activation of osteoclast precursors resulting in osteoclastic bone resorption, followed by osteoblastic bone formation to repair the defect (Figure 14.1).

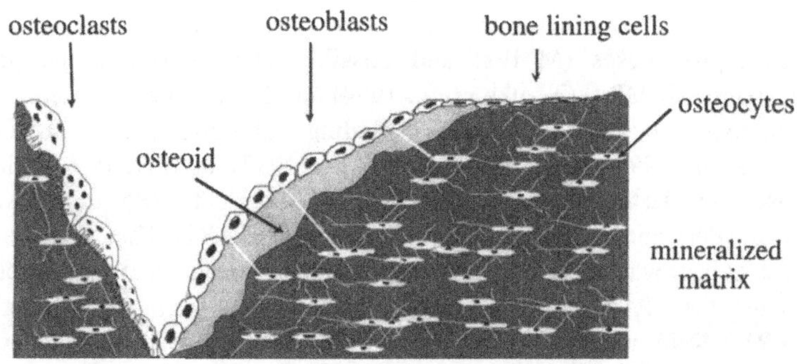

Figure 14.1. Bone remodeling. The bone remodeling sequence involves the activation of osteoclast precursors resulting in osteoclastic bone resorption, followed by osteoblastic bone formation to repair the defect (from Evis Daci, Sophie van Cromphaut and Roger Bouillon, Mechanisms influencing bone metabolism in chronic illness, *Hormone Research*, 2002, 58 (Suppl. 1): 44–51, reprinted with permission of S. Karger AG, Medical and Scientific Publishers, Basel).

In order to maintain bone mass, the processes of bone resorption and formation must be tightly coupled: Packets of bone removed during resorption must be replaced during formation. Any imbalance between these processes will result in bone disease, as is the case for postmenopausal osteoporosis, in which bone resorption exceeds bone formation resulting in bone loss. Local cytokine control mechanisms are so prominent in this process that there are a number of possible mechanisms responsible for mediating the (un)coupling process. In a recent morphological study of this process, Everts *et al.* (2002), using a number of *in vitro* models to examine Howship's lacunae, identified the bone lining cells cytologically at sites of resorption, and showed that these cells actually engulf the collagen fragments on the bone surface after the osteoclasts have resorbed. They showed also that a central role in this process was played by MMPs—most likely matrix-bound, and perhaps activated by plasmin generated through the action of PAs. Participation of the PA system in local bone remodeling events such as this would be consistent with its roles in other tissues.

4. Regulation of the PA/Plasmin System in Bone Cells

4.1. Peptide Hormones and Cyclic AMP

The bone-resorbing agents parathyroid hormone (PTH) and prostaglandin E_2 (PGE_2) have been shown to stimulate PA activity in normal and malignant rat osteoblast cultures via a cAMP-dependent mechanism, in that the effect of each was enhanced when administered with a phosphodiesterase inhibitor, and the same response was obtained with forskolin treatment (Hamilton *et al.*, 1984, 1985; Allan *et al.*, 1986). The increased activity was identified as tPA by fibrin autography, with detectable uPA activity in culture medium after treatment with PTH (Fukumoto *et al.*, 1992a). Hoekman *et al.* (1991), using a spectrophotometric enzyme assay for PA, also found both tPA and uPA were increased following PTH treatment of normal and malignant osteoblasts. Therefore, it appears likely that the osteoblasts synthesize and secrete single chain uPA (scuPA), which requires proteolytic activation. Upon secretion of scuPA from the cell it can bind to its cellular receptor and be activated. The PTH-induced increase in PA activity in both normal and malignant osteoblasts was found to be closely paralleled by a decrease in mRNA and protein for PAI-1, also a cAMP-dependent process (Fukumoto *et al.*, 1992a). It is likely that hormonal regulation of PA activity by PTH is achieved predominantly by changes in the synthesis and secretion of PAI-1 rather than by a direct modulation of uPA or tPA mRNA and protein, since the latter are relatively small effects. PGE_2 appears to exert its stimulatory effect on PA activity in primary osteoblast-like cells largely through modulation of tPA and uPA (Allan *et al.*, 1995). mRNA for tPA, uPA, and uPAR are all enhanced following PGE_2 treatment. In these cells steady state PAI-1 mRNA is low, so PAI-1 modulation may not contribute significantly to the overall level of activity, but when PAI-1 mRNA levels are stimulated by treatment of the cells with transforming growth factor beta (TGFβ), simultaneous treatment with PGE_2 inhibits the increase due to TGFβ alone. In bone, where osteoblasts are regulated by numerous hormones, cytokines, and growth factors, PGE_2 inhibition of PAI-1 may contribute significantly to PA activity.

4.2. Glucocorticoids and $1,25(OH)_2$ Vitamin D_3

The glucocorticoids dexamethasone, corticosterone, and hydroxycortisone suppressed both basal and $1,25(OH)_2$ vitamin D_3 ($1,25(OH)_2D_3$)-stimulated PA activity in normal and malignant osteoblasts (Hamilton *et al.*, 1984, 1985), while progesterone and estradiol were ineffective. The dexamethasone effect was associated with an increase in PAI-1 mRNA levels but no change in uPA or tPA mRNA levels (Fukumoto *et al.*, 1992b). PAI-1 protein levels were also elevated, and therefore dexamethasone would appear to exert its inhibitory effect on PA activity solely by increasing PAI-1 mRNA and protein levels. Dexamethasone, as well as two other steroids, prednisolone and deflazacort, which also increase PAI-1 protein (Martin *et al.*, 1993) and mRNA, all inhibited PN-1 mRNA (Allan and Martin, unpublished observations).

$1,25(OH)_2$ D_3 was found to stimulate PA activity in normal and malignant osteoblasts by a cAMP-independent mechanism. Fukumoto *et al.* (1994) demonstrated that $1,25(OH)_2D_3$, like PTH, stimulated PA activity largely through inhibiting PAI-1 mRNA and protein, with only small changes in both uPA and tPA mRNA. The similarity of the PTH and $1,25(OH)_2D_3$ effects despite their different signaling systems in osteoblastic cells, is reflective of their similar effects on the expression of RANKL and OPG in similar cells (Suda *et al.*, 1999).

4.3. Cytokines and Growth Factors

Various growth factors and cytokines modulate one or more components of the PA/plasmin system. Their effects are likely to be of importance in bone since many are synthesized and secreted by osteoblasts which may also have receptors for them. TGFβ, an important and abundant bone protein, decreased PA activity in normal and malignant osteoblasts (Pfielschifter *et al.*, 1990; Allan *et al.*, 1991) and stimulated PAI-1 mRNA levels and protein in the same cells (Allan *et al.*, 1991). It was surprising that although TGFβ inhibited PA activity, uPA and uPAR steady-state mRNA levels were increased, but there was no effect on tPA mRNA levels (Allan *et al.*, 1991; Allan and Martin, unpublished observations). Furthermore, although PA activity in conditioned medium and cell lysates from TGFβ-treated osteoblasts was inhibited, fibrin zymography revealed a time-dependent stimulation of uPA activity, particularly in the cell extracts, an effect possibly due to uPA binding to its cell membrane receptor. Similar responses to those of TGFβ were shown by leukaemia inhibitory factor (LIF), which is synthesized and secreted by osteoblasts, which also have receptors for it. LIF was found to inhibit PA activity in both the normal and malignant osteoblasts, accompanied by an increase in PAI-1 protein and a transient increase in PAI-1 mRNA (Allan *et al.*, 1990). Other bone cytokines and growth factors that have been reported to stimulate PA activity in osteoblast cultures are epidermal growth factor (Hamilton *et al.*, 1984, 1985; Cheng *et al.*, 1991; Laschinger *et al.*, 1991; Pfeilschifter *et al.*, 1992), acidic and basic fibroblast growth factors, and platelet-derived growth factor (Evans *et al.*, 1990; Cheng *et al.*, 1991; Hoekman *et al.*, 1991; Laschinger *et al.*, 1991; Fawthrop *et al.*, 1992; De Bart *et al.*, 1995), thrombin (Allan and Martin, 1996), while there are conflicting reports on whether the agents interleukin-1 (IL-1) and TNFα modulate PA activity (Cheng *et al.*, 1991; Hoekman *et al.*, 1991).

5. Potential Roles of the PA/Plasmin System in Bone

The participation of the PA/plasmin system in tissue turnover and remodeling in so many sites in development, as well as in physiological and pathological processes, together with the tight regulation of the components of the system demonstrated in bone cells, have led to a number of suggested roles in bone biology (Verheijen, 1997), but direct evidence for an important role is limited.

5.1. Bone Resorption

Bone remodeling involves several processes such as tissue degradation and renewal, proenzyme and growth factor activation, and cell motility. Studies in other systems suggest that the PA/plasmin pathway can be involved in such processes in a plasmin-dependent or -independent way. The former includes effects related to the proteolytic activity of plasmin to directly degrade several noncollagenous components of the extracellular matrix, or to activate proMMPs and participate in collagen breakdown (Figure 14.2). In agreement with this hypothesis, uPA-generated plasmin participates in the degradation of a bone-like extracellular matrix by human osteosarcoma cells or uPA-transfected mouse fibroblasts (De Bart *et al.*, 1995; Ronday *et al.*, 1997).

Degradation by collagenase activity of the osteoid layer, universally present on bone surfaces except in areas of active bone resorption (Raina, 1972), is suggested to be important for the initiation of the resorption process, because it allows the osteoclasts to contact bone mineral and to become activated (Chambers and Fuller, 1985; Chambers *et al.*, 1985; Vaananen, 1996a), and there may be plasmin dependence of this process. Although Leloup *et al.*, (1994) did not find evidence for plasmin involvement in the resorption process in explanted fetal metatarsal bones, that organ culture system might not be the most appropriate in which to test the possibility. Further strong support for the involvement of collagenase activity in the resorption process comes from the observations that bone-resorbing agents induce collagenase synthesis in osteoblasts both *in vitro* and *in vivo* (Heath *et al.*, 1984; Meikle *et al.*, 1992; Zhao *et al.*, 1999). Most strikingly though, this hypothesis is

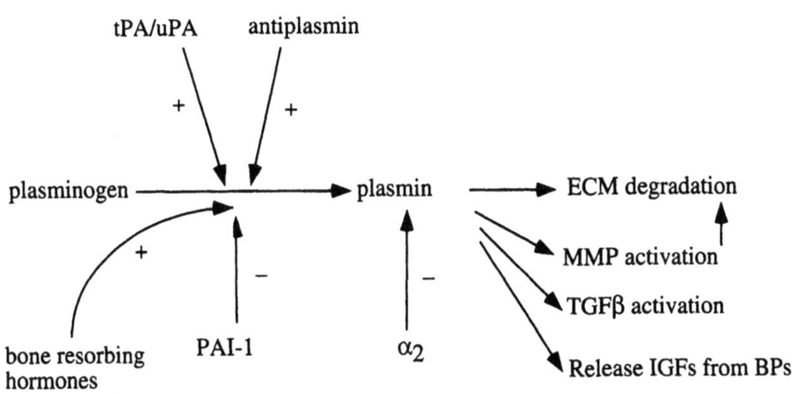

Figure 14.2. The plasminogen activator/plasmin pathway and its potential involvement in bone metabolism. ECM = extracellular matrix, BPs = binding proteins.

consistent with the observations made in transgenic mice whose type I collagen is rendered resistant to collagenase cleavage (Zhao *et al.*, 1999). The calvaria of these mice are extremely thick and have almost no marrow cavity. Moreover, the collagenase-resistant mice show a markedly blunted response to the bone-resorptive effects of PTH (Zhao *et al.*, 1999). In addition, removal of the organic matrix remaining in the resorption lacuna after mineral solubilization, is a component of the ongoing resorption process (Vaananen, 1996). As indicated earlier, the recent study of Everts *et al.* (2002) is informative of this process, in that they showed in a morphological study that bone lining cells are central to the process of digestion and degradation of collagen in Howship's lacunae after osteoclastic resorption. The process requires participation of MMP's, most likely MT-MMP, and may be plasmin-dependent.

5.2. Studies in Genetically Manipulated Mice

The nature of the PA/plasmin system, with its several contributing components, control by, hormones and cytokines, and participation in localized tissue processes make it likely that resolution of its roles in many tissues, especially in bone, will require studies in genetically manipulated mice, or in cells derived from such mice. Interesting information is beginning to be derived from such an approach, first in studying several aspects of bone resorption, including osteoclast formation, mineral dissolution, and degradation of the organic matrix (Daci *et al.*, 1999). These different processes were assayed *in vitro* using cells derived from mice with either tPA (tPA−/−), uPA (uPA−/−), PAI-1 (PAI-1−/−) inactivation, or with a combined inactivation (tPA−/−:uPA−/−) and compared with wild-type (WT) mice (Carmeliet *et al.*, 1993a, 1994). First, osteoclast formation, assessed by investigating the number and characteristics of tartrate-resistant acid phosphatase-positive multinucleated cells formed in cocultures of primary osteoblasts and bone marrow cells treated with 1,25-$(OH)_2D_3$, was no different among the cell types obtained from different knockout animals (Daci *et al.*, 1999). Second, dentine resorption, an assay for osteoclast activity, was not affected by the combined deficiency of both tPA and uPA, an observation that still does not exclude participation of plasmin in the digestion of the thin collagen layer on the bone surface, because that would not be present in the dentine slices. Of most interest, however, was the finding that osteoblastic cells from tPA−/−:uPA−/− mice or from uPA−/− mice had significantly reduced ability to degrade nonmineralized bone-like matrix, as compared to WT cells (Figure 14.3A) (Daci *et al.*, unpublished observations). In contrast, matrix degradation by PAI-1−/− primary osteoblasts was significantly increased (Figure 14.3B). The participation of the plasminogen activators in matrix degradation was plasminogen-dependent, did not require active MMPs, and resulted in proteolytic degradation of the noncollagenous components of the extracellular bone matrix (Daci *et al.*, unpublished observations), indicating that the process involves the direct proteolysis of the extracellular matrix components by plasmin.

These data suggest that the PA/plasmin system is not required for osteoclast formation, and perhaps not for the resorption of the mineral phase, but is involved in the removal of noncollagenous proteins present in the nonmineralized bone matrix. This might fit very well with the mechanism proposed by Everts *et al.* (2002) for "cleaning" collagen fragments out of bone resorption pits, a process requiring MT-MMP participation. Plasmin has a broad proteolytic activity and is able to cleave extracellular matrix molecules. The "classical" plasmin substrate is fibrin, but, in the context of extracellular matrix

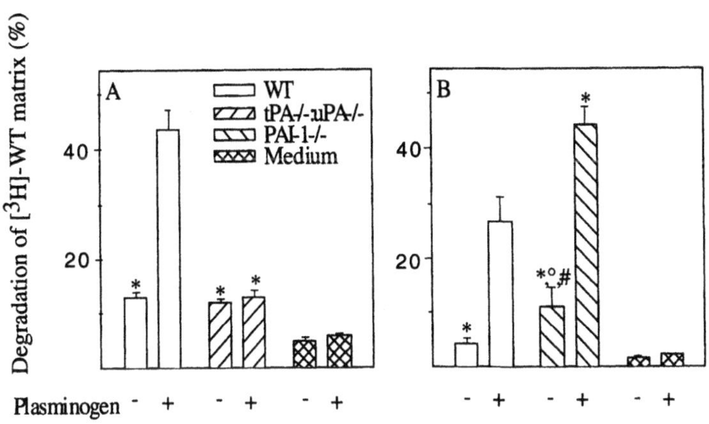

Figure 14.3. Plasmin(ogen)-dependent degradation of nonmineralized bone-like matrix by primary osteoblasts cultured on matrix from WT cells. Matrix degradation was calculated as the percentage of radioactivity released from [^{3}H]-WT matrix over the total amount of radioactivity. *p<0.05 vs. WT with plasminogen; °p<0.05 vs. PAI-1 with plasminogen; #p<0.05 vs. WT without plasminogen.

proteolysis, it is clear that several components of the extracellular matrix such as fibronectin, laminin, and the protein core of proteoglycans, can be degraded by this enzyme (Mignatti and Rifkin, 1993; Clark and Murphy, 1999). Of particular relevance to bone may be the ability of plasmin to cleave osteocalcin and release it from hydroxyapatite (Novak *et al.*, 1997). In addition, uPA-generated plasmin is a likely pathological activator of several zymogen MMPs (Carmeliet *et al.*, 1997a), which play an important role in extracellular matrix degradation.

The second study of the bone phenotype of genetically manipulated mice is that of PAI-1 −/− mice in a condition of increased bone turnover and bone loss due to estrogen deficiency (Daci *et al.*, 2000). This was undertaken since PAI-1 inactivation has little effect on other systems in basal conditions, but results in pathology after challenge (Carmeliet *et al.*, 1993b, 1997b). A link between PAI-1 and estrogen deficiency has been suggested by epidemiological studies showing increased plasma levels of PAI-1 in postmenopausal women, and their down regulation following hormone replacement therapy (Meilahn *et al.*, 1996, Nasr *et al.*, 1998). How this is reflected in bone is not known, since PAI-1 activity in bone has not been measured. Estrogen deficiency is the most important factor in the pathogenesis of postmenopausal bone loss, which is characterized by increased osteoclastic bone resorption and impaired compensatory bone formation. In a study of bone metabolism in mice, which used pQCT histomorphometric analysis to assess bone growth and remodeling, these processes were found to proceed normally in mice with inactivation of the *PAI-1* gene, indicating that the role of PAI-1 in basal bone metabolism was not critical. Nevertheless, in estrogen deficiency, PAI-1 deficiency protects from the increased bone turnover and trabecular bone loss induced by estrogen deficiency, without affecting cortical bone loss, an effect which might result either from a proteolytic or a non-proteolytic action of PAI-1. For example, PAI-1 can bind to vitronectin and block the binding of the integrin $\alpha_v\beta_3$ to its ligands. This interaction, which has been shown to inhibit the migration of smooth muscle cells (Stefansson and Lawrence, 1996), may also affect the behaviour of bone cells. These observations suggest a specific role for PAI-1 in the regulation of trabecular bone turnover during estrogen deficiency, and one which requires further investigation.

5.3. Bone Formation

Activation of growth factors by plasmin could also provide a crucial role in bone formation for the PA/plasmin pathway in osteoblasts, as has been proposed for TGFβ activation in other models (Sato *et al.*, 1990; Laschinger *et al.*, 1991; Flaumenhaft *et al.*, 1992). Latent TGFβ, produced by osteoblasts, has been shown to be activated by plasmin (Allan *et al.*, 1991); therefore, if the activation of latent TGFβ by plasmin were a physiological event, activation could be regulated in a targeted manner, by osteotropic factors and local factors, which influence the PA/plasmin pathway in bone. We have extended the early observation in finding that treatment of cultures of a rat osteogenic sarcoma cell line with PTH and plasminogen resulted in a 6-fold increase in the concentration of active TGFβ and that the increase was plasminogen-dependent (Yee *et al.*, 1993). Neither agent alone had any effect. Furthermore, in a wound healing assay it was found that PTH and plasminogen together led to an increase in the migration of neonatal mouse calvarial osteoblasts into the denuded area, and that the increase could be prevented by the addition of anti-TGFβ antibody. None of the agents alone had any effect on the migration of osteoblasts. In these experiments PA activity would be elevated as a result of decreased synthesis of PAI-1 (Fukumoto *et al.*, 1992a), and would therefore be available to activate plasminogen, either in the medium or bound to cell surface receptors. The plasmin would in turn activate latent TGFβ secreted by the osteoblasts, or in the case of bone *in vivo*, released from resorbed matrix. Once activated, TGFβ could induce the synthesis of PAI-1, which would inhibit PA activity, and further TGFβ activation would cease. Indeed the data presented in Allan *et al.* (1995) supports this hypothesis. In this way TGFβ could control its own activation from the precursor, protein-bound form. An analogous function of locally produced PA production has been proposed by Campbell *et al.* (1992) since it has been found that plasmin, generated from both tPA and uPA activity in human osteosarcoma cells, can dissociate insulin-like growth factor-1 (IGF-1) from the binding proteins, which render it inactive. The IGF-1 thus generated is then available to exert its biological functions. It has also been shown in bone cell and organ culture that PTH treatment promotes IGF-1 production, and this has been suggested as a factor contributing to the anabolic effect of PTH (Canalis *et al.*,1989). Consistent with such a possibility, Mijakoski *et al.* (2001) found that in mice rendered null for the *IGF-1* gene, treatment with PTH had no effect on bone mineral density or any bone formation parameters by histomorphometry. WT mice on the other hand exhibited the expected substantial response to PTH.

Any resolution of a place for the PA/plasmin pathway in anabolic responses in bone through growth factor activation, will require similar studies in mice with the appropriate genes ablated.

References

Allan, E.H. and Martin, T.J. (1995). Prostaglandin E_2 regulates production of plasminogen activator isoenzymes, urokinase recptor, and plasminogen activator inhibitor-1 in primary cultures of rat calvarial osteoblasts. *J. Cell. Physiol.* **165**, 521–529.

Allan, E.H. and Martin, T.J. (1996). Receptor-mediated effect of thrombin on plasminogen activator inhibitor-1 synthesis in rat osteoblast-like cells. *Fibrinolysis* **10**, 285–293.

Allan, E.H. and Martin, T.J. (1996). Prostaglandin E_2 regulates production of plasminogen activator isozyures, urokinase receptor, calvarial osteoblasts. *J. Cell. Physiol.* **165**, 521–529.

Allan, E.H., Hamilton J.A., Medcalf, R.L., Kubota, M., and Martin T.J. (1986). Cyclic-AMP-dependent and -independent effects on tissue-type plasminogen activator in osteogenic sarcoma cells; evidence from phosphodiesterase inhibition and parathyroid hormone antagonists. *Biochim. Biophys. Acta* **888**, 199–207.

Allan, E.H., Hilton, D.J., Brown, M.A., Evely, R.S., Yumita, S., Medcalf, D., Gough, N.M., Ng, K.W., Nicola, N.A., and Martin, T.J. (1990). Osteoblast display receptors for and responses to leukemia-inhibitory factor. *J. Cell. Physiol.* **145**, 110–119.

Allan, E.H., Zeheb, R., Gelehrter, T.D., Heaton, J.H., Fukumoto, S., Yee, J.A., and Martin, T.J. (1991). Transforming growth factor beta inhibits plasminogen activator (PA) activity and stimulates production of urokinase-type PA, PA inhibitor-1 mRNA, and protein in rat osteoblast-like cells. *J. Cell. Physiol.* **149**, 34–43.

Baron, R. (1999). Anatomy and ultrastructure of bone. In M.J. Favus (ed.) *Primer on the Metabolic Bone Diseases and Disorders of Mineral Metabolism.* Lippincott Williams & Wilkins, Philadelphia, PA, pp. 3–10.

Boskey, A.L. (1999). Mineralization, structure and function of bone. In M.J. Seibel, S.P. Robins, and J.P. Bilezikian (eds) *Dynamics of Bone and Cartilage Metabolism.* Academic Press, San Diego, CA, pp. 153–164.

Burger, E.H. and Klein-Nulend, J. (1999). Mechanotransduction in bone—role of the lacuno-canalicular network. *FASEB J.* **13** Suppl, S101–112.

Campbell, P.G., Novak, J.F., Yanosick, T.B., and McMaster, J.H. (1992). Involvement of the plasmin system in dissociation of the insulin-like growth factor-binding protein complex. *Endocrinology* **130**, 1401–1412.

Canalis, E., Centrella, M., Burck, W., and McCarthy, T.L. (1989). Insulin like growth factor 1 mediates selective anabolic effects of parathyroid hormone in bone cultures. *J. Clin. Invest.* **83**, 60–65.

Carmeliet, P., Kieckens, L., Schoonjans, L., Ream, B., Vannuffelen, A., Prendergast, G., Cole, M., Bronson, R., Collen, D., and Mulligan, R.C. (1993a). Plasminogen activator inhibitor-1 gene-deficient mice. I. Generation by homologous recombination and characterization. *J. Clin. Invest.* **92**, 2746–2755.

Carmeliet, P., Stassen, J.M., Schoonjans, L., Ream, B., Vandenoord, J.J., Demol, M., Mulligan, R.C., and Collen, D. (1993b). Plasminogen activator inhibitor-1 gene-deficient mice. II. Effects on hemostasis, thrombosis, and thrombolysis. *J. Clin. Invest.* **92**, 2756–2760.

Carmeliet, P., Schoonjans, L., Kieckens, L., Ream, B., Degen, J., Bronson, R., De Vos R., Vandenoord, J.J., Collen, D., and Mulligan, R.C. (1994). Physiological consequences of loss of plasminogen activator gene function in mice. *Nature* **368**, 419–424.

Carmeliet, P., Moons, L., Lijnen, R., Baes, M., Lemaitre, V., Tipping, P., Drew, A., Eeckhout, Y., Shapiro, S., Lupu, F., and Collen, D. (1997a). Urokinase-generated plasmin activates matrix metalloproteinases during aneurysm formation. *Nat. Genet.* **17**, 439–444.

Carmeliet, P., Moons, L., Lijnen, R., Janssens, S., Lupu, F., Collen, D., and Gerard, R.D. (1997b). Inhibitory role of plasminogen activator inhibitor 1 in arterial wound healing and neointima formation. *Circulation* **96**, 3180–3191.

Chambers, T.J. and Fuller, K. (1985). Bone cells predispose endosteal surfaces to resorption by exposure of bone mineral to osteoclast contact. *J. Cell Sci.* **261**, 9972–9978.

Chambers, T.J., Darby, J.A., and Fuller, K. (1985). Mammalian collagenase predisposes bone surfaces to osteoclastic resorption. *Cell Tiss. Res.* **241**, 671–675.

Cheng, S.-L., Shen, V., and Peck, W.A. (1991). Regulation of plasminogen activator and plasminogen activator inhibitor production by growth factors and cytokines in rat calvarial cells. *Calcif. Tiss. Int.* **49**, 321–327.

Clark I.M. and Murphy, G. (1999). Regulation of plasminogen activator and plasminogen activator inhibition production by growth factors and cytokines in rat calvarial cells. *Calcif. Tiss. Int.* **49**, 321–327.

Daci, E., Udagawa, N., Martin, T.J., Bouillon, R., and Carmeliet, G. (1999). The role of the plasminogen system in bone resorption *in vitro*. *J. Bone Min. Res.* **14**, 946–952.

Daci, E., Verstuyf, A., Moermans, K., Bouillon R., and Carmeliet, G. (2000). Mice lacking the plasminogen activator inhibitor 1 are protected from trabecular bone loss induced by estrogen deficiency. *J. Bone Min. Res.* **15**, 1510–1516.

De Bart, A.C.W., Quax, P.H.A., Löwik, C.W.G.M., and Verheijen, J.H. (1995). Regulation of plasminogen activation, matrix metalloproteinases and urokinase-type plasminogen activator-mediated extracellular matrix degradation in human osteosarcoma cell line MG63 by interleukin-1 alpha. *J. Bone Min. Res.* **10**, 1374–1384.

Dempster, D.W. (1999). New concepts in bone remodeling. In M.J. Seibel, S.P. Robins, and J.P. Bilezikian (eds) *Dynamics of Bone and Cartilage Metabolism*. Academic Press, San Diego, CA, pp. 261–273.

Ducy, P. and Karsenty, G. (1995). Two distinct osteoblast-specific *cis*-acting elements control expression of a mouse osteocalcin gene. *Mol. Cell. Biol.* **15**, 1858–1869.

Ducy, P., Zhang, R., Geoffroy, V., Ridall, A.L., and Karsenty, G. (1997). Osf2/Cbfa1: A transcriptional activator of osteoblast differentiation. *Cell* **89**, 747–754.

Ducy, P., Starbuck, M., Priemel, M., Shen, J., Pinero, G., Geoffroy, V., Amling, M., and Karsenty, G. (1999). A Cbfa1-dependent genetic pathway controls bone formation beyond embryonic development. *Genes Dev.* **13**, 1025–1036.

Evans, D.B., Bunning, R.A.D., and Russell, R.G.G. (1990). The effects of recombinant human interleukin-1β on cellular proliferation and production of prostaglandin E_2, plasminogen activator, osteocalcin and alkaline phosphatase by osteoblast-like cells derived from human bone. *Biochem. Biophys. Res. Commun.* **66**, 208–216.

Everts, V., Delaisse, J.M., Korper., W., Jansen, D.G., Tigchebar-Gutter, W., Saftig, P., and Beertsen W. (2002). The bone living cell: Its role in cleaning Howship's lacunae and initiating bone formation. *J. Bone Min. Res.* **17**, 77–90.

Fawthrop, F.W., Oyajobi, B.O., Bunning, R.A.D., and Russell, R.G.C. (1992). The effect of transforming growth factor β on the plasminogen activator activity in normal-human osteoblast-like cells and a human osteosarcoma cell line MG-63. *J. Bone Min. Res.* **7**, 1363–1371.

Flaumenhaft, R., Abe, M., Mignatti, P., and Rifkin, D.B. (1992). Basic fibroblast growth factor-induced activation of latent transforming growth factor β in endothelial cells: Regulation of plasminogen activator activity. *J. Cell Biol.* **118**, 901–909.

Fukumoto, S., Allan, E.H., Yee, J.A., Gelehrter, T.D., and Martin, T.J. (1992a). Plasminogen activator regulation in osteoblasts: Parathyroid inhibition of type-1 plasminogen activator inhibitor and its mRNA. *J. Cell. Physiol.* **152**, 346–355.

Fukumoto, S., Allan, E.H., Zeheb, R., Gelehrter, T.D., and Martin, T.J. (1992b). Glucocorticoid regulation of plasminogen activator inhibitor-1 messenger ribonucleic acid and protein in normal and malignant rat osteoblasts. *Endocrinology* **130**, 797–804.

Fukumoto, S., Allan, E.H., and Martin, T.J. (1994). Regulation of plasminogen activator inhibitor-1 (PAI-1) expression by 1,25-dihydroxyvitamin D_3 in normal and malignant rat osteoblasts. *Biochim. Biophys. Acta* **1201**, 223–228.

Goto, T., Kiyoshima, T., Moroi, R., Tsukuba, T., Nishimura, Y., Himeno, M., Yamamoto, K., and Tanaka, T. (1994). Localization of cathepsins B, D, and L in the rat osteoclast by immuno-light and -electron microscopy. *Histochemistry* **101**, 33–40.

Grills, B.L., Gallagher, J.A., Allan, E.H., Yumita, S., and Martin, T.J. (1990). Identification of plasminogen activator in osteoclasts. *J. Bone Min. Res.* **5**, 499–505.

Hamilton, J.A., Lingelbach, S.R., Partridge, N.C., and Martin, T.J. (1984). Stimulation of plasminogen activator release in osteoblast-like cells by bone resorbing-hormones. *Biochem. Biophys. Res. Commun.* **122**, 230–236.

Hamilton, J.A., Lingelbach, S.R., Partridge, N.C., and Martin, T.J. (1985). Regulation of plasminogen activator production by bone-resorbing hormones in normal and malignant osteoblasts. *Endocrinology* **116**, 2186–2191.

Heath, J.K., Atkinson, S.J., Meikle, M.C., and Reynolds, J.J. (1984). Mouse osteoblasts synthesize collagenase in response to bone resorbing agents. *Biochim. Biophys. Acta* **802**, 151–154.

Hoekman, K., Löwik, C.W.G.M., Ruit, M., Bijuoet, O.L.M., Verheijen, J.H., and Papadopoulous, S.E. (1991). Regulation of production of plasminogen activators by bone resorption enhancing and inhibiting factors in three types of osteoblast-like cells. *Bone Min.* **14**, 189–204.

Inaoka, T., Bilbe, G., Ishibashi, O., Tezuka, K., Kumegawa, M., and Kokubo, T. (1995). Molecular cloning of human cDNA for cathepsin K: Novel cysteine proteinase predominantly expressed in bone. *Biochem. Biophys. Res. Commun.* **206**, 89–96.

Karsenty, G. (1999). The genetic transformation of bone biology. *Genes Dev.* **13**, 3037–3051.

Komori, T., Yagi, H., Nomura, S., Yamaguchi, A., Sasaki, K., Deguchi, K., Shimizu, Y., Bronson, R.T., Gao, Y.H., Inada, M., Sato, M., Okamoto, R., Kitamura, Y., Yoshiki, S., and Kishimoto, T. (1997). Targeted disruption of Cbfa1 results in a complete lack of bone formation owing to maturational arrest of osteoblasts. *Cell* **89**, 755–764.

Laschinger, C.A., Bellows, C.G., and Wasi, S. (1991). Modulation of plasminogen activators and plasminogen activator inhibitors by TGFβ, IL-1α and EGF in foetal rat calvarial cells at different times of culture. *Bone Min.* **13**, 23–34.

Leloup, G., Delaisse, J.M., and Vaes, G. (1994). Relationship of the plasminogen activator/plasmin cascade to osteoclast invasion and mineral resorption in explanted fetal metatarsal bones. *J. Bone. Min. Res.* **9**, 891–902.

MacDonald, B.R. and Gowen, M. (1993). The cell biology of bone. *Baillieres Clin. Rheumatol.* **7**, 421–443.

Martin, T.J., Allan, E.H., and Heath, J.H. (1993). Cellular actions of glucocorticoids: Studies in bone cells. In G. Christiansen and B. Rus (eds) *Fourth International Symposium on Osteoporosis.* Handelstrykkerief, Aalborg, Aps., Denmark, pp. 509–511.

Meikle, M.C., Bord, S., Hembry, R.M., Compston, J., Croucher, P.I., and Reynolds, J.J. (1992). Human osteoblasts in culture synthesize collagenase and other matrix metalloproteinases in response to osteotropic hormones and cytokines. *J. Cell Sci.* **103**, 1093–1099.

Meilahn, E.N., Cauley, J.A., Tracy, R.P., Macy, E.O., Gutai, J.P., and Kuller, L.H. (1996). Association of sex hormones and adiposity with plasma levels of fibrinogen and PAI-1 in postmenopausal women. *Am. J. Epidemiol.* **143**, 159–166.

Mignatti, P. and Rifkin, D.B. (1993). Biology and biochemistry of proteinases in tumor invasion. *Physiol. Rev.* **73**, 161–195.

Miyakoshi, N., Kasukawa, Y., Linkhart, T.A., Baylink, D.J., and Mohan, S. (2001). Evidence that anabolic effects of PTH on bone require IGF-1 in growing mice. *Endocrinology* **142**, 4349–4356.

Mundlos, S., Otto, F., Mundlos C., Mulliken, J.B., Aylsworth, A.S., Albright, S., Lindhout, D., Cole, W.G., Henn, W., Knoll, J.H., Owen, M.J., Mertelsmann, R., Zabel, B.U., and Olsen, B.R. (1997). Mutations involving the transcription factor CBFA1 cause cleidocranial dysplasia. *Cell* **89**, 773–779.

Mundy, G.R. (1999). Bone remodelling. In M.J. Focus (ed.) *Primers on the Metabolic Bone Diseases and Disorders of Mineral Metabolism.* Lippinott Williams & Wilkins, Philadephia, PA, pp. 30–38.

Nakashima, K., Zhou, X., Kunkel, G., Zhang, Z., Deng, J.M., Behringer, R.R., and de Crombrugghe, B. (2002). The novel zinc finger-containing transcription factor oxterix is required for osteoblast differentiation and bone formation. *Cell* **108**, 17–29.

Nasr, A. and Breckwoldt, M. (1998). Estrogen replacement therapy and cardiovascular protection: Lipid mechanisms are the tip of an iceberg. *Gynecol. Endocrinol.* **12**, 43–59.

Novak, J.F., Hayes, J.D., and Nishimoto, S.K. (1997) Plasmin-mediated proteolysis of osteocalcin. *J. Bone Min. Res.* **12**, 1035–1042.

Otto, F., Thornell, A.P., Crompton, T., Denzel, A., Gilmour, K.C., Rosewell, I.R., Stamp, G.W., Beddington, R.S., Mundlos, S., Olsen, B.R., Selby, P.B., and Owen, M.J. (1997). Cbfa1, a candidate gene for cleidocranial dysplasia syndrome, is essential for osteoblast differentiation and bone development. *Cell* **89**, 765–771.

Pfeilschifter, J., Erdmann, J., Schmidt, W., Naumann, A., Minne, H.W., and Ziegler, R. (1990). Differential regulation of plasminogen activator and plasminogen activator inhibitor by osteotropic factors in primary cultures of mature osteoblasts and osteoblast precursors. *Endocrinology* **126**, 703–711.

Pfeilschifter, J., Krempien, R., Nauman, A., Gronwald, R.G.K., Hoppa, J., and Zeiglar, R. (1992). Differential effects of platelet-derived growth factor isoforms on plasminogen activator activity in foetal rat osteoblasts due to isoform-specific receptor functions. *Endocrinology* **130**, 2059–2066.

Raina, V. (1972). Normal osteoid tissue. *J. Clin. Pathol.* **25**, 229–232.

Ronday, H.K., Smits, H.H., Quax, P.H.A., Van Der Pluijm, G., Löwik, C.W.G.M., Breedveld, F.C., and Verheijen, J.H. (1997). Bone matrix degradation by the plasminogen activation system. Possible mechanism of bone destruction in arthritis. *Br. J. Rheumatol.* **36**, 9–15.

Roodman, G.D. (1996). Advances in bone biology: The osteoclast. *Endocrinol. Rev.* **17**, 308–332.

Sato, T., del Carmen, Ovejero, M., Hou, P., Heegaard, A.M., Kumegawa, M., Foged, N.T., and Delaisse, J.M. (1997). Identification of the membrane-type matrix metalloproteinase MT1-MMP in osteoclasts. *J. Cell Sci.* **110** (Pt 5), 589–596.

Sato, Y., Tsuboi, R., Lyons, R., Moses, H., and Rifkin, D.B. (1990). Characterization of the activation of latent transforming growth factor β by co-cultures of endothelial cells and pericytes or smooth muscle cells: A self regulating system. *J. Cell Biol.* **111**, 757–763.

Stefansson, S. and Lawrance, D.A. (1996). The serpin PAI-1 inhibits cell migration by blocking intergin $\alpha_v\beta_3$ binding to vitronectin. *Nature* **383**, 441–443.

Suda, T., Takahashi, N., and Martin, T.J. (1992). Modulation of osteoclast differentiation. *Endocrinol. Rev.* **13**, 66–80.

Suda, T., Takahashi, N., Udagawa, N., Jimi, E., Gillespie, M.T., and Martin, T.J. (1999). Modulation of osteoclast differentiation and function by the new members of the tumor necrosis receptor and ligand families. *Endocrinol. Rev.* **20**, 345–357.

Tezuka, K., Nemoto, K., Tezuka, Y., Sato, T., Ikeda, Y., Kobori, M., Kawashima, H., Eguchi, H., Hakeda, Y., and Kumegawa, M. (1994). Identification of matrix metalloproteinase 9 in rabbit osteoclasts. *J. Biol. Chem.* **269**, 15006–15009.

Vaananen, H.K. (1996). Osteoclast function: Biology and mechanisms. In J.P. Bilezikian, L.G. Raisz, and G.A. Rodan (eds) *Principles of Bone Biology*. Academic Press, San Diego, CA, pp. 103–113.

Vaananen, H.K., Salo, J., and Lehenkari, P. (1996). Mechanism of osteoclast-mediated bone resorption. *J. Bone Min. Metab.* **14**, 187–192.

Verheijen, J.H. (1997). Bone matrix degradation by the plasminogen activation system. Possible mechanism of bone destruction in arthritis. *Br. J. Rheumatol.* **36**, 9–15.

Yang, J.N., Allan, E.H., Anderson, G.I., Martin, T.J., and Minkin, C. (1997). Plasminogen activator system in osteoclasts. *J. Bone Min. Res.* **12**, 761–768.

Yee, J.A., Yan, L., Dominguez, J.C., Allan, E.H., and Martin, T.J. (1993). Plasminogen-dependent activation of latent transforming growth factor beta (TGFβ) by growing cultures of osteoblast-like cells. *J. Cell. Physiol.* **157**, 528–534.

Zhao, W., Byrne, M.H., Boyce, B.F., and Krane, S.M. (1999). Bone resorption induced by parathyroid hormone is strikingly diminished in collagenase-resistant mutant mice. *J. Clin. Invest.* **103**, 517–524.

15

Plasminogen Activators in CNS Physiology and Disease

Tambet Teesalu, Andres Kulla, Toomas Asser, Aadu Simisker, and Antti Vaheri

1. Introduction

Extracellular proteolysis represents a potent and irreversible mechanism of modulating cell–cell and cell–matrix interactions and remodeling structural components of tissues. Understanding of the role of extracellular proteolysis has evolved during the last decade from a somewhat simplistic concept of proteases being important for breaking down/remodeling mechanical tissue barriers of the extracellular matrix (ECM) into a more complex view of proteases affecting indirectly many aspects of cell physiology, from cell division and differentiation to programmed cell death (Basbaum and Werb, 1996; Werb, 1997). Not surprisingly, proteases and their regulatory molecules are involved in a wide range of physiological processes of tissue growth and remodeling, as well as in pathological conditions such as tumorigenesis and tissue-destructive diseases (Werb *et al.*, 1999; Andreasen *et al.*, 2000).

Plasminogen activators are extracellular serine proteases that catalyze the conversion of the proteolytically inactive zymogen, plasminogen, into the broad-spectrum serine protease plasmin (de Vries *et al.*, 1996). There are two molecularly and functionally distinct plasminogen activators, tissue-type plasminogen activator (tPA) and urokinase-type plasminogen activator (uPA) (Lijnen and Collen, 1988). Due to ubiquitous supply of plasminogen and inhibitors, the availability of plasminogen activators is the key limiting factor of the plasminogen cascade. Expression of plasminogen activators is controlled mostly at the transcriptional level, although some evidence exists also for posttranscriptional control mechanisms. Both plasminogen activators can be extracellularly cleaved into two-chain disulfide-bridged forms, for uPA this cleavage is required for proteolytic activity

Tambet Teesalu, Aadu Simisker, and Antti Vaheri • Haartman Institute, University of Helsinki, POB 21, FIN-00014 Helsinki, Finland.
Tambet Teesalu and Andres Kulla • Department of Pathology and Neuropathology, Tartu University Clinics, Estonia.
Tambet Teesalu, Toomas Asser, and Aadu Simisker • Department of Neurology, University of Tartu, Estonia.

Plasminogen: Structure, Activation, and Regulation, edited by David M. Waisman. Kluwer Academic / Plenum Publishers, New York, 2003.

(Conese and Blasi, 1995). The activity of tPA is strongly enhanced by its interaction with fibrin, and the primary role of tPA is considered to be participation in the fibrin clot surveillance (Stubbs *et al.*, 1998). uPA interacts with its high-affinity cell-surface receptor and focalizes plasmin activity pericellularily to allow cell-directed matrix remodeling and cell migration (Blasi, 1999). Following secretion, the metabolism and tissue localization of PAs are regulated by a whole set of functionally related molecules: inhibitors (plasminogen activator inhibitor type-1 and 2, neuroserpin, α_2-macroglobulin), receptors, and co-receptors (Andreasen *et al.*, 2000). Vascular endothelium is the main source of endogenous tPA in the adult. In addition tPA is synthesized and released by cells not involved in intravascular fibrinolysis, such as neurons and glial cells of the CNS. During the last decade, an intense research effort has focused on the PA system in the context of the central nervous system: neurogenesis, neurophysiology and diverse nervous system pathologies, such as stroke, epilepsy, Alzheimer's disease and CNS tumors (Seeds *et al.*, 1992; Strickland, 2001).

Rapid progress in deciphering the function of plasminogen activators in the CNS has been made possible by methodological advances such as (a) generation of gene-deficient and transgenic mice with changes in PA expression levels and/or localizations in the central nervous system, (b) development of improved animal models of neurological disease, (c) establishment of improved cell/organ culture systems, and (d) availability of new specific pharmacological inhibitors of proteases.

This review will summarize the current knowledge on the role of plasminogen activation system in the developing and adult CNS as well as in several neuropathologies.

2. Expression and Function of Plasminogen Activators in the Normal CNS

2.1. Plasminogen Activators in Developing and Adult CNS

Plasminogen Activators in Neurogenesis

In the early 1980s, advances in understanding of biochemistry and regulation of proteolytic enzymes made researchers to consider these molecules as potential players in the neural development. A variety of cultured neuronal and glial cells from the peripheral and central nervous system were found to release serine proteinases (Moonen *et al.*, 1982; Rogister *et al.*, 1988). In neurons, the plasminogen activators were found to be secreted from distal processes and growth cones, consistent with concept of proteases mediating neurite outgrowth, pathfinding, target recognition, and synaptogenesis (Neuman *et al.*, 1989; Pittman and Buettner, 1989; Pittman *et al.*, 1989)

Development of the mouse nervous system begins with the formation of the neural plate around embryonic day 8 (E8). The neural plate forms the neural tube, which eventually forms the mature CNS. This process of development of the nervous system is associated with extensive structural and cellular changes occurring within a short time window. *In situ* hybridization analysis was used to map the expression of plasminogen activators through murine prenatal development (Sumi *et al.*, 1992; Friedman and Seeds, 1994). In early developing CNS (E10.5–E13.5) tPA is expressed by migrating neuroepithelial cells in the ventricular zone of basal mesencephalon, metencephalon, and myelencephalon.

In addition, tPA transcripts are highly expressed in tegmentum and basal rhombencephalon. From E15, a strong expression of tPA transcripts appeared in the regions that develop into pons and medulla, and in the forming cerebellum. Also olfactory neurons, but not support cells, scored positive for the expression of tPA transcripts. Expression of uPA in the developing CNS appears to be widespread as well. Most identifiable neurons appear to express uPA mRNA from an early stage in their development, and this expression appears to coincide with the onset of axogenesis and persists to later stages.

In neonatal rodent brain, prominent neuronal migration takes place in the developing cerebellum where large numbers of newly formed granule neurons migrate inwards from the surface germinative zone. These germinative neurons turn on the expression of tPA gene as they leave the external granule layer (Krystosek and Seeds, 1981; Friedman and Seeds, 1995). The tPA expression of granule neurons peaks on postnatal day 7 (P7) and expression in this cell population persists, although at a lower level, in adulthood. Zymographic analysis of the PA activity on the cerebellar extracts demonstrated the presence of active tPA in a pattern compatible with the mRNA data.

Mapping of the embryonic expression has supported the dominant hypothesis that PAs facilitate neuronal migration and neuronal outgrowth, probably by degrading cell–matrix adhesions and matrix molecules.

Extracellular proteolysis catalyzed by plasminogen activators is not confined to ontogeny, but contributes also to adult nervous system physiology. In the adult CNS, tPA is the major plasminogen activator (Sappino *et al.*, 1993; Davies *et al.*, 1998), although uPA is also expressed (Masos and Miskin, 1996). tPA transcripts are found in neurons in most brain regions, with high levels of enzymatically active tPA found in discrete areas, in particular in the hippocampus and hypothalamus (Sappino *et al.*, 1993). The reason for such uncoupling of tPA catalytic activity and mRNA distribution patterns might be due to post-transcriptional control or differential expression of tPA regulatory molecules, such as its neuronally derived inhibitor neuroserpin (Krueger *et al.*, 1997).

Recently, we have performed a systematic characterization of tPA expression in the human CNS using a set of tissue arrays representing 36 brain/spinal cord localizations (Teesalu *et al.*, manuscript in preparation). Widespread and strong neuronal expression of tPA transcripts and antigen was detected as early as 1 month postnatally, and it persisted through adulthood. The cells expressing the highest levels of tPA transcripts were the neurons of the entire neocortical mantle, thalamus, amygdala, and hippocampal pyramidal neurons, whereas white matter, rich in bundles of myelinated axons, remained unlabeled. Strong tPA expression was also detected in the spinal cord neurons with particularly prominent expression in ventral horn motoneurons, as well as in large neurons of dorsal root ganglia. Whereas in all individual cases tPA was detected in neurons, in some brains also micro- and astroglial expression of tPA was detected. Glial tPA expression appeared to be associated with gliosis due to chronic disease or history of alcohol abuse. Also, tPA antigen was present in a subpopulation of endothelial cells. We also immunostained CNS arrays with antibodies to the major substrate of tPA, plasminogen. Plasminogen/plasmin immunoreactivity was present in neurons in a number of locations, and tPA and plasminogen/plasmin double immunostaining indicated that the two proteins co-localize in the vesicular compartment in the neurons. These experiments document the widespread presence of tPA stored in the neurons of the human brain. In these cells the role of tPA under physiological conditions is probably to participate in the modulation of synaptic microremodeling and neuronal plasticity. The presence of pre-synthesized tPA in the human CNS could mean that under pathological conditions, this protease could be massively released into the extracellular space, thus being one of the primary effectors that lead to neuronal sensitization and apoptosis.

2.2. Role of Plasminogen Activators in Neuronal Migration, Plasticity, and Learning

The functional significance of precisely coordinated expression of plasminogen activators in developing and adult CNS has been assessed by many complementary approaches, most importantly using gene-inactivated mice (Seeds *et al.*, 1996; Strickland, 2001). The initial finding of no obvious neuroanatomic and/or neurological phenotypes in tPA and uPA null mutant mice seriously undermined the concept of these proteases as important players in neurogenesis and physiology. However, in subsequent years a number of more subtle but insightful CNS phenotypes have been described, especially for tPA-null mice (Table 15.1).

Table 15.1. Neurophysiological Phenotypes Resulting from Genetic Modulation of PA Expression in the CNS

Genotype	Processes analysed	Phenotype(s)	Reference
tPA−/−	L-LTP	Glutamergic homosynaptic L-LTP replaced by heterosynaptic L-LTP involving also GABAergic transmission	Frey *et al.* (1996)
tPA−/−	L-LTP/learning	Reduction of L-LTP in both Schaffer collateral and the mossy fiber pathways Impaired two way avoidance test	Huang *et al.* (1996)
tPA−/−	L-LTP/learning	Reduction of L-LTP in striatal neurons Impaired performance in context conditioning and two-way avoidance tests	Calabresi *et al.* (2000)
tPA overexpression	LTP/learning	Increased and prolonged LTP Improved performance in spatial orientation learning tasks	Madani *et al.* (1999)
tPA−/−	Hippocampal mossy fiber outgrowth	Decreased mossy fiber outgrowth, aberrant processing of DSD-1-PG/phosphacan	Wu *et al.* (2000)
uPA overexpression	Learning	Impaired spatial, olfactory and taste aversion learning	Meiri *et al.* (1994)
tPA−/−	Granule neuron migration	Retarded rate of granule neuron migration in postnatal cerebellum	Seeds *et al.* (1999)
tPA−/−	Reverse occlusion plasticity in the visual cortex	Inhibited recovery of cortical function and thalamic neuron size during reverse occlusion	Muller and Griesinger (1998)

2.2.1. Neuronal Migration

tPA is actively secreted from cultured granule neurons of the early postnatal mouse cerebellum and PA and plasmin inhibitors substantially inhibited migration of external granular layer neurons in organotypic culture (Moonen *et al.*, 1982; Seeds *et al.*, 1992). Recently, a fine description of cerebellar neuroanatomy of tPA-null mice was performed (Seeds *et al.*, 1999). During the most active phase of granule cell migration, there were about twice more migratory granule cells in transit in the molecular layer in tPA-null mice than in wild-type counterparts. Also, the migration speed of granule cells in cerebellar slice cultures was reduced by about half in tPA null mice. These data suggested that tPA is an important, although not indispensable mediator of granule cell migration. These data also served as a good demonstration of the concept that the absence of an end-point phenotype in mutant mice does not mean that the gene product is not involved in the process of interest.

2.2.2. Synaptic Plasticity

It is well established that the learning process involves structural remodeling of synaptic connections (Ruegg, 2001). Alterations in the structure of synaptic connections take place in a variety of memory processes ranging in complexity from those produced by elementary forms of non-associative learning in invertebrates to those produced by higher-order associative tasks in higher vertebrates. It is likely that the mechanisms of structural changes accompanying long-term memory are the same as those responsible for de novo synapse formation and synapse pruning during development (Bailey and Kandel, 1993). tPA expression in the adult mouse brain in structures associated with memory and learning (hippocampus and cerebellum) served as a first evidence suggesting that tPA may be involved in synaptic plasticity and microremodeling accompanying learning processes (Sappino *et al.*, 1993). Furthermore, systematic study of serine protease mRNAs in the rodent hippocampus identified tPA as the most abundant (Davies *et al.*, 1998). The role of tPA (and uPA) has been tested in different paradigms of learning and remodeling of synaptic connections (long-term potentiation, motor learning, monocular deprivation).

tPA is induced as an immediate-early gene accompanying long-term potentiation (LTP) (Qian *et al.*, 1993), a form of synaptic plasticity that is thought to be involved in the storage of long-term memory for explicit forms of learning. tPA expression is required for LTP as was demonstrated by finding that tPA-null mice exhibit a different form of late-LTP (L-LTP) characterized by an *N*-methyl-D-aspartate (NMDA) receptor-dependent modification of GABA transmission in the CA1 region of the hippocampus (Frey *et al.*, 1996). tPA-null mice completely lack homosynaptic L-LTP seen in wild-type animals and have instead heterosynaptic form of L-LTP, which appears to be functionally equivalent to normal L-LTP. Addition of exogenous tPA was found to facilitate L-LTP (Baranes *et al.*, 1998). Analysis of transgenic animals constitutively expressing elevated levels of active tPA in post-natal neurons under control of the Thy1.2 promoter provided further support for tPA in synaptic plasticity and learning (Madani *et al.*, 1999). These mice appeared healthy and exhibited no obvious neurological differences from wild-type counterparts, or alteration in neuroanatomical structure of the brain. Interestingly, electrophysiological studies indicated that tPA overexpression is associated with selective enhancement of paired-pulse facilitation and LTP in the hippocampus, and that the mice exhibit enhanced performance in hippocampus-dependent learning task (hidden platform water maze). These data indicate that neuronal overexpression of tPA is accompanied by improvement in learning ability and memory storage that may be related to an effect of extracellular proteolysis on hippocampal plasticity. A number of other studies have used tPA-null mice to address *in vivo* various behavioral paradigms. Whereas there appears to be some controversy about the contribution of tPA in various hippocampus-dependent memory tests, there is consensus that tPA-deficient animals have impaired response to negative stimuli in two-way avoidance tests that rely on striatum and/or amygdala (Huang *et al.*, 1996; Calabresi *et al.*, 2000). The precise mechanism of action of tPA in memory storage is not understood. It is possible, that secreted tPA participates in synaptic microremodeling by activating neuronally expressed plasminogen and subsequent degradation of a number of extracellular substrates, such as N-CAM and laminin (Chen and Strickland, 1997; Endo *et al.*, 1998). Also, tPA may affect the synaptic status through modulating the profile of active growth factors (Rifkin *et al.*, 1997) and through cleavage of cell surface receptors such as the NR1 subunit of the NMDA receptor (Nicole *et al.*, 2001).

Learning of complex motor skills requires synaptic remodeling of Purkinje cell inputs (Ito, 1993). In rats trained to perform complex motor tasks there is an increased number of synapses of Purkinje cells in the molecular layer of cerebellar cortex. tPA mRNA and antigen are rapidly elevated in Purkinje cells and in granule cell layer after learning a complex motor task coherent with a role of tPA in the formation of the new synapses (Seeds *et al.*, 1995).

Further support to the participation of tPA in the synaptic microremodeling comes from the finding that intracortical blockade of tissue plasminogen activator or plasmin selectively prevents recovery of cortical function and thalamic neuron size during reverse occlusion, without affecting the monocular deprivation response (Muller and Griesinger, 1998).

Much less is known about uPA in the learning and synaptic plasticity. Transgenic mice overexpressing uPA in the cortex and limbic system have normal sensory and motor capabilities, however, they perform poorly in spatial, olfactory, and aversion learning tasks (Meiri *et al.*, 1994).

Collectively, these studies are compatible with tPA playing an important role in activity-dependent synaptic plasticity. Control of tPA activity in these physiological processes appears to be precisely regulated at the levels of transcription, secretion, and extracellular inhibition of tPA.

3. Plasminogen Activators in CNS Pathologies

3.1. Excitotoxicity

Excitotoxic neuronal degeneration is initiated by overexcitation of neurons by high levels of extracellular glutamate and subsequent massive depolarization, calcium influx, and cell death (Shaw, 1993). Experimentally, excitotoxic neuronal death can be induced by systemic or local injection of glutamate receptor antagonists (e.g., kainate). Importantly, such experimentally induced excitotoxic neuronal death is though to be a paradigm of many acute and progressive diseases in the CNS (Meldrum and Garthwaite, 1990).

Recent pioneering research by S. Strickland and coworkers has implicated PA/plasmin system as a key downstream effector contributing to kainate-induced excitotoxic neuronal damage in the hippocampus reviewed by (Strickland, 2001). tPA appears to be a key effector in the proteolytic cascade leading to excitotoxic neurodegeneration and tPA-deficient mice are resistant to kainate-induced excitotoxicity in the hippocampus (Tsirka *et al.*, 1995). Injection of proteolytically active tPA sensitizes tPA-null mice for hippocampal neuronal loss, suggesting that for excitotoxicity there was an acute (as opposed to developmental) requirement of tPA. In contrast, injection of the tPA inhibitor PAI-1 rendered wild-type mice resistant to kainate-induced neurotoxicity (Tsirka *et al.*, 1996). tPA participates in the excitotoxic cascade by converting plasminogen to plasmin, as demonstrated by the finding that plasminogen-deficient mice displayed a similar degree of resistance to neurodegeneration as tPA-null mice (Tsirka *et al.*, 1997). What is the mechanism of tPA/plasmin mediated neurodegeneration? Resistance of plasminogen/fibrinogen double-null mice to excitotoxicity unequivocally demonstrated, that fibrin is not the critical end substrate of the cascade (Tsirka *et al.*, 1997). Instead, systematic analysis of a number of other ECM molecules after excitotoxin injection demonstrated the disappearance of laminin immunoreactivity immediately prior to neuronal degeneration, whereas laminin immunoreactivity remained unchanged after treatments that block excitotoxicity (e.g., injection of α_2-antiplasmin) (Chen and Strickland, 1997). Laminins are large heterotrimeric glycoproteins that can exist in up to 11 isoforms (Patton, 2000). The precise spectrum and functional roles of hippocampal laminins need further study.

During characterization of the role of tPA in excitotoxic cascade it was observed that proteolytically inactive tPA elicited activation response in microglia (Rogove *et al.*, 1999). tPA could thus serve as a critical paracrine signal for microglial activation and neuroinflammation.

The experiments with gene-inactivated mice led Strickland and coworkers to propose a model for neuronal degeneration in which tPA/plasmin mediated ECM degradation and loss of neuron–matrix interaction leads to sensitizing the neurons to genetic program of cell suicide, apoptosis (Figure 15.1) (Strickland, 2001). tPA and plasminogen are known to be co-expressed in hippocampal pyramidal neurons; also microglial expression of tPA in hippocampus has been reported. It seems likely that in neurons, a pool of pre-synthesized tPA is stored for regulated release during processes associated with plasticity, such as synaptic microremodeling. Massive membrane depolarization during excitotoxic injury leads to uncontrolled secretion of tPA and plasminogen, and degradation of ECM molecules as laminins that serve as neuronal survival factors. The concept that adhesive interactions are critical for cell survival has been described for other cell types and tissues. Apoptosis following loss-of-contact has been termed anoikis (loosely derived from Greek "homelessness") (Frisch and Screaton, 2001).

3.2. Ischemic Stroke

During ischemic stroke, lack of vascular perfusion due to a blood clot or rupture of a cerebral blood vessel, leads to hypoxia and excitotoxic neuronal death. tPA is a major fibrin-directed plasminogen activator and is widely used in thrombolytic therapies. In the case of thrombotic stroke, therapeutic administration of tPA has been applied to restore vascular perfusion, and intravenous tPA thrombolytic treatment is clinically effective if started within 3 hr after onset of stroke symptoms (1995).

There has been intense controversy about potential neurological risks associated with intravenous tPA administration during thrombolytic therapy of myocardial infarction

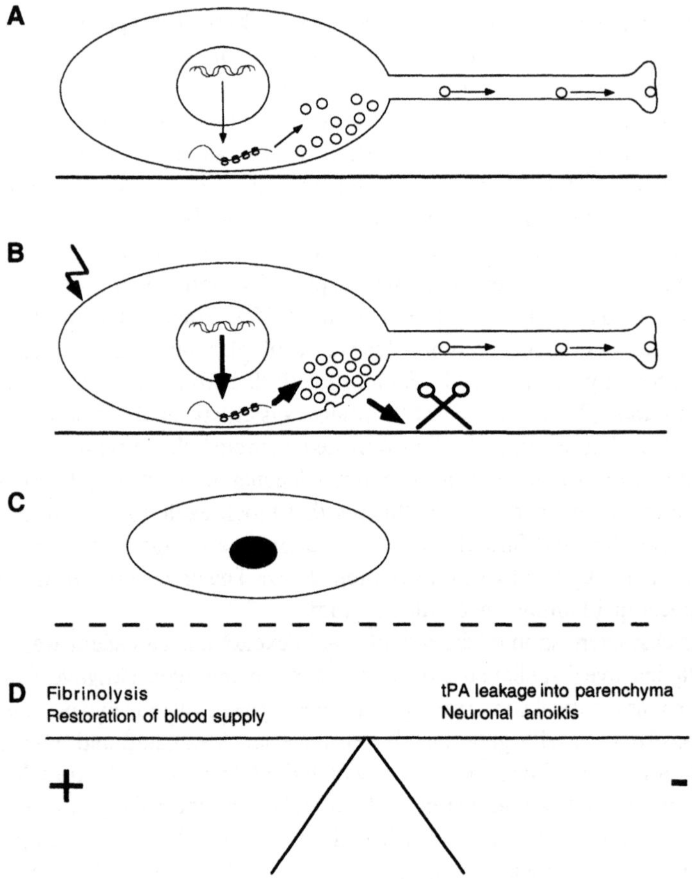

Figure 15.1. Possible mechanism of excitotoxin-induced neuronal degeneration. (A) Under basal conditions, low amount of tPA is synthesized and released from neurons to enable synaptic microremodeling and neuronal plasticity. (B) Upon injury, membrane depolarization leads to rapid release of pre-made tPA into the extracellular milieu, and induction of tPA gene expression. Extracellularly, tPA activity leads to plasmin generation and degradation of the underlying extracellular matrix. (C) Degradation of the matrix leads to loss of neuron anchorage to the matrix and initiates a program of cell suicide (apoptosis, also termed, in the case of loss of matrix attachment, anoikis). (D) Dual role of therapeutic intravenous administration of tPA in thrombotic stroke patients. tPA-catalyzed plasminogen activation is beneficial for thrombolysis and restoration of blood supply, however, increases the risk of leakage of tPA into brain parenchyma and initiation of neuronal apoptosis program.

and stroke. Indeed, exogenous tPA could penetrate the blood–brain-barrier, and contribute to a pathway leading to neuronal anoikis as discussed above. Use of experimental animals to dissect the role of tPA in stroke has produced mixed results. In stroke models not associated with blood clot formation (ligation of middle cerebral artery or insertion of siliconized thread into the artery), tPA appears to be a pathogenic factor (Wang *et al.*, 1998). In these models tPA-null mice showed 50% less damage than wild-type counterparts, and intravenous administration of tPA, similar to thrombolytic therapy in humans, led to an increased size of infarction. In contrast, in the models with blood clot formation (e.g., insertion of non-siliconized thread into the cerebral artery), tPA is a beneficial factor mediating fibrinolysis and reperfusion (Kilic *et al.*, 1999; Tabrizi *et al.*, 1999). Thus, tPA has a dual role: it is beneficial for fibrinolysis and results in reperfusion of an occluded blood vessel, and it contributes to neuronal damage upon its leakage into brain parenchyma through damaged blood–brain-barrier. These data suggest that an optimal agent for stroke therapy would be a highly fibrin-dependent protease not exhibiting activity towards other ECM molecules.

The importance of the tPA/plasminogen axis for neuronal ischemic damage has been further underlined by analysis of dynamics of the main inhibitor of tPA in the CNS, neuroserpin, upon ischemia. Immunoreactivity of neuroserpin was prominently upregulated in the area surrounding ischemic core (penumbra) as early as 6 hr after stroke and remained elevated for up to 7 days (Yepes *et al.*, 2000). Intracerebral perfusion of neuroserpin decreased the damaged area by 64%, this was associated by a decrease in tPA activity to baseline levels and significant reduction of uPA activity in the infarction area. Neuroserpin injection also appeared to reduce apoptosis in the area of penumbra. It will be of great interest to follow evolution of the idea of neuroserpin as a naturally occurring neuroprotective molecule, and its therapeutic use.

3.3. Neurodegenerative Diseases

3.3.1. Alzheimer's Disease

Neocortical deposits of amyloid-β (Aβ), a 4 kDa fragment of amyloid precursor protein (APP), are a prominent hallmark of the leading form of dementia, Alzheimer's disease (AD) (Selkoe, 2001). The amyloid cascade hypothesis postulates that cerebral accumulation of Aβ is an early, invariant, and necessary event in the genesis of AD. Aβ is generated from APP by endoproteolytic processing involving β-and γ-secretase (Walter *et al.*, 2001). The alternative processing pathway involves cleavage by α-secretase in the center of the Aβ domain, generating non-amyloidogenic peptide termed p3. Unlike the Aβ pathway, which appears to operate in the secretory and endosomal compartments, α-secretase is active in glycolipid-enriched membrane rafts. Interestingly, plasminogen and plasmin also localize to neuronal rafts, and addition of exogenous tPA results in enhanced α-cleavage of APP (Periz and Fortini, 2000; Ledesma *et al.*, 2000). It is not known, whether plasmin itself carries out α-cleavage of the APP, or whether it participates via activation of extracellular matrix metalloproteinases (Parvathy *et al.*, 1999) known to be capable of α-cleavage.

Plasminogen system is not only involved in APP processing, but it appears to be capable of mediating degradation of preformed extracellular neurotoxic fibrillar Aβ. It has been demonstrated that aggregated, neurotoxic form of Aβ increases the levels of transcripts of tPA and uPA in primary cortical cultures and in transgenic mice overexpressing Aβ (Tucker *et al.*, 2000a,b). Aβ also appears to act *in vitro* as an activating co-factor for tPA in the same way as fibrin does (Van Nostrand and Porter, 1999). Physiological concentrations of plasmin cleave Aβ *in vitro* at multiple sites and block its neurotoxicity (Tucker *et al.*, 2000a,b). Consistent with the hypothesized beneficial role of plasmin for preventing plaque deposition and promoting plaque lysis, reduced plasmin activity has been documented in affected cortical extracts (Ledesma *et al.*, 2000). A decrease in plasmin activity may reflect a shift in protease/inhibitor balance due to inflammatory response elicited by Aβ deposition. Clearly, understanding the precise role of the PA system in relation to Aβ turnover and AD pathogenesis requires further functional experiments, for example, generating and analysis of Aβ-overexpressing mice deficient in genes of the plasminogen activation system.

3.3.2. Creutzfeldt-Jakob Disease

Prion diseases, including bovine spongioformic encephalopathy (BSE) and human Creutzfeldt-Jakob disease (CJD), represent a group of degenerative diseases of the nervous system. Similarly to AD, CJD is characterized by the deposition of insoluble β-pleated sheet peptides in the extracellular spaces of grey matter in the brain (Prusiner, 1998). In the case of CJD, this protein is a modified form of the normal prion protein PrPC (cellular prion protein), which is produced by all mammals and is present in the brain, particularly on the surface of nerve cells. The normal and the malignant prion protein have the same chemical composition and differ from each other exclusively in their conformation. Plasminogen appears to bind strongly to PrPSc but not to PrPC, as demonstrated by a series of binding studies (Fischer *et al.*, 2000; Maissen *et al.*, 2001). In these experiments, magnetic beads were coated with plasminogen and mixed together with brain material from prion-infected mice. While PrPSc attached to the plasminogen beads, no normal prion protein PrPC attached. This selective binding property could serve as the basis for a diagnostic test aimed at differentiating and diagnosing BSE or CJD. It remains to be seen, if and how PrPSc–plasminogen interaction is related to pathogenesis of CJD.

3.4. Neuroinflammatory Diseases

Multiple sclerosis (MS) is a common, chronic demyelinating disease of the central nervous system, thought to be of autoimmune pathogenesis, whose etiology may involve genetic, viral, and immunological factors. MS features autoimmune inflammatory attack against the myelin insulation of neurons. T cells sensitized against myelin self-antigens secrete tumor necrosis factor, cytokines, prostaglandins, and other inflammatory mediators that strip away the myelin and sometimes destroy the axons (Trapp *et al.*, 1999).

Extracellular proteolytic enzymes are implied as pathogenic factors in the MS and its animal model experimental autoimmune encephalomyelitis (EAE) (Cuzner and Opdenakker, 1999). Increased levels of matrix metalloproteinases (MMPs) have been found in the cerebrospinal fluid in MS and EAE, and these enzymes are upregulated in situ in a pattern suggesting an active role in inflammation and demyelination (Gijbels *et al.*, 1992, 1994). In addition, plasmin inhibitors were shown to suppress the development of EAE (Brosnan *et al.*, 1980). Our findings of elevated tPA antigen levels in the cerebrospinal fluid of MS patients and *in situ* upregulation of tPA in MS lesions are coherent with involvement of this plasminogen activation system in the mechanism of pathogenesis of MS (Akenami *et al.*, 1996, 1999). In an independent study, prominent upregulation of uPA, uPAR, and PAI-1 was detected in mononuclear cells of perivascular cuffs and macrophages in the lesion parenchyma (Gveric *et al.*, 2001). In the latter study, tPA appeared down-regulated and it colocalized with fibrin(ogen) in damaged axons.

It is likely that the pathophysiological process of progression of MS involves a concerted action of several enzymes of different proteolytic pathways. Recently, we have described expression and activity patterns of proteases of plasminogen activators and MMP systems as well as several functionally related molecules (inhibitors, receptors) in the CNS during acute EAE in Balb/c mice (Teesalu *et al.*, 2001). Expression of several molecules of the plasminogen activation (PA) system and several members of the MMP family of extracellular proteases and their inhibitors was analyzed by *in situ* hybridization. During clinical EAE, the PA system was upregulated in the spinal cords at several levels.

Induction of expression of tPA and PAI-1 transcripts was detected in activated astrocytes in the white matter. Inflammatory cells scored positive for the expression high affinity receptor of uPA, uPAR. *In situ* zymography demonstrated the presence of increased tPA and uPA activities in the areas of the inflammatory damage. Accumulation of fibrin, fibronectin, and vitronectin immunoreactivity was seen in perivascular matrices of symptomatic animals.

In addition, transcription of the membrane-type MT1-MMP and metalloelastase (in inflammatory cells), and tissue inhibitor of matrix metalloproteinases type 1 (TIMP-1) (in activated astrocytes) was induced during EAE. Increased gelatinolytic activity was detected at the sites of inflammatory cell accumulation by *in situ* zymography of fluorescently labeled gelatin; SDS–PAGE zymography identified the upregulated gelatinolytic activity as gelatinase B.

The above analysis demonstrated a parallel upregulation of two key systems of extracellular proteolysis—the plasminogen activator system and MMPs—both in MS and during EAE in mice. Regulated expression of the components of PA system warrants inhibitor/knockout-based functional studies on the effect of inhibition of the PA system on the progression of inflammatory demyelination. Plasminogen activators and MMP systems have been demonstrated to cooperate functionally in the process of wound healing in mice (Lund *et al.*, 1999). Similarly, the concurrent induction of the two systems in MS and EAE suggests that treatments aiming at modulating the expression of both PAs and MMPs simultaneously may prove more efficient than therapeutic strategies focusing on a single system of extracellular proteolysis.

3.5. CNS Malignancy

Interactions between the cancer cells and adjacent stromal cells and ECM critically affect the behavior of tumors and their clinical outcome. Extracellular microenvironment will eventually determine the extent of spread and invasion of tumor cells, vascular supply of tumor, and bioavailability of trophic/inhibitory factors. Extracellular proteolysis represents a direct and irreversible means of changing cellular microenvironment. Extracellular proteases of PA and MMP families are central players in the progression of tumor invasion and metastasis (Yamamoto *et al.*, 1994a,b,c; Koblinski *et al.*, 2000). These enzymatic systems are of critical importance for cancer cell migration and invasion, as well as for tumor-directed matrix remodeling in general (angiogenesis, desmoplasia). A number of tumors are positive for uPA and its cell surface receptor uPAR, and the levels of uPA and uPAR are prognostic markers for many types of tumors (Duffy, 1996; Andreasen *et al.*, 2000). uPAR is a pleiotrophic molecule and a key regulator of proteolysis on tumor cell surface. uPAR acts to focalize uPA activity (e.g., at cell-substratum adhesion sites) (Myohanen *et al.*, 1993), serves as an adhesion receptor to ECM protein vitronectin, and acts as a cell surface chemokine (Blasi, 1997, 1999).

Malignant gliomas comprise the majority of primary intracranial brain tumors. Most of the gliomas, including glioblastomas, have constituent cells with recognizable astrocytic properties. Gliomas are characterized by local invasive growth, tissue destruction at the border between tumor and normal brain tissue, and intense angiogenesis. Diffusely infiltrative and angiogenic nature of malignant gliomas represents a major obstacle to their successful surgical control and results in extremely poor prognosis. uPA and uPAR are expressed in many glioma subtypes, and their expression level by tumor cells appears to correlate with higher grades of gliomas (Mohanam *et al.*, 1994; Yamamoto *et al.*, 1994a,b,c; Bhattacharya *et al.*, 2001). In contrast, tPA is downregulated in malignant gliomas (Sawaya *et al.*, 1991). In high-grade gliomas (glioblastomas and anaplastic astrocytomas) uPA and uPAR immunoreactivities are localized mainly at tumor cells and microvascular endothelial cells, especially at sites of angiogenesis and at invasion front. The same areas also exhibit immunoreactivity for PAI-1 (Sawaya *et al.*, 1995) and are positive for the permissive migration substrate vitronectin (Sawaya *et al.*, 1995). Functional relevance of uPAR expression to the mechanism of pathogenesis of gliomas has been tested in cell culture and experimental animals. The invasiveness of glioblastoma cells *in vitro* is reduced by their exposure to anti-uPAR antibodies *in vitro*; also antisense downregulation of uPAR in a glioblastoma cell line results in decreased invasive behavior both *in vitro* and *in vivo* (Go *et al.*, 1997). Interestingly, uPAR appears to be a survival factor for glioblastoma cells, and uPAR antisense downregulation is associated with apoptosis (Kin *et al.*, 2000). uPA/uPAR system may represent in high-grade gliomas a key checkpoint for tumor-directed tissue remodeling (Figure 15.2). Pericellularily focalized uPA activity may result in proteolytic

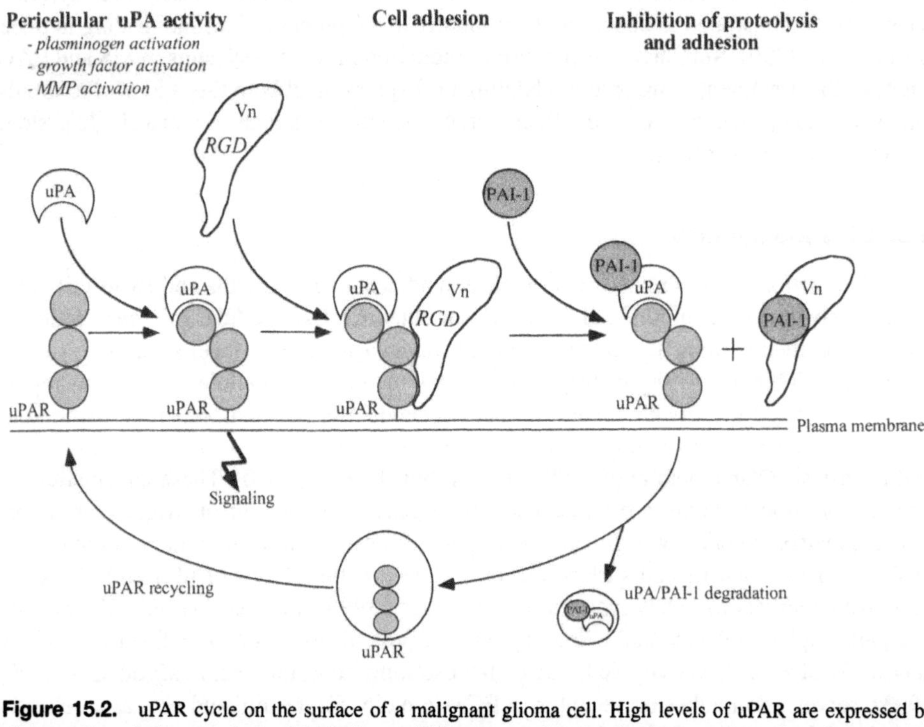

Figure 15.2. uPAR cycle on the surface of a malignant glioma cell. High levels of uPAR are expressed by glioma cells at the invasive front. uPA binding to uPAR activates the plasminogen cascade on the cell surface and results in focalized ECM remodeling and activation of growth factors such as TGF-β. Ligand binding to uPAR initiates intracellular signaling cascade that may further contribute to glioma cell malignancy. N-terminal somatomedin B domain of vitronectin binds both uPAR and PAI-1, which compete for binding. PAI-1 also competes with integrins for binding to vitronectin. uPAR has a dual role acting as uPA receptor and as a cell surface receptor of vitronectin, and PAI-1 acts to block both cell adhesion to vitronectin and pericellular plasminogen activation. uPA-PAI-1 complexes are internalized and degraded, whereas uPAR recycles back to cell surface, where it is available for another cycle.

cascades leading to MMP (e.g., gelatinase B) activation (Hahn Dantona *et al.*, 1999), to ECM remodeling and to activation of latent growth factors, such as HGF-SF and TGFβ (Rifkin *et al.*, 1999). In addition, uPA–uPAR interaction is known to initiate in many cell types an intracellular signaling cascade and uPAR is known to act as a specific cell surface receptor for vitronectin (Blasi, 1997). The presence of abundant PAI-1 in malignant gliomas may render the uPA–uPAR–vitronectin system even more dynamic, and further contribute to cell invasion and tissue remodeling.

4. Conclusions and Future Perspectives

PAs are expressed by many types of CNS-resident cells in developing and adult tissues. In the CNS, PAs are involved in a diverse range of processes, from neuronal migration to glioma invasion and from synaptic microremodeling to microglia activation.

tPA is the major PA in the adult CNS, where it is synthesized primarily by neurons, and in some instances by glial cells. Neuronal secretion of tPA is activity-dependent, and its primary physiological role in neurons is to catalyze plasminogen activation to allow neuronal plasticity. The presence of a pool of tPA protein in a secretory compartment of neurons poses inherent risk of neuronal damage in the case of uncontrolled neuronal stimulation, for example, during stroke. Experiments with null mice identify tPA as a major effector in the pathway of excitotoxic neurodegeneration. This finding is of great interest, since excitotoxic damage is involved in most neurological diseases. Currently, an important challenge lies in systematic mapping of the expression and activity of plasminogen activators and functionally related molecules in the human CNS diseases. Such "guilt by association" type of data, combined with cell culture studies, will constitute an essential component in the validation of the relevance of PA system to pathologies in which it has been suggested to be involved based on animal experimentation.

In recent years, a number of new extracellular CNS-expressed proteases and their functionally related molecules have been identified. It is likely that CNS physiology and disease involves a concerted action of several proteolytic enzymes, and therapeutic strategies of interfering with several activities simultaneously might prove much more effective than targeting a single enzyme.

5. Acknowledgments

We would like to thank all the members of plasminogen activation group of Haartman Institute for their support. The original work was supported by Academy of Finland, grant 4334 from Estonian Science Foundation and Finnish Cancer Societies.

References

Akenami, F.O., Siren, V., Koskiniemi, M., Siimes, M.A., Teravainen, H., and Vaheri, A. (1996). Cerebrospinal fluid activity of tissue plasminogen activator in patients with neurological diseases. *J. Clin. Pathol.* 49, 577–580.

Akenami, F.O., Siren, V., Wessman, M., Koskiniemi, M., and Vaheri, A. (1999). Tissue plasminogen activator gene expression in multiple sclerosis brain tissue. *J. Neurol. Sci.* **165**, 71–76.

Andreasen, P.A., Egelund, R., and Petersen, H.H. (2000). The plasminogen activation system in tumor growth, invasion, and metastasis. *Cell. Mol. Life Sci.* **57**, 25–40.

Bailey, C.H. and Kandel, E.R. (1993). Structural changes accompanying memory storage. *Annu. Rev. Physiol.* **55**, 397–426.

Baranes, D., Lederfein, D., Huang, Y.Y., Chen, M., Bailey, C.H., and Kandel, E.R. (1998). Tissue plasminogen activator contributes to the late phase of LTP and to synaptic growth in the hippocampal mossy fiber pathway. *Neuron.* **21**, 813–825.

Basbaum, C.B. and Werb, Z. (1996). Focalized proteolysis: Spatial and temporal regulation of extracellular matrix degradation at the cell surface. *Curr. Opin. Cell Biol.* **8**, 731–738.

Bhattacharya, A., Lakka, S.S., Mohanam, S., Boyd, D., and Rao, J.S. (2001). Regulation of the urokinase-type plasminogen activator receptor gene in different grades of human glioma cell lines. *Clin. Cancer Res.* **7**, 267–76.

Blasi, F. (1997). uPA, uPAR, PAI-1: Key intersection of proteolytic, adhesive and chemotactic highways? *Immunol. Today.* **18**, 415–417.

Blasi, F. (1999). The urokinase receptor. A cell surface, regulated chemokine. *APMIS* **107**, 96–101.

Brosnan, C.F., Cammer, W. Norton, W.T., and Bloom, B.R. (1980). Proteinase inhibitors suppress the development of experimental allergic encephalomyelitis. *Nature* **285**, 235–7.

Calabresi, P., Napolitano, M., Centonze, D., Marfia, G.A., Gubellini, P., Teule, M.A., Berretta, N., Bernardi, G., Frati, L., Tolu, M., and Gulino, A. (2000). Tissue plasminogen activator controls multiple forms of synaptic plasticity and memory. *Eur. J. Neurosci.* **12**, 1002–1012.

Chen, Z.L. and Strickland, S. (1997). Neuronal death in the hippocampus is promoted by plasmin-catalyzed degradation of laminin. *Cell* **91**, 917–925.

Conese, M. and Blasi, F. (1995). Urokinase/urokinase receptor system: Internalization/degradation of urokinase-serpin complexes: Mechanism and regulation. *Biol. Chem. Hoppe Seyler.* **376**, 143–155.

Cuzner, M.L. and Opdenakker, G. (1999). Plasminogen activators and matrix metalloproteases, mediators of extracellular proteolysis in inflammatory demyelination of the central nervous system. *J. Neuroimmunol.* **94**, 1–14.

Davies, B.J., Pickard, B.S., Steel, M., Morris, R.G., and Lathe, R. (1998). Serine proteases in rodent hippocampus. *J. Biol. Chem.* **273**, 23004–23011.

de Vries, T.J., van_Muijen, G.N., and Ruiter, D.J. (1996). The plasminogen activation system in tumour invasion and metastasis. *Pathol. Res. Pract.* **192**, 718–733.

Duffy, M.J. (1996). Proteases as prognostic markers in cancer. *Clin. Cancer. Res.* **2**, 613–618.

Endo, A., Nagai, N., Urano, T., Ihara, H., Takada, Y., Hashimoto, K., and Takada, A. (1998). Proteoloysis of highly polysialylated NCAM by the tissue plasminogen activator-plasmin system in rats. *Neurosci Lett.* **246**, 37–40.

Fischer, M.B., Roeckl, C., Parizek, P., Schwarz, H.P., and Aguzzi, A. (2000). Binding of disease-associated prion protein to plasminogen. *Nature* **408**, 479–483.

Frey, U., Muller, M., and Kuhl, D. (1996). A different form of long-lasting potentiation revealed in tissue plasminogen activator mutant mice. *J. Neurosci.* **16**, 2057–2063.

Friedman, G.C. and Seeds, N.W. (1994). Tissue plasminogen activator expression in the embryonic nervous system. *Brain Res. Dev. Brain Res.* **81**, 41–49.

Friedman, G.C. and Seeds, N.W. (1995). Tissue plasminogen activator mRNA expression in granule neurons coincides with their migration in the developing cerebellum. *J. Comp. Neurol.* **360**, 658–670.

Frisch, S.M. and Screaton, R.A. (2001). Anoikis mechanisms. *Curr. Opin. Cell Biol.* **13**, 555–562.

Gijbels, K., Galardy, R.E., and Steinman, L. (1994). Reversal of experimental autoimmune encephalomyelitis with a hydroxamate inhibitor of matrix metalloproteases. *J. Clin. Invest.* **94**, 2177–2182.

Gijbels, K., Masure, S., Carton, H., and Opdenakker, G. (1992). Gelatinase in the cerebrospinal fluid of patients with multiple sclerosis and other inflammatory neurological disorders. *J. Neuroimmunol.* **41**, 29–34.

Go, Y., Chintala, S.K., Mohanam, S., Gokaslan, Z., Venkaiah, B., Bjerkvig, R., Oka, K., Nicolson, G.L., Sawaya, R., and Rao, J.S. (1997). Inhibition of *in vivo* tumorigenicity and invasiveness of a human glioblastoma cell line transfected with antisense uPAR vectors. *Clin. Exp. Metastasis.* **15**, 440–446.

Gveric, D., Hanemaaijer, R., Newcombe, J., van_Lent, N.A., Sier, C.F., and Cuzner, M.L. (2001). Plasminogen activators in multiple sclerosis lesions: Implications for the inflammatory response and axonal damage. *Brain* **124**, 1978–88.

Hahn Dantona, E., Ramos_DeSimone, N., Sipley, J., Nagase, H., French, D.L., and Quigley, J.P. (1999). Activation of proMMP-9 by a plasmin/MMP-3 cascade in a tumor cell model. Regulation by tissue inhibitors of metalloproteinases. *Ann. N. Y. Acad. Sci.* **878**, 372–387.

Huang, Y.Y., Bach, M.E., Lipp, H.P., Zhuo, M., Wolfer, D.P., Hawkins, R.D., Schoonjans, L., Kandel, E.R., Godfraind, J.M., Mulligan, R., Collen, D., and Carmeliet, P. (1996). Mice lacking the gene encoding tissue-type plasminogen activator show a selective interference with late-phase long-term potentiation in both Schaffer collateral and mossy fiber pathways. *Proc. Natl. Acad. Sci. USA*. **93**, 8699–8704.

Ito, M. (1993). Synaptic plasticity in the cerebellar cortex and its role in motor learning. *Can. J. Neurol. Sci.* **20**(Suppl. 3), S70–S74.

Kilic, E., Hermann, D.M., and Hossmann, K.A. (1999). Recombinant tissue plasminogen activator reduces infarct size after reversible thread occlusion of middle cerebral artery in mice. *Neuroreport* **10**, 107–111.

Kin, Y., Chintala, S.K., Go, Y., Sawaya, R., Mohanam, S., Kyritsis, A.P., and Rao, J.S. (2000). A novel role for the urokinase-type plasminogen activator receptor in apoptosis of malignant gliomas. *Int. J. Oncol.* **17**, 61–65.

Koblinski, J.E., Ahram, M., and Sloane, B.F. (2000). Unraveling the role of proteases in cancer. *Clin. Chim. Acta.* **291**, 113–135.

Krueger, S.R., Ghisu, G.P., Cinelli, P., Gschwend, T.P., Osterwalder, T., Wolfer, D.P., and Sonderegger, P. (1997). Expression of neuroserpin, an inhibitor of tissue plasminogen activator, in the developing and adult nervous system of the mouse. *J. Neurosci.* **17**, 8984–8996.

Krystosek, A. and Seeds, N.W. (1981). Plasminogen activator secretion by granule neurons in cultures of developing cerebellum. *Proc. Natl. Acad. Sci. USA* **78**, 7810–7814.

Ledesma, M.D., Da_Silva, J.S., Crassaerts, K., Delacourte, A., De_Strooper, B., and Dotti, C.G. (2000). Brain plasmin enhances APP alpha-cleavage and Abeta degradation and is reduced in Alzheimer's disease brains. *EMBO Rep.* **1**, 530–535.

Lijnen, H.R. and Collen, D. (1988). Mechanisms of plasminogen activation by mammalian plasminogen activators. *Enzyme* **40**, 90–96.

Lund, L.R., Romer, J., Bugge, T.H., Nielsen, B.S., Frandsen, T.L., Degen, J.L., Stephens, R.W., and Dano, K. (1999). Functional overlap between two classes of matrix-degrading proteases in wound healing. *EMBO J.* **18**, 4645–4656.

Madani, R., Hulo, S., Toni, N., Madani, H., Steimer, T., Muller, D., and Vassalli, J.D. (1999). Enhanced hippocampal long-term potentiation and learning by increased neuronal expression of tissue-type plasminogen activator in transgenic mice. *EMBO J.* **18**, 3007–3012.

Maissen, M., Roeckl, C., Glatzel, M., Goldmann, W., and Aguzzi, A. (2001). Plasminogen binds to disease-associated prion protein of multiple species. *Lancet* **357**, 2026–2028.

Masos, T. and Miskin, R. (1996). Localization of urokinase-type plasminogen activator mRNA in the adult mouse brain. *Brain Res. Mol. Brain Res.* **35**, 139–148.

Meiri, N., Masos, T., Rosenblum, K., Miskin, R., and Dudai, Y. (1994). Overexpression of urokinase-type plasminogen activator in transgenic mice is correlated with impaired learning. *Proc. Natl. Acad. Sci. USA.* **91**, 3196–3200.

Meldrum, B. and Garthwaite, J. (1990). Excitatory amino acid neurotoxicity and neurodegenerative disease. *Trends Pharmacol. Sci.* **11**, 379–387.

Mohanam, S., Sawaya, R.E., Yamamoto, M., Bruner, J.M., Nicholson, G.L., and Rao, J.S. (1994). Proteolysis and invasiveness of brain tumors: Role of urokinase-type plasminogen activator receptor. *J. Neurooncol.* **22**, 153–160.

Moonen, G., Grau_Wagemans, M.P., and Selak, I. (1982). Plasminogen activator-plasmin system and neuronal migration. *Nature* **298**, 753–755.

Muller, C.M. and Griesinger, C.B. (1998). Tissue plasminogen activator mediates reverse occlusion plasticity in visual cortex. *Nat. Neurosci.* **1**, 47–53.

Myohanen, H.T., Stephens, R.W., Hedman, K., Tapiovaara, H., Ronne, E., Hoyer-Hansen, G., Dano, K., and Vaheri, A. (1993). Distribution and lateral mobility of the urokinase-receptor complex at the cell surface. *J. Histochem. Cytochem.* **41**, 1291–1301.

Neuman, T., Stephens, R.W., Salonen, E.M., Timmusk, T., and Vaheri, A. (1989). Induction of morphological differentiation of human neuroblastoma cells is accompanied by induction of tissue-type plasminogen activator. *J. Neurosci. Res.* **23**, 274–281

Nicole, O., Docagne, F., Ali, C., Margaill, I., Carmeliet, P., MacKenzie, E.T., Vivien, D., and Buisson, A. (2001). The proteolytic activity of tissue-plasminogen activator enhances NMDA receptor-mediated signaling. *Nat. Med.* **7**, 59–64.

Parvathy, S., Hussain, I., Karran, E.H., Turner, A.J., and Hooper, N.M. (1999). Cleavage of Alzheimer's amyloid precursor protein by alpha-secretase occurs at the surface of neuronal cells. *Biochemistry* **38**, 9728–9734.

Patton, B.L. (2000). Laminins of the neuromuscular system. *Microsc. Res. Tech.* **51**, 247–261.

Periz, G. and Fortini, M.E. (2000). Proteolysis in Alzheimer's disease. Can plasmin tip the balance? *EMBO Rep.* **1**, 477–478.

Pittman, R.N. and Buettner, H.M. (1989). Degradation of extracellular matrix by neuronal proteases. *Dev. Neurosci.* **11**, 361–375.

Pittman, R.N., Ivins, J.K., and Buettner, H.M. (1989). Neuronal plasminogen activators: Cell surface binding sites and involvement in neurite outgrowth. *J. Neurosci.* **9**, 4269–4686.

Prusiner, S.B. (1998). The prion diseases. *Brain Pathol.* **8**, 499–513.

Qian, Z., Gilbert, M.E., Colicos, M.A., Kandel, E.R., and Kuhl, D. (1993). Tissue-plasminogen activator is induced as an immediate-early gene during seizure, kindling and long-term potentiation. *Nature* **361**, 453–457.

Rifkin, D.B., Gleizes, P.E., Harpel, J., Nunes, I., Munger, J., Mazzieri, R., and Noguera, I. (1997). Plasminogen/plasminogen activator and growth factor activation. *Ciba Found. Symp.* **212**, 105–115; discussion 116–118.

Rifkin, D.B., Mazzieri, R., Munger, J.S., Noguera, I., and Sung, J. (1999). Proteolytic control of growth factor availability. *APMIS* **107**, 80–85.

Rogister, B., Leprince, P., Pettmann, B., Labourdette, G., Sensenbrenner, M., and Moonen, G. (1988). Brain basic fibroblast growth factor stimulates the release of plasminogen activators by newborn rat cultured astroglial cells. *Neurosci. Lett.* **91**, 321–326.

Rogove, A.D., Siao, C., Keyt, B., Strickland, S., and Tsirka, S.E. (1999). Activation of microglia reveals a non-proteolytic cytokine function for tissue plasminogen activator in the central nervous system. *J. Cell Sci.* **112** (Pt 22), 4007–4016.

Ruegg, M.A. (2001). Molecules involved in the formation of synaptic connections in muscle and brain. *Matrix Biol.* **20**, 3–12.

Sappino, A.P., Madani, R., Huarte, J., Belin, D., Kiss, J.Z., Wohlwend, A., and Vassalli, J.D. (1993). Extracellular proteolysis in the adult murine brain. *J. Clin. Invest.* **92**, 679–685.

Sawaya, R., Ramo, O.J., Shi, M.L., and Mandybur, G. (1991). Biological significance of tissue plasminogen activator content in brain tumors. *J. Neurosurg.* **74**, 480–486.

Sawaya, R., Yamamoto, M., Ramo, O.J., Shi, M.L., Rayford, A., and Rao, J.S. (1995). Plasminogen activator inhibitor-1 in brain tumors: Relation to malignancy and necrosis. *Neurosurgery* **36**, 375–380; discussion 380–381.

Seeds, N.W., Verrall, S., Friedman, G., Hayden, S., Gadotti, D., Haffke, S., Christensen, K., Gardner, B., McGuire, P., and Krystosek, A. (1992). Plasminogen activators and plasminogen activator inhibitors in neural development. *Ann. N. Y. Acad. Sci.* **667**, 32–40.

Seeds, N.W., Basham, M.E., and Haffke, S.P. (1999). Neuronal migration is retarded in mice lacking the tissue plasminogen activator gene. *Proc. Natl. Acad. Sci. USA.* **96**, 14118–14123.

Seeds, N.W., Friedman, G., Hayden, S., Thewke, D., Haffke, S., McGuire, P., and Krystosek, A. (1996). Plasminogen activators and their interaction with extracellular matrix in neural development, plasticity and regeneration. **8**, 405–412.

Seeds, N.W., Verrall, S., Friedman, G., Hayden, S., Gadotti, D., Haffke, S., Christensen, K., Gardner, B., McGuire, P., and Krystosek, A. (1992). Plasminogen activators and plasminogen activator inhibitors in neural development. *Ann. N. Y. Acad. Sci.* **667**, 32–40.

Seeds, N.W., Williams, B.L., and Bickford, P. C. (1995). Tissue plasminogen activator induction in Purkinje neurons after cerebellar motor learning. *Science* **270**, 1992–1994.

Selkoe, D.J. (2001). Alzheimer's disease: Genes, proteins, and therapy. *Physiol. Rev.* **81**, 741–766.

Shaw, P.J. (1993). Excitatory amino acid receptors, excitotoxicity, and the human nervous system. *Curr. Opin. Neurol. Neurosurg.* **6**, 414–422.

Strickland, S. (2001). Tissue plasminogen activator in nervous system function and dysfunction. *Thromb. Haemost.* **86**, 138–143.

Stubbs, M.T., Renatus, M., and Bode, W. (1998). An active zymogen: Unravelling the mystery of tissue-type plasminogen activator. *Biol. Chem.* **379**, 95–103.

Sumi, Y., Dent, M.A., Owen, D.E., Seeley, P.J., and Morris, R.J. (1992). The expression of tissue and urokinase-type plasminogen activators in neural development suggests different modes of proteolytic involvement in neuronal growth. *Development* **116**, 625–637.

Tabrizi, P., Wang, L., Seeds, N., McComb, J.G., Yamada, S., Griffin, J.H., Carmeliet, P., Weiss, M.H., and Zlokovic, B.V. (1999). Tissue plasminogen activator (tPA) deficiency exacerbates cerebrovascular fibrin deposition and brain injury in a murine stroke model: Studies in tPA-deficient mice and wild-type mice on a matched genetic background. *Arterioscler. Thromb. Vasc. Biol.* **19**, 2801–2806.

Teesalu, T., Hinkkanen, A.E., and Vaheri, A. (2001). Coordinated induction of extracellular proteolysis systems during experimental autoimmune encephalomyelitis in mice. *Am. J. Pathol.* **159**, 2227–2237.

The National Institute of Neurological Disorders and Stroke. (1995). Tissue plasminogen activator for acute ischemic stroke. rt-PA Stroke Study Group. *N. Engl. J. Med.* **333**, 1581–1587.

Trapp, B.D., Bo, L., Mork, S., and Chang, A. (1999). Pathogenesis of tissue injury in MS lesions. *J. Neuroimmunol.* **98**, 49–56.

Tsirka, S.E., Bugge, T.H., Degen, J.L., and Strickland, S. (1997). Neuronal death in the central nervous system demonstrates a non-fibrin substrate for plasmin. *Proc. Natl. Acad. Sci. USA.* **94**, 9779–9781.

Tsirka, S.E., Gualandris, A., Amaral, D.G., and Strickland, S. (1995). Excitotoxin-induced neuronal degeneration and seizure are mediated by tissue plasminogen activator. *Nature* **377**, 340–344.

Tsirka, S.E., Rogove, A.D., and Strickland, S. (1996). Neuronal cell death and tPA. *Nature* **384**, 123–124.

Tsirka, S.E., Rogove, A.D., Bugge, T.H., Degen, J.L., and Strickland, S. (1997). An extracellular proteolytic cascade promotes neuronal degeneration in the mouse hippocampus. *J. Neurosci.* **17**, 543–552.

Tucker, H.M., Kihiko, M., Caldwell, J.N., Wright, S., Kawarabayashi, T., Price, D., Walker, D., Scheff, S., McGillis, J.P., Rydel, R.E., and Estus, S. (2000a). The plasmin system is induced by and degrades amyloid-beta aggregates. *J. Neurosci.* **20**, 3937–3946.

Tucker, H.M., Kihiko_Ehmann, M., Wright, S., Rydel, R.E., and Estus, S. (2000b). Tissue plasminogen activator requires plasminogen to modulate amyloid-beta neurotoxicity and deposition. *J. Neurochem.* **75**, 2172–2177.

Van Nostrand, W.E. and Porter, M. (1999). Plasmin cleavage of the amyloid beta-protein: Alteration of secondary structure and stimulation of tissue plasminogen activator activity. *Biochemistry* **38**, 11570–11576.

Walter, J., Kaether, C., Steiner, H., and Haass, C. (2001). The cell biology of Alzheimer's disease: Uncovering the secrets of secretases. *Curr. Opin. Neurobiol.* **11**, 585–590.

Wang, Y.F., Tsirka, S.E., Strickland, S., Stieg, P.E., Soriano, S.G., and Lipton, S.A. (1998). Tissue plasminogen activator (tPA) increases neuronal damage after focal cerebral ischemia in wild-type and tPA-deficient mice. *Nat. Med.* **4**, 228–231.

Werb, Z. (1997). ECM and cell surface proteolysis: Regulating cellular ecology. *Cell* **91**, 439–442.

Werb, Z., Vu, T.H., Rinkenberger, J.L., and Coussens, L.M. (1999). Matrix-degrading proteases and angiogenesis during development and tumor formation. *APMIS* **107**, 11–18.

Wu, Y.P., Siao, C.J., Lu, W., Sung, T.C., Frohman, M.A., Milev, P., Bugge, T.H., Degen, J.L., Levine, J.M., Margolis, R.U., and Tsirka, S.E. (2000). The tissue plasminogen activator (tPA)/plasmin extracellular proteolytic system regulates seizure-induced hippocampal mossy fiber outgrowth through a proteoglycan substrate. *J. Cell Biol.* **148**, 1295–1304.

Yamamoto, M., Sawaya, R., Mohanam, S., Rao, V.H., Bruner, J.M., Nicolson, G.L., and Rao, J.S. (1994a). Expression and localization of urokinase-type plasminogen activator receptor in human gliomas. *Cancer Res.* **54**, 5016–5020.

Yamamoto, M., Sawaya, R., Mohanam, S., Bindal, A.K., Bruner, J.M., Oka, K., Rao, V.H., Tomonaga, M., Nicolson, G.L., and Rao, J.S. (1994b). Expression and localization of urokinase-type plasminogen activator in human astrocytomas *in vivo*. *Cancer Res.* **54**, 3656–3661.

Yamamoto, M., Sawaya, R., Mohanam, S., Rao, V.H., Bruner, J.M., Nicolson, G.L., Ohshima, K., and Rao, J.S. (1994c). Activities, localizations, and roles of serine proteases and their inhibitors in human brain tumor progression. *J. Neurooncol.* **22**, 139–151.

Yepes, M., Sandkvist, M., Wong, M.K., Coleman, T.A., Smith, E., Cohan, S.L., and Lawrence, D.A. (2000). Neuroserpin reduces cerebral infarct volume and protects neurons from ischemia-induced apoptosis. *Blood* **96**, 569–576.

16

Role of the Plasminogen Activator–Plasmin System in Angiogenesis

Michael S. Pepper

1. Introduction

It is firmly established that angiogenesis is an absolute requirement for the growth of normal and neoplastic tissues (Pepper, 1997). Many biological processes, including angiogenesis, depend on tightly controlled interactions between cells and the extracellular matrix (ECM). These interactions are mediated (a) by integral membrane proteins including integrins, which provide a link between the ECM and the cytoskeleton; and (b) by extracellular proteinases and their inhibitors, which mediate focal degradation of components of the ECM, some of which (e.g., the fibrillar collagens) are highly resistant to broad spectrum proteases. Most of the relevant extracellular proteolytic enzymes belong to one of two families: the serine proteases, in particular the plasminogen activator (PA)/plasmin system, and the matrix metalloproteinases (MMPs) (Mignatti and Rifkin, 1996; Pepper, 2001a; Vu and Werb, 2000).

Like most other biological processes, angiogenesis is the result of subtle and often complex interactions between regulatory and effector molecules. In order to facilitate its analysis, it is useful to divide angiogenesis into a phase of activation (sprouting) and a phase of resolution. The phase of activation encompasses (a) increased vascular permeability and extravascular fibrin deposition; (b) vessel wall disassembly; (c) basement membrane degradation; (d) cell migration and ECM invasion; (e) endothelial cell proliferation; and (f) capillary lumen formation. The phase of resolution includes (a) inhibition of endothelial cell proliferation; (b) cessation of cell migration; (c) basement membrane reconstitution; (d) junctional complex maturation; and (e) vessel wall assembly, including recruitment and differentiation of smooth muscle cells and pericytes. Implicit in the definition of the resolution phase is the establishment of blood flow in the newly formed vessel (Pepper, 1997). Extracellular proteolysis has been implicated in many of these processes including basement membrane degradation, cell migration/ECM invasion, and capillary lumen formation (Table 16.1) (Pepper *et al.*, 1996).

Michael S. Pepper • Department of Morphology, University Medical Center, Geneva, Switzerland.

Plasminogen: Structure, Activation, and Regulation, edited by David M. Waisman. Kluwer Academic/Plenum Publishers, New York, 2003.

Table 16.1. Extracellular Proteolysis and Angiogenesis

Breaching of mechanical barriers
Basement membrane degradation
Cell migration and ECM invasion
Lumen formation
Regulation of cytokine activity
Activation of latent TGF-β
Release of matrix-bound bFGF, VEGF
Release of membrane-anchored TNF-α
Inactivation of MCP-3 (and generation of a chemokine antagonist)
Generation of bioactive degradation products
ECM components
Plasminogen $\rightarrow$ angiostatin, kringle 5 (inhibitory)
Collagen IV $\rightarrow$ arresten, canstatin, tumstatin (inhibitory)
Collagen XV $\rightarrow$ restin (inhibitory)
Collagen XVIII $\rightarrow$ endostatin (inhibitory)
Thrombospondin (inhibitory)
Hyaluronan (stimulatory)
SPARC (stimulatory and inhibitory)
Other molecules
Prolactin (inhibitory)
Platelet factor 4 (inhibitory)
MMP-2 $\rightarrow$ PEX (inhibitory)
Antithrombin (inhibitory)
Calreticulin $\rightarrow$ vasostatin (inhibitory)
High molecular weight kininogen domain 5 $\rightarrow$ kininostatin (inhibitory)

The classic descriptions by Clark and Clark in 1939 of new blood vessel formation in transparent chambers in the rabbit ear, clearly demonstrate the production of fibrinolytic activity by growing capillary sprouts:

> As new blood vessels advanced ... the fibrin outside them dissolved, leaving a clear space surrounding the new vessels which was filled at times with fluid but more commonly with a transparent gelatinous substance containing occasional blood cells, macrophages and fibroblasts. Although long, tapering, solid tips might be seen extending directly into the fibrin, which was still present beyond the advancing capillary network, the new capillaries, and even capillaries in which the circulation had not yet started, but which possessed a lumen and a rouleaux of erythrocytes, were all surrounded by a characteristic clear space from which fibrin had disappeared.

A similar phenomenon of fibrin dissolution was described in studies on acute inflammation induced by subcutaneous implantation of a fibrin clot in the rat (Kwaan and Astrup, 1964). Capillary sprouts were associated with "empty zones" in the fibrin clot, which was indicative of fibrin degradation. Fibrinolytic activity decreased progressively in the later stages of the inflammatory process, as the fibrin clot was replaced by connective tissue. These two papers were amongst the first to describe the presence of extracellular proteolytic activity in the endothelial pericellular environment during angiogenesis. They also highlighted the existence of a temporal sequence of matrix dissolution and assembly, in which fibrin provides a temporary matrix scaffold for migrating endothelial cells, which with time is replaced by a more mature collagenous matrix.

Considerable progress has since been made in understanding the mechanisms of angiogenesis. Thus, angiogenesis is controlled by the net balance between molecules which have positive and negative regulatory activity (Pepper, 1997). With the exception of angiogenesis, which occurs in response to tissue injury or in female reproductive organs, endothelial cell turnover in the healthy adult organism is very low. The maintenance of endothelial quiescence is thought to be due to the presence of endogenous negative regulators, since positive regulators are frequently detected in adult tissues in which there is apparently no angiogenesis. The converse is also true, namely that positive and negative regulators often coexist in tissues in which endothelial cell turnover is increased. This has led to the notion of the "angiogenic switch," in which endothelial activation status is determined by a balance between positive and negative regulators: In activated (angiogenic) endothelium, positive regulators predominate, whereas endothelial quiescence is achieved and maintained by the dominance of negative regulators (Figure 16.1). The switch involves either the induction of a positive regulator and/or the loss of a negative regulator (Bouck, 1990; Hanahan and Folkman, 1996). Used initially in the context of tumor progression to describe the passage from the prevascular to the vascular phase, the notion of the "switch" can also be applied to developmental, physiological, and pathological angiogenesis.

Positive regulators of angiogenesis include, but are not limited to, the vascular endothelial growth factor (VEGF) and fibroblast growth factor (FGF) families of cytokines and their tyrosine kinase receptors (Dvorak *et al.*, 1999; Ferrara, 1999; Gerwins *et al.*, 2000). Interestingly, it has been observed that VEGF and basic FGF (bFGF or FGF-2) induce a synergistic angiogenic response, both *in vitro* (Pepper *et al.*, 1992a; Goto *et al.*, 1993), and *in vivo* (Asahara *et al.*, 1995). A number of endogenous negative regulators have been described, including extracellular proteases inhibitors, thrombospondins-1 and -2, and bioactive fragments of ECM and other molecules (see below) (Cao, 1999; Keshet and Ben-Sasson, 1999; Pepper, 2001b).

The notion of the "balance" applied to positive and negative regulators of angiogenesis can also be applied to extracellular proteolysis. The resulting notion of the "proteolytic

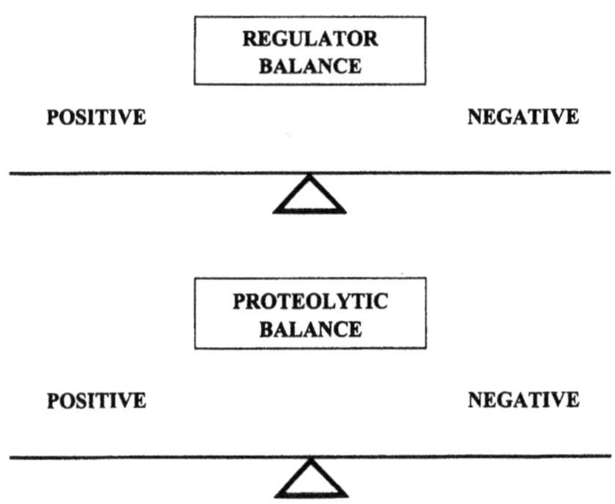

Figure 16.1. Angiogenesis—a question of balance at multiple levels. Angiogenesis is controlled by a balance between molecules with positive (e.g., cytokines VEGF and bFGF) and negative (e.g., angiostatin, thrombospondin) regulatory activity, as well as by proteases (PA–plasmin axis and MMPs) and their inhibitors (PAIs and tissue inhibitors of MMPs).

balance" (Pepper and Montesano, 1990; Pepper *et al.*, 1996), which highlights the importance of a precise protease–antiprotease equilibrium, has become one of the cornerstones on which our understanding of the molecular mechanisms of angiogenesis is based. Thus, although it has consistently been observed that endothelial cell migration and invasion are associated with increased proteolysis, protease inhibitors play an important albeit permissive role during angiogenesis by preserving matrix integrity (Figure 16.1). For example, in wound healing and inflammatory skin lesions, protease inhibitors have been localized to stromal cells surrounding vessels in regions of active vascular proliferation, where they may serve to prevent uncontrolled matrix destruction (Saarialho-Kere *et al.*, 1994). Excessive proteolysis is incompatible with normal angiogenesis, since it results either in destruction of the matrix scaffold, which is required for invasion, or in the formation of aberrant vascular structures (Montesano *et al.*, 1990). In the absence of sufficient proteolysis on the other hand, invasion and lumen formation are inhibited (Pepper *et al.*, 1990).

Another important notion is that of spatial restriction of extracellular proteolysis. The objective of spatial restriction is to amplify cell-surface activity while at the same time limiting proteolysis to the immediate pericellular environment. This is achieved through the local production of protease inhibitors as outlined above, as well as through the assembly of components of one or more protease systems on the cell surface, which brings the enzymes into close proximity to one another. The result is a highly efficient proteolytic cascade on the cell surface, which increases the efficiency of invasion without destroying the matrix scaffold, which is required for cell migration (e.g., see Mazzieri *et al.*, 1997).

In addition to its requirement for matrix invasion and capillary morphogenesis, extracellular proteolysis has been implicated in the regulation of cytokine activity (Table 16.1). Examples include activation of latent transforming growth factor-β (TGF-β) (Flaumenhaft and Rifkin, 1992), release of matrix-bound VEGF (Houck *et al.*, 1992; Bergers *et al.*, 2000), release of membrane-anchored tumor necrosis factor-α (TNF-α) (Black *et al.*, 1997), and inactivation of MCP-3, which concomitantly results in the generation of a chemokine antagonist (McQuibban *et al.*, 2000).

Finally, extracellular proteolysis has been implicated in the generation of proteolytic fragments of ECM and other molecules, a number of which have been reported to have angioregulatory activity, either positive or negative (Table 16.1) (Sage, 1997). Much attention has been focused on angiostatin and endostatin, which are negative regulators derived from plasminogen and collagen XVIII, respectively (O'Reilly *et al.*, 1994, 1997). Fragments of collagen IV, which are also negative regulators, include arresten, from the alpha 1 chain (Colorado *et al.*, 2000), canstatin, from the alpha 2 chain (Kamphaus *et al.*, 2000), and tumstatin, from the alpha 3 chain (Maeshima *et al.*, 2000a,b). Other ECM proteins from which fragments with angioregulatory activity have been derived include thrombospondin (inhibitory), hyaluronan (stimulatory), SPARC (stimulatory and inhibitory), and collagen XV/restin (inhibitory) (Ramchandran *et al.*, 1999). Non-ECM molecules from which fragments with angioinhibitory activity have been derived include prolactin, platelet factor 4, MMP-2 (PEX—see Table 16.1), calreticulin (vasostatin) (Pike *et al.*, 1998), and high molecular weight kininogen domain 5 (kininostatin) (Colman *et al.*, 2000). Finally, a cleaved form of antithrombin has also been shown to have anti-angiogenic activity (O'Reilly *et al.*, 1999). It should be pointed out however that despite a large amount of information which has accumulated on their anti-angiogenic activity, a degree of caution is still required when interpreting the effects of many of these molecules, including the mechanisms and precise molecular nature of their biological activity.

2. Role of the PA–Plasmin System in Angiogenesis

Plasmin is a broad-spectrum protease of tryptic specificity, which either directly, or indirectly through the activation of certain proMMPs, is presumed to hydrolyze many extracellular proteins, the most notable of which is fibrin (Figure 16.2). Urokinase-type PA (uPA) and tissue-type PA (tPA), the principal activators of plasminogen, are the products of separate genes, which are regulated independently by a variety of factors including hormones and cytokines. Like plasmin, uPA and tPA are serine proteases of tryptic specificity, but in contrast to plasmin, they have a very restricted substrate specificity, namely plasminogen. uPA (or more precisely its inactive precursor pro-uPA) is secreted as a soluble protein that binds with high affinity to a specific glycosyl phosphatidyl inositol-anchored cell-surface receptor (uPA receptor—uPAR), which is present on a variety of cells including endothelial cells. Plasminogen and plasmin also associate with plasma membranes, and colocalization of uPA and plasminogen on the cell surface increases the efficiency of plasminogen activation and subsequent plasmin-dependent proteolysis. The existence of multiple specific physiological inhibitors of both plasmin (i.e., α_2-antiplasmin) and PAs (i.e., PA inhibitors-1 and -2 or PAI-1 and PAI-2) provides additional points of regulation along this protease cascade.

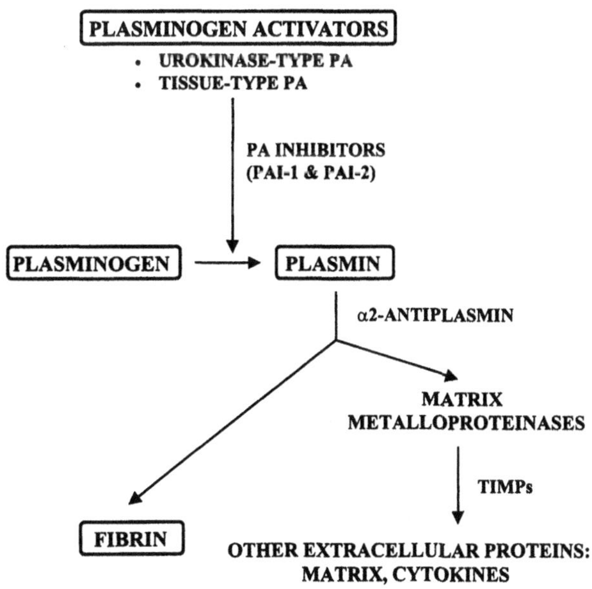

Figure 16.2. The PA–plasmin system. Schematic representation of interactions between uPA, tPA, and activation of plasminogen to plasmin (uPA is bound to a GPI-linked cell-surface receptor—not shown). The principal activity of plasmin is fibrinolysis. Plasmin also activates latent MMPs, which in turn degrade components of the ECM as well as other extracellular proteins (including cytokines). Physiological inhibitors of this system include PAI-1 and PAI-2, α2-antiplasmin, and TIMPs (tissue inhibitors of MMPs or TIMPs).

The specificity, activity, and localization of serine protease inhibitors can be regulated by the ECM and other glycoproteins (Andreasen *et al.*, 2000; Collen, 1999).

With regard to localization studies, the overall picture is that uPA, uPAR, and PAI-1 are not expressed by quiescent endothelium. tPA on the other hand has been detected in the quiescent endothelium of a small percentage of arterioles, venules, and vasa vasorum of normal human tissues. In contrast, uPA, uPAR, and PAI-1 are all expressed during angiogenesis *in vivo* in a variety of settings. uPA and uPAR appear to be expressed by endothelial cells, and depending on the situation, PAI-1 is expressed either by endothelial cells themselves, or by non-endothelial (stromal and epithelial) cells, where it is thought to play a role in the preservation of matrix integrity (Pepper *et al.*, 1996; Carmeliet and Collen, 2000). Little is known about tPA expression during angiogenesis.

These *in vivo* observations are amply supported by results obtained *in vitro*. That cultured endothelial cells (which are closer to activated than quiescent endothelium) produce uPA, uPAR, tPA, and PAI-1 has been known for some time, and their regulation by bFGF and VEGF has been described (Mignatti and Rifkin, 1996; Pepper *et al.*, 1996). It should be pointed out however that endothelial cells of human origin respond less well than do cells of bovine origin with regard to induction of PAs by bFGF and VEGF. The reason for the lack of response may be related to the chronic stimulation of these cells by components of the endothelial growth supplements (FGFs and probably VEGF) used to culture the cells. The response to VEGF also appears to be dependent on the vascular bed from which the cells were isolated. Thus, VEGF induces uPA and tPA in endothelial cells derived from the microvasculature but not in cells derived from the aorta (Cavallaro *et al.*, 2001; Pepper *et al.*, 2001). This may have important implications for the regulation of angiogenesis, which traditionally has been considered to occur from preexisting vessels of the microvasculature. Hypoxia, which is a major stimulus for angiogenesis, has been reported to increase uPAR (Kroon *et al.*, 2000) and PAI-1 (Uchiyama *et al.*, 2000) in endothelial cells. Induction of PAI-1 by hypoxia is mediated through binding of the transcription factor HIF-1 to hypoxia responsive elements in the PAI-1 promoter (Fink *et al.*, 2002). We have previously demonstrated that uPA, uPAR, and PAI-1 are increased in endothelial cells in a two-dimensional wound-induced migration assay, and that both migration and protease/inhibitor induction are dependent on endogenous bFGF (Pepper *et al.*, 1987, 1992b, 1993). Using a coculture system of endothelial cells and fibroblasts, PAI-1 mRNA and promoter activity were induced only in those fibroblasts apposed to sprouting but not resting endothelium (Bacharach *et al.*, 1998). The authors reason that this may be of importance during sprouting, which constitutes the only period during which endothelial cells establish direct contact with fibroblasts. They suggest the paracrine induction of PAI-1 provides a mechanism to negate excessive pericellular proteolysis.

Angiogenesis results in the formation of new vessels in the blood vascular system. A comparable process, called lymphangiogenesis, is responsible for the formation of new vessels in the lymphatic system. Lymphangiogenesis has traditionally been overshadowed by the greater emphasis placed on angiogenesis. This is due in part to the lack of identification of lymphangiogenic factors, as well as suitable markers with which to distinguish blood from lymphatic vascular endothelium. This scenario is changing rapidly following the discovery of the first lymphangiogenic factors (VEGF-C and -D) as well as a limited number of lymphatic endothelium-specific markers (Pepper, 2001c). In adult tissues, new lymphatic capillaries grow by sprouting as extensions from preexisting lymphatics, much in the same way as new blood capillaries arise by sprouting from the preexisting microvasculature during angiogenesis. Although very little is known about the mechanisms of lymphangiogenesis, it has been demonstrated that VEGF, VEGF-C, and bFGF are capable of stimulating uPA, uPAR, tPA, and PAI-1 in bovine large vessel lymphatic endothelial cells (Pepper *et al.*, 1994, 1998; Liu and He, 1997; Leak *et al.*, 2000). The relevance of these *in vitro* observations to lymphangiogenesis *in vivo* remains to be determined.

The generation of mice deficient in components of the PA–plasmin system (Carmeliet and Collen, 2000) provided an opportunity to assess the role of individual components of this system during angiogenesis *in vivo*. Previous studies, both *in vivo* and *in vitro*, had pointed to a crucial role for the PA–plasmin system during angiogenesis. Unexpectedly however, it was found that embryonic and postnatal development were unperturbed in these mice, at least from the perspective of an angiogenesis-associated phenotype. A seminal study on wound healing in plasminogen-deficient mice clearly defined a role for the PA–plasmin system during keratinocyte migration *in vivo* (Rømer *et al.*, 1996). However, although keratinocyte migration was severly impaired, there were no overt qualitative differences in the granulation tissue (inflammatory cell infiltration, fibroblast migration and neovascularization) between wild-type and plasminogen −/− mice, suggesting that in this setting, angiogenesis is not PA–plasmin-dependent.

Although developmental and wound-healing-associated angiogenesis appear to be unaffected in uPA, uPAR, tPA, PAI-1, and plasminogen null mice, several *in vivo* studies have demonstrated a requirement for this system during angiogenesis in other settings. For example, tumor-associated angiogenesis (vascular density) was inhibited in an athymic mouse model utilizing a human prostate carcinoma cell line, PC-3, stably transfected with human PAI-1 (Soff *et al.*, 1995). Although this study pointed to a role for PAs in tumor angiogenesis, it provided no mechanistic information, including the cell-type specificity of the effect. More precise information has been provided in a murine model of myocardial infarction, in which post-infarction myocardial revascularization (angiogenesis) was severely impaired in uPA −/− mice (Heymans *et al.*, 1999). Since this phenotype could not be rescued by exogenously added VEGF, and since endothelial and smooth muscle cells expressed uPA during revascularization, it was concluded that the defect in angiogenesis was due to impaired uPA-dependent endothelial cell invasion. Interestingly, when assayed in the rabbit cornea or on the chick chorioallantoic membrane (CAM), uPA itself had a pro-angiogenic effect (Berman *et al.*, 1982; Goldfarb *et al.*, 1986; Ribatti *et al.*, 1999). In the CAM, this could be ascribed to the release of endogenous bFGF (Ribatti *et al.*, 1999).

2.1. uPA/uPAR Interactions are Required for Angiogenesis

The importance of the interaction between uPA and uPAR during angiogenesis has been demonstrated in a number of *in vivo* systems. Thus, bFGF-induced angiogenesis in subcutaneously injected Matrigel was inhibited by a fusion protein consisting of the receptor-binding amino-terminal fragment (ATF) of uPA and the Fc portion of human IgG, which functions as a high affinity uPAR antagonist (Min *et al.*, 1996). Similarly, in syngeneic and xenograft murine tumor models, adenovirally delivered ATF specifically inhibited tumor angiogenesis (Li *et al.*, 1998). In a homologous immunocompetent rat model using tumor cells transfected with a mutant murine uPA that retains receptor binding but not proteolytic activity, microvessel density was markedly reduced (Evans *et al.*, 1997).

Tumor angiogenesis in several murine models was also inhibited by an 8-mer peptide derived from the non-receptor-binding region of uPA (Guo *et al.*, 2000; Mishima *et al.*, 2000). These studies defined a novel epitope in uPA involved in the uPA/uPAR interaction.

2.2. PAI-1 is Required for Angiogenesis

Localization studies have demonstrated that PAI-1 is strongly expressed in endothelial cells in a number of tumor types (Pepper *et al.*, 1996). Studies in PAI-1 null mice have revealed an absolute requirement for PAI-1 in tumor-induced angiogenesis. In one model, tumor and stromal cells are separated by the interposition of a type I collagen gel, which allows independent assessment of exogenously added tumor cell as well as host stromal cell invasion (Bajou *et al.*, 1998, 2001). The absence of PAI-1 (but not of uPA, uPAR, or tPA) from the host markedly impaired tumor cell invasion and host-derived neovascularization. Reversal of this phenotype was achieved by adenovirally delivered human PAI-1 to PAI-1 null mice. In wild-type mice, PAI-1 expression in the stroma could be localized to endothelial and non-endothelial cells. Interestingly, although uPA was localized to newly formed blood vessels in wild-type mice, this was replaced by tPA in mice deficient in uPA, suggesting the capacity for compensation between the PAs. The requirement for host PAI-1 (but not uPA) during tumor angiogenesis has also been demonstrated in a fibrosarcoma model in gene inactivation mice (Gutierrez *et al.*, 2000). bFGF-induced corneal angiogenesis was similarly impeded in PAI-1- but not in uPA-deficient mice.

Further studies showed that tumor growth was inhibited in PAI-1 null mice and stimulated in mice which overexpressed PAI-1 (McMahon *et al.*, 2001), and it has been reported that exogenously added PAI-1 "at therapeutic concentrations" is able to inhibit bFGF-induced angiogenesis in the chorioallantoic membrane (Stefansson *et al.*, 2001). Taken together, these observations raised the possibility that the response to PAI-1 might be biphasic or follow a bell-shaped curve. Using a rat aortic ring model of *ex vivo* angiogenesis, the pro- and anti-angiogenic effects of PAI-1 have indeed been shown to be dose dependent (Devy *et al.*, 2002). It is possible therefore to conclude that angiogenesis is inefficient in the absence of PAI-1, that at an optimal concentration of PAI-1 angiogenesis is maximal, and that at supraphysiologic concentrations, PAI-1 inhibits angiogenesis.

Two potential mechanisms may explain why PAI-1 is essential for angiogenesis. First, by protecting the ECM against excessive degradation, PAI-1 may serve to maintain the matrix scaffold required for endothelial cell migration and tube formation (Montesano *et al.*, 1987) (Figure 16.3). Second, a complex series of interactions have been described between PAI-1, uPAR, integrins and the ECM component, vitronectin (Loskutoff *et al.*, 1999; Ossowski and Aguirre-Ghiso, 2000; Preissner *et al.*, 2000). Alterations in PAI-1 expression and activity would therefore be expected to alter the adhesive, migratory, and growth properties of endothelial cells, which in turn would regulate angiogenesis. To address this question directly, adenovirus was used to deliver two mutant forms of PAI-1 to PAI-1-deficient mice using the collagen gel interposition model described above (Bajou *et al.*, 2001). The PAI-1 mutants either (a) bound vitronectin normally but failed to inhibit uPA or tPA or (b) inhibited PAs normally but had negligable binding to vitronectin. Angiogenesis was restored by the PAI-1 mutant which retained PA inhibitory activity but not by the mutant which bound vitronectin but failed to inhibit the PAs, demonstrating that the requirement for PAI-1 in this model is due to inhibition of excessive proteolysis rather

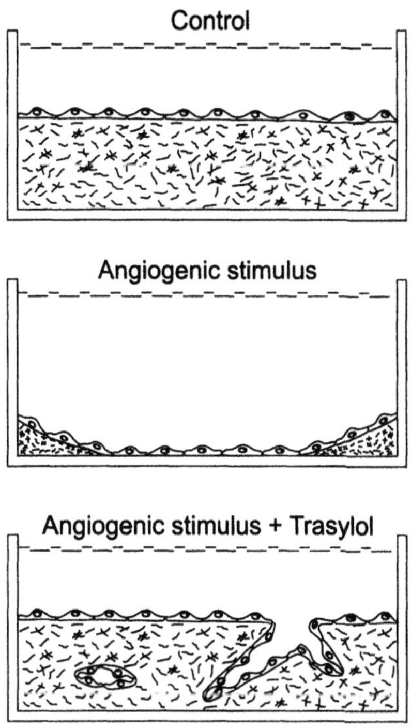

Figure 16.3. Protease inhibitors play an important permissive role during angiogenesis by preserving ECM integrity. When endothelial cells are grown to confluence on the surface of a three-dimensional fibrin gel, they form a monolayer on the surface of the gel and do not invade the underlying matrix (Control). With time, the fibrin gel is progressively lysed, indicating the production of unopposed fibrinolytic activity by the cells (not shown). Upon addition of an angiogenic stimulus, fibrinolysis is massively accelerated, and the three-dimensional matrix scaffold is rapidly destroyed, thereby precluding endothelial cell invasion and tube formation. In contrast, when a broad spectrum serine protease inhibitor such as Trasylol (aprotonin) is added to the cells, matrix integrity is preserved, which in turn allows the cells to invade the underlying matrix within which they form capillary-like tubes. See Montesano *et al.* (1987) for experimental details.

than inhibition of cellular adhesion to vitronectin. These findings were confirmed using the rat aortic ring model of *ex vivo* angiogenesis (Devy *et al.*, 2002). In a separate study using a variety of PAI-1 mutants, the inhibitory activity of PAI-1 on bFGF-induced angiogenesis in the CAM could be ascribed both to PAI-1's anti-protease activity as well as its vitronectin-binding capacity (Stefansson *et al.*, 2001).

2.3. Does tPA Play a Role in Angiogenesis?

Little is known about the role of tPA in angiogenesis. *In vivo*, tPA antigen is localized principally to the endothelium of pre-capillary arterioles and post-capillary venules (Levin and del Zoppo, 1994), the latter of which, as morphological studies have revealed, is the region of the vascular bed from which new blood vessels are formed during angiogenesis. tPA is also present in adventitial vasa vasorum, where it is thought to be produced by the sympathetic nerves that innervate these vessels (Jiang *et al.*, 2000). The presence of tPA in angiogenic endothelium has yet to be described.

In vitro, tPA has been implicated in capillary-like tube formation on matrices constituted of type I collagen, Matrigel and the explanted amnion (Mignatti *et al.*, 1989; Yasunaga *et al.*, 1989; Sato *et al.*, 1993; Ito *et al.*, 1995; Schnaper *et al.*, 1995). The requirement for tPA in endothelial cell migration *in vitro* has been demonstrated in some studies (Mawatari *et al.*, 1991) but not in others (Zablocki *et al.*, 2000). Interestingly, the combination of bFGF and VEGF induces tPA mRNA and enzyme activity in a synergistic manner in cultured endothelial cells (Pepper *et al.*, 2001). In this study, immunofluorescence analysis revealed that cell-associated tPA was colocalized with von Willebrand factor, a marker for Weibel–Palade bodies, which are regulated, endothelial cell-specific storage granules. A combination of bFGF and VEGF increased the number of tPA-positive cells as well as the number of tPA-positive granules per cell. Angiostatin, an endogenous inhibitor of angiogenesis, has been shown to inhibit tPA-catalyzed plasminogen activation in endothelial cells (Stack *et al.*, 1999).

At present one can only speculate on the function of tPA in Weibel–Palade bodies in the context of angiogenesis. An attractive role would be to protect newly forming blood vessels from thrombotic occlusion. Indeed, stagnant blood in the forming dead-end of new blood vessels would be particularly prone to coagulation, first because of the potential exposure of blood to the ECM (and in particular to collagen) in the hyperpermeable and rapidly remodeling migrating sprout; and second, because activated blood coagulation would be less easily washed away. To overcome this problem, locally generated thrombin would result in the acute release of tPA from Weibel–Palade bodies (Datta *et al.*, 1999; Rosnoblet *et al.*, 1999), which in turn would aid in the removal of fibrin deposits.

2.4. MMP Requirement for Fibrinolysis During Angiogenesis

Traditionally, physiological fibrinolysis has been ascribed to the PA–plasmin system, and studies in null mice have confirmed this notion (Collen, 1999). Surprisingly however, no overt phenotype related to aberrant angiogenesis has been observed in these mice, either during development or wound healing (see page 7). This raised the possibility that other protease systems might assume a fibrinolytic role during angiogenesis, that is, that in contrast to other cell types, endothelial cells invade fibrin matrices in a plasmin-independent manner (Montesano *et al.*, 1987). This hypothesis has been confirmed in an *ex vivo* model in which tissue fragments from null mice are embedded in a three-dimensional matrix (Hiraoka *et al.*, 1998). Thus, using inhibitors of serine, cysteine, and aspartate proteinases, it was found that endothelial cell invasion and capillary morphogenesis progressed unabated. In contrast, these processes were completely inhibited by natural and synthetic MMP inhibitors. When a series of MMPs including MMPs-1, -2, -3 and a soluble form of MT1-MMP were assessed for their fibrinolytic activity in a cell-free assay, all had the capacity to solubilize fibrin gels, and this could be completely blocked by MMP inhibitors. Quantitatively, the most potent fibrinolytic activity was ascribed to MT1-MMP (Hiraoka *et al.*, 1998). Further studies identified MT2- and MT3-MMPs as alternate membrane-anchored MMPs that drive fibrin invasive activity (Hotary *et al.*, 2002). The requirement for MT1-MMP in both collagen and fibrin gel invasion during *in vitro* angiogenesis has recently been reported (Galvez *et al.*, 2001). Furthermore, MT1-MMP expression and activity were upregulated on migrating endothelial cells.

These unexpected and highly novel observations suggest that when endothelial cells invade a provisional fibrin ECM scaffold, this occurs in a plasmin-independent manner, and may require the activity of MMPs such as MT1-MMP. These observations have potentially far-reaching mechanistic and therapeutic implications.

2.5. Angiostatin and other Angiostatic Derivatives of Plasminogen

Angiostatin is an internal proteolytically derived fragment of plasminogen spanning the first four kringle domains (O'Reilly *et al.*, 1994). It has been suggested that, dependent on the type of tumor, either tumor cells themselves or tumor-associated macrophages express enzymatic activity which is capable of converting inactive plasminogen into the angiogenesis-inhibiting proteolytic fragment, angiostatin (reviewed by Pepper, 2001b; Moser *et al.*, 2002). Thus, metalloelastase (MMP-12), gelatinase B (MMP-9), stromelysin-1 (MMP-3), matrilysin (MMP-7), cathepsin D, and prostate-specific antigen are capable of hydrolyzing plasminogen to generate angiostatin. An angiostatin-like molecule can also be generated by plasmin autodigestion (Gately *et al.*, 1996, 1997; Stathakis *et al.*, 1997, 1999; Fitzpatrick *et al.*, 2000; Kassam *et al.*, 2001).

Angiostatin reduces endothelial cell number in proliferation assays, inhibits endothelial cell migration, and prevents endothelial cell invasion and tube formation in three-dimensional *in vitro* angiogenesis assays (reviewed by Pepper, 2001b; Moser *et al.*, 2002). Angiostatin induces endothelial cell apoptosis (Claesson-Welsh *et al.*, 1998; Lucas *et al.*, 1998). Angiostatin has recently been shown to bind specifically to endothelial cell integrin $\alpha v\beta 3$ (Tarui *et al.*, 2001). This integrin is expressed by angiogenic endothelial cells, and has been shown to be important for angiogenesis in certain settings. Most of the *in vitro* effects of angiostatin described above have been observed on endothelial cells of human and bovine origin. Angiostatin has been reported to bind ATP synthase on the surface of human endothelial cells (Moser *et al.*, 1999). Angiostatin has also been shown to directly inhibit neutrophil chemotaxis as well as neutrophil-mediated angiogenesis (Benelli *et al.*, 2002).

In vivo, angiostatin inhibits spontaneous angiogenesis in the CAM (O'Reilly *et al.*, 1994) as well as bFGF-induced corneal angiogenesis (Gately *et al.*, 1996; Morikawa *et al.*, 2000). Recombinant angiostatin also inhibits primary tumor growth as well as angiogenesis-dependent growth of metastases in mice (Cao *et al.*, 1996; O'Reilly *et al.*, 1996; Gately *et al.*, 1997; Sim *et al.*, 1997; Wu *et al.*, 1997; Kirsch *et al.*, 1998). Angiostatin derived from stably transfected tumor cells (Cao *et al.*, 1998) as well as adenovirus- or adeno-associated virus (AAV)-delivered angiostatin (Griscelli *et al.*, 1998; Gyorffy *et al.*, 2001a; Ma *et al.*, 2002a,b), inhibits primary and metastatic tumor growth and tumor angiogenesis. AAV-delivered angiostatin was effective when administered either intratumorally or intramuscularly (Ma *et al.*, 2002a,b). The angiostatin cDNA, delivered both local-and systemically by liposomes, has been shown to inhibit primary and secondary tumor formation (Sacco *et al.*, 2000, 2001; Rodolfo *et al.*, 2001). Recombinant angiostatin has been shown to reduce VEGF and bFGF levels in primary tumors (Kirsch *et al.*, 1998). Angiostatin gene transfer has also been shown to be effective in reducing collagen-induced arthritis (Kim *et al.*, 2002) as well as choroidal neovascularization (Lai *et al.*, 2001) and oxygen-induced retinopathy in mice (Drixler *et al.*, 2001).

When administered systemically, angiostatin does not induce toxicity or interfere with organ vascularization and growth (Drixler *et al.*, 2001). Interestingly, angiostatin has recently been shown to impair anastomotic wound healing in mice (te Velde *et al.*, 2002). However, on discontinuation of anti-angiogenic therapy, normal anastomotic healing was promptly restored.

A combination of angiostatin and ionizing radiation has proven to be significantly more efficaceous than either therapy alone for inhibiting tumor growth (Gorski *et al.*, 1998; Mauceri *et al.*, 1998; Griscelli *et al.*, 2000). Synergistic inhibition of primary tumor growth has been observed with combined angiostatin and endostatin therapy (Yokoyama *et al.*, 2000; Scappaticci *et al.*, 2001). A synergistic cytotoxic effect was observed with angiostatin and photodynamic therapy on endothelial but not non-endothelial cells (Renno *et al.*, 2000), and the inhibitory activity of angiostatin delivered via a gene therapy approach was potentiated when combined with the suicidal thymidine kinase gene (Ma *et al.*, 2002b). Finally, it has recently been suggested that the combination of angiostatin and IL-12 gene therapy can effectively shrink a solid tumor and vaccinate the animal against rechallenge (Gyorffy *et al.*, 2001b). The rationale for this therapy is to limit tumor size by attacking the vasculature with angiostatin, thereby allowing IL-12 to mount a T cell-specific response against tumor antigens.

Angiostatin spans the first four kringle domains of plasminogen. However, angioinhibitory activity has also been shown to reside in other plasminogen kringles, either alone or in combination, most notably in kringle 5 or kringles 1–5 (Cao *et al.*, 1997; Ji *et al.*, 1998a,b). It has recently been shown that normal and tumor cells can convert plasminogen into a 22 kDa fragment named p22, which consists of amino acid residues 78–180 of plasminogen and therefore includes the first plasminogen kringle (residues 84–162) as well as additional N- and C-terminal residues (Kwon *et al.*, 2001). p22 inhibited proliferation of endothelial but not non-endothelial cells, prevented vascular growth in the CAM, and suppressed the growth of murine Lewis lung carcinoma metastatic foci.

The tumor microvasculature is an important target in strategies aimed at inhibiting tumor growth. Angiostatin apparently does not induce systemic cytotoxicity, which suggests that its antitumor effect is mediated by specific targeting of angiogenic endothelial cells. However, with the exception of ATP synthase, which may serve as a cell-surface binding site for angiostatin, little is known about the molecules with which this inhibitor interacts on the cell surface. Similarly, the signaling pathways which lead to its angioinhibitory effect have not been described. Therefore, despite the large amount of preclinical information which has accumulated on its antitumor activity, a degree of caution is still required when interpreting the effects of this molecule.

3. Requirement for the PA–Plasmin System in Polyoma Virus Middle T Oncogene-induced Vascular Tumor Formation

Endothelial cells are the primary target of the transforming ablilty of the polyomavirus middle T (PymT) oncogene. Tumors induced by PymT (called endotheliomas) have been used as a model system for endothelial cell tumors. Although significant morphological differences exist between endotheliomas and human vascular tumors (including hemangiomas), the PymT system does provide a good functional model for hemangiomas because (a) the tumors are endothelial specific and organ non-specific, (b) there appears to be rapid uncontrolled endothelial cell proliferation, (c) mice develop many signs of the Kasabach–Meritt syndrome (i.e., thrombocytopenia, anemia, and associated splenomegaly), which may be a life-threatening complication of hemangiomas in humans. It has also been suggested that endotheliomas may serve as a model system for vascular malformations.

This is due to the cavernous nature of the lesions, which is a characteristic feature of cavernous angiomas in humans. Finally, since recruited host cells are an important component of these tumors, endotheliomas have also been used as a model system for Kaposi's sarcoma (Pepper *et al.*, 1997).

PymT-transformed endothelial cells (endothelioma or End. cells), which can be isolated from endotheliomas with relative ease, have the potential to grow indefinitely, and retain many important features of differentiated endothelium. Endotheliomas can be induced not only by PymT itself, but also following subcutaneous or intraperitoneal injection of End. cells. The following is a working hypothesis for endothelioma formation following subcutaneous End. cell injection (Pepper *et al.*, 1997). End. cells invade and integrate into host blood vessels leading to their rupture and subsequent haemorrhage. This in turn is followed by rapid recruitment of host endothelial and non-endothelial cells. Recruited endothelial cells form new capillary blood vessels which feed the growing cystic tumor. Recruitment is dependent on the continuous presence of End. cells, and mainly involves host cell migration with low levels of proliferation. The signal for host endothelial cell recruitment is not known, but probably originates from the injected End. cells as well as from the intense inflammatory reaction observed at the periphery of the haemorrhagic cysts. It is likely that in addition to host cell recruitment, proliferating End. cells themselves contribute significantly to tumor expansion during the later phases of tumor growth. Taraboletti *et al.* have partially characterized an End. cell-derived motility factor which may be responsible for recruitment of host endothelial cells (Taraboletti *et al.*, 1993; Vergani *et al.*, 2001).

An *in vitro* model has been developed to study the autonomous morphogenetic behaviour of End. cells (Montesano *et al.*, 1990). This consists of growing End. cells in suspension in three-dimensional fibrin gels. In fibrin gels, End. cells proliferate rapidly, and after 10–14 days form large cystic structures lined by a monolayer of endothelial cells. These cyst-like structures are strikingly reminiscent of the cavernous endotheliomas which develop *in vivo*. Under the same experimental conditions, non-transformed endothelial cells from a variety of species and organs form elongated branching tube-like structures; in no instance are cyst-like structures seen. The fibrin gel model, in which normal vascular morphogenesis is disrupted, that is, formation of endothelial-lined cyst- rather than tube-like structures, provides a powerful *in vitro* system for investigating the mechanisms of protease-dependent tumor formation.

With regard to the PA–plasmin system and endothelioma formation, End. cells display increased PA activity when compared to non-transformed endothelial cells, which can be accounted for both by an increase in uPA, and in some cell lines by a concominant decrease in PAI-1 (Montesano *et al.*, 1990; Dubois-Stringfellow *et al.*, 1994; Besser *et al.*, 1995; Primo *et al.*, 2000). The mechanism for this increase is not known, although it has been shown that PymT increases uPA gene activation without influencing uPA mRNA stability (Besser *et al.*, 1995). Taken together with the observation that transformed endothelial cells form cyst-like structures when embedded in fibrin gels, these findings demonstrate that excessive proteolysis is associated with aberrant vascular morphogenesis. Reduction of this activity by inhibiting plasmin restores normal morphogenetic properties to these cells, that is, the formation of capillary-like tubes. These findings clearly implicate increased PA–plasmin-mediated proteolysis in aberrant vascular morphogenesis *in vitro* (Montesano *et al.*, 1990).

Based on these *in vitro* observations, it was hypothesized that the PA–plasmin system would also be required for endothelioma formation *in vivo*. This was tested directly in uPA-, tPA-, PAI-1- and plasminogen-deficient mice. Essentially, tumor formation following viral infection/transduction was unaffected in the different genotypes, suggesting that the primary transformation process is PA–plasmin-independent. However, in tumors induced by injection of wild-type End. cell, tumor incidence and latency were increased in uPA −/− and double transgenic uPA : tPA −/− mice, but not in tPA−/− or PAI-1−/−mice. Tumor incidence and latency were also increased when uPA−/− and uPA : tPA −/− End. cells were injected into wild-type mice (Sabapathy *et al.*, 1997). These genetic studies demonstrated that components of the PA–plasmin system, and in particular uPA, are critical for tumor growth *in vivo*. With respect to the *in vitro* morphogenetic behaviour of End. cells derived from protease knockout mice, while PA activity was necessary, it did not appear to be sufficient for cyst formation (Sabapathy *et al.*, 1997).

The plasminogen-dependent and -independent proteolytic activities of wild-type and mutant End. cells have been determined (Lijnen *et al.*, 1997). Thus, plasminogen activation with wild-type and tPA −/− cells was comparable, but was reduced 4-fold with uPA −/− cells and increased 3-fold with PAI-1 −/− cells. Lysis of an ^{125}I-labeled matrix in the presence of plasminogen was comparable in wild-type, tPA −/−, and PAI-1 −/− cells, but was significantly reduced in uPA −/− cells. Lysis of a human umbilical vein-derived endothelial cell ^{3}H-proline-labeled ECM in the presence of plasminogen was comparable between wild-type, tPA −/−, and PAI-1 −/− cells, but was virtually abolished in uPA −/− cells. These findings provide a mechanism for the reduction in tumor growth seen in the absence of uPA (Sabapathy *et al.*, 1997).

Finally, using the *in vitro* fibrin gel model, we have infrequently observed the formation of cyst-like structures by End. cells deficient in both uPA and tPA (unpublished observation). Clonal populations of these cells have been obtained, in which cyst formation can be completely inhibited by synthetic MMP inhibitors (unpublished observation). This implicates another protease system(s) (possibly MMPs) in the fibrinolysis required for cyst formation, and is in accord with the emerging notion of MMP-mediated fibrinolysis in endothelial cells.

4. Conclusions and Perspectives

Increased understanding of the mechanisms of angiogenesis has opened up novel and exciting potential therapeutic avenues, and considerable benefit can be envisaged in the clinical setting from manipulating angiogenesis, either positively or negatively (Pepper, 1997; Griffioen and Molema, 2000).

The studies on the PA–plasmin system cited in this chapter provide ample evidence for the requirement for this system during angiogenesis *in vivo*. In particular, a role for uPAR and PAI-1 has been clearly defined, with uPA playing a more ancillary role. The counterintuitive possibility that inhibition of PAI-1 might serve to inhibit angiogenesis is one that merits further investigation. To date, little is known about the role of tPA in angiogenesis.

The tumor vasculature has become a key target in the war against cancer (Eatock *et al.*, 2000). Increased understanding of the complex role of the PA–plasmin system in the regulation of angiogenesis may in the future provide a rational basis for the development of anti-proteolysis-based therapeutic strategies for cancer and other angiogenesis-associated diseases.

Acknowledgments

Work in the author's laboratory has been supported principally by the Swiss National Science Foundation and the Department of Public Instruction of the Canton of Geneva.

References

Andreasen, P.A., Egelund, R., and Petersen, H.H. (2000). The plasminogen activation system in tumor growth, invasion, and metastasis. *Cell. Mol. Life Sci.* **57**, 25–40.

Asahara, T., Bauters, C., Zheng, L.P., Takeshita, S., Bunting, S., Ferrara, N., Symes, J.F., and Isner, J.M. (1995). Synergistic effect of vascular endothelial growth factor and basic fibroblast growth factor on angiogenesis *in vivo*. *Circulation* **92**(Suppl. II), II-365–II-371.

Bacharach, E., Itin, A., and Keshet, E. (1998). Apposition-dependent induction of plasminogen activator inhibitor type-1 expression: A mechanism for balancing pericellular proteolysis during angiogenesis. *Blood* **92**, 939–945.

Bajou, K., Masson, V., Gerard, R.D., Schmitt, P.M., Albert, V., Praus, M., Lund, L.R., Frandsen, T.L., Brunner, N., Dano, K., Fusenig, N.E., Weidle, U., Carmeliet, G., Loskutoff, D., Collen, D., Carmeliet, P., Foidart, J.M., and Noël, A. (2001).The plasminogen activator inhibitor PAI-1 controls *in vivo* tumor vascularization by interaction with proteases, not vitronectin: Implications for antiangiogenic strategies. *J. Cell Biol.* **152**, 777–784.

Bajou, K., Noel, A., Gerard, R.D., Masson, V., Brunner, N., Holst-Hansen, C., Skobe, M., Fusenig, N.E., Carmeliet, P., Collen, D., and Foidart, J.M. (1998). Absence of host plasminogen activator inhibitor 1 prevents cancer invasion and vascularization. *Nat. Med.* **4**, 923–928.

Benelli, R., Morini, M., Carrozzino, F., Ferrari, N., Minghelli, S., Santi, L., Cassatella, M., Noonan, D.M., and Albini, A. (2002). Neutrophils as a key cellular target for angiostatin: Implications for regulation of angiogenesis and inflammation. *FASEB J.* **16**, 267–269.

Bergers, G., Brekken, R., McMahon, G., Vu, T.H., Itoh, T., Tamaki, K., Tanzawa, K., Thorpe, P., Itohara, S., Werb, Z., and Hanahan, D. (2000). Matrix metalloproteinase-9 triggers the angiogenic switch during carcinogenesis. *Nat. Cell Biol.* **2**, 737–744.

Berman, M., Winthrop, S., Ausprunk, D., Rose, J., Langer, R., and Gage, J. (1982). Plasminogen activator (urokinase) causes vascularization of the cornea. *Invest. Ophthalmol. Vis. Sci.* **22**, 191–199.

Besser, D., Urich, M., Sakaue, M., Messerschmitt, A., Ballmer-Hofer, K., and Nagamine, Y. (1995). Urokinase-type plasminogen activator gene regulation by polyomavirus middle-T antigen. *Oncogene* **11**, 2383–2391.

Black, R.A., Rauch, C.T., Kozlosky, C.J., Peschon, J.J., Slack, J.L., Wolfson, M.F., Castner, B.J., Stocking, K.L., Reddy, P., Srinivasan, S., Nelson, N., Boiani, N., Schooley, K.A., Gerhart, M., Davis, R., Fitzner, J.N., Johnson, R.S., Paxton, R.J., March, C.J., and Cerretti, D.P. (1997). A metalloproteinase disintegrin that releases tumour-necrosis factor-alpha from cells. *Nature* **385**, 729–733.

Bouck, N. (1990). Tumor angiogenesis: The role of oncogenes and tumor suppressor genes. *Cancer Cells* **2**, 179–185.

Cao, Y. (1999). Therapeutic potentials of angiostatin in the treatment of cancer. *Haematologica* **84**, 643–650.

Cao, Y., Chen, A., An, S.S.A., Ji, R.W., Davidson, D., Cao, Y., and Llinás, M. (1997). Kringle 5 of plasminogen is a novel inhibitor of endothelial cell growth. *J. Biol. Chem.* **272**, 22924–22928.

Cao, Y., Ji, R.W., Davidson, D., Schaller, J., Marti, D., Söhndel, S., McCance, S.G., O'Reilly, M.S., Llinás, M., and Folkman, J. (1996). Kringle domains of human angiostatin. *J. Biol. Chem.* **271**, 29461–29467.

Cao, Y., O'Reilly, M.S., Marshall, B., Flynn, E., Ji, R.W., and Folkman, J. (1998). Expression of angiostatin cDNA in a murine fibrosarcoma suppresses primary tumor growth and produces long-term dormancy of metastases. *J. Clin. Invest.* **101**, 1055–1063.

Carmeliet, P., and Collen, D. (2000). Transgenic mouse models in angiogenesis and cardiovascular disease. *J. Pathol.* **190**, 387–405.

Cavallaro, U., Tenan, M., Castelli, V., Perilli, A, Maggiano, N., Van Meir, E., Montesano, R., Soria, M.R., and Pepper M.S. (2001). The response of bovine endothelial cells to FGF-2 and VEGF is dependent on their site of origin: Relevance to the regulation of angiogenesis. *J. Cell Biochem.* **82**, 619–633.

Claesson-Welsh, L., Welsh, M., Ito, N., Anand-Apte, B., Soker, S., Zetter, B., O'Reilly, M., and Folkman, J. (1998). Angiostatin induces endothelial cell apoptosis and activation of focal adhesion kinase independently of the integrin-binding motif RGD. *Proc. Natl Acad. Sci. USA* **95**, 5579–5583.

Clark, E.R. and Clark, E.L. (1939). Microscopic observations on the growth of blood capillaries in the living mammal. *Am. J. Anat.* **64**, 251–301.

Collen, D. (1999). The plasminogen (fibrinolytic) system. *Thromb. Hemost.* **82**, 259–270.

Colman, R.W., Jameson, B.A., Lin, Y., Johnson, D., and Mousa, S.A. (2000). Domain 5 of high molecular weight kininogen (kininostatin) down-regulates endothelial cell proliferation and migration and inhibits angiogenesis. *Blood* **95**, 543–550.

Colorado, P.C., Torre, A., Kamphaus, G., Maeshima, Y., Hopfer, H., Takahashi, K., Volk, R., Zamborsky, E.D., Herman, S., Sarkar, P.K., Ericksen, M.B., Dhanabal, M., Simons, M., Post, M., Kufe, D.W., Weichselbaum, R.R., Sukhatme, V.P., and Kalluri, R. (2000). Anti-angiogenic cues from vascular basement membrane collagen. *Cancer Res.* **60**, 2520–2526.

Datta, Y.H., Youssoufian, H., Marks, P.W., and Ewenstein, B.M. (1999). Targeting of a heterologous protein to a regulated secretion pathway in cultured endothelial cells. *Blood* **94**, 2696–703.

Devy, L., Blacher, S., Grignet-Debrus, C., Bajou, K., Masson, V., Gerard, R.D., Gils, A., Carmeliet, G., Carmeliet, P., Declerck, P.J., Noel, A., and Foidart, J.M. (2002). The pro- or antiangiogenic effect of plasminogen activator inhibitor 1 is dose dependent. *FASEB J.* **16**, 147–154.

Drixler, T.A., Rinkes, I.H., Ritchie, E.D., Treffers, F.W., van Vroonhoven, T.J., Gebbink, M.F., and Voest, E.E. (2001). Angiostatin inhibits pathological but not physiological retinal angiogenesis. *Invest. Ophthalmol. Vis. Sci.* **42**, 3325–3330.

Dubois-Stringfellow, N., Jonczyk, J., and Bautch, V.L. (1994). Perturbations in the fibrinolytic pathway abolish cyst formation but not capillary-like organization of cultured murine endothelial cells. *Blood* **83**, 3206–3217.

Dvorak, H.F., Nagy, J.A., Feng, D., Brown, L.F., and Dvorak, A.M. (1999). Vascular permeability factor/ vascular endothelial growth factor and the significance of microvascular hyperpermeability in angiogenesis. *Curr. Top. Microbiol. Immunol.* **237**, 97–132.

Eatock, M.M., Schätzlein, A., and Kaye, S.B. (2000). Tumour vasculature as a target for anticancer therapy. *Cancer Treatment Rev.* **26**, 191–204.

Evans, C.P., Elfman, F., Parangi, S., Conn, M., Cunha, G., and Shuman, M.A. (1997). Inhibition of prostate cancer neovascularization and growth by urokinase-plasminogen activator receptor blockade. *Cancer Res.* **57**, 3594–3599.

Ferrara, N. (1999). Vascular endothelial growth factor: Molecular and biological aspects. *Curr. Top. Microbiol. Immunol.* **237**, 1–30.

Fink, T., Kazlauskas, A., Poellinger, L., Ebbesen, P., and Zachar, V. (2002). Identification of a tightly regulated hypoxia-response element in the promoter of human plasminogen activator inhibitor-1. *Blood* **99**, 2077–2083.

Fitzpatrick, S.L., Kassam, G., Choi, K.S., Kang, H.M., Fogg, D.K., and Waisman, D.M. (2000). Regulation of plasmin activity by annexin II tetramer. *Biochemistry* **39**, 1021–1028.

Flaumenhaft, R. and Rifkin, D.B. (1992). The extracellular regulation of growth factor action. *Mol. Biol. Cell* **3**, 1057–1065.

Galvez, B.G., Matias-Roman, S., Albar, J.P., Sanchez-Madrid, F., and Arroyo, A.G. (2001). Membrane type 1-matrix metalloproteinase is activated during migration of human endothelial cells and modulates endothelial motility and matrix remodeling. *J. Biol. Chem.* **276**, 37591–37500.

Gately, S., Twardowski, P., Stack, M.S., Cundiff, D.L., Grella, D., Castellino, F.J., Enghild, J., Kwaan, H.C., Lee, F., Kramer, R.A., Volpert, O., Bouck, N., and Soff, G.A. (1997). The mechanism of cancer-mediated conversion of plasminogen to the angiogenesis inhibitor, angiostatin. *Proc. Natl. Acad. Sci. USA* **94**, 10868–10872.

Gately, S., Twardowski, P., Stack, M.S., Patrick, M., Boggio, L., Cundiff, D.L., Schapner, H.W., Madison, L., Volpert, O., Bouck, N., Enghild, J., Kwaan, H.C., and Soff, G.A. (1996). Human prostate carcinoma cells express enzymatic activity that converts human plasminogen to the angiogenesis inhibitior, angiostatin. *Cancer Res.* **56**, 4887–4890.

Gerwins, P., Skoldenberg, E., and Claesson-Welsh, L. (2000). Function of fibroblast growth factors and vascular endothelial growth factors and their receptors in angiogenesis. *Crit. Rev. Oncol. Hematol.* **34**, 185–194.

Goldfarb, R.H., Ziche, M., Murano, G., and Liotta, L.A. (1986). Plasminogen activators (urokinase) mediate neovascularization: Possible role in tumor angiogenesis. *Semin. Thromb. Hemost.* **12**, 337–338.

Gorski, D.H., Mauceri, H.J., Salloum, R.M., Gately, S., Hellman, S., Beckett, M.A., Sukhatme, V.P., Soff, G.A., Kufe, D.W., and Weichselbaum, R.R. (1998). Potentiation of the antitumor effect of ionizing radiation by brief concomitant exposures to angiostatin. *Cancer Res.* **58**, 5686–5689.

Goto, F., Goto, K., Weindel, K., and Folkman, J. (1993). Synergistic effects of vascular endothelial growth factor and basic fibroblast growth factor on the proliferation and cord formation of bovine capillary endothelial cells within collagen gels. *Lab. Invest.* **69**, 508–517.

Griffioen, A.W. and Molema, G. (2000). Angiogenesis: Potentials for pharmacologic intervention in the treatment of cancer, cardiovascular diseases, and chronic inflammation. *Pharmacol. Rev.* **52**, 237–268.

Griscelli, F., Li, H., Bennaceur-Griscelli, A., Soria, J., Opolon, P., Soria, C., Perricaudet, M., Yeh, P., and Lu, H. (1998). Angiostatin gene transfer: Inhibition of tumor growth *in vivo* by blockage of endothelial cell proliferation associated with a mitosis arrest. *Proc. Natl. Acad. Sci. USA* **95**, 6367–6372.

Griscelli, F., Li, H., Cheong, C., Opolon, P., Bennaceur-Griscelli, A., Vassal, G., Soria, J., Soria, C., Lu, H., Perricaudet, M., and Yeh, P. (2000). Combined effects of radiotherapy and angiostatin gene therapy in glioma tumor model. *Proc. Natl. Acad. Sci. USA* **97**, 6698–6703.

Guo, Y., Higazi, A.A., Arakelian, A., Sachais, B.S., Cines, D., Goldfarb, R.H,, Jones, T.R., Kwaan, H., Mazar, A.P., and Rabbani, S.A. (2000). A peptide derived from the nonreceptor binding region of urokinase plasminogen activator (uPA) inhibits tumor progression and angiogenesis and induces tumor cell death *in vivo*. *FASEB J.* **14**, 1400–1410.

Gutierrez, L.S., Schulman, A., Brito-Robinson, T., Noria, F., Ploplis, V.A., and Castellino, F.J. (2000). Tumor development is retarded in mice lacking the gene for urokinase-type plasminogen activator or its inhibitor, plasminogen activator inhibitor-1. *Cancer Res.* **60**, 5839–5847.

Gyorffy, S., Palmer, K., and Gauldie, J. (2001a). Adenoviral vector expressing murine angiostatin inhibits a model of breast cancer metastatic growth in the lungs of mice. *Am. J. Pathol.* **159**, 1137–1147.

Gyorffy, S., Palmer, K., Podor, T.J., Hitt, M., and Gauldie, J. (2001b). Combined treatment of a murine breast cancer model with type 5 adenovirus vectors expressing murine angiostatin and IL-12: A role for combined anti-angiogenesis and immunotherapy. *J. Immunol.* **166**, 6212–6217.

Hanahan, D. and Folkman, J. (1996). Patterns and emerging mechanisms of the angiogenic switch during tumorigenesis. *Cell* **86**, 353–364.

Heidtmann, H.H., Nettelbeck, D.M., Mingels, A., Jäger, R., Welker, H.G., and Knotermann, R.E. (1999). Generation of angiostatin-like fragments from plasminogen by prostate-specific antigen. *Br. J. Cancer* **81**, 1269–1273.

Heymans, S., Luttun, A., Nuyens, D., Theilmeier, G., Creemers, E., Moons, L., Dyspersin, G.D., Cleutjens, J.P., Shipley, M., Angellilo, A., Levi, M., Nube, O., Baker, A., Keshet, E., Lupu, F., Herbert, J.M., Smits, J.F., Shapiro, S.D., Base, M., Borgers, M., Collen, D., Daemen, M.J., and Carmeliet, P. (1999). Inhibition of plasminogen activators or matrix metalloproteinases prevents cardiac rupture but impairs therapeutic angiogenesis and causes cardiac failure. *Nat. Med.* **5**, 1135–1142.

Hiraoka, N., Allen, E., Appel, I.J., Gyetko, M.R., and Weiss, S.J. (1998). Matrix metalloproteinases regulate neovascularization by acting as pericellular fibrinolysins. *Cell* **95**, 365–377.

Hotary, K.B., Yana, I., Sabeh, F., Li, X.-Y., Holmbeck, K., Birkendal-Hansen, H., Allen, E.D., Hiraoka, N., and Weiss, S.J. (2002). Matrix metaloproteinases (MMPs) regulate fibrin invasive activity via MT1-MMP-dependent and -independent processes. *J. Exp. Med.* **195**, 295–308.

Houck, K.A., Leung, D.W., Rowland, A.M., Winer, J., and Ferrara, N. (1992). Dual regulation of vascular endothelial growth factor bioavailability by genetic and proteolytic mechanisms. *J. Biol. Chem.* **267**, 26031–26037.

Ito, K.I., Ryuto, M., Ushiro, S., Ono, M., Sugenoya, A., Kuraoka, A., Shibata, Y., and Kuwano, M. (1995). Expression of tissue-type plasminogen activator and its inhibitor couples with development of capillary network by human microvascular endothelial cells on Matrigel. *J. Cell Physiol.* **162**, 213–224.

Ji, W.R., Barrientos, L.G., Llinás, M., Gray, H., Villarreal, X., DeFord, M.E., Castellino, F.J., Kramer, R.A., and Trail, P.A. (1998a). Selective inhibition by kringle 5 of human plasminogen on endothelial cell migration, an important process in angiogenesis. *Biochem. Biophy. Res. Commun.* **247**, 414–419.

Ji, W.R., Castellino, F.J., Chang, Y., Deford, M.E., Gray, H., Villarreal, X., Eghtedarzadeh, K., Marti, D.N., Llinas, M., Schaller, J., Kramer, R.A., and Trail, P.A. (1998b). Characterization of kringle domains of angiostatin as antagonists of endothelial cell migration, an important process in angiogenesis. *FASEB J.* **12**, 1731–1738.

Jiang, X., Wang, Y., Hand, A.R., Gillies, C., Cone, R.E. and O'Rourke, J. (2000). Presence of tissue plasminogen activator (t-PA) in the adventitial sympathetic nerves that innervate small arteries: Morphologic evidence for a neural fibrinolysis. *Fibrinol. Proteol.* **14**, 35–26,.

Kamphaus, G.D., Colorado, P.C., Panka, D.J., Hopfer, H., Ramchandran, R., Torre, A., Maeshima, Y., Mier, J.W., Sukhatme, V.P., and Kalluri, R. (2000). Canstatin, a novel matrix-derived inhibitor of angiogenesis and tumor growth. *J. Biol. Chem.* **275**, 1209–1215.

Kassam, G., Kwon, M., Yoon, C.S., Graham, K.S., Young, M.K., Gluck, S., and Waisman, D.M. (March, 23 2001). Purification and characterization of A61. An angiostatin-like plasminogen fragment produced by plasmin autodigestion in the absence of sulfhydryl donors. *J. Biol. Chem.* **276**(12), 8924–8933.

Keshet, E. and Ben-Sasson, S.A. (1999). Anticancer drug targets: Approaching angiogenesis. *J. Clin. Invest.* **104**, 1497–1501.

Kim, J.M., Ho, S.H., Park, E.J., Hahn, W., Cho, H., Jeong, J.G., Lee, Y.W., and Kim, S. (2002). Angiostatin gene transfer as an effective treatment strategy in murine collagen-induced arthritis. *Arthritis Rheum.* **46**, 793–801.

Kirsch, M., Strasser, J., Allende, R., Bello, L., Zhang, J., and Black, P.McL. (1998). Angiostatin suppresses malignant glioma growth *in vivo*. *Cancer Res.* **58**, 4654–4659.

Kroon, M.E., Koolwijk, P., van Der Vecht, B., and van Hinsbergh, V.W.M. (2000). Urokinase receptor expression on human microvascular endothelial cells is increased by hypoxia: Implications for capillary-like tube formation in a fibrin matrix. *Blood* **96**, 2775–2783.

Kwaan, H.C. and Astrup, T. (1964). Fibrinolytic activity of reparative connective tissue. *J. Path. Bact.* **87**, 409–414.

Kwon, M., Yoon, C.S., Fitzpatrick, S., Kassam, G., Graham, K.S., Young, M.K., and Waisman, D.M. (2001). p22 is a novel plasminogen fragment with antiangiogenic activity. *Biochemistry* **40**, 13246–13253.

Lai, C.C., Wu, W.C., Chen, S.L., Xiao, X., Tsai, T.C., Huan, S.J., Chen, T.L., Tsai, R.J., and Tsao, Y.P. (2001). Suppression of choroidal neovascularization by adeno-associated virus vector expressing angiostatin. *Invest. Ophthalmol. Vis. Sci.* **42**, 2401–2407.

Leak, L.V., Saunders, M., Day, A.A., and Jones, M. (2000). Stimulation of plasminogen activator and inhibitor in the lymphatic endothelium. *Microvasc. Res.* **60**, 201–211.

Levin, E.G. and del Zoppo, G.J. (1994). Localization of tissue plasminogen activator in the endothelium of a limited number of vessels. *Am. J. Pathol.* **144**, 855–861.

Li, H., Lu, H., Griscelli, F., Opolon, P., Sun, L.Q., Ragot, T., Legrand, Y., Belin, D., Soria, J., Soria, C., Perricaudet, M., and Yeh, P. (1998). Adenovirus-mediated delivery of a uPA/uPAR antagonist suppresses angiogenesis-dependent tumor growth and dissemination in mice. *Gene Ther.* **5**, 1105–1113.

Lijnen, H.R., Wagner, E.F., and Collen, D. (1997). Plasminogen-dependent and -independent proteolytic activity of murine endothelioma cells with targeted inactivation of fibrinolytic genes. *Thromb. Hemost.* **77**, 362–367.

Liu, N.-F. and He, Q.-L. (1997). The regulatory effects of cytokines on lymphatic angiogenesis. *Lymphology.* **30**, 3–12.

Loskutoff, D.J., Curriden, S.A., Hu, G., and Deng, G. (1999). Regulation of cell adhesion by PAI-1. *APMIS* **107**, 54–61.

Lucas, R., Holmgren, L., Garcia, I., Jiminez, B., Mandriota, S.J., Borlat, F., Sim, B.K.L., Wu, Z., Grau, G.E., Shing, Y., Soff, G.A., Bouck, N., and Pepper, M.S. (1998). Multiple forms of angiostatin induce apoptosis in endothelial cells. *Blood* **92**, 4730–4741.

Ma, H.I., Guo, P., Li, J., Lin, S.Z., Chiang, Y.H., Xiao, X., and Cheng, S.Y. (2002a). Suppression of intracranial human glioma growth after intramuscular administration of an adeno-associated viral vector expressing angiostatin. *Cancer Res.* **62**, 756–763.

Ma, H.I., Lin, S.Z., Chiang, Y.H., Li, J., Chen, S.L., Tsao, Y.P., and Xiao, X. (2002b). Intratumoral gene therapy of malignant brain tumor in a rat model with angiostatin delivered by adeno-associated viral (AAV) vector. *Gene Ther.* **9**, 2–11.

Maeshima, Y., Colorado, P.C., and Kalluri, R. (2000a). Two RGD-independent alpha vbeta 3 integrin binding sites on tumstatin regulate distinct anti-tumor properties. *J. Biol. Chem.* **275**, 23745–23750.

Maeshima, Y., Colorado, P.C., Torre, A., Holthaus, K.A., Grunkemeyer, J.A., Ericksen, M.B., Hopfer, H., Xiao, Y., Stillman, I.E., and Kalluri, R. (2000b). Distinct antitumor properties of a type IV collagen domain derived from basement membrane. *J. Biol. Chem.* **275**, 21340–21348.

Mauceri, H.J., Hanna, N.N., Beckett, M.A., Gorski, D.H., Staba, M.J., Stellato, K.A., Bigelow, K., Heimann, R., Gately, S., Dhanabal, M., Soff, G.A., Sukhatme, V.P., Kufe, D.W., and Weichselbaum, R.R. (1998). Combined effects of angiostatin and ionizing radiation in antitumor therapy. *Nature* **394**, 287–291.

Mawatari, M., Okamura, K., Matsuda, T., Hamanaka, R., Mizoguchi, H., Higashio, K., Kohno, K., and Kuwano, M. (1991). Tumor necrosis factor and epidermal growth factor modulate migration of human microvascular endothelial cells and production of tissue-type plasminogen activator and its inhibitor. *Exp. Cell Res.* **192**, 574–580.

Mazzieri, R., Masiero, L., Zanetta, L., Monea, S., Onisto, M., Garbisa, S., and Mignatti, P. (1997). Control of type IV collagenase activity by components of the urokinase-plasmin system: A regulatory mechanism with cell-bound reactants. *EMBO J.* **16**, 2319–2332.

McMahon, G.A., Petitclerc, E., Stefansson, S., Smith, E., Wong, M.K., Westrick, R.J., Ginsburg, D., Brooks, P.C., and Lawrence, D.A. (2001). Plasminogen activator inhibitor-1 regulates tumor growth and angiogenesis. *J. Biol. Chem.* **276**, 33964–33968.

McQuibban, G.A., Gong, J.H., Tam, E.M., McCulloch, C.A., Clark-Lewis, I., and Overall, C.M. (2000). Inflammation dampened by gelatinase A cleavage of monocyte chemoattractant protein-3. *Science* **289**, 1202–1206.

Mignatti, P. and Rifkin, D.B. (1996). Plasminogen activators and matrix metalloproteinases in angiogenesis. *Enzyme Protein* **49**, 117–137.

Mignatti, P., Tsuboi, R., Robbins, E., and Rifkin, D.B. (1989). *In vitro* angiogenesis on the human amniotic membrane: Requirement for basic fibroblast growth factor-induced proteinases. *J. Cell Biol.* **108**, 671–682.

Min, H.Y., Doyle, L.V., Vitt, C.R., Zandonella, L., Stratton-Thomas, J.R., Shuman, M.A., and Rosenberg, S. (1996). Urokinase receptor antagonists inhibit angiogenesis and primary tumor growth in syngeneic mice. *Cancer Res.* **56**, 2428–2433.

Mishima, K., Mazar, A.P., Gown, A., Skelly, M., Ji, X.D., Wang, X.D., Jones, T.R., Cavenee, W.K., and Huang, H.J. (2000). A peptide derived from the non-receptor-binding region of urokinase plasminogen activator inhibits glioblastoma growth and angiogenesis *in vivo* in combination with cisplatin. *Proc. Natl. Acad. Sci. USA* **97**, 8484–8489.

Montesano, R., Pepper, M.S., Möhle-Steinlein, U., Risau, W., Wagner, E.F., and Orci, L. (1990). Increased proteolytic activity is responsible for the aberrant morphogenetic behavior of endothelial cells expressing middle T oncogene. *Cell* **62**, 435–445.

Montesano, R., Pepper, M.S., Vassalli, J.D., and Orci L. (1987). Phorbol ester induces cultured endothelial cells to invade a fibrin matrix in the presence of fibrinolytic inhibitors. *J. Cell Physiol.* **132**, 509–516.

Morikawa, W., Yamamoto, K., Ishikawa, S., Takemoto, S., Ono, M., Fukushi, J., Naito, S., Nozaki, C., Iwanaga, S., and Kuwano, M. (2000). Angiostatin generated by cathepsin D secreted by human prostate carcinoma cells. *J. Biol. Chem.* **275**, 38912–38920.

Moser, T.L., Stack, M.S., Asplin, I., Enghild, J.J., Hojrup, P., Everitt, L., Hubchak, S., Schapner, H.W., and Pizzo, S.V. (1999). Angiostatin binds ATP synthase on the surface of human endothelial cells. *Proc. Natl. Acad. Sci.* USA **96**, 2811–2816.

Moser, T.L., Stack, M.S., Wahl, M.L., and Pizzo, S.V. (2002).The mechanism of action of angiostatin: Can you teach an old dog new tricks? *Thromb. Haemost.* **87**, 394–401.

O'Reilly, M.S., Boehm, T., Shing, Y., Fukai, N., Vasios, G., Lane, W.S., Flynn, E., Birkhead, J.R., Olsen, B.R., and Folkman, J. (1997). Endostatin: An endogenous inhibitor of angiogenesis and tumor growth. *Cell* **88**, 277–285.

O'Reilly, M.S., Holmgren, L., Chen, C., and Folkman, J. (1996). Angiostatin induces and sustains dormancy of human primary tumors in mice. *Nature Med.* **2**, 689–692.

O'Reilly, M.S., Holmgren, L., Shing, Y., Chen, C., Rosenthal, R.A., Moses, M., Lane, W.S., Cao, Y., Sage, E.H., and Folkman, J. (1994). Angiostatin: A novel angiogenesis inhibitor that mediates the suppression of metastases by a Lewis lung carcinoma. *Cell* **79**, 315–328.

O'Reilly, M.S., Pirie-Shepherd, S., Lane, W.S., and Folkman, J. (1999). Antiangiogenic activity of the cleaved conformation of the serpin antithrombin. *Science* **285**, 1926–1928.

Ossowski, L. and Aguirre-Ghiso, J.A. (2000). Urokinase receptor and integrin partnership: Coordination of signalling for cell adhesion, migration and growth. *Curr. Opin. Cell Biol.* **12**, 613–620.

Pepper, M.S. (1997). Manipulating angiogenesis: From basic science to the bedside. *Arterioscl. Thromb. Vasc. Biol.* **17**, 605–619.

Pepper, M.S. (2001a). Role of the matrix metalloproteinase and plasminogen activator-plasmin systems in angiogenesis. *Arterioscler. Thromb. Vasc. Biol.* **21**, 1104–1117.

Pepper, M.S. (2001b) Extracellular proteolysis and angiogenesis. *Thromb. Hemost.* **86**, 346–355.

Pepper, M.S. (2001c) Lymphangiogenesis and tumor metastasis: Myth or reality. *Clin. Cancer Res.* **7**, 462–468.

Pepper, M.S., Belin, D., Montesano, R., Orci, L., and Vassalli, J.D. (1990). Transforming growth factor-beta 1 modulates basic fibroblast growth factor-induced proteolytic and angiogenic properties of endothelial cells *in vitro. J. Cell Biol.* **111**, 743–755.

Pepper, M.S., Ferrara, N., Orci, L., and Montesano, R. (1992a). Potent synergism between vascular endothelial growth factor and basic fibroblast growth factor in the induction of angiogenesis *in vitro. Biochem. Biophys. Res. Commun.* **189**, 824–831.

Pepper, M.S., Mandriota, S.J., Jeltsch, M., Kumar, V., and Alitalo, K. (1998). Vascular endothelial growth factor (VEGF)-C synergises with basic fibroblast growth factor and VEGF in the induction of angiogenesis *in vitro*, and alters endothelial cell proteolytic properties. *J. Cell Physiol.* **177**, 439–452.

Pepper, M.S. and Montesano, R. (1990). Proteolytic balance and capillary morphogenesis. *Cell Differ. Dev.* **32**, 319–327.

Pepper, M.S., Montesano, R., Mandriota, S.J., Orci, L., and Vassalli, J.D. (1996). Angiogenesis: A paradigm for balanced extracellular proteolysis during cell migration and morphogenesis. *Enzyme Protein* **49**, 138–162.

Pepper, M.S., Rosnoblet, C., Di Sanza, C., and Kruithof, E..K.O. (2001). Synergistic induction of tissue-type plasminogen activator (tPA) by vascular endothelial growth factor and basic fibroblast growth factor and localization of tPA to Weibel–Palade bodies in bovine microvascular endothelial cells. *Thromb. Hemost.* **86**, 702–709.

Pepper, M.S., Sappino, A.-P., Montesano, R., Orci, L., and Vassalli, J.-D. (1992b). Plasminogen activator inhibitor-1 is induced in migrating endothelial cells. *J. Cell Physiol.* **153**, 129–139.

Pepper, M.S., Sappino, A.-P., Stocklin, R., Montesano, R., Orci, L., and Vassalli, J.-D. (1993). Upregulation of urokinase receptor expression on migrating endothelial cells. *J. Cell Biol.* **122**, 673–684.

Pepper, M.S., Tacchini-Cottier, F., Sabapathy, K., Montesano, R., and Wagner, E.F. (1997). Endothelial cells transformed by polyomavirus middle T oncogene: A model for hemangiomas and other vascular tumours. In R. Bicknell, C.E. Lewis and N. Ferrara (eds) *Tumour angiogenesis*. Oxford University Press, Oxford, pp. 309–331.

Pepper, M.S., Vassalli, J.-D., Montesano, R., and Orci, L. (1987). Urokinase-type plasminogen activator is induced in migrating capillary endothelial cells. *J. Cell Biol.* **105**, 2535–2541.

Pepper, M.S., Wasi, S., Ferrara, N., Orci, L., and Montesano, R. (1994). *In vitro* angiogenic and proteolytic properties of bovine lymphatic endothelial cells. *Exp. Cell Res.* **210**, 298–305.

Pike, S.E., Yao, L., Jones, K.D., Cherney, B., Appella, E., Sakaguchi, K., Nakhasi, H,, Teruya-Feldstein, J., Wirth, P., Gupta, G., and Tosato, G. (1998). Vasostatin, a calreticulin fragment, inhibits angiogenesis and suppresses tumor growth. *J. Exp Med.* **188**, 2349–2356.

Preissner, K.T., Kanse, S.M., and May, A.E. (2000). Urokinase receptor: A molecular organizer in cellular communication. *Curr. Opin. Cell Biol.* **12**, 621–628.

Primo, L., Roca, C., Ferrandi, C., Lanfrancone, L., and Bussolino, F. (2000). Human endothelial cells expressing polyoma middle T induce tumors. *Oncogene* **19**, 3632–3641.

Ramchandran, R., Dhanabal, M., Volk, R., Waterman, M.J.F., Segal, M., Lu, H., Knebelmann, B., and Sukhatme, V.P. (1999). Antiangiogenic activity of restin, NC10 domain of human collagen XV: Comparison to endostatin. *Biochem. Biophys. Res. Commun.* **255**, 735–739.

Renno, R.Z., Delori, F.C., Holzer, R.A., Gragoudas, E.S., and Miller, J.W. (2000). Photodynamic therapy using Lu-Tex induces apoptosis *in vitro*, and its effect is potentiated by angiostatin in retinal capillary endothelial cells. *Invest. Ophthalmol. Vis. Sci.* **41**, 3963–3971.

Ribatti, D., Leali, D., Vacca, A., Giuliani, R., Gualandris, A., Roncali, L., Nolli, M.L., and Presta, M. (1999). *In vivo* angiogenic activity of urokinase: Role of endogenous fibroblast growth factor-2. *J. Cell Sci.* **112**, 4213–4221.

Rodolfo, M., Cato, E.M., Soldati, S., Ceruti, R., Asioli, M., Scanziani, E., Vezzoni, P., Parmiani, G., and Sacco, M.G. (2001). Growth of human melanoma xenografts is suppressed by systemic angiostatin gene therapy. *Cancer Gene Ther.* **8**, 491–496.

Rømer, J., Bugge, T.H., Pyke, C., Lund, L.R., Flick, M.J., Degan, L.J., and Danø, K. (1996). Impaired wound healing in mice with a disrupted plasminogen gene. *Nat. Med.* **2**, 287–292.

Rosnoblet, C., Vischer, U.M., Gerard, R.D., Irminger, J.C., Halban, P.A., and Kruithof, E.K.O. (1999). Storage of tissue-type plasminogen activator in Weibel–Palade bodies of human endothelial cells. *Arterioscler. Thromb. Vasc. Biol.* **19**, 1796–1803.

Saarialho-Kere, U.K., Pentland, A.P., Birkedal-Hansen, H., Parks, W.C., and Welgus, H.G. (1994). Distinct populations of basal keratinocytes express stromelysin-1 and stromelysin-2 in chronic wounds. *J. Clin. Invest.* **94**, 79–88.

Sabapathy, K.T., Pepper, M.S., Kiefer, F., Möhle-Steinlein, U., Tacchini-Cottier, F., Fetka, I,, Breier, G., Carmeliet, P., Risau, W., Montesano, R., and Wagner, E.F. (1997). Polyoma middle T-induced vascular tumor formation: The role of the plasminogen activator/plasmin system. *J. Cell Biol.* **137**, 953–963.

Sacco, M.G., Caniatti, M., Cato, E.M., Frattini, A., Chiesa, G., Ceruti, R., Adorni, F., Zecca, L., Scanziani, E., and Vezzoni, P. (2000). Liposome-delivered angiostatin strongly inhibits tumor growth and metastatization in a transgenic model of spontaneous breast cancer. *Cancer Res.* **60**, 2660–2665.

Sacco, M.G., Cato, E.M., Ceruti, R., Soldati, S., Indraccolo, S., Caniatti, M., Scanziani, E., and Vezzoni, P. (2001). Systemic gene therapy with anti-angiogenic factors inhibits spontaneous breast tumor growth and metastasis in MMTVneu transgenic mice. *Gene Ther.* **8**, 67–70.

Sage, E.H. (1997). Pieces of eight: Bioactive fragments of extracellular proteins as regulators of angiogenesis. *Trends Cell Biol.* **7**, 182–186.

Sato, Y., Okamura, K., Morimoto, A., Hamanaka, R., Hamaguchi, K., Shimada, T., Ono, M., Kohno, K., Sakata, T., and Kuwano, M. (1993). Indispensable role of tissue-type plasminogen activator in growth factor-dependent tube formation of human microvascular endothelial cells *in vitro*. *Exp. Cell Res.* **204**, 223–229.

Scappaticci, F.A., Smith, R., Pathak, A., Schloss, D., Lum, B., Cao, Y., Johnson, F., Engleman, E.G., and Nolan, G.P. (2001). Combination angiostatin and endostatin gene transfer induces synergistic antiangiogenic activity *in vitro* and antitumor efficacy in leukemia and solid tumors in mice. *Mol. Ther.* **3**, 186–196.

Schnaper, H.W., Barnathan, E.S., Mazar, A., Maheshwari, S., Ellis, S., Cortez, S.L., Baricos, W.H., and Kleinman, H.K. (1995). Plasminogen activators augment endothelial cell organization *in vitro* by two distinct pathways. *J. Cell. Physiol.* **165**, 107–118.

Sim, B.K.L., O'Reilly, M.S., Liang, H., Fortier, A.H., He, W., Madsen, J.W., Lapcevich, R., and Nacy, C.A. (1997). A recombinant human angiostatin protein inhibits experimental primary and metastatic cancer. *Cancer Res.* **57**, 1329–1334.

Soff, G.A., Sanderowitz, J., Gately, S., Verrusio, E., Weiss, I., Brem, S., and Kwaan, H.C. (1995). Expression of plasminogen activator inhibitor type 1 by human prostate carcinoma cells inhibits primary tumor growth, tumor-associated angiogenesis, and metastasis to lung and liver in an athymic mouse model. *J. Clin. Invest.* **96**, 2593–2600.

Stack, M.S., Gately, S., Bafeti, L.M., Enghild, J.J., and Soff, G.A. (1999). Angiostatin inhibits endothelial and melanoma cellular invasion by blocking matrix-enhanced plasminogen activation. *Biochem. J.* **340**, 77–84.

Stathakis, P., Fitzgerald, M., Matthias, L.J., Chesterman, C.N., and Hogg, P.J. (1997). Generation of angiostatin by reduction and proteolysis of plasmin. Catalysis by a plasmin reductase secreted by cultured cells. *J. Biol. Chem.* **272**, 20641–20645.

Stathakis, P., Lay, A.J., Fitzgerald, M., Schlieker, C., Matthias, L.J., and Hogg, P.J. (1999). Angiostatin formation involves disulfide bond reduction and proteolysis in kringle 5 of plasmin. *J. Biol. Chem.* **274**, 8910–8916.

Stefansson, S., Petitclerc, E., Wong, M.K., McMahon, G.A., Brooks, P.C., and Lawrence, D.A. (2001). Inhibition of angiogenesis *in vivo* by plasminogen activator inhibitor-1. *J. Biol. Chem.* **276**, 8135–8141.

Taraboletti, G., Belotti, D., Dejana, E., Mantovani, A., and Giavazzi, R. (1993). Endothelial cell migration and invasiveness are induced by a soluble factor produced by murine endothelioma cells transformed by polyoma virus middle T oncogene. *Cancer Res.* **53**, 3812–3816

Tarui, T., Miles, L.A., and Takada, Y. (2001). Specific interaction of angiostatin with integrin alpha(v)beta(3) in endothelial cells. *J. Biol. Chem.* **276**, 39562–39568.

te Velde, E.A., Voest, E.E., van Gorp, J.M., Verheem, A., Hagendoorn, J., Gebbink, M.F., and Borel Rinkes, I.H. (2002). Adverse effects of the antiangiogenic agent angiostatin on the healing of experimental colonic anastomoses. *Ann. Surg. Oncol.* **9**, 303–309.

Uchiyama, T., Kurabayashi, M., Ohyama, Y., Utsugi, T., Akuzawa, N., Sato, M., Tomono, S., Kawazu, S., and Nagai, R. (2000). Hypoxia induces transcription of the plasminogen activator inhibitor-1 gene through genistein-sensitive tyrosine kinase pathways in vascular endothelial cells. *Arterioscler. Thromb. Vasc. Biol.* **20**, 1155–1161.

Vergani, V., Garofalo, A., Bani, M.R., Borsotti, P., Parker, M.P., Drudis, T., Mazzarol, G., Viale, G., Giavazzi, R., Stetler-Stevenson, W.G., and Taraboletti, G. (2001). Inhibition of matrix metalloproteinases by over-expression of tissue inhibitor of metalloproteinase-2 inhibits the growth of experimental hemangiomas. *Int. J. Cancer.* **91**, 241–247.

Vu, T.H. and Werb, Z. (2000). Matrix metalloproteinases: Effectors of development and normal physiology. *Genes Dev.* **14**, 2123–2133.

Wu, Z., O'Reilly, M.S., Folkman, J., and Shing, Y. (1997). Suppression of tumor growth with recombinant murine angiostatin. *Biochem. Biophys. Res. Commun.* **236**, 651–654.

Yasunaga, C., Nakashima, Y., and Sueishi, K. (1989). A role of fibrinolytic activity in angiogenesis. Quantitative assay using *in vitro* model. *Lab. Invest.* **61**, 698–704.

Yokoyama, Y., Dhanabal, M., Griffioen, A.W., Sukhatme, V.P., and Ramakrishnan, S. (2000). Synergy between angiostatin and endostatin: Inhibition of ovarian cancer growth. *Cancer Res.* **60**, 2190–2196.

Zablocki, D.K., Rade, J.J., and Alevriadou, B.R. (2000). Adenovirus-mediated expression of tissue plasminogen activator does not alter endothelial cell proliferation and invasion. *In Vitro Cell Dev. Biol. Anim.* **36**, 625–628.

Index

GPSR Compliance

The European Union's (EU) General Product Safety Regulation (GPSR) is a set of rules that requires consumer products to be safe and our obligations to ensure this.

If you have any concerns about our products, you can contact us on ProductSafety@springernature.com

In case Publisher is established outside the EU, the EU authorized representative is:

Springer Nature Customer Service Center GmbH
Europaplatz 3
69115 Heidelberg, Germany

Batch number: 10607920

Printed by Printforce, the Netherlands